Climate Ethics

Climate Ethics

Essential Readings

Edited by

Stephen M. Gardiner, Simon Caney, Dale Jamieson, and Henry Shue

2010

OXFORD
UNIVERSITY PRESS

Oxford University Press, Inc., publishes works that further
Oxford University's objective of excellence
in research, scholarship, and education.

Oxford New York
Auckland Cape Town Dar es Salaam Hong Kong Karachi
Kuala Lumpur Madrid Melbourne Mexico City Nairobi
New Delhi Shanghai Taipei Toronto

With offices in
Argentina Austria Brazil Chile Czech Republic France Greece
Guatemala Hungary Italy Japan Poland Portugal Singapore
South Korea Switzerland Thailand Turkey Ukraine Vietnam

Published by Oxford University Press, Inc.
198 Madison Avenue, New York, New York 10016

www.oup.com

Library of Congress Cataloging-in-Publication Data

Climate ethics : essential readings / edited by Stephen M. Gardiner...[et al.].
p. cm.
Includes bibliographical references and index.
ISBN 978–0–19–539962–2; 978–0–19–539961–5 (pbk.)
1. Environmental ethics. 2. Global warming—Moral and ethical aspects. 3. Climatic changes—Moral and ethical aspects. 4. Environmental responsibility. I. Gardiner, Stephen Mark.
GE42.C555 2010
179'.1—dc22 2009027303

9 8 7 6 5 4 3 2 1

Printed in the United States of America
on acid-free paper

For those who will come after us;
and for Brian Barry,
who showed us the importance of thinking seriously about them

Foreword

R. K. Pachauri

The international debate on climate change has thus far focused essentially on the scientific and technical aspects or the economic implications of burden sharing to reduce the emissions of greenhouse gasses. Since the Eighth Conference of the Parties held in New Delhi in 2002, the debate has also extended to cover relevant aspects of the impacts of climate change and related adaptation options. It is only now that some focused concern is being provided to the ethical dimensions of this subject as well as the future of actions that are defined by a comprehensive assessment of climate change in its entirety. The Fourth Assessment Report (AR4) of the Intergovernmental Panel on Climate Change (IPCC) has provided considerable information and analysis on the following:

- The differential nature of the impacts of climate change and the vulnerability of some of the poorest and most economically disadvantaged societies in the world.
- The existing handicaps and weak capacity of governments in coming up with adaptation measures for meeting the projected impacts in some of these vulnerable regions and societies.
- The attractiveness of mitigation measures, particularly in the developed countries where a large range of co-benefits can be identified with mitigation actions.

Against this background, it would appear inexplicable that there is a lack of adequate action to deal with the global challenge of climate change, both in respect of adaptation measures in the developing countries and stringent mitigation in the developed world. This clearly raises the issue of ethics, because given the historical responsibility for cumulative emissions of greenhouse gasses in the developed world, as a result of which high levels of prosperity have been attained in these countries, inaction implies a neglect of overall responsibility and ethical considerations. On the other hand, the IPCC's projections of future impacts of climate change in the small island states, low lying coastal areas and food deficient and water stressed regions of the world (which are essentially spread across several developing countries), inaction by the global community takes on an ethical significance as well.

This book has brought in fresh perspectives on the ethical dimensions of climate change, articulated and presented by a distinguished set of authors. A publication of this nature could not have been more timely and relevant. Leaders in various countries as well as the negotiators working towards a global agreement would do well to study the material presented in this book, because that might help to lift their visions and perspectives above self-centered and short-sighted considerations that appear dominant at present. The current approach hardly does justice to the interests of future generations and the preservation of natural resources on this planet. Ethics demand proper attention to issues of intra- as well as inter-generational equity. As citizens of spaceship earth, it is essential for us to keep these considerations in view, and I compliment the authors for the excellent material provided in this book on such a vitally important subject.

Preface

Climate change poses a severe challenge to current institutions and ways of life. The idea that this challenge involves ethics is not unfamiliar. In 2006, for example, Al Gore infamously declared that climate change "is not a political issue so much as a moral issue," and Gordon Brown (now U.K. prime minister) said that "the developed world has a moral duty to tackle climate change." Still, it remains true that other ways of talking about climate change—especially scientific, economic, and geopolitical ways—dominate the current discussion.

Much might be said about this (see chapters 1 through 4). But surely one reason for the marginalization of ethics is that moving beyond general pronouncements about its relevance poses major challenges of its own, both political and intellectual. The intellectual challenge is formidable, and philosophers have not been particularly swift to meet it. Although some have been writing about climate change since the late 1980s, as of January 2009, the *Philosopher's Index* listed only about 100 articles under "climate change" and "global warming," most of them recent. By contrast, there were more than 700 listings for "informed consent" and more than 1,000 for "euthanasia." Part of the reason for the reluctance of philosophers to enter the discussion is the complexity of the scientific, economic, and legal issues involved (see chapter 1). But there may also be reasons of academic culture. In his book *Morality's Progress,* Dale Jamieson reports one of his colleagues asking some years ago, "How can you write on something that no one else has written about?" This is a sobering question. Being among the first to enter a new area of philosophical inquiry is daunting. For one thing, one must make difficult decisions about how and where to start and how to divide up the terrain; for another, making these choices turns one into cannon fodder for the next generation and so poses professional risks. (Much better to be the cannon than the fodder.) Still, when the stakes are so high, it is necessary to begin.

With this apologia in mind, this collection brings together what we see as core papers from those foolish (or brave) enough to make a beginning, the first generation of philosophers working on climate change. The aim is to capture much of the best work so far, work that is currently dispersed across two decades and many venues. We hope that this is a service to future scholars and students of these issues and also to those outside philosophy who would like a window into what has been done.

There are, naturally, a few caveats. First, we would like to have included more papers and from a wider variety of disciplines. We did so in the first draft, but we were advised (no doubt quite rightly) that no one but ourselves would buy a 700-page book on this topic. Second, we acknowledge slight embarrassment at including two or three papers from each of us. Our only defenses here are to say that there really was very little out there when we were pulling the collection together (in early 2008), and that whenever one of us tried to cut our own contributions down, the others resisted. Third, we are aware that a number of important issues have been ignored. This is partly because of constraints of space but also because the literature on these topics is still underdeveloped. We hope that the obviousness of the lacunae will prompt others to fill them. Hopefully, the obviousness of the lacunae will prompt others to fill them. If so, this will fit well with the general aim of this collection, which is to help spur the wider debate. Finally, we note that all of the papers included appear in their original versions. In particular, we have largely resisted the urge to revise their content in light of political and scientific changes. This is partly because major "updates" would probably have resulted in substantially new papers, and the history of the discussion would be lost. But it is also because one of the surprising facts about twenty years of climate policy is how little has changed. Rather than hide that fact, we recommend it for further study.

I'd like to thank a few people who made this volume possible. Janice Moskalik and Jason Benchimol were outstanding in pulling together the permissions and bibliography. Peter Ohlin, our editor at Oxford University Press, was enthusiastic, supportive, good-humored, and efficient in guiding us through the process. The Department of Philosophy and the Program on Values in Society at the University of Washington also helped out in numerous ways. The mainstays of the project were, of course, my coeditors, Dale, Henry, and Simon. I am grateful to them for their guidance and enthusiasm and also for initially being willing to take an off-the-cuff idea and run with it. Dale deserves special thanks for extra help with much of the behind-the-scenes work and the difficult decisions. Although I suspect that the collection would look different if any one of us had done it alone, I do believe that it would not have been better. Finally, I'd like to thank Lynn, Ben, and Matthew for putting up with the final push with their usual love and grace. They were always there, reminding me why this is important.

I close with a thought about where we now stand. On the day that I was composing this preface, some of my students gave a presentation to our class on a well-known article of Henry Shue's, originally published in 1993 (chapter 11 here). They concluded by quoting a recent report from the *Washington Post* summarizing the United Nations climate negotiations in Poznan in December 2008. The report said: "the core questions—how much industrialized countries will slash their emissions, what they expect in return from major emerging economies, and what they will do to help poorer countries pursue low-carbon development—remain untouched." As the presenters pointed out, these were the questions Henry tried to address back in the early 1990s, and the continued relevance of his seminal discussion is in some ways depressing. More depressing still, in my view, is the fact that global emissions have risen by roughly 30 percent since Henry wrote, amid the heady atmosphere of the Rio Earth Summit. My suspicion is that this shows that we need more than just the usual scientific and economic rhetoric on climate change, which so far has failed to move us. Perhaps then we should challenge ourselves to do more than merely pay lip service to the idea that climate change is a moral issue and seek to engage more fully with the ethics of a warming world. We hope that this collection is a good place to start.

—Stephen M. Gardiner
Seattle, February 2009

Postscript (January 2010): We look over the final proofs of the collection in early 2010, just after an international meeting in Copenhagen that was orginally intended to provide a successor to the Kyoto Protocol that achieved so little that it was widely declared a "disaster." Hopefully, the following chapters can help to explain this sad state of affairs and why it matters, and also contribute to a resolution.

Acknowledgments and Credits

Stephen M. Gardiner, "Ethics and Global Climate Change," originally appeared in *Ethics* 114, (2004): 555–600. Copyright © 2004 by the University of Chicago.

Nicholas Stern, "The Economics of Climate Change," originally appeared in *American Economic Review* 98.2 (2008): 1–37. Reprinted with permission of the author and the American Economic Association.

Dale Jamieson, "Ethics, Public Policy, and Global Warming," originally appeared in *Science, Technology, Human Values* 17.2 (1992): 139–153. Reprinted with permission of Sage Publications.

Stephen M. Gardiner, "A Perfect Moral Storm: Climate Change, Intergenerational Ethics, and the Problem of Moral Corruption," originally appeared in *Environmental Values* 15 (2006): 397–413. Reprinted with permission of the White Horse Press.

Henry Shue, "Global Environment and International Inequality," originally appeared in *International Affairs* 75.3 (1999): 531–545. Reprinted with the permission of Blackwell Publishing.

Derek Parfit, "Energy Policy and the Further Future: The Identity Problem," originally appeared in Douglas MacLean and Peter Brown, eds. *Energy and the Future* (Totowa, N.J.: Rowman & Allanheld, 1983): 166–179. Reprinted with permission of the author and Rowman & Littlefield.

Simon Caney, "Cosmopolitan Justice, Responsibility, and Global Climate Change," originally appeared in *Leiden Journal of International Law* (2005): 747–775. Reprinted with permission of *Leiden Journal of International Law.*

Henry Shue, "Deadly Delays, Saving Opportunities: Creating a More Dangerous World?" is original to this volume.

Simon Caney, "Climate Change, Human Rights, and Moral Thresholds," originally appeared in Stephen Humphreys, ed., *Human Rights and Climate Change* (Cambridge, U.K.: Cambridge University Press, 2010): 69–70. Reprinted with permission of Cambridge University Press.

Peter Singer, "One Atmosphere," originally appeared in his *One World: The Ethics of Globalization* (New Haven, Conn.: Yale University Press, 2002). © 2002 Yale University Press. Reprinted with permission of Yale University Press.

Henry Shue, "Subsistence Emissions and Luxury Emissions," originally appeared in *Law & Policy* 15. 1 (1993): 39–59. Reprinted with the permission of Blackwell Publishing.

Paul Baer, with Tom Athanasiou, Sivan Kartha, and Eric Kemp-Benedict, "Greenhouse Development Rights: A Framework for Climate Protection That Is 'More Fair' Than Equal Per Capita Emissions Rights" is original to this volume.

Robert E. Goodin, "Selling Environmental Indulgences," originally appeared in *Kyklos* 47 (1994): 573–596. Reprinted with the permission of Blackwell Publishing.

Paul Baer, "Adaptation: Who Pays Whom?" originally appeared in W. Adger, J. Paavola, S. Huq, and M. Mace, eds., *Fairness in Adaptation to Climate Change* (Cambridge, Mass.: MIT Press, 2006): 131–153. Copyright © 2006 by the Massachusetts Institute of Technology. Reprinted with the permission of MIT Press.

Dale Jamieson, "Adaptation, Mitigation, and Justice," originally appeared in Walter Sinnott-Armstrong and Richard Howarth, eds., *Perspectives on Climate Change* (Amsterdam: Elsevier, 2005): 221–253. Reprinted with permission of Emerald Publishing Group Limited.

Stephen M. Gardiner, "Is 'Arming the Future' with Geoengineering Really the Lesser Evil? Some Doubts about the Ethics of Intentionally Manipulating the Climate System" is original to this volume.

Dale Jamieson, "When Utilitarians Should Be Virtue Theorists," originally appeared in *Utilitas* 19.2 (2007): 160–183. Reprinted with permission of Cambridge University Press.

Walter Sinnott-Armstrong, "It's Not *My* Fault," originally appeared in Walter Sinnott-Armstrong and Richard Howarth, eds., *Perspectives on Climate Change* (Amsterdam: Elsevier, 2005): 221–253. Reprinted with permission of Emerald Publishing Group Limited.

Contents

Contributors

Paul Baer is an interdisciplinary scholar specializing in climate change and climate policy, with a focus on issues of equity and uncertainty. He holds a PhD from the Energy and Resources Group at University of California, Berkeley, and is currently a postdoctoral scholar at the Woods Institute for the Environment at Stanford University as well as research director of EcoEquity.

Simon Caney is professor in political theory and tutorial fellow in politics at Magdalen College, Oxford. He has published articles on climate change, global and intergenerational justice, poverty, rights, and cosmopolitanism in politics, philosophy, and law journals. He is the author of *Justice beyond Borders* (2005). He is currently working on a book titled *On Cosmopolitanism* (under contract to Oxford University Press) and a book titled *Global Justice and Climate Change* (coauthored with Derek Bell and also under contract to Oxford University Press). He holds an ESRC Leadership Fellowship on Climate Change (2008–2011).

Stephen M. Gardiner is associate professor of philosophy at the University of Washington in Seattle. His main research interests are in ethical theory, political philosophy, environmental ethics, and Aristotle's ethics. He is the author of *A Perfect Moral Storm* (under contract to Oxford University Press) and the editor of *Virtue Ethics, Old and New* (2005).

Robert E. Goodin is Distinguished Professor of Social and Political Theory and of Philosophy at the Research School of Social Sciences, Australian National University. He is founding editor of the *Journal of Political Philosophy,* general editor of the 10-volume series of *Oxford Handbooks of Political Science,* and author of, most recently, *Innovating Democracy* (2008).

Dale Jamieson teaches environmental studies, philosophy, and law at New York University. His most recent book is *Ethics and the Environment: An Introduction* (2008), and he is currently at work on a book about the moral and political dimensions of climate change.

Derek Parfit has been a fellow of All Souls College, Oxford, since 1967. He also teaches as a regular visiting professor to the philosophy departments of Rutgers, New York University, and Harvard. His first book, *Reasons and Persons,* was published by Oxford University Press in 1984. A second book, *On What Matters,* is nearly complete and will also be published by Oxford.

Henry Shue is a senior research fellow at the Centre for International Studies at the University of Oxford.

Peter Singer has taught at Oxford, La Trobe University, and Monash University. Since 1999, he has been Ira W. DeCamp Professor of Bioethics at the University Center for Human Values at Princeton University. From 2005 on, he has also held the part-time position of laureate professor at the University of Melbourne, in the Centre for Applied Philosophy and Public Ethics.

Walter Sinnott-Armstrong is Chauncey Stillman Professor of Practical Ethics in the Philosophy Department and the Kenan Institute for Ethics at Duke University. He is currently vice chair of the Board of Officers of the American Philosophical Association and codirector of the MacArthur Law and Neuroscience Program. He has published extensively on ethics (theoretical and applied), philosophy of law, epistemology, philosophy of religion, and informal logic. His research now focuses on empirical moral psychology as well as law and neuroscience.

Nicholas Stern holds the IG Patel Chair in Economics and Government at the London School of Economics and is the author of *The Global Deal: Climate Change and the Creation of a New Era of Progress and Prosperity* (2009). He is also the first chair of the Grantham Research Institute on Climate Change. Lord Stern was adviser to the U.K. government on the economics of climate change and development from 2005 to 2007 and head of the Stern Review on the Economics of Climate Change. He was also head of the Government Economic Service, 2003–2007; second permanent secretary to Her Majesty's Treasury, 2003–2005; and director of policy and research for the Prime Minister's Commission for Africa, 2004–2005. He was chief economist and senior vice president at the World Bank from 2000 to 2003.

Part I

Introductory Overview

Ethics and Global Climate Change

Stephen M. Gardiner

Very few moral philosophers have written on climate change.[1] This is puzzling, for several reasons. First, many politicians and policy makers claim that climate change is not only the most serious environmental problem currently facing the world but also one of the most important international problems per se.[2] Second, many of those working in other disciplines describe climate change as fundamentally an ethical issue.[3] Third, the problem is theoretically challenging, both in itself and in virtue of the wider issues it raises.[4] Indeed, some have even gone so far as to suggest that successfully addressing climate change will require a fundamental paradigm shift in ethics (Jamieson 1992, p. 292).

Arguably, then, there is a strong presumption that moral philosophers should be taking climate change seriously. So why the neglect? In my view, the most plausible explanation is that study of climate change is necessarily interdisciplinary, crossing boundaries between (at least) science, economics, law, and international relations.

This fact not only creates an obstacle to philosophical work (since amassing the relevant information is both time-consuming and intellectually demanding) but also makes it tempting to assume that climate change is essentially an issue for others to resolve. Both factors contribute to the current malaise—and not just within philosophy but in the wider community, too.

My aims in this survey, then, are twofold. First, I try to overcome the interdisciplinary obstacle to some extent, by making the climate-change issue more accessible to both philosophers and nonphilosophers alike. Second, by drawing attention to the ethical dimensions of the climate change problem, I make the case that the temptation to defer to experts in other disciplines should be resisted. Climate change is fundamentally an ethical issue. As such, it should be of serious concern to both moral philosophers and humanity at large.

The interdisciplinary nature of the climate-change problem once prompted John Broome to imply that a truly comprehensive survey of the relevant literature would be impossible (Broome 1992, p. viii). I do not attempt the impossible. Instead, I present an overview of the most major and recent work relevant to philosophical discussion. Inevitably, this overview is to some extent selective and opinionated. Still, I hope that it will help to reduce the interdisciplinary obstacles to philosophical work on climate change, by giving both philosophers and the public more generally some sense of what has been said so far and what might be at stake. In my view, the ethics of global climate change is still very much in its infancy. I hope that this small contribution will encourage its development.

I. Terminology

"While global warming has catastrophic communications attached to it, climate change sounds a more controllable and less emotional challenge."

—From a memo by strategist Frank Luntz recommending that Republicans adopt the new terminology (Lee 2003)

Potential confusion about the climate-change problem begins even with the terms used to describe it: from *greenhouse effect* to *global warming* to the more recently favored *climate change*.[5] To begin with, many people spoke of "the greenhouse effect." This refers to the basic physical mechanism behind projected changes in the climate system.[6] Some atmospheric gases (called greenhouse gases, or GHG) have asymmetric interactions with radiation of different frequencies: just like glass in a conventional greenhouse, they allow short-wave incoming solar radiation through but reflect some of the earth's outgoing long-wave radiation back to the surface. This creates "a partial blanketing effect," which causes the temperature at the surface to be higher than would otherwise be the case (Houghton 1997, pp. 11–12). Humans are increasing the atmospheric concentrations of these gases through industrialization. This would, other things being equal, be expected to result in an overall warming effect.

The basic greenhouse mechanism is both well understood and uncontroversial. Still, the term *greenhouse effect* remains unsatisfactory to describe the problem at hand. There are two reasons. First, there is a purely natural greenhouse effect, without which the earth would be much colder than it is now.[7] Hence, it is not accurate to say that "the greenhouse effect" as such is a problem; in fact, the reverse is true: without some greenhouse effect, the earth would be much less hospitable for life as we know it. The real problem is the enhanced, human-induced greenhouse effect. Second, it is not the greenhouse effect in isolation that causes the climate problem. Whether an increase in the concentration of greenhouse gases does in fact cause the warming we would otherwise expect depends on how the immediate effects of an increase in low-frequency radiation play out in the overall climate system. But that system is complex, and its details are not very well understood.

For a while, then, the term *global warming* was favored. This term captures the point that it is the effects of increased levels of greenhouse gases that are of concern. However, it also has its limitations. In particular, it highlights a specific effect, higher temperatures, and thus suggests a one-dimensional problem. But while it is true that rising temperature has been a locus for concern about increasing human emissions of greenhouse gases, it is not true that temperature as such defines either the core problem or even (arguably) its most important aspects. Consider, for example, the following. First, a higher global temperature does not in itself constitute the most important impact of climate change. Indeed, considered in isolation, there might be no particular reason to prefer the world as it is now to one several degrees warmer.[8] However, second, this thought is liable to be misleading. For presumably, if one is imagining a warmer world and thinking that it might be appealing, one is envisioning the planet as it might be in a stable, equilibrium state at the higher level, where humans, animals, and plants have harmoniously adapted to higher temperatures. But the problem posed by current human behavior is not of this kind. The primary concern of many scientists is that an enhanced greenhouse effect puts extra energy into the earth's climate system and so creates an imbalance. Hence, most of the unease about present climate change has been brought about because it seems that change is occurring at an unprecedented rate, that any equilibrium position is likely to be thousands, perhaps tens or hundreds of thousands, of years off, and that existing species are unlikely to be able to adapt quickly and easily under such conditions. Third, although it is at present unlikely, it is still possible that temperature might go down as a result of the increase in atmospheric greenhouse-gas concentrations. But this does not cast any doubt on the serious nature of the problem. This is partly because a rapid and unprecedented lowering of temperature would have similar kinds of adverse effects on human and nonhuman life and health as a rapid warming and partly because the effects most likely to cause cooling (such as a shutdown of the thermohaline circulation, or THC, which supports the Gulf Stream current to northern Europe, as discussed in the next section) may well be catastrophic even in relation to the other projected effects of global warming.

For all of these reasons, current discussion tends to be carried out under the heading *climate change*. This term captures the fact that it is interference in the climate system itself that is the crucial issue, not what the particular effects of that interference turn out to be. The fundamental problem is that it is now possible for humans to alter the underlying dynamics of the planet's climate and therefore the basic life-support system for themselves and all other forms of life on earth. Whether the alteration of these dynamics is most conveniently tracked in terms of increasing, declining, or even stable temperatures is of subsidiary interest in comparison with the actual changes in the climate itself and their consequences for human, and nonhuman, life.[9]

II. Climate Science

"Almost no one would deny that in principle, our actions and policies should be informed by our best scientific judgments, and it is hard to deny that our best scientific judgments about climate change are expressed in the IPCC reports."

Dale Jamieson (1998, p. 116) (for a dissenting view, see Michaels and Balling 2000, chap. 11)

"Recent scientific evidence shows that major and widespread climate changes have occurred with startling speed. . . . Climate models typically underestimate the size, speed, and extent of those changes. . . . Climate surprises are to be expected."

—U.S. National Research Council, Committee on Abrupt Climate Change (2002, p. 1)

What do we know about climate change? In 1988, the Intergovernmental Panel on Climate Change (IPCC) was jointly established by the World Meteorological Association and the United Nations Environment Program to provide member governments with state-of-the-art assessments of "the science, the impacts, and the economics of—and the options for mitigating and/or adapting to—climate change" (IPCC 2001b, p. vii).[10] The IPCC has, accordingly, submitted three comprehensive reports, in 1990, 1995, and 2001.[11] The results have remained fairly consistent across all three reports, although the level of confidence in those results has increased.[12] The main findings of the 2001 report are as follows.

The IPCC begins with an account of patterns of climate change observed so far. On temperature, it reports: "The global average surface temperature has increased over the 20th century by about 0.6 °C"; "Globally, it is very likely[13] that the 1990s was the warmest decade and 1998 the warmest year in the instrumental record, since 1861"; and "The increase in temperature in the 20th century is likely to have been the largest of any century during the past 1,000 years" (IPCC 2001b, p. 152). For other phenomena, the IPCC says that snow cover and ice extent have decreased, global average sea level has risen, and ocean heat content has increased. It also cites evidence for increases in the amount of precipitation in some regions; the frequency of heavy precipitation events; cloud cover in some latitudes; and the frequency, persistence, and intensity of El Niño phenomenon.[14]

The IPCC also surveys the literature on relevant human activities. It concludes that since preindustrial times (1750 is the usual benchmark), humans have altered "the atmosphere in ways that are expected to affect the climate" by markedly increasing the concentrations of greenhouse gases (IPCC 2001b, p. 154). The main culprit is carbon dioxide,[15] for which "the concentration has increased by 31% since 1750"; "the present CO_2 concentration has not been exceeded during the past 420,000 years and likely not during the past 20 million years"; and "the current rate of increase is unprecedented during at least the past 20,000 years...at about 1.5 ppm [parts per million] (0.4%) per year" (IPCC 2001b, p. 155). The main anthropogenic sources of CO_2 are the burning of fossil fuels (about 75 percent) and changes in land-use patterns (principally, deforestation). Of secondary importance is methane, where the present atmospheric concentration "has increased by...151% since 1750; and has not been exceeded during the past 420,000 years," and "slightly more than half of current...emissions are anthropogenic (e.g., use of fossil fuels, cattle, rice agriculture and landfills)" (IPCC 2001b, pp. 156–157). Molecule for molecule, methane is a more potent greenhouse gas than carbon dioxide. Still, because CO_2 lasts much longer in the atmosphere (most between 5 to 200 years, but 10 to 15 percent over 10,000 years—as opposed to methane's 12 years)[16] it is more important.[17]

The IPCC also tries to predict future climate. To do so, it uses computer models to simulate a variety of different possible future scenarios, incorporating different assumptions about economic growth, world population, and technological change. The basic results are as follows. First, carbon dioxide emissions resulting from the burning of fossil fuels are "virtually certain to be the dominant influence on the trends in atmospheric CO_2 concentration during the 21st century," and by 2100, that concentration should be 90–250 percent above preindustrial levels (of 280 parts per million), at 540–970 parts per million (IPCC 2001b, pp. 158–159). Second, if this occurs, the full range of model scenarios predicts that surface temperature will increase by 1.4°C to 5.8°C over the century. The IPCC states that this is not only a much larger projected rate of warming than that observed during the 20th century but one "very likely...without precedent during at least the last 10,000 years."[18] Third, models indicate that "stabilisation of atmospheric CO_2 concentrations at 450, 650, or 1,000 ppm would require global anthropogenic CO_2 emissions to drop below 1990 levels, within a few decades, about a century, or about two centuries, respectively, and continue to decrease steadily thereafter. Eventually CO_2 emissions would need to decline to *a very small fraction* of current emissions" (IPCC 2001b, p. 160; emphasis added).

As alarming as the IPCC predictions are, we should also pay attention to the fact that they might be overly optimistic. Some authors argue that the current climate models typically underestimate the potential for nonlinear threshold effects (U.S. National Research Council 2002; Gagosian 2003). One well-known threat of this sort is the potential collapse of the West Antarctic Ice Sheet (WAIS), which would eventually raise global sea levels by four to six meters. But the recent (2002–2003) literature registers even greater concern about a lesser-known issue: the possibility of a weakening or shutdown of the deep circulation system that drives the world's ocean currents. This system, known as the Ocean Conveyor, distributes "vast quantities of heat around our planet, and thus plays a fundamental role in governing Earth's climate [and] in the distribution of life-sustaining water" (Gagosian 2003, p. 4).

The Ocean Conveyor has been called the climate's Achilles' heel (Broecker 1997), because it appears to be a major threshold phenomenon. There are two grounds for concern. First, there is strong evidence that in the past, the Conveyor has slowed, and slowed very quickly, with significant climatic consequences. One such event, 12,700 years ago, saw a drop in temperatures in the North Atlantic region of around 5 °C in a single decade. This apparently caused icebergs to spread as far south as the coast of Portugal and has been linked to widespread global drought. Second, the operation of the Conveyor is governed by factors that can be affected by climate change. In particular, the world's currents are driven by the sinking of a large volume of salty water in the North Atlantic region. But this process can be disrupted by an influx of fresh water, which both dilutes the salty water and can create a lid over it, restricting heat flow to the atmosphere.

The possibility of dramatic climate shifts of this sort complicates the picture of a global-warming world in several ways. First, it suggests that gradual warming at the global level could cause, and coexist with, dramatic cooling in some regions. (Among other things, this has serious ramifications for our ability to plan for future changes.) Second, it envisages that the major losers from climate change may not be the usual suspects, the less developed countries (LDCs). For it is the rich countries bordering the North Atlantic that are particularly vulnerable to Conveyor shifts. Climate models predict that "the North Atlantic region would cool 3 to 5 degrees Celsius if conveyor circulation were totally disrupted," producing winters "twice as cold as the worst winters on record in the eastern United States in the past century" for a period of up to a century (Gagosian 2003, p. 7).

The IPCC does not emphasize the problem of the Ocean Conveyor. For one thing, although it acknowledges that most models predict a weakening of the Conveyor during the 21st century, it emphasizes that such changes are projected to be offset by the more general warming; for another, it suggests that a complete shutdown is unlikely during the 21st century (though increasingly likely thereafter) (IPCC 2001b, p. 16). Hence, the IPCC's attitude is relatively complacent. Still, it is not clear what justifies such complacency. On the one hand, even if the threshold will not be reached for 100 years, this is still a matter of serious concern for future generations, since once the underlying processes that will breach it are in motion, it will be difficult, if not impossible, to reverse them. On the other hand, the current models of thermohaline circulation are not very robust, primarily because scientists simply do not know where the threshold is. And some models do predict complete shutdown within a range that overlaps with IPCC projections for the 21st century (IPCC 2001b, p. 440).[19]

III. Scientific Uncertainty

"Scientists aren't any time soon going to give politicians some magic answer. Policy makers for a long, long time are going to have to deal with a situation where it's not clear what the costs and benefits are, where lots of people disagree about them, and they can't wait until everything is resolved."

—Robert J. Lampert, senior scientist and expert in risk analysis at the RAND Corporation (Revkin 2001b)

"Should the public come to believe that the scientific issues are settled, their views about global warming will change accordingly. Therefore, you need to continue to make the lack of scientific certainty a primary issue."

—Frank Luntz (Lee 2003)

"It is sometimes argued that the uncertainty of the scientist's predictions is a reason for not acting at present, and that we should wait until some further research has been concluded. This argument is poor economics."

—John Broome (1992, p. 17)

Politically, the most common objection raised to action on climate change is that of scientific uncertainty.[20] In this section, I will explain why most writers on the subject believe this objection to be a red herring.

The first thing to note is that, at least in economics, *uncertainty* is a technical term, to be distinguished from *risk*. In the technical sense, a risk involves a known, or reliably estimable, probability that a certain set of outcomes may occur, whereas an uncertainty arises when such probabilities are not available. So to say that there is scientific uncertainty surrounding global warming is to claim that we do not know, and cannot reliably estimate, the probability that climate change will occur or its extent if it does occur.

This distinction is useful, because the first problem with the objection from scientific uncertainty is that the IPCC does not seem to view global warming as uncertain in the technical sense. As we have seen, the 2001 Scientific Assessment explicitly assigns probabilities to its main climate predictions, making the situation one of risk rather than uncertainty. Furthermore, these probabilities are of considerable magnitude. For example, the IPCC says that it is "very likely" that in the 21st century, there will be "higher maximum temperatures and more hot days over nearly all land areas" (IPCC 2001b, p. 162), by which it means a probability of 90 to 99 percent (IPCC 2001b, p. 152, n. 7). Given that many of the effects assigned high probabilities are associated with significant costs, they would seem to justify some kinds of action.

But perhaps the idea is that the IPCC's probability statements are not reliable, so that we should ignore them,[21] treat the situation as genuinely uncertain, and hence refuse to act. Still, there is a difficulty. To an important extent, some kind of uncertainty "is an inherent part of the problem" (Broome 1992, p. 18). Arguably, if we knew exactly what was going to happen, to whom, and whose emissions would cause it, the problem might be more easily addressed;[22] at the very least, it would have a very different shape. Hence, to refuse to act because of uncertainty is either to refuse to accept the global-warming problem as it is (insisting that it be turned into a more respectable form of problem before one will address it) or else to endorse the principle that to do nothing is the appropriate response to uncertainty. The former is a head-in-the-sand approach and clearly unacceptable, but the latter is also dubious and does not fit our usual practice.

The third, and perhaps most crucial, point to make about the problem of uncertainty is that it is important not to overplay it. For one thing, many decisions we have to make in life, including many important decisions, are also subject to considerable uncertainties.[23] For another, all uncertainties are not created equal. On the one hand, the reason I am unable to assign probabilities may be that I know absolutely nothing about the situation[24] or else that I have only one past instance to go on. But I might also be uncertain in circumstances where I have considerable information.[25]

Now it seems clear that uncertainty in the first kind of case is worse than uncertainty in the second—and potentially more paralyzing. Furthermore, and this is the crucial point, it seems reasonably clear that scientific uncertainty about global warming is of the second kind. As Donald Brown argues: "A lot of climate change science has never been in question,... many of the elements of global warming are not seriously challenged even by the scientific skeptics, and...the issues of scientific certainty most discussed by climate skeptics usually deal with the magnitude and timing of climate change, not with whether global warming is a real threat" (Brown 2002, p. 102).[26] To see this, let us briefly examine a number of sources of uncertainty about global warming.

The first concerns the direct empirical evidence for anthropogenic warming itself. This has two main aspects. First, systematic global temperature records, based on measurements of air temperature on land and surface-water temperature measurements at sea, exist only from 1860,[27] and satellite-based measurements are available only from 1979. The direct evidence for recent warming comes from the former. But skeptics suggest that the satellite measurements do not match the surface readings and do not provide evidence for warming.[28] Second, there is no well-defined baseline from which to measure change.[29] While it is true that the last couple of decades have been the warmest in human history, it is also true that the long-term climate record displays significant short-term variability and that, even accounting for this, climate seems to have been remarkably stable since the end of the last Ice Age, 10,000 years ago, as compared with the preceding 100,000 years.[30] Hence, global temperatures have fluctuated considerably over the long-term record, and it is clear that these fluctuations have been naturally caused.[31]

The skeptics are right, then, when they assert that the observational temperature record is a weak data set and that the long-term history of the climate is such that even if the data were more robust, we would be rash to conclude that humans are causing it solely on this basis.[32] Still, it would be a mistake to infer too much from the truth of these claims. It would be equally rash to dismiss the possibility of warming on these grounds. For, even though it might be true that the empirical evidence is consistent with there being no anthropogenic warming, it is also true that it provides just the kind of record we would expect if there were a real global-warming problem.

This paradox is caused by the fact that our epistemological position with respect to climate change is intrinsically very difficult; it may simply be impossible to confirm climate change empirically from this position. This is because our basic situation may be a bit like that of a coach who is asked whether the current performance of a 15-year-old athlete shows that she will reach the highest level of her sport. Suppose the coach has the best evidence that she can have. It will still only be evidence for a 15-year-old. It will be at most consistent with reaching the highest level. It cannot be taken as a certain prediction. But that does not mean it is no prediction at all or that it is worthless. It is simply the best prediction she is currently in a position to make.

Fortunately, for the climate-change problem, the concern with the empirical record is not the end of the matter. The temperature record is far from our only evidence for warming. Instead, we also have strong theoretical grounds for concern. First, the basic physical and chemical mechanisms that give rise to a potential global-warming effect are well understood. In particular, there is no scientific controversy over the claims (a) that in itself a higher concentration of greenhouse gas molecules in the upper atmosphere would cause more heat to be retained by the earth and less radiated out into the solar system so that, other things being equal, such an increase would cause global temperatures to rise; and (b) that human activities since the industrial revolution have significantly increased the atmospheric concentration of greenhouse gases. Hence, everyone agrees that the basic circumstances are such that an enhanced greenhouse effect is to be expected.[33]

Second, the scientific dispute, insofar as there is one, concerns the high level of complexity of the global climate system, given which there are the other mechanisms that might be in play to moderate such an effect. The contentious issue here is whether there might be negative feedbacks that either sharply reduce or negate the effects of higher levels of greenhouse gases or even reduce the amount of them present in the atmosphere. However, current climate models suggest that most related factors will likely exhibit positive feedbacks (water vapor, snow, and ice),[34] while others have both positive and negative feedbacks whose net effect is unclear (e.g., clouds, ocean currents). Hence, there is genuine scientific uncertainty. But this does not by itself justify a skeptical position about action on climate change. For there may be no more reason to assume that we will be saved by unexpectedly large negative feedbacks than that the warming effect will be much worse than we would otherwise anticipate, as a result of unexpectedly large positive feedbacks.[35]

This is the basic scientific situation. However, three further aspects of uncertainty are worth mentioning. First, the conclusions about feedback are also open to doubt because considerable uncertainties remain about the performance of the models. In particular, they are not completely reliable against past data.[36] This is to be expected, because the climate is a highly complex system that is not very well understood.[37] Still, it clouds the overall picture.[38] Second, as mentioned earlier, the current models tend to assume that atmospheric feedbacks scale linearly with surface warming, and they do not adequately account for possible threshold effects, such as the possible collapse of the West Antarctic Ice Sheet. Hence, they may underestimate the potential risks from global warming. Finally, there is a great deal of uncertainty about the distribution of climate change. Moreover, the focus on global temperature tends to conceal considerable variation within years and across regions. Still, although it is very difficult to predict which regions will suffer most and in what ways, such evidence as there is suggests that, at least in the medium term, the impact will be heaviest in the tropical and subtropical regions (where most of the LDCs are) and lighter in the temperate regions (where most of the richer countries are).

In conclusion, there are substantial uncertainties surrounding both the direct empirical evidence for warming and our theoretical understanding of the overall climate system. But these uncertainties cut both ways. In particular, while it is certainly conceivable (though, at present, unlikely) that the climate-change problem will turn out to be chimerical, it is also possible that global warming will turn out to be much worse than anyone has yet anticipated. More importantly, the really vital issue concerns not the presence of scientific uncertainty but rather how we decide what to do under such circumstances. To this issue we now turn.

IV. Economics

"Economic analyses clearly show that it will be far more expensive to cut CO_2 emissions radically than to pay the costs of adaptation to the increased temperatures."
—Bjørn Lomborg (2001, p. 318)

"Cost-benefit analysis, when faced with uncertainties as big as these, would simply be self-deception. And in any case, it could not be a successful exercise, because the issue is too poorly understood, and too little accommodated in the current economic theory."
—John Broome (1992, p. 19)

As it turns out, many recent skeptics no longer cite scientific uncertainty as their reason for resisting action on climate change. Instead, they claim to accept the reality of human-induced climate change but argue that there is a strong economic rationale for refusing to act.[39] Prevention, they insist, is more expensive than adaptation; hence, both present and future generations would be better off if we simply accepted that there will be climate change and tried to live with it. Furthermore, they assert, money that might be spent on prevention would be better spent helping the world's poor. I will consider the first of these arguments in this section and the second argument later on.

Several attempts have been made to model the economic implications of climate change.[40] Politically prominent among these is the DICE model proposed by Yale economist William Nordhaus. The DICE model is an integrated assessment (IA) model. IA models combine the essential elements of biophysical and economic systems in an attempt to understand the impact of climate and economic policies on one another. Typically, such models aim to find a climate policy that will maximize the social-welfare function. And many give the surprising result that only limited abatement should occur in the next 20 to 30 years, since the costs of current reductions are too high in comparison with the benefits.[41] Hence, proponents of these models argue that based on economic costs, the developed world (and the United States in particular) should focus on adaptation rather than abatement. This is the argument embraced by Lomborg, who cites Nordhaus's work as his inspiration.

1. The Cost Argument

A full response to Lomborg's proposal requires addressing both the argument about costs and the more general argument for an adaptation, rather than mitigation, strategy. Let us begin with the cost argument.

The first point to make is that, even if Nordhaus's calculations were reliable, the costs of climate-change mitigation do not seem unmanageable. As Thomas Schelling puts it:

> The costs in reduced productivity are estimated at two percent of GNP forever. Two percent of GNP seems politically unmanageable in many countries. Still, if one plots the curve of US per capita GNP over the coming century with and without the two percent permanent loss, the difference is about the thickness of a line drawn with a number two pencil, and the doubled per capita income that would have been achieved by 2060 is reached in 2062. If someone could wave a wand and phase in, over a few years, a climate-mitigation program that depressed our GNP by two percent in perpetuity, no one would notice the difference. (Schelling 1997)

Even Lomborg agrees with this. He not only cites the 2 percent figure with approval but adds, "there is no way that the cost [of stabilizing abatement measures] will send us to the poorhouse" (Lomborg 2001, p. 323).[42]

The second point is that Nordhaus's work is extremely controversial. Some claim that his model is simplistic, both in itself and, especially, relative to the climate models.[43] Indeed, one commentator goes so far as to say that "the model is extremely simple—so simple that I once, during a debate, dubbed it a toy model" (Gundermann 2002, p. 150). Others offer rival models that endorse the exact opposite of Nordhaus's conclusion: that strong action now (in the form of substantial carbon taxes, etc.) would be more beneficial in the long term than waiting, even perhaps if global warming does not actually transpire (e.g., Costanza 1996; De Leo et al. 2001; Woodward and Bishop 1997).

Part of the reason such disputes arise is that the models embody some very questionable assumptions.[44] Some are specific to Nordhaus (e.g., Gundermann 2002, p. 154). But others are the result of two more general kinds of difficulty.

The first is practical. There are severe informational problems involved in any reliable cost-benefit analysis for climate change. In particular, over the time scale relevant for climate change, "society is bound to be radically transformed in ways which are utterly unpredictable to us now," and these changes will themselves be affected by climate (Broome 1992, p. 10; see also Jamieson 1992, pp. 288–289).[45] Broome, for example, argues that fine-grained cost-benefit analyses are simply not possible for climate change.

The second kind of difficulty, of more interest to ethicists perhaps, is that there are some basic philosophical problems inherent in the methods of conventional economic analysis. I will mention just two prominent examples.

One concerns the standard economic treatments of intergenerational issues. Economists typically employ a social discount rate (SDR) of 2 to 10 percent for future costs[46] (Lomborg uses 5 percent; Nordhaus 3 to 6 percent).[47] But this raises two serious concerns. The first is that, for the short- to medium-term effects of climate change (say, over 10 to 50 years), model results can be extremely sensitive to the rate chosen. For example, Shultz and Kasting claim that the choice of SDR makes the rest of the climate-change model largely irrelevant in Nordhaus's model, and variations in the SDR make a huge difference to model results more generally (Schultz and Kasting 1997, cited by Gundermann 2002, p. 147). The other concern is that when the SDR is positive, all but the most catastrophic costs disappear after a number of decades, and even these become minimal over very long time periods.[48] This has serious consequences for the intergenerational ethics of climate change. As John Broome puts it: "It is people who are now children and people who are not yet born who will reap most of the benefits of any project that mitigates the effects of global warming. Most of the benefits of such a project will therefore be ignored by the consumer-price method of project evaluation. It follows that this method is quite useless for assessing such long-term projects. This is my main reason for rejecting it [for climate change]" (Broome 1992, p. 72).[49]

The second philosophical problem inherent in conventional economic analysis is that it cannot adequately capture all of the relevant costs and benefits. The obvious cases here are costs to nonhumans (such as animals, plants, species, and ecosystems) and noneconomic costs to humans, such as aesthetic costs (Sagoff 1998; Schmidtz 2001). But there is also concern that conventional economic analysis cannot adequately take into account costs with special features, such as irreversible and nonsubstitutable damages, that are especially associated with climate change (Shogren and Toman 2000; Costanza 1996).[50]

We can conclude, then, that there are strong reasons to be skeptical about Lomborg's cost argument in particular and about the reliability of fine-grained economic analyses of climate change more generally. Still, John Broome argues that two things can be said with some confidence: first, the specific effects of climate change "are very uncertain," where (as argued in the previous section) "this by itself has important consequences for the work that needs to be done"; and second, these effects "will certainly be long lived, almost certainly large, probably bad, and possibly disastrous" (Broome 1992, p. 12). To these claims, we might add that at 2 percent of world production, the estimated costs of stabilizing emissions do not seem obviously prohibitive.

2. The Adaptation Argument

We can now turn to the more general argument that instead of reducing emissions, we should pursue a policy of trying to adapt to the effects of climate change.[51] The first thing to note about this argument is that adaptation measures will clearly need to be part of any sensible climate policy, because we are already committed to some warming as a result of past emissions, and almost all of the proposed abatement strategies envisage that overall global emissions will continue to rise for at least the next few decades, committing us to even more.[52] Hence, the choice cannot be seen as being one between abatement and adaptation, since advocates of abatement generally support a combination of strategies. The real issue is rather whether adaptation should be our only strategy, so that abatement is ignored (Jamieson, 2005).

If this is the proposal, several points can be made about it. First, we should beware of making the case for adaptation a self-fulfilling prophecy. For example, it is true that the existing capital stock in the United States made it difficult for America to meet its original Kyoto target for 2008–2012.[53] But it is also true that a significant amount of this capital was invested after the United States committed itself to stabilizing emissions at the Rio Earth Summit of 1992. Furthermore, matters will only get worse. As of 2003, the Bush administration's energy plan called for building 1,300 new power plants in the next 20 years, boosting supply (and thereby emissions) by more than 30 percent.

Second, the comparison between abatement and adaptation costs looks straightforward but is not. In particular, we have to bear in mind the different kinds of economic costs at stake in each case. On the one hand, suppose we allow global warming to continue unchecked. What will we be adapting to? Chances are, we will experience both a range of general gradual climatic changes and an increase in severe weather and climate events. On the other hand, if we go for abatement, we will also be adapting but this time to increases in tax rates on (or decreases in permits for) carbon emissions.[54] But there is a world of difference between these kinds of adaptation: in the first case, we would be dealing with sudden, unpredictable, large-scale impacts descending at random on particular individuals, communities, regions, and industries and visiting them with pure, unrecoverable costs,[55] whereas in the second, we would be addressing gradual, predictable, incremental impacts, phased in so as to make adaptation easier.[56] Surely, adaptation in the second kind of case is, other things being equal, preferable to that in the first.[57]

Third, any reasonable abatement strategy would need to be phased in gradually, and it is well documented that many economically beneficial energy savings could be introduced immediately, using existing technologies.[58] These facts suggest that the adaptation argument is largely irrelevant to what to do now. The first steps that need to be taken would be economically beneficial, not costly. Yet opponents of action on climate change do not want to do even this much.

V. Risk Management and the Precautionary Principle

"The risk assessment process ... is as much policy and politics as it is science. A typical risk assessment relies on at least 50 different assumptions about exposure, dose-response, and relationships between animals and humans. The modeling of uncertainty also depends on assumptions. Two risk assessments conducted on the same problem can vary widely in results."

—Carolyn Raffensberger and Joel Tickner (1999, p. 2)

As serious as they are, these largely technical worries about conventional economic analysis are not the only reasons to be wary of any economic solution to the climate-change problem. Some writers suggest that exclusive reliance on economic analysis would be problematic even if all of the numbers were in, since the climate problem is ultimately one of values, not efficiency. As Dale Jamieson puts it, its "fundamental questions" concern "how we ought to live, what kinds of societies we want, and how we should relate to nature and other forms of life" (Jamieson 1992, p. 290).

But the problem may not be just that climate change raises issues of value. It may also show that our existing values are insufficient to the task. Jamieson, for example, offers the following argument. First, he asserts that our present values evolved relatively recently, in "low-population-density and low-technology societies, with seemingly unlimited access to land and other resources." Then he claims that these values include as a central component an account of responsibility that "presupposes that harms and their causes are individual, that they can be readily identified, and that they are local in time and space." Third, he argues that problems such as climate change fit none of these criteria. Hence, he concludes, a new value system is needed (Jamieson 1992, pp. 291–292).[59]

How, then, should we proceed? Some authors advocate a rethinking of our basic moral practices. For example, Jamieson claims that we must switch our focus away from approaches (such as those of contemporary economics) that concentrate on "calculating probable outcomes" and instead foster and develop a set of "twenty-first century virtues," including "humility, courage, . . . moderation," "simplicity and conservatism" (Jamieson 1992, p. 294).

Other climate-change theorists, however, are less radical. For example, Henry Shue employs the traditional notions of a "No Harm Principle" and rights to physical security (Shue 1999a, p. 43). He points out that even in the absence of certainty about the exact impacts of climate change, there is a real moral problem posed by subjecting future generations to the risk of severe harms. This implies a motive for action in spite of the scientific and economic uncertainties. Similarly, many policy makers appeal to the "precautionary principle,"[60] which is now popular in international law and politics[61] and receives one of its canonical statements in the United Nations Framework Convention on Climate Change (1992).[62] The exact formulation of the precautionary principle is controversial, but one standard version is the Wingspread Statement, which reads: "When an activity raises threats of harm to human health or the environment, precautionary measures should be taken even if some cause and effect relationships are not fully established scientifically" (Wingspread Statement 1998).

Both "no harm" principles and the precautionary principle are, however, controversial. "No harm" principles are often criticized for being either obscure or overly conservative when taken literally. The precautionary principle generates similar objections; its critics say that it is vacuous, extreme, and irrational.[63] Still, I would argue that, at least in the case of the precautionary principle, many of these initial objections can be overcome (Gardiner 2006). In particular, a core use of the precautionary principle can be captured by restricting its application to those situations that satisfy John Rawls's criteria for the application of a maximin principle: the parties lack, or have good reason to doubt, relevant probability information; they care little for potential gains; and they face unacceptable outcomes (Rawls 1999, p. 134). And this core use escapes the initial, standard objections.[64]

More importantly for current purposes, I would also claim that a reasonable case can be made that climate change satisfies the conditions for the core precautionary principle (Gardiner 2004a). First, many of the predicted outcomes from climate change seem severe, and some are catastrophic. Hence, there are grounds for saying that there are unacceptable outcomes. Second, as we have seen, for gradual change, either the probabilities of significant damage from climate change are high, or else we do not know the probabilities; and for abrupt change, the probabilities are unknown. Finally, given widespread endorsement of the view that stabilizing emissions would impose a cost of "only" 2 percent of world production, one might claim that we care little about the potential gains—at least relative to the possibly catastrophic costs.

There is reason to believe, then, that the endorsement by many policy makers of some form of precautionary or "no harm" approach is reasonable for climate change. But exactly which "precautionary measures" should be taken? One obvious first step is that those changes in present energy consumption that would have short-term, as well as long-term, economic benefits should be made immediately. In addition, we should begin acting on low-cost emissions-saving measures as soon as possible. Beyond that, it is difficult to say exactly how we should strike a balance between the needs of the present and those of the future. Clearly, this is an area where further thought is urgently needed.

Still, it is perhaps worthwhile to close this section with one speculative opinion about how we should direct our efforts. By focusing on the possibility of extreme events, and considering the available science, Brian O'Neill and Michael Oppenheimer suggest in a recent article in *Science* that "taking a precautionary approach because of the very large uncertainties, a limit of 2°C above 1990 global average temperature is justified to protect [the West Antarctic Ice Sheet]. To avert shutdown of the [Thermohaline circulation], we define a limit of 3°C warming over 100 years" (O'Neill and Oppenheimer 2002). It is not clear how robust these assertions are. Still, they suggest a reasonable starting point for discussion. On the assumption that these outcomes are unacceptable and given the IPCC projections of a warming of between 1.4°C and 5.8°C over the century, both claims appear to justify significant immediate action on greenhouse-gas stabilization.[65]

VI. Responsibility for the Past

"I'll tell you one thing I'm not going to do is I'm not going to let the United States carry the burden for cleaning up the world's air, like the Kyoto Treaty would have done. China and India were exempted from that treaty. I think we need to be more even-handed."

—George W. Bush, from the second televised presidential debate of 2000 (Singer 2002, p. 30)

"Even in an emergency one pawns the jewellery before selling the blankets. . . . Whatever justice may positively require, it does not permit that poor nations be told to sell their blankets [compromise their development strategies] in order that the rich nations keep their jewellery [continue their unsustainable lifestyles]."

—Henry Shue (1992, p. 397; quoted by Grubb 1995, p. 478)

"To demand that [the developing countries] act first is patently unfair and would not even warrant serious debate were it not the position of a superpower."

—Paul Harris (2003)

Suppose that action on climate change is morally required. Whose responsibility is it? The core ethical issue here concerns how to allocate the costs and benefits of greenhouse-gas emissions and abatement.[66] On this issue, there is a surprising convergence of philosophical writers on the subject: they are virtually unanimous in their conclusion that the developed countries should take the lead role in bearing the costs of climate change, while the less developed countries should be allowed to increase emissions for the foreseeable future.[67]

Still, agreement on the fact of responsibility masks some notable differences about its justification, form, and extent, so it is worth assessing the competing accounts in more detail. The first issue to be considered is that of "backward-looking considerations."[68] The facts are that developed countries are responsible for a very large percentage of historical emissions, whereas the costs likely to be imposed by those emissions are expected to be disproportionately visited on the poorer countries (IPCC 1995, p. 94).[69] This suggests two approaches. First, one might invoke historical principles of justice that require that one "clean up one's own mess." This suggests that the industrialized countries should bear the costs imposed by their past emissions.[70] Second, one might characterize the earth's capacity to absorb man-made emissions of carbon dioxide as a common resource, or sink (Traxler 2002, p. 120),[71] and claim that, since this capacity is limited, a question of justice arises about how its use should be allocated (Singer 2002, pp. 31–32).[72] On this approach, the obvious argument to be made is that the developed countries have largely exhausted the capacity in the process of industrializing and so have, in effect, denied other countries the opportunity to use "their shares." On this view, justice seems to require that the developed countries compensate the less developed for this overuse.

It is worth observing two facts about these two approaches. First, they are distinct. On the one hand, the historical principle requires compensation for damage inflicted by one party on another and does not presume that there is a common resource; on the other, the sink consideration crucially relies on the presence of a common resource and does not presume that any (further) damage is caused to the disenfranchised beyond their being deprived of an opportunity for use.[73] Second, they are compatible. One could maintain that a party deprived of its share of a common resource ought to be compensated both for that and for the fact that material harm has been inflicted on it as a direct result of the deprivation.[74]

Offhand, the backward-looking considerations seem weighty. However, many writers suggest that in practice, they should be ignored.[75] One justification that is offered is that until comparatively recently, the developed countries were ignorant of the effects of their emissions on the climate and so should not be held accountable for past emissions (or at least those prior to 1990, when the IPCC issued its first report).[76] This consideration seems to me far from decisive, because it is not clear how far the ignorance defense extends.[77] On the one hand, in the case of the historical principle, if the harm inflicted on the world's poor is severe, and if they lack the means to defend themselves against it, it seems odd to say that the rich nations have no obligation to assist, especially when they could do so relatively easily and are in such a position largely because of their previous causal role. On the other hand, in the case of the sink consideration, if you deprive me of my share of an important resource, perhaps one necessary to my very survival, it seems odd to say that you have no obligation to assist because you were ignorant of what you were doing at the time. This is especially so if your overuse both effectively denies me the means of extricating myself from the problem you have created and further reduces the likelihood of fair outcomes on this and other issues (Shue 1992).[78]

A second justification for ignoring past emissions is that taking the past into account is impractical. For example, Martino Traxler claims that any agreement that incorporates backward-looking considerations would require "a prior international agreement on what constitutes international distributive justice and then an agreement on how to translate these considerations into practical allocations" and that, given that "such an agreement is [un] likely in our lifetime," insisting on it "would amount to putting off any implementation concerning climate change indefinitely" (Traxler 2002, p. 128). Furthermore, he asserts that climate change takes the form of a commons problem and so poses a significant problem of defection:[79] "Each nation is (let us hope) genuinely concerned with this problem, but each nation is also aware that it is in its interest not to contribute or do its share, regardless of what other countries do.... In short, in the absence of the appropriate international coercive muscle, defection, however unjust it may be, is just too tempting" (Traxler 2002, p. 122).

Though rarely spelled out, such pragmatic concerns seem to influence a number of writers. Still, I am not convinced—at least, by Traxler's arguments. For one thing, I do not see why a complete background understanding of international justice is required, especially just to get started.[80] For another, I am not sure that defection is quite the problem, or at least has the implications, that Traxler suggests. In particular, Traxler's argument seems to go something like this: since there is no external coercive body, countries must be motivated not to defect from an agreement; but (rich) countries will be motivated to defect if they are asked to carry the costs of their past (mis)behavior; therefore, past behavior cannot be considered, otherwise (rich) countries will defect. But this reasoning is questionable, on several grounds. First, it seems likely that if past behavior is not considered, then the poor countries will defect. Since, in the long run, their cooperation is required, this would suggest that Traxler's proposal is at least as impractical as anyone else's.[81] Second, it is not clear that no external coercive instruments exist. Trade and travel sanctions, for example, are a possibility and have precedents. Third, the need for such sanctions (and indeed, the problem of defection in general) is not brought on purely by including the issue of backward-looking considerations in negotiation, nor is it removed by their absence. So it seems arbitrary to disallow such considerations on this basis. Finally, Traxler's argument seems to assume (first) that the only truly urgent issue that needs to be addressed with respect to climate change is that of future emissions growth and (second) that this issue is important enough that concerns about (a) the costs of climate change to which we are already committed and (b) the problem of inequity in the proceeds from those emissions (e.g., that the rich countries may have, in effect, stolen rights to develop from the poorer countries) can be completely ignored. But such claims seem controversial.[82]

The arguments in favor of ignoring past emissions are, then, unconvincing. Hence, contrary to many writers on this subject, I conclude that we should not ignore the presumption that past emissions pose an issue of justice that is both practically and theoretically important. Since this has the effect of increasing the obligations of the developed nations, it strengthens the case for saying that these countries bear a special responsibility for dealing with the climate-change problem.

VII. Allocating Future Emissions

"The central argument for equal per capita rights is that the atmosphere is a global commons, whose use and preservation are essential to human well being."
—Paul Baer (2002, p. 401)

"Much like self-defense may excuse the commission of an injury or even a murder, so their necessity for our subsistence may excuse our indispensable current emissions and the resulting future infliction of harm they cause."
—Martino Traxler (2002, p. 107)

Let us now turn to the issue of how to allocate future emissions. Here I cannot survey all of the proposals that have been made, but I will consider four prominent suggestions.[83]

1. Equal Per Capita Entitlements

The most obvious initial proposal is that some acceptable overall level of anthropogenic greenhouse emissions should be determined and then that this should be divided equally among the world's population, to produce equal per capita entitlements to emissions.[84] This proposal seems intuitive but would have a radical redistributive effect. Consider the following illustration. Singer points out that stabilizing carbon emissions at current levels would give a per capita rate of roughly one metric ton per year. But actual emissions in the rich countries are substantially in excess of this: the United States is at more than five metric tons per capita; and Japan, Australia, and western Europe are all in a range from 1.6 to 4.2 metric tons per capita (with most below three). India and China, on the other hand, are significantly below their per capita allocation (at 0.29 and 0.76 metric ton, respectively).[85] Thus, Singer suggests (against Bush's claim at the beginning of the previous section), an "even-handed approach" implies that India and China should be allowed increases in emissions, while the United States should take a massive cut (Singer 2002, pp. 39–40).[86]

Two main concerns have been raised about the per capita proposa1.[87] The first is that it might encourage population growth, through giving countries an incentive to maximize their population in order to receive more emissions credits (Jamieson 2001, p. 301).[88] But this concern is easily addressed: most proponents of a per capita entitlement propose indexing population figures for each country to a certain time. For example, Jamieson proposes a 1990 baseline (relevant because of the initial IPCC report), whereas Singer proposes 2050 (to avoid punishing countries with younger populations at present). The second concern is more serious. The per capita proposal does not take into account the fact that emissions may play very different roles in people's lives. In particular, some emissions are used to produce luxury items, whereas others are necessary for most people's survival.

2. Rights to Subsistence Emissions

This concern is the basis for the second proposal on how to allocate emissions rights. Henry Shue argues that people should have inalienable rights to the minimum emissions necessary to their survival or to some minimal quality of life.[89] This proposal has several implications. First, it suggests that there might be moral constraints on the limitation of emissions, so that establishing a global emissions ceiling will not be simply a matter for climatologists or even economists. If some emissions are deemed morally essential, then they may have to be guaranteed even if this leads to an overall allocation above some scientific optimum. Traxler is explicit about why this is the case. Even if subsistence emissions cause harm, they can be morally excusable, because "they present their potential emitters with such a hard choice between avoiding a harm today or avoiding a harm in the future" that they are morally akin to self-defense.[90] Second, the proposal suggests that actual emissions entitlements may not be equal for all individuals and may vary over time. For the benefits that can actually be drawn from a given quantity of greenhouse-gas emissions vary with the existing technology, and the necessity of them depends on the available alternatives. But both vary by region and will no doubt evolve in the future, partly in response to emissions regulation. Third, as Shue says, the guaranteed-minimum principle does not imply that allocation of any remaining emissions rights above those necessary for subsistence must be made on a per capita basis. The guaranteed-minimum view is distinct from a more robust egalitarian position that demands equality of a good at all levels of its consumption (Shue 1995a, pp. 387–388); hence, above the minimum, some other criterion might be adopted.

The guaranteed-minimum approach has considerable theoretical appeal. However, there are two reasons to be cautious about it. First, determining what counts as a "subsistence emission" is a difficult matter, both in theory and in practice. For example, Traxler defines subsistence emissions in terms of physiologically and socially necessary emissions but characterizes social necessity as "what a society needs or finds indispensable in order to survive" (Traxler 2002, p. 106). But this is problematic. For one thing, much depends on how societies define what they find indispensable. (It is hard not to recall the first President Bush's comment, back in 1992, that "the American way of life is not up for negotiation.") For another, and perhaps more importantly, there is something procedurally odd about the proposal. It appears to envisage that the climate-change problem can be resolved by appealing to some notion of social necessity that is independent of, and not open to, moral assessment. But this seems somehow backward. After all, several influential writers argue that part of the challenge of climate change is the deep questions it raises about how we should live and what kinds of societies we ought to have (Jamieson 1992, p. 290; IPCC 2001a, 1.4; questioned by Lomborg 2001, pp. 318–322).

Second, in practice, the guaranteed approach may not differ from the per capita principle and yet may lack the practical advantages of that approach. On the first issue, given the foregoing point, it is hard to see individuals agreeing on an equal division of basic emissions entitlements that does anything less than exhaust the maximum permissible on other (climatological and intergenerational) grounds, and it is easy to see them being tempted to overshoot it. Furthermore, determining an adequate minimum may turn out to be almost the same task as (a) deciding what an appropriate ceiling would be and then (b) assigning per capita rights to the emissions it allows. For (a) would also require a view about what constitutes an acceptable form of life and how many emissions are necessary to sustain it. On the second issue, the subsistence emissions proposal carries political risks that the per capita proposal does not, or at least not to the same extent. For one thing, the claim that subsistence emissions are nonnegotiable seems problematic given the first point (above) that there is nothing to stop some people claiming that almost any emission is essential to their way of life. For another, the claim that nonsubsistence emissions need not be distributed equally may lead some in developed countries to argue that what is required to satisfy the subsistence constraint is extremely minimal and that emissions above that level should be either grandfathered or distributed on other terms favorable to those with existing fossil-fuel-intensive economies. But this would mean that developing countries might be denied the opportunity to develop, without any compensation.

3. Priority to the Least Well-Off

The third proposal I wish to consider offers a different justification for departing from the per capita principle, namely that such a departure might maximally (or at least disproportionately) benefit the least well-off.[91] The obvious version of this argument suggests, again, that the rich countries should carry the costs of dealing with global warming, and the LDCs should be offered generous economic assistance.[92] But there are also less obvious versions, some of which may be attributable to some global-warming skeptics.

The first is offered by Bjørn Lomborg, who claims that the climate-change problem is ultimately reduced to the question of whether to help poor inhabitants of the poor countries now or their richer descendants later. And he argues that the right answer is to help now, since the present poor are both poorer and more easily helped. Kyoto, he says, "will likely cost at least $150 billion a year, and possibly much more," whereas "just $70–80 billion a year could give all Third World inhabitants access to the basics like health, education, water and sanitation" (Lomborg 2001, p. 322).

But this argument is far from compelling. For one thing, it seems falsely to assume that helping the poor now and acting on climate change are mutually exclusive alternatives (Grubb 1995, p. 473, n. 25).[93] For another, it seems to show a giant leap of political optimism. If their past record is anything to go by, the rich countries are even less likely to contribute large sums of money to help the world's poor directly than they are to do so to combat climate change (Singer 2002, pp. 26–27).

A second kind of priority argument may underlie former President Bush's proposal of a "greenhouse gas intensity approach," which sought to index emissions to economic activity.[94] Bush suggested reducing the amount of greenhouse gas per unit of U.S. GDP by 18 percent in 10 years, saying that "economic growth is the solution, not the problem," and that "the United States wants to foster economic growth in the developing world, including the world's poorest nations" (Singer 2002, p. 43). Hence, he seemed to appeal to a Rawlsian principle.

Peter Singer, however, claims that there are two serious problems with this argument. First, it faces a considerable burden of proof: it must show that U.S. economic activity makes the poor not only better off but maximally so. Second, this burden cannot be met: not only do CIA figures show the United States "well above average in emissions per head it produces in proportion to per capita GDP,"[95] but "the vast majority of the goods and services that the US produces—89 per cent of them—are consumed in the US" (Singer 2002, pp. 44–45). This, Singer argues, strongly suggests that the world's poor would be better off if the majority of the economic activity the United States undertakes (with its current share of world emissions) occurred elsewhere.

4. Fair Chore Division

A final proposal superficially resembles the equal-intensity principle but is advocated for very different reasons. Martino Traxler proposes a "fair chore division," which equalizes the marginal costs of those aiming to prevent climate change. Such a proposal, he claims, is politically expedient, in that it (a) provides each nation in the global commons with "no stronger reasons to defect from doing its (fair) share than it gives any other nation" and so (b) places "the most moral pressure possible on each nation to do its part" (Traxler 2002, p. 129).

Unfortunately, it is not clear that Traxler's proposal achieves the ends he sets for it. First, by itself, (a) does not seem a promising way to escape a traditional commons or prisoner's dilemma situation. What is crucial in such situations is the magnitude of the benefits of defecting relative to those of cooperating; whether the relative benefits are equally large for all players is of much less importance.[96] Second, this implies that (b) must be the crucial claim, but (b) is also dubious in this context. Traxler explicitly rules out backward-looking considerations on practical grounds. But this means ignoring the previous emissions of the rich countries, the extent to which those emissions have effectively denied the LDCs "their share" of fossil-fuel-based development in the future, and the damages that will be disproportionately visited on the LDCs because of those emissions. So, it is hard to see why the LDCs will experience "maximum moral pressure" to comply. Third, equal-marginal-costs approaches are puzzling for a more theoretical reason. In general, equality-of-marginal-welfare approaches suffer from the intuitive defect that they take no account of the overall level of welfare of each individual. Hence, under certain conditions, they might license taking large amounts from the poor (if they are so badly off anyway that changes for the worse make little difference), while leaving the rich relatively untouched (if they are so used to a life of luxury that they suffer greatly from even small losses).[97] Now, Traxler's own approach does not fall into this trap, but this is because he advocates that costs should be measured not in terms of preferences or economic performance but rather in terms of subsistence, near-subsistence, and luxury emissions. Thus, his view is that the rich countries should have to give up all of their luxury emissions before anyone else need consider giving up subsistence and near-subsistence emissions. But this raises a new concern.[98] In practice, it means that Traxler's equal-burdens proposal actually demands massive action from the rich countries before the poor countries are required to do anything at all (if indeed they ever are). And however laudable, or indeed morally right, such a course of action might be, it is hard to see it as securing the politically stable agreement that Traxler craves, or, at least, it is hard to see it as more likely to do so than the alternatives. So, the equal-marginal-costs approach seems to undercut its own rationale.

VIII. What Has the World Done? The Kyoto Deal

"This has been a disgraceful performance. It is the single worst failure of political leadership that I have seen in my lifetime."

—Al Gore, then a U.S. senator, criticizing the first Bush administration's performance in Rio (Hopgood 1998, p. 199)

"The system is made in America, and the Americans aren't part of it."

—David Doniger, former Kyoto negotiator and director of climate programs for the Natural Resources Defense Council (Pohl 2003)

We have seen that there is a great deal of convergence on the issue of who has primary responsibility to act on climate change. The most defensible accounts of fairness and climate change suggest that the rich countries should bear the brunt, and perhaps even the entirety, of the costs. What, then, has the world done?

The current international effort to combat climate change has come in three main phases. The first came to fruition at the Rio Earth Summit of 1992. There, the countries of the world committed themselves to the Framework Convention on Climate Change (FCCC), which required "stabilization of greenhouse gas concentrations in the atmosphere at a level that would prevent dangerous anthropogenic interference with the climate system" and endorsed a principle of "common but differentiated responsibilities," according to which the richer, industrialized nations (listed under "Annex I" in the agreement) would take the lead in cutting emissions, while the less developed countries would pursue their own development and take significant action only in the future.[99] In line with the FCCC, many of the rich countries (including the United States, the European Union, Japan, Canada, Australia, New Zealand, and Norway) announced that they would voluntarily stabilize their emissions at 1990 levels by 2000.

Unfortunately, it soon became clear that merely voluntary measures were ineffective. As it turned out, most of those who had made declarations did nothing meaningful to try to live up to them, and their emissions continued to rise without constraint.[100] Thus, a second phase ensued. In a meeting in Berlin in 1995, it was agreed that the parties should accept binding constraints on their emissions, and this was subsequently achieved in Japan in 1997, with the negotiation of the Kyoto Protocol.[101] This agreement initially appeared to be a notable success, in that it required the Annex I countries to reduce emissions to roughly 5 percent below 1990 levels between 2008 and 2012. But it also contained two major compromises on the goal of limiting overall emissions, in that it allowed countries to count forests as sinks and to meet their commitments through buying unused capacity from others through permit trading.

The promise of Kyoto turned out to be short-lived. First, it proved so difficult to thrash out the details that a subsequent meeting, in the Hague in November 2000, broke down amid angry recriminations. Second, in March 2001, the Bush administration withdrew U.S. support, effectively killing the Kyoto agreement. Or so most people thought. As it turned out, the U.S. withdrawal did not cause immediate collapse. Instead, during the remainder of 2001, in meetings in Bonn and Marrakesh, a third phase began in which a full agreement was negotiated, with the European Union, Russia, and Japan playing prominent roles,[102] and sent to participating governments for ratification. Many nations swiftly ratified, including the European Union, Japan, and Canada, so that, at the time of writing (2003), the Kyoto Treaty needs only ratification by Russia to pass into international law.[103]

On the surface, then, the effort to combat global climate change looks a little bruised but still on track. But this appearance may be deceptive. There is good reason to think that the Kyoto Treaty is deeply flawed, both in its substance and in its background assumptions (Barrett 2003; Gardiner 2004). Let us begin with two substantive criticisms.

The first is that Kyoto currently does very little to limit emissions. Initial projections suggested that the Bonn-Marrakesh agreement would reduce emissions for participants by roughly 2 percent on 1990 levels, down from the 5 percent initially envisaged by the original Kyoto agreement (Ott 2001). But recent research suggests that such large concessions were made in the period from Kyoto to Marrakesh that (a) even full compliance by its signatories would result in an overall increase in their emissions of 9 percent above 2000 levels by the end of the first commitment period, and (b) if slow economic growth persisted, this would actually match or exceed projected business-as-usual emissions (Babiker et al. 2002). Coupled with emissions growth in the LDCs, this means that there will be another substantial global increase by 2012.[104] This is nothing short of astounding, given that by then, we will be "celebrating" 20 years since the Earth Summit (Gardiner 2004).

It is worth pausing to consider potential objections to this criticism. Some would argue that, even if it achieves very little, the current agreement is to be valued either procedurally (as a necessary first step),[105] symbolically (for showing that some kind of agreement is possible),[106] geopolitically (for showing that the rest of the world can act without the United States),[107] or as simply the best that is possible under current conditions (Athanasiou and Baer 2001, 2002, p. 24). There is something to be said for these views. The current Kyoto Protocol sets targets only for 2008–2012, and these targets are intended as only the first of many rounds of abatement measures. Kyoto's enthusiasts anticipate that the level of cuts will be deepened and their coverage expanded (to include the developing countries) as subsequent targets for new periods are negotiated.[108]

Nevertheless, I remain skeptical. This is partly because of the history of climate negotiations in general and the current U.S. energy policy in particular and partly because I do not think future generations will see reason to thank us for symbolism rather than action. But the main reason is that there are clear ways in which the world could have done better (Gardiner 2004).

This leads us to the second substantive criticism of Kyoto: that it contains no effective compliance mechanism. This criticism arises because, although the Bonn-Marrakesh agreement allows for reasonably serious punishments for those who fail to reach their targets,[109] these punishments cannot be enforced.[110] The envisioned treaty has been set up so that countries have several ways to avoid being penalized. On the one hand, enforcement is not binding on any country that fails to ratify the amendment necessary to punish it (Barrett 2003, p. 386).[111] On the other hand, the penalties take the form of more demanding targets in the next decade's commitment period—but parties can take this into account when negotiating their targets for that commitment period, and in any case, a country is free to exit the treaty with one year's notice, three years after the treaty has entered into force for it (FCCC, article 25).[112]

The compliance mechanisms for Kyoto are thus weak. Some would object to this, saying that they are as strong as is possible under current institutions.[113] But I argue that this is both misleading and, to some extent, irrelevant. It is misleading because other agreements have more serious, external sanctions (e.g., the Montreal Protocol on ozone depletion allows for trade sanctions) and also because matters of compliance are notoriously difficult in international relations, leading some to suggest that it is only the easy, and comparatively trivial, agreements that get made. It is somewhat irrelevant because part of what is at stake with climate change is whether we have institutions capable of responding to such global and long-term threats (Gardiner 2004).

Kyoto is also flawed in its background assumptions. Consider the following three examples. First, the agreement assumes a "two-track" approach, whereby an acceptable deal on climate can be made without addressing the wider issue of international justice. But this, Shue argues, represents a compound injustice to the poor nations, whose bargaining power on climate change is reduced by existing injustice (Shue 1992, p. 373). Furthermore, this injustice appears to be manifest, in that the treaty directly addresses only the costs of preventing future climate change and only indirectly (and minimally) addresses the costs of coping with climate change to which we are already committed (Shue 1992, p. 384).[114] Second, the Bonn-Marrakesh deal eschews enforcement mechanisms external to the climate-change issue, such as trade sanctions. Given the apparent fragility of such a commitment on the part of the participant countries, this is probably disastrous. Third, Kyoto takes as its priority the issue of cost-effectiveness. As several authors point out, this tends to shift the focus of negotiations away from the important ethical issues and (paradoxically) to tend to make the agreement less, rather than more, practica1.[115]

Why is Kyoto such a failure? The reasons are no doubt complex and include the political role of energy interests, confusion about scientific uncertainties and economic costs, and the inadequacies of the international system. But two further factors have also been emphasized in the literature, and I will mention them in closing. The first is the role of the United States, which with 4 percent of the world's population emits roughly 25 percent of global greenhouse gases. From the early stages, and on the most important issues, the United States effectively molded the agreement to its will, persistently objecting when other countries tried to make it stronger. But then it abandoned the treaty, seemingly repudiating even those parts on which it had previously agreed. This behavior has been heavily criticized for being seriously unethical (e.g., Brown 2002; Harris 2000a).[116] Indeed, Singer even goes so far as to suggest that it is so unethical that the moral case for economic sanctions against the United States (and other countries that have refused to act on climate change) is stronger than it was for apartheid South Africa, since the South African regime, as horrible as it was, harmed only its own citizens, whereas the United States harms citizens of other countries.

The second reason behind Kyoto's failure is its intergenerational aspect. Most analyses describe the climate-change problem in intragenerational, game-theoretic terms, as a prisoner's dilemma (Barrett 2003, p. 368; Danielson 1993, pp. 95–96; Soroos 1997, pp. 260–261) or battle-of-the-sexes problem (Waldron 1990).[117] But I have argued that the more important dimension of climate change may be its intergenerational aspect (Gardiner 2001). Roughly speaking, the point is that climate change is caused primarily by fossil-fuel use. Burning fossil fuels has two main consequences: on the one hand, it produces substantial benefits through the production of energy; on the other hand, it exposes humanity to the risk of large, and perhaps catastrophic, costs from climate change. But these costs and benefits accrue to different groups: the benefits arise primarily in the short to medium term and so are received by the present generation, but the costs fall largely in the long term, on future generations. This suggests a worrying scenario. For one thing, as long as high energy use is (or is perceived to be) strongly connected to self-interest, the present generation will have strong egoistic reasons to ignore the worst aspects of climate change. For another, this problem is iterated: it arises anew for each subsequent generation as it gains the power to decide whether or not to act. This suggests that the global-warming problem has a seriously tragic structure. I have argued that it is this background fact that most readily explains the Kyoto debacle (Gardiner 2004).[118]

IX. Conclusion

This chapter has been intended as something of a primer. Its aim is to encourage and facilitate wider engagement by ethicists with the issue of global climate change.[119] At the outset, I offered some general reasons to explain why philosophers should be more interested in climate change. In closing, I would like to offer one more. I have suggested that climate change poses some difficult ethical and philosophical problems. Partly as a consequence of this, the public and political debate surrounding climate change is often simplistic, misleading, and awash in conceptual confusion. Moral philosophers should see this as a call to arms. Philosophical clarity is urgently needed. Given the importance of the problem, let us hope that the call is answered quickly.

Acknowledgments

For support during an early stage of this work, I am very grateful to the University of Melbourne Division of the ARC Special Research Centre for Applied Philosophy and Public Ethics and to the University of Canterbury, New Zealand. For helpful discussion, I would like to thank Chrisoula Andreou, Paul Baer, Roger Crisp, David Frame, Leslie Francis, Dale Jamieson, David Nobes, and especially the reviewers for *Ethics*. I am especially grateful to Robert Goodin for both suggesting and encouraging this project.

Notes

1. Prominent exceptions include John Broome (Broome 1992), Dale Jamieson (including Jamieson 1990, 1991, 1992, 1996, 1998, 2001, 2005), Henry Shue (Shue 1992, 1993, 1994, 1995a, 1995b, 1996, 1999a, 1999b, 2004), and an early anthology (Coward and Hurka 1993). Recently, a few others have joined the fray. Gardiner (2004b), Singer (2002), and Traxler (2002) all write specifically about climate change; and Francis (2003), Gardiner (2001), and Green (2002) discuss issues in global ethics more generally but take climate change as their lead example. (Moellendorf 2002 contains a short but substantive discussion.) There are also brief overviews in two recent collections (Hood 2003, Shue 2001). There is rather more work by nonphilosophers. Grubb (1995) is something of a classic. Also worth reading are Athanasiou and Baer 2002; Baer 2002; Harris 2000a, 2001; Holden 1996, 2002; Intergovernmental Panel on Climate Change (IPCC) 1995; Lomborg 2001; Paterson 1996, 2001; Pinguelli-Rosa and Munasinghe 2002; and Victor 2001. Brown 2002 provides a very readable introduction, aimed at a general audience.

2. Such claims are made by both liberals (such as former U.S. President Bill Clinton and Britain's former environment minister Michael Meacher) and conservatives (U.S. Senator Chuck Hagel and the Bush administration's first EPA director, Christine Todd Whitman). See Johansen 2002, pp. 2, 93; and Lomborg 2001, p. 258.

3. For example, the most authoritative report on the subject begins: "Natural, technical, and social sciences can provide essential information and evidence needed for decisions on what constitutes 'dangerous anthropogenic interference with the climate system.' At the same time, *such decisions are value judgments* determined through socio-political processes, taking into account considerations such as development, equity, and sustainability, as well as uncertainties and risk" (IPCC 2001b, p. 2, emphasis added). See also Grubb 1995, p. 473.

4. For example, I argue that climate change is an instance of a severe and underappreciated intergenerational problem (Gardiner 2001).

5. Sometimes skeptics suggest that the terminological change is suspicious. Recently, however, most have embraced it.

6. It is perhaps worth pointing out that the global-warming problem is distinct from the problem of stratospheric ozone depletion. Ozone depletion is principally caused by man-made chlorofluorocarbons (CFCs) and has as its main effect the ozone "hole" in the southern hemisphere, which increases the intensity of radiation dangerous to human health through incidence of skin cancer. These compounds are currently regulated by the Montreal Protocol, apparently with some success. Since some of them are also potent greenhouse gases, their regulation is to be welcomed from the point of view of global warming. However, their main replacements, hydrochlorofluorocarbons (HCFCs) and hydrofluorocarbons (HFCs) are also greenhouse gases, although they are less

potent and less long-lived than CFCs. There is an agreement to phase out HCFCs by 2030, but the concentration of such compounds remains a concern from the point of view of global warming. (See Houghton 1997, pp. 35–38. Houghton's book provides an excellent overview of the science. Also worth reading is Alley 2000.)

7. Houghton calculates that the average temperature at the earth's surface without the natural greenhouse effect would be −6°C. With the natural effect, it is about 15°C (Houghton 1997, pp. 11–12).

8. Skeptics sometimes correctly point out that the earth has been much warmer in previous periods of its history. They might also note, however, that we were not around during those times, that the climate has been extremely stable during the rise of civilization, and that we have never been subject to climate changes as swift, or of such a magnitude, as those projected by the IPCC.

9. It is perhaps worth noting that *climate change* is not yet the perfect term. For one thing, it may turn out that there are other ways in which humans can profoundly alter global climate than through greenhouse gases; for another, much of our concern with climate change would remain even if it turned out to have a natural source.

10. It should be noted that IPCC processes are politicized in several ways. For one thing, the scientific membership is decided by participant governments, which nominate their representatives. For another, the most important part of each report (the Summary for Policymakers, or SPM) is approved by member governments on a line-by-line, consensus basis (although this is not true of the scientific reports themselves). The latter procedure in particular is vigorously attacked both by skeptics (see, e.g., Lomborg 2001, p. 319, who complains that the IPCC toughened the language of the 2001 SPM for political reasons) and by nonskeptics (many of whom believe that the consensus necessary for the SPMs substantially weakens the claims that would be justified based on the fuller scientific reports). Since they were the subject of intense negotiation, I have repeated the precise wording of the IPCC statements here, rather than paraphrasing.

11. *2009 Update*: The 2007 report appeared three years after this article was orginally published. Its basic message is the same, albeit expressed with greater confidence. For example, in 2007, the IPCC upgrade its assessment of the claim that "most of the observed warming over the last 50 years has been due to the increase in greenhouse gas concentrations" from "likley" to "very likely" (meaning a probability of 90 percent or more). The IPCC's main conclusions have been endorsed by all major scientific bodies, including the National Academy of Sciences, the American Meterological Society, the Amercian Geophysical Union, and the American Association for the Advancement of Science.

12. The U.S. National Academy of Science's Committee on the Science of Climate Change reviewed the issue in 2001, at the request of the Bush administration, and found itself in general agreement with the IPCC.

13. The IPCC's scientific report defines likelihoods in terms of probabilities. Its definitions are as follows: virtually certain (greater than 99 percent chance that a result is true), very likely (90–99 percent chance), likely (66–90 percent chance), of medium likelihood (33–66 percent chance), unlikely (10–33 percent chance), very unlikely (1–10 percent chance), and exceptionally unlikely (less than 1 percent chance). See IPCC 2001b, p. 152, n. 7.

14. Some phenomena that are sometimes cited as a source of concern are reported not to have shown a change yet. These include tropical storm intensity and frequency; the frequency of tornados, thunder, and hail; and the extent of Antarctic sea ice (IPCC 2001b, p. 154).

15. Water vapor is the main atmospheric greenhouse gas, but humans have been doing little to increase its concentration. However, the IPCC does report that one expected consequence of global warming would be an increase in water-vapor concentration as a positive feedback.

16. For this reason, David Victor argues that methane emissions do not raise the same issues of intergenerational justice as CO_2 emissions, for most of the warming effects of the former will be visited in the short to medium term on the present and next generation (Victor 2001).

17. Other, but less significant, contributing factors include nitrous oxide, halocarbons, aerosols, and natural factors (including variations in solar output) (IPCC 2001b, p. 157).

18. Furthermore, the temperature rise is not evenly spread. Models suggest that it is "very likely" that the land will warm more quickly, and more so in the northern hemisphere. In fact, northern North America and Asia are projected to exceed the global average "by more than 40 percent." Based on these temperature results, over the course of the 21st century, the IPCC predicts increases in global average water-vapor concentration and precipitation, mean sea level, maximum and minimum temperatures, the number of hot days, and the risk of drought and decreases in the

day-night temperature range and (in the northern hemisphere) in snow cover and sea ice (IPCC 2001b, pp. 161–163).

19. *2009 Update*: Scientific concern about this specific tipping point seems to have diminished of late. But many facets of the work remain controversial. Though most agree that the past events occured and were accompanied by a slowdown, there is disagreement about the extent of the climatic impacts, how they might be relevent to predicting future climate change, and whether we are indeed seeing signs of such change already. Still, much of this controversy concerns *when* we might expect a change, not whether there will be one if global warming continues well into the future. Models do predict a point "beyond which the thermohaline circulation cannot be maitained." But there is a disagreement about what conditions are necessary to trigger this. On the one hand, many scientists apparently believe that it requires warming of 4–5 degrees Celsius, and that we will not experience that this century (IPCC 2007; Schiermeier 2006, 257). On the other hand, some say that the range goes lower, to 3–5 degrees, and that some simulations "clearly pass a THC tipping point this century" (Lenton et al. 2008, 1789–1790). From an ethical point of view, we should note that even a small probability of collapse this century is a matter for concern, and (more importantly) that it is not clear why we should put so much emphasis on whether it may come before or after 2100.

20. See, e.g., former White House spokesman Ari Fleischer, as quoted by Traxler 2002, p. 105.

21. There is some case for this. It is not clear how the IPCC generates its "probability" estimates (Reilly et al. 2001).

22. For example, using ozone depletion and deforestation as his case studies, Rado Dimitrov argues that the crucial variable in resolving global environmental problems is knowledge of their cross-border consequences, rather than of their extent and causes, since this "facilitates utility calculations and the formation of interests" (Dimitrov 2003, p. 123).

23. For example, suppose I am weighing a job offer in a distant city. Suppose also that one major consideration in my decision is what kind of life my 18-month-old son will have. The information I have about this is riddled with uncertainty. I know that my current location offers many advantages as a place for children to grow up (e.g., the schools are good, the society values children, there are lots of wholesome activities available) but some considerable disadvantages (e.g., great distances from other family members, a high youth-suicide rate). But I have no idea how these various factors might affect my son (particularly since I can only guess at this stage what his personality might turn out to be). So I am in a situation of uncertainty.

24. For example, suppose that the position I've been offered is on the other side of the world in New Zealand. Suppose also that I have never been to New Zealand, nor do I know anyone who has. I might be completely bereft of information on which to base a decision. (These days, of course, I have the Internet, the local library, and Amazon.com. But pity the situation of the early settlers.)

25. For example, suppose I'm considering the job offer again, but now I'm thinking about whether my 15-year-old daughter will like the move. This time, I do have considerable information about her personality, preferences, goals, and aspirations. But this does not mean that there is not considerable uncertainty about how good the move would be for her. Suppose, I know that the most important thing from her point of view is having very close friends. I also know that she is good at making friends, but I don't know whether suitable friends will present themselves.

26. According to Brown, these facts have been obscured in the American mind by aggressive propaganda campaigns by some business interests and the media's tendency to run "for and against" articles (and so overrepresent the views of skeptics).

27. There are also notable issues within this data set, especially in comparing different instruments used and in a possible locational bias in favor of urban areas, which have quite likely warmed during the period as a result of industrialization.

28. *2009 Update*: This worry has substantially diminished since Wigley et al. 2006 reported that the discrepancy rested on data errors which have now been corrected.

29. There is, of course, an important presumption here. Dale Jamieson points out that the very idea of climate change presupposes a paradigm of stability versus change, and this brings with it a need to distinguish signal from noise (see Jamieson 1991, pp. 319–321).

30. According to data largely from Arctic ice cores, in the last 10,000 years, the variation in average global temperatures was less than 1°C; in the preceding 100,000 years, variations were sometimes experienced of up to 5°C or 6°C in less than 100 years (Houghton 1997, chap. 4; United Nations Environment Program 1999, sheet 8).

31. A significant and poorly understood factor here is energy output from the sun (although

fluctuations caused by variations in the earth's orbit are better known).

32. Interestingly, this does not imply that we should not have a policy to limit emissions. Since a prolonged natural warming would be just as disastrous for current patterns of human life on the planet as artificially induced warming, it could turn out that some abatement of projected anthropogenic emissions would be justified as a counteracting measure.

33. I have pointed out elsewhere that the potential gains from carbon emissions are far from exhausted, given the low per capita rates in most parts of the world. Hence, even if global warming were not yet occurring, we would, other things being equal, expect it at some time in the future, as global emissions rise (Gardiner 2004).

34. These may amplify the direct warming by a factor of two or three (United Nations Environment Program 1999, sheet 7).

35. In particular, there is no reason to assume that our planet's atmosphere is robustly stable in the face of different inputs. The atmosphere of Venus, for example, has undergone a runaway greenhouse effect. (It is easy to forget that what we are dealing with fundamentally is a band of gases around the earth that is just a few miles wide.)

36. For an overview, see Edwards et al. 2007.

37. David Frame has suggested to me that the problem has more to do with the models being tuned to fit the current and recent climate record and that the lingering errors might result from the omission from the models of processes such as fully interactive biogeochemical and cryosphere cycles.

38. The IPCC is sometimes criticized for now positing a wider projection range in its latest report than before. This suggests expanding uncertainty. But it is worth noting that the IPCC range is not, as might be expected, a statistical measure, capturing error bars. Instead, it encompasses a cluster of model results. (Leading climate scientists such as Stephen Schneider have criticized the IPCC for being misleading here and so leaving themselves open to political manipulation.)

39. See, e.g., Lomborg 2001, p. 317 (although Lomborg does argue elsewhere in the chapter that the IPCC overstates both the temperature effect and the importance of the likely consequences).

40. The models and their results are summarized in Mabey et al. 1997, chap. 3.

41. Nordhaus claims that even the Kyoto controls are much too aggressive. For why this might be surprising, see the later discussion of the Kyoto Protocol.

42. Peter Singer adds that with global emissions trading, Lomborg's own figures suggest that Kyoto would be a net economic benefit (Singer 2002, p. 27). Lomborg's argument, of course, is that even though this is true, the investment would be better placed elsewhere, in direct aid to poor countries (Lomborg 2001, p. 322).

43. It is worth noting that there is a serious paradox for at least some skeptics here. Some are both very skeptical and demanding on the standards they impose on predictive models from climatology but not at all cautious about the power of the economic models on which they choose to focus. But this should be surprising. For, without wishing in any way to be derogatory about contemporary macroeconomics, it has at least as dubious a status as a predictive science as climatology, if not worse. Hence, if one is going to be quite so critical of the IPCC consensus on climate change as some skeptics are, one should be even-handed in one's approach to the economic models (Gundermann 2002, p. 154).

44. For example, many models (including Nordhaus's) do not take into account indirect social and environmental costs and benefits not associated with production. But some claim that benefits of this sort might actually outweigh the direct costs of abatement (see, e.g., De Leo et al. 2001, pp. 478–479).

45. Jamieson is particularly concerned about climate effects. He says that the regional effects are varied and uncertain; predicting human behavior will be difficult since the impacts will affect a wide range of social, economic, and political activities; we have limited understanding of the global economy; and there will be complex feedbacks among different economic sectors.

46. Discounting is "a method used by economists to determine the dollar value today of costs and benefits in the future. Future monetary values are weighted by a value < 1, or 'discounted'" (Toman 2001, p. 267). The SDR is the rate of discounting: "Typically, any benefit (or cost), B (or C), accruing in T years' time is recorded as having a 'present' value, PV of…" (Pearce 1993, p. 54).

47. For philosophical objections to the SDR, see Parfit (1985, app. F). A (partial) reply is to be found in Broome (1994). However, Broome explicitly denies that a positive SDR should be used for climate change (see Broome 1992, pp. 60, 72).

48. Alex Dubgaard makes the point with an example. Suppose that Denmark needs to be evacuated because of flooding. Current real estate value in Denmark is estimated at about $238 billion

(U.S.). If a discount rate of 5 percent is applied, then over 500 years, the same real estate would be worth just $6. Hence, "If they do not enlarge their property in the meantime, the loss of all real estate in Denmark would be compensated if, today, we make a saving equivalent to half a barbequed chicken with potato fritters." He calls such a conclusion obviously absurd (Dubgaard 2002, pp. 200–201).

49. This quotation refers specifically to the consumer-price method. But Broome also rejects other ways of generating a positive discount rate for future generations in the case of climate change (Broome 1992, chap. 3) and, indeed, specifically endorses a discount rate of zero in this context (Broome 1992, p. 108).

50. Economists tend to operate under the assumption that all goods are readily substitutable for one another, so that in principle, any one kind of good (such as clean air or blankets) can be substituted for any other kind (such as jewelry). But this seems dubious in general and, in the case of environmental quality, to embody a significant value judgment that is not widely shared. Good starting points for discussion of such philosophical issues might be Adler and Posner 2001; and Chang 1997.

51. This argument received political prominence at a meeting in Delhi in 2002, where it was promoted by the United States and India (Revkin 2002; Harding 2002).

52. This is why the IPCC and others speak of further emissions reductions as "mitigation" rather than prevention.

53. Victor argues that, given an actual 12 percent rise in U.S. emissions from 1990 to 1999 and a projected further 10 percent rise to 2008, the Kyoto requirement of a 7 percent cut on 1990 levels amounts to a 30 percent cut overall from projected emissions. He adds, "Compliance with a sharp 30% cut would force the premature disposal of some of the 'capital stock' of energy equipment and retard significant parts of the US economy. Electricity power generation is especially vulnerable. About half of US electric power is supplied by coal, which is the most greenhouse gas intensive of all fossil fuels. *The time to implement easy changes has already passed.* About four-fifths of the US generating capacity that will electrify 2010 will already have been built by the end of the year 2000" (Victor 2001, pp. 3–4, emphasis added).

54. Of course, in reality, the contrast between the two scenarios is not so stark. Since we are already committed to some warming as a result of past emissions, it is not true that we can completely shield ourselves from the possibility of unpredictable impacts. But we can shield ourselves to some extent from unpredictable impacts from our future emissions.

55. One effect of this would be to introduce new and more widespread costs. For example, since the impacts are unpredictable, all prudent agents will insure against them, so that some will spend money on emergency services and flood walls that they do not need. This contrasts with an abatement strategy, where the direct costs are incurred only by those responsible for excessive emissions.

56. Not only do we avoid the unnecessary costs mentioned above, but costs in the second case can be distributed in a rational fashion over the sources of the problem and may even generate revenue (through taxation or the price of permits), which could be used to alleviate the effects of warming to which we are already committed or for other socially beneficial purposes.

57. There is something of a paradox here in the attitudes of some commentators, who appear to have great faith in the ability of the market to adapt in the first case but not in the second. It is not clear what could justify such a prejudice. Commenting on some early works by Nordhaus and Beckerman, Broome says that they are "evidently assuming that human life is by now fairly independent of the natural world.... I find this assumption too complacent" (Broome 1992, p. 25, n. 31.)

58. There are many ways in which developed countries waste energy, and thereby carbon emissions, through inefficient practices. For example, the most fuel-efficient cars and trucks/sport-utility vehicles available in the United States are capable of 66 and 29 miles per gallon, respectively, on the open highway; the least efficient are capable of 14 and 16 miles per gallon (U.S. Environmental Protection Agency 2003). Furthermore, in recent years, manufacturers in the United States have actually stopped making the most fuel-efficient cars, as such vehicles have been crowded out of the marketplace by sport-utility vehicles. Hence, average fuel efficiency has declined (Heavenrich and Hellman 2000). Less markedly, substantial energy savings could be made simply by switching to the most efficient currently available models of washing machines, hot water heaters, and the like.

59. In a later article, Jamieson's position seems more modest. He suggests that there are two moral and legal paradigms associated with responsibility in the Western tradition: a causal paradigm and an "ability to benefit or prevent harm" paradigm. He then argues that the former founders with climate change; but the latter, which he associates with the utilitarian tradition, does not. See Jamieson 1998, pp. 116–117.

60. The literature on the precautionary principle is voluminous, though mostly written by nonphilosophers, and a thorough treatment of it would require a separate article. Two representative collections are O'Riordan, Cameron, and Jordan 2001; Raffensberger and Tickner 1999. Haller 2002 is a recent philosophical study of related issues, with some emphasis on climate change.

61. Versions appear in the Third North Sea Conference (1990) and the Ozone Layer Protocol (1987); they are also endorsed by major institutions, such as the UN Environment Program (1989), the European Union in its environment policy (1994), and the U.S. President's Council on Sustainable Development (1996). See Raffensberger 1999.

62. Some take the precautionary principle to be equivalent to a "do no harm" principle and to have roots in the Hippocratic Oath (see, e.g., Ozonoff 1999, p. 100).

63. In a recent piece in the *New York Times,* a self-described "former Reagan administration trade hawk" asserted: "Without any scientific grounds, but on the basis of the so-called precautionary principle—that is, if we can't prove absolutely that it is harmless, let's ban it—the [European] Union has prevented genetically modified food from the United States from entering its markets" (Prestowitz 2003). For more measured, philosophical criticisms, see Soule 2000; Manson 2002.

64. I would also argue that it renders many objections made to the principle in practical contexts misguided: instead of calling into doubt the reasonableness of the precautionary principle itself, critics are often arguing that the conditions for its application are not met.

65. O'Neill and Oppenheimer (2002) suggest stabilization at 450 parts per million of carbon dioxide, which would require a peak in global emissions between 2010 and 2020.

66. Shue usefully distinguishes four issues of distributive fairness here: how to allocate the costs of preventing avoidable change; how to allocate the costs of coping with change that will not be avoided; the background allocation of wealth that would allow fair bargaining about such issues; and the allocation of the gases themselves, both in the long run and during any period of transition to it (Shue 1993, p. 40).

67. Some try to account for the convergence. For example, Peter Singer claims that it arises because the facts of climate change are such that all of the major traditional lines of thought about justice in ethical theory point to the same conclusion (Singer 2002); Henry Shue argues that three "commonsense principles of fairness, none of them dependent upon controversial theories of justice," all support the position (Shue 1999b, p. 531); and Wesley and Peterson believe that the United States should accept heavier burdens because they are justified by "at least four of Ross's prima facie duties" (see Wesley and Peterson 1999, p. 191).

68. The term is from Traxler. Singer calls them "historical." Shue objects to that label, preferring to use a fault-based and no-fault distinction. (He argues that no-fault principles are not necessarily ahistorical: an ability to pay principle might emerge from a historical analysis; Shue 1993, p. 52.)

69. Singer cites Hayes and Smith 1993, chap. 2, table 2.4, which says that, even from 1950 to 1986, the United States, with about 5 percent of world population, was responsible for 30 percent of cumulative emissions, while India, with 17 percent of world population, was responsible for less than 2 percent. (Another study suggests that the developed world is responsible for 85.9 percent of the increase in atmospheric concentration of carbon dioxide since 1800; see Grubler and Fujii 1991, cited by Neumayer 2000, p. 190; and IPCC 1995, p. 94.) Furthermore, Singer says that "at present rates of emissions . . . including . . . changes in land use . . . contributions of the developing nations to the atmospheric stock of GHG will not equal the built-up contributions of developed nations until about 2038. If we adjust . . . for population—per person contributions . . . —the answer is: not for at least another century" (Singer 2002, pp. 36–37).

70. This approach is reflected in the conventional environmental "polluter pays" principle and in Shue's first "commonsense principle" of equity (Shue 1999b, p. 534). (Shue suggests that his principle is wider than "polluter pays," since he claims that the latter is exclusively forward-looking, demanding only that future pollution costs should be reflected in prices. But many writers seem to use "polluter pays" in a wider sense than this.)

71. Shue characterizes the issue as one of an international regime imposing a ceiling on

emissions and thereby creating an issue of justice, through making emissions a zero-sum good (see Shue 1995b, p. 385).

72. Singer suggests that it is this feature of the problem that renders the Lockean Proviso, of leaving "enough and as good" for others, inoperative under the circumstances for climate change.

73. Traxler suggests that they produce "very much the same results" (Traxler 2002, p. 120). But this might not turn out to be the case. For example, I might be responsible for some of the costs of upkeep of a common resource, so that the compensation due to me for a given level of pollution might be less than if there were no common property involved; or use of the resource might necessarily involve some imposed costs, of which I am expected to bear a fair share. Neither would be true on the other principle.

74. A further point to be made about the approaches is that they are potentially rebuttable. In particular, proponents of historical accounts of appropriation generally suggest that due compensation is typically paid, in the form of the increased standard of living for all that the appropriation allows. Singer, however, argues that such arguments will not work for climate change. For one thing, he says, the poor do not benefit from the increased productivity of the rich, industrialized world—"they cannot afford to buy its products"—and, if natural disasters ensue, they may even be made substantially worse off by it (Singer 2002, pp. 33–34). For another, he claims that the benefits received by the rich are wildly disproportionate. Singer dismisses Adam Smith's argument that there is an invisible hand at work so that, although the rich take the "most precious" things, "they consume little more than the poor [and] divide with the poor the produce of all their improvements." Instead, Singer claims, there is nothing even close to an equal distribution of the benefits of greenhouse-gas emissions, because "the average American... uses more than fifteen times as much of the global atmospheric sink as the average Indian" and so effectively deprives the poor of the opportunity to develop along the same lines (see Singer 2002, pp. 34–35). Shue argues that "whatever benefits the LDCs have received, they have mostly been charged for" (Shue 1999b, p. 535).

75. Other considerations are discussed by Beckerman and Pasek (1995), Neumayer (2000), Shue (1993, pp. 44–45), and Grubb (1995, p. 491).

76. Singer and Jamieson both want to ignore emissions prior to 1990, and both mention ignorance as a relevant factor. However, their endorsement of the ignorance defense is lukewarm, and this may indicate that they are more concerned with practicality. Singer suggests that there is a "strong case" for backward-looking principles but imagines that the poor countries might "generously" overlook it (Singer 2002, pp. 38–39, 48). Jamieson argues that emissions prior to 1990 are at least not morally equivalent to those after, because they do not amount to an intentional effort to deprive the poor of their share (Jamieson 2001, p. 301).

77. It is perhaps worth noticing that U.S. tort law allows for circumstances of strict liability—in which a party causing harm is liable for damages even when not guilty of negligence—and that this concept has been successfully upheld in several environmental cases and employed in environmental legislation.

78. According to Shue, far from being irrelevant, backward-looking considerations exacerbate the problems through creating compound injustice.

79. I will comment on the appropriateness of describing the climate-change problem in this way toward the end of the chapter.

80. One reason comes from historical precedent. Thomas Schelling argues that our one experience with redistribution of this magnitude is the post–World War II Marshall Plan. In that case, "there was never a formula... there were not even criteria; there were 'considerations'... every country made its claim for aid on whatever grounds it chose," and the process was governed by a system of "multilateral reciprocal scrutiny," where the recipient nations cross-examined each other's claims until they came to a consensus on how to divide the money allocated or faced arbitration from a two-person committee. Though not perfect, such a procedure did at least prove workable (Schelling 1997).

81. This concern is exacerbated by the fact that the principle of "differentiated responsibilities" was explicitly agreed to long ago, under the Framework Convention for Climate Change, and ratified by all of the major governments. So LDCs would have a procedural as well as several substantive reasons to defect.

82. It should also be clear that to restrict concern to future emissions growth has the effect of addressing only the single issue that matters to the rich countries. Again, this heightens the risk of poor-country defection.

83. For critiques of some other possibilities, see Baer 2002 and Jamieson 2001.

84. Versions of this proposal are made by Agarwal and Narain 1991; Jamieson 2001; Singer 2002, pp. 39–40; and Baer 2002. Politically, it is also advocated by China, India, and most of the LDCs.

85. Agarwal, Narain, and Sharma point out that "in 1996, one U.S. citizen emitted as much as... 19 Indians, 30 Pakistanis, 107 Bangladeshis... and 269 Nepalis" (Agarwal, Narain, and Sharma 1999, p. 107).

86. This is even without taking into account the historical issues. The IPCC 1995 report says: "If the total CO_2 absorption were assigned on an equal per capita basis, most developing countries are in fact 'in credit'—their cumulative emissions are smaller than the global average per capita absorption, and so on this basis their past contribution is not merely small but actually negative" (IPCC 1995, p. 94).

87. Other issues include the need, in practice, to assign the rights to countries rather than to individuals and the need for large transfers of resources from rich countries to poor. The former undermines the egalitarianism of the proposal, since governments might have other objectives; the latter may undermine its political feasibility. For discussion, see Baer 2002, pp. 402–4; and Beckerman and Pasek 2001, p. 183.

88. Singer suggests merely that it will give nations insufficient incentives to combat population growth and that this is an issue because under a fixed ceiling, such growth effectively reduces other countries' shares (Singer 2002, p. 40). But note that whether there is an incentive to increase population is an empirical issue, involving more than one factor: while it is true that the growing country's allocation will go up, that country will then have an extra person to look after. So, a larger population is desirable only if an extra person "costs" notably less than the emissions allotment.

89. Shue views the "maintain an adequate minimum" requirement as a no-fault principle, therefore having the advantage that no inquiry needs to be conducted to see who is to blame. (Resources are to be generated through an "ability to pay" criterion.) See Shue 1993, pp. 53–54. Moellendorf endorses an "ability to pay" criterion as a no-fault principle, but only to the extent that the rich countries should pay 40 percent of the costs, which is equivalent to their current percentage of global emissions; see Moellendorf 2002, p. 100. Traxler accepts Henry Shue's argument for the importance of subsistence emissions but argues that the difference between subsistence and luxury emissions is one of degree and that a fair allocation of costs would involve a "fair chore division" among nations based on their marginal costs. See below.

90. Traxler does admit that those committing the harm have an obligation to minimize the damage inflicted on others and may still owe compensation for the damage they cause (Traxler 2002, pp. 107–108).

91. I have in mind both the Rawlsian requirement of fairness, captured in his famous Difference Principle, and the milder views of present-day "prioritarians." For the former, see Rawls 1999; for the latter, see Parfit 1997 and, for climate change in particular, Beckerman and Pasek 2001.

92. Offhand, one would expect utilitarian approaches to recommend the same thing, based on global inequalities in welfare and diminishing marginal returns to utility. But two things make the utilitarian approach difficult. The first is logistical: calculating the maximally happiness-inducing climate policy seems to be impossible. The second is ethical: the rich might claim that they have become so used to emissions-intensive lifestyles that they will suffer more from losing them than the poor will from being denied access to them and, hence, should be required to sacrifice less. Singer claims that the logistical problem can be dealt with by treating the other distributive criteria as secondary principles to utilitarianism and that there is no ethical problem, since the rich have a legitimate concern but one that can be accommodated by allowing them to buy emissions permits from the poor (Singer 2002, pp. 45–48). Beckerman and Pasek are more pessimistic (1995, p. 406).

93. Lomborg himself seems to recognize the criticism at the end of his chapter (Lomborg 2001, p. 324).

94. This would give the United States a larger share of global emissions than per capita principles, since it has a large share of the global economy. Raul A. Estrada-Oyuela suggests a more complex, international "standard of efficiency for work performed approach," with different criteria for different economic sectors (Estrada-Oyuela 2002, p. 44).

95. It is worth noting that the "per capita" clause makes all the difference. Developed countries typically produce more GDP per unit of energy than LDCs; see Jamieson 2001, p. 295.

96. For a discussion of the commons in reference to climate change, see Gardiner 2001.

97. This kind of point is made by Amartya Sen in a classic piece (Sen 1980).

98. One might also object that there are plenty of rich people in poor countries and poor people in

rich countries, so that it doesn't seem fair to deny some rich people (those in rich countries) their luxuries, while leaving the luxuries of others (the rich in poor countries) untouched.

99. Articles 2 and 3.1, FCCC. This treaty was later ratified by all of the major players, including the United States.

100. The United States, for example, posted a 12 percent increase for the decade. Only the European Union looked likely to succeed, but this was merely because, by a fortuitous coincidence, the United Kingdom and Germany posted sharp reductions in emissions for economic reasons unrelated to climate change.

101. The best guide to the Kyoto agreement is Grubb et al. 1999. Also very informative is Victor 2001. On the role played by ethical considerations in international environmental agreements in general, see Albin 2001.

102. The latter two countries won substantial concessions on their targets, with a further weakening of the overall goal.

103. *2009 Update*: Russia ultimately ratified in November 2004, and the Protocol went into effect in February 2005. For some time, ratification was far from a foregone conclusion. President Putin promised in 2002 to have the process under way by the begininning of 2003, but by October 2003, this had still not occurred. Many commentators had initially assumed that Russia would be eager to ratify, since the economic collapse following the end of communism had reduced its own emissions and therefore appeared to give it a large surplus of permits to sell once the Kyoto targets were in place. However, some Russian leaders expressed doubts about this scenario. For example, in October 2003, Andrei Illarionov, an advisor to President Putin on economic policy, was widely reported to oppose Russian participation, saying that it would "doom Russia to poverty, weakness and backwardness" (Hirsch 2003; Brown 2003, p. 13). Pravda reported that Russia was ultimately "forced to ratify the Kyoto Protocol" in order to advance its membership in the World Trade Organization (Pravda 2004).

104. Grubb suggests that non–Annex I emissions will grow by 114 percent during the period and that (even if the United States had been included in Kyoto) this would have led to a global emissions rise of 31 percent above 1990 levels; see Grubb et al. 1999, p. 156. A 2003 United Nations report anticipated that developed-country emissions will increase by 8 percent from 2000 to 2010 (http://www.usinfo.state.gov/topical/climate/03060501.htm, June 3, 2003).

105. For example, Eileen Claussen, the president of the Pew Center on Global Climate Change, concedes that "the protocol does not do much of anything for the atmosphere" but goes on to say that "you've got to get a framework in place before you can take more than relatively small steps" (Revkin 2002). See also DeSombre 2004.

106. For example, Kate Hampton of Friends of the Earth said when the Bonn deal was made: "The Kyoto Protocol is still alive. That in itself is a triumph. But the price of success has been high. It has been heavily diluted" (Clover 2001).

107. For example, Jennifer Morgan of the World Wildlife Fund said in Bonn: "The agreement reached today is a geopolitical earthquake. Other countries have demonstrated their independence from the Bush administration on the world's most critical environmental problem" (Kettle and Brown 2001).

108. Grubb et al. 2003 is one broadly optimistic assessment.

109. It allows for parties who do not meet their targets in a given period to be assigned penalties in terms of tougher targets in subsequent periods (subject to a multiple of 1.3 times the original missed amount) and to have their ability to trade emissions suspended (United Nations Framework Convention on Climate Change 2002, decision 24/CP.7, p. 75).

110. My reasons for skepticism here all have to do with the particular format of the Kyoto Treaty. But some claim that it is also true that countries cannot be forced to keep to their international agreements (Barrett 1990, p. 75).

111. Article 18 of the Kyoto Protocol requires that the enforcement of compliance rules be approved by amendment to the Protocol. But Article 20 allows that such an amendment would be binding only on those parties that ratify the amendment.

112. For more extensive discussions, see Barrett 2003, pp. 384–386; and Gardiner 2004b.

113. For example, Doniger called it "by far the strongest environmental treaty that's ever been drafted, from the beginning to the end, from the soup of measuring emissions to the nuts of the compliance regime.... The parties have reached complete agreement on what's an infraction, how you decide a case and what are the penalties. That's as good as it gets in international relations" (Revkin 2001a).

114. Kyoto allows for help with coping through its Clean Development Mechanism (CDM) and Joint Implementation (JI) programs.

115. For the first claim, see Brown (2002). Victor makes the second claim in relation to Kyoto's provisions for international permit trading, saying that "under international law... it is not possible to create the institutional conditions that are necessary for an international tradable permit system to operate effectively" (Victor 2001, p. xiii). Shue makes both claims in his objections to the workings of the CDM and JI (Shue, in press).

116. Harris argued in 2000 that the Clinton administration had not in fact repudiated "common but differentiated responsibilities" but merely wanted something ("virtually anything") that indicated that the LDCs would aim to limit their projected future emissions (Harris 2000b, p. 239).

117. A battle-of-the-sexes analysis is also briefly suggested by some remarks of Mabey et al. (1997, pp. 356–359, 409–410) and, for the specific issue of ratification of the Kyoto Protocol, by Barrett (1998, pp. 36–37). Against this, I have argued that the intragenerational problem is more likely a prisoner's dilemma and that we have reason to treat it as if it were if there is any doubt (Gardiner 2001).

118. A theoretical analysis of the intergenerational problem is to be found in Gardiner 2003. Other intergenerational problems relevant to global warming include Derek Parfit's infamous Non-Identity Problem (Parfit 1985; Page 1999).

119. This has the paradoxical consequence that if it succeeds, this survey will soon appear obsolete and simplistic.

References

Adler, Matthew D., and Eric A. Posner, eds. 2001. *Cost-Benefit Analysis: Legal, Economic and Philosophical Perspectives.* Chicago: University of Chicago Press.

Agarwal, Anil, and Sunita Narain. 1991. *Global Warming in an Unequal World: A Case of Environmental Colonialism.* New Delhi: Centre for Science and Environment.

Agarwal, Anil, Sunita Narain, and Anju Sharma, eds. 1999. *Global Environmental Negotiations, Vol. 1, Green Politics.* New Delhi: Centre for Science and Environment.

Albin, Cecilia. 2001. *Justice and Fairness in International Negotiation.* Cambridge, U.K.: Cambridge University Press.

Alley, Richard. 2000. *The Two Mile Time Machine: Ice Cores, Abrupt Climate Change, and Our Future.* Princeton, N.J.: Princeton University Press.

Athanasiou, Tom, and Paul Baer. 2001. "Climate Change after Marrakesh: Should Environmentalists Still Support Kyoto?" Earthscape Update, December. Available at http://www.earthscape.org/p1/att02/att02.html.

———. 2002. *Dead Heat: Global Justice and Global Warming.* New York: Seven Stories Press.

Babiker, Mustapha H., Henry D. Jacoby, John M. Reilly, and David M. Reiner. 2002. "The Evolution of a Climate Regime: Kyoto to Marrakech and Beyond." *Environmental Science and Policy* 5: 195–206.

Baer, Paul. 2002. "Equity, Greenhouse Gas Emissions, and Global Common Resources." In *Climate Change Policy: A Survey,* ed. Stephen H. Schneider, Armin Rosencranz, and John O. Niles (Washington, D.C.: Island Press), pp. 393–408.

Barrett, Scott. 1990. "The Problem of Global Environmental Protection." *Oxford Review of Economic Policy* 6: 68–79.

———. 1998. "The Political Economy of the Kyoto Protocol." *Oxford Review of Economic Policy* 14: 20–39.

———. 2003. *Environment and Statecraft.* Oxford: Oxford University Press.

Beckerman, Wilfred, and Joanna Pasek. 1995. "The Equitable International Allocation of Tradable Carbon Emission Permits." *Global Environmental Change* 5: 405–413.

———. 2001. *Justice, Posterity and the Environment.* Oxford: Oxford University Press.

Broecker, Wallace S. 1997. "Thermohaline Circulation, the Achilles' Heel of Our Climate System: Will Man-Made CO_2 Upset the Current Balance?" *Science* 278 (November 28): 1582–1588.

Broome, John. 1992. *Counting the Cost of Global Warming.* Isle of Harris, U.K.: White Horse Press.

———. 1994. "Discounting the Future." *Philosophy & Public Affairs* 23: 128–156. Reprinted in *Ethics Out of Economics,* ed. John Broome (Cambridge, U.K.: Cambridge University Press, 1999), pp. 44–67. References are to the later version.

Brown, Donald. 2002. *American Heat: Ethical Problems with the United States' Response to Global Warming.* Lanham, Md.: Rowman & Littlefield.

Brown, Paulo, 2003. "Russia Urged to Rescue Kyoto Pact." *Guardian,* February 26.

Chang, Ruth, ed. 1997. *Incommensurability, Incomparability and Practical Reason.* Cambridge, Mass.: Harvard University Press.

Clover, Charles. 2001. "Pollution Deal Leaves US Cold." *Daily Telegraph,* July 24.

Committee on the Science of Climate Change. 2001. *Climate Change Science: An Analysis of Some Key Questions.* Washington, D.C.: National Academy Press.

Costanza, Robert. 1996. "Review of *Managing the Commons: The Economics of Climate Change,* by William D. Nordhaus." *Environment and Development Economics* 1: 381–384.

Coward, Harold, and Thomas Hurka, eds. 1993. *Ethics and Climate Change: The Greenhouse Effect.* Waterloo, Ont.: Wilfrid Laurier Press.

Danielson, Peter. 1993. "Personal Responsibility." In *Ethics and Climate Change: The Greenhouse Effect,* ed. Harold Coward and Thomas Hurka (Waterloo, Ont.: Wilfrid Laurier Press, 1993), pp. 81–98.

De Leo, Giulio, L. Rizzi, A. Caizzi, and M. Gatto. 2001. "The Economic Benefits of the Kyoto Protocol." *Nature* 413 (October 4): 478–479.

DeSombre, Elizabeth R. 2004. "Global Warming: More Common Than Tragic." *Ethics and International Affairs* 18: 41–46.

Dimitrov, R. 2003. "Knowledge, Power and Interests in Environmental Regime Formation." *International Studies Quarterly* 47: 123–150.

Dubgaard, Alex. 2002. "Sustainability, Discounting, and the Precautionary Principle." In *Sceptical Questions and Sustainable Answers,* by Danish Ecological Council (Copenhagen: Danish Ecological Council), pp. 196–202.

Earth Negotiations Bulletin. 2003. "COP-9 Final." International Institute for Sustainable Development. Available at http://www.iisd.ca/linkages/climate/cop9.

Edwards, Tamsin L., Michel Crucifi, and Sandy P. Harrison. 2007. "Using the past to constrain the future: how the palaeorecord can improve estimates of global waming." *Progress in Physical Geography* (31)5: 481–500. DOI: 10.1177/0309133307083295.

Estrada-Oyuela, Raul A. 2002. "Equity and Climate Change." In *Ethics, Equity and International Negotiations on Climate Change,* ed. Luiz Pinguelli-Rosa and Mohan Munasinghe (Cheltenham, U.K.: Edward Elgar), pp. 36–46.

Francis, Leslie Pickering. 2003. "Global Systemic Problems and Interconnected Duties." *Environmental Ethics* 25: 115–128.

Gagosian, Robert. 2003. "Abrupt Climate Change: Should We Be Worried?" Woods Hole Oceanographic Institute. Available at http://www.whoi.edu/institutes/occi/hottopics_climatechange.html.

Gardiner, Stephen M. 2001. "The Real Tragedy of the Commons." *Philosophy & Public Affairs* 30: 387–416.

———. 2003. "The Pure Intergenerational Problem." *Monist* 86: 481–500.

———. 2004. "The Global Warming Tragedy and the Dangerous Illusion of the Kyoto Protocol." *Ethics and International Affairs* 18: 23–39.

———. 2006. "A Core Precautionary Principle," *Journal of Political Philosophy* 14(1): 33–60.

Green, Michael. 2002. "Institutional Responsibility for Global Problems." *Philosophical Topics* 30: 1–28.

Grubb, Michael. 1995. "Seeking Fair Weather: Ethics and the International Debate on Climate Change." *International Affairs* 71: 463–496.

Grubb, Michael, with Christian Vrolijk and Duncan Brack. 1999. *The Kyoto Protocol: A Guide and Assessment.* London: Royal Institute of International Affairs.

Grubb, Michael, Tom Brewer, Benito Muller, John Drexhage, Kirsty Hamilton, Taishi Sugiyama, and Takao Aiba, eds. 2003. "A Strategic Assessment of the Kyoto-Marrakesh System: Synthesis Report." Sustainable Development Programme Briefing Paper 6, Royal Institute of International Affairs, London. Available at http://www.riia.org.

Grubler, A., and Y. Fujii. 1991. "Inter-generational and Spatial Equity Issues of Carbon Accounts." *Energy* 16: 1397–1416.

Gundermann, Jesper. 2002. "Discourse in the Greenhouse." In *Sceptical Questions and Sustainable Answers,* by Danish Ecological Council (Copenhagen: Danish Ecological Council), pp. 139–164.

Haller, Stephen F. 2002. *Apocalypse Soon? Wagering on Warnings of Global Catastrophe.* Montreal: McGill-Queens.

Harding, Luke. 2002. "Just So Much Hot Air." *Guardian,* October 31.

Harris, Paul, ed. 2000a. *Climate Change and American Foreign Policy.* New York: St. Martin's Press.

———. 2000b. "International Norms of Responsibility and US Climate Change Policy." In *Climate Change and American Foreign Policy,* ed. Paul Harris (New York: St. Martin's Press), pp. 225–240.

———. 2001. *International Equity and Global Environmental Politics.* Aldershot, U.K.: Ashgate.

———. 2003. "Fairness, Responsibility, and Climate Change." *Ethics and International Affairs* 17: 149–156.

Hayes, Peter, and Kirk Smith, eds. 1993. *The Global Greenhouse Regime: Who Pays?* London: Earthscan.

Heavenrich, Robert, and Karl Hellman. 2000. "Light-Duty Automotive Technology Trends 1975 through 2000." EPA420-S-00–003. Available at http://www.epa.gov/otaq/cert/mpg/fetrends/s00003.pdf.

Hirsch, Tim. 2003. "Climate Talks End without Result." BBC News, October 3. Available at http://www.news.bbc.co.uk/1/hi/sci/tech/3163030.stm.

Holden, Barry, ed. 1996. *The Ethical Dimensions of Climate Change.* Basingstoke, U.K.: Macmillan.

———. 2002. *Democracy and Global Warming.* London: Continuum.

Hood, Robert. 2003. "Global Warming." In *A Companion to Applied Ethics,* ed. R. G. Frey and Christopher Wellman (Oxford: Blackwell), pp. 674–684.

Hopgood, Stephen. 1998. *American Foreign Policy and the Power of the State.* Oxford: Oxford University Press.

Houghton, John. 1997. *Global Warming: The Complete Briefing,* 2d ed. Cambridge, U.K.: Cambridge University Press.

IPCC (Intergovernmental Panel on Climate Change). 1995. *Climate Change 1995: Economic and Social Dimensions of Climate Change.* Cambridge, U.K.: Cambridge University Press.

———. 2001a. *Climate Change 2001: Mitigation.* Cambridge, U.K.: Cambridge University Press. Available at http://www.ipcc.ch.

———. 2001b. *Climate Change 2001: Synthesis Report.* Cambridge, U.K.: Cambridge University Press. Available at http://www.ipcc.ch.

———. 2007. *Climate Change 2007: The Physical Science Basis.* Cambridge, U.K.: Cambridge University Press. Available at http://www.ipcc.ch.

Jamieson, Dale. 1990. "Managing the Future: Public Policy, Scientific Uncertainty, and Global Warming." In *Upstream/Downstream: Essays in Environmental Ethics,* ed. D. Scherer (Philadelphia: Temple University Press), pp. 67–89.

———. 1991. "The Epistemology of Climate Change: Some Morals for Managers." *Society and Natural Resources* 4: 319–329.

———. 1992. "Ethics, Public Policy and Global Warming." *Science, Technology and Human Values* 17: 139–153. Reprinted in Dale Jamieson, *Morality's Progress* (Oxford: Oxford University Press, 2003). References are to the later version. Also chapter 3 in this volume.

———. 1996. "Ethics and Intentional Climate Change." *Climatic Change* 33: 323–336.

———. 1998. "Global Responsibilities: Ethics, Public Health and Global Environmental Change." *Indiana Journal of Global Legal Studies* 5: 99–119.

———. 2001. "Climate Change and Global Environmental Justice." In *Changing the Atmosphere: Expert Knowledge and Global Environmental Governance,* ed. P. Edwards and C. Miller (Cambridge, Mass.: MIT Press), pp. 287–307.

———. 2005. "Adaptation, Mitigation, and Justice." In *Perspectives on Climate Change: Science, Economics, Politics, Ethics,* ed. Walter Sinnott-Armstrong and Richard Howarth. (Amsterdam: Elsevier). Also chapter 15 in this volume.

Johansen, Bruce. 2002. *The Global Warming Desk Reference.* Westport, Conn.: Greenwood.

Kettle, Martin, and Paul Brown. 2001. "US Stands Defiant Despite Isolation in Climate Debate." *Guardian,* July 24.

Lee, Jennifer. 2003. "GOP Changes Environmental Message." *Seattle Times,* March 2.

Lenton, Timothy, Hermann Held, Elmar Kriegler, Jim Hall, Wolfgang Lucht, Stefan Rahmsdorf, and Hans Joachim Schnellnhuber. 2008. "Tipping Points in the Earth's Climate System." *Proceedings of the National Academies of Sciences* 105(6): 1786–1793.

Lomborg, Bjørn. 2001. "Global Warming." In *The Sceptical Environmentalist,* by Bjørn Lomborg (Cambridge, U.K.: Cambridge University Press), pp. 258–324.

Mabey, Nick, Stephen Hall, Claire Smith, and Sujata Gupta. 1997. *Argument in the Greenhouse: The International Economics of Controlling Global Warming.* London: Routledge.

Manson, Neil A. 2002. "Formulating the Precautionary Principle." *Environmental Ethics* 24: 263–274.

Michaels, Patrick, and Robert Balling Jr. 2000. *The Satanic Gases: Clearing the Air about Global Warming.* Washington, D.C.: Cato Institute.

Moellendorf, Darrell. 2002. *Cosmopolitan Justice.* Boulder, Colo.: Westview.

Neumayer, Eric. 2000. "In Defence of Historical Accountability for Greenhouse Gas Emissions." *Ecological Economics* 33: 185–192.

Nicholls, N., G. V. Gruza, J. Jouzel, T. R. Karl, L. A. Ogallo, and D. E. Parker. 1996. "Observed

Climate Variability and Change." In *Climate Change 1995: The Science of Climate Change,* ed. J. T. Houghton, L. G. M. Filho, B. A. Callander, N. Harris, A. Kattenberg, and K. Maskell (Cambridge, U.K.: Cambridge University Press), pp. 133–192.

O'Neill, Brian C., and Michael Oppenheimer. 2002. "Dangerous Climate Impacts and the Kyoto Protocol." *Science* 296 (June 14): 1971–1972.

O'Riordan, T., J. Cameron, and A. Jordan, eds. 2001. *Reinterpreting the Precautionary Principle.* London: Cameron & May.

Ott, Hermann. 2001. "Climate Policy after the Marrakesh Accords: From Legislation to Implementation." Available at http://www.wupperinst.org/download/Ott-after-marrakesh.pdf. Published as *Global Climate: Yearbook of International Law* (Oxford: Oxford University Press).

Ozonoff, David. 1999. "The Precautionary Approach as a Screening Device." In *Protecting Public Health and the Environment: Implementing the Precautionary Principle,* ed. Carolyn Raffensberger and Joel Tickner (Washington, D.C.: Island Press), pp. 100–105.

Page, Edward. 1999. "Intergenerational Justice and Climate Change." *Political Studies* 47: 53–66.

Parfit, Derek. 1985. *Reasons and Persons.* Oxford: Oxford University Press.

———. 1997. "Equality and Priority." *Ratio* 10: 202–221.

Paterson, Matthew. 1996. *Global Warming and Global Politics.* London: Routledge.

———. 2001. "Principles of Justice in the Context of Global Climate Change." In *International Relations and Global Climate Change,* ed. Urs Luterbacher and Detlef Sprinz (Cambridge, Mass.: MIT Press), pp. 119–126.

Pearce, David. 1993. *Economic Values and the Natural World.* London: Earthscan.

Pinguelli-Rosa, Luiz, and Mohan Munasinghe, eds. 2002. *Ethics, Equity and International Negotiations on Climate Change.* Cheltenham, U.K.: Edward Elgar.

Pohl, Otto. 2003. "US Left Out of Emissions Trading." *New York Times,* April 10.

Pravda 2004. "Russia forcted to ratify Kyoto Protocol to become WTO member." October 26. Available at http://english.pravada.ru/russia/politics/7274-kyoto-0.

Prestowitz, Clyde. 2003. "Don't Pester Europe on Genetically Modified Food." *New York Times,* January 25.

Raffensberger, Carolyn. 1999. "Uses of the Precautionary Principle in International Treaties and Agreements." Available at http://www.biotech-info.net/treaties_and_agreements.html.

Raffensberger, Carolyn, and Joel Tickner, eds. 1999. *Protecting Public Health and the Environment: Implementing the Precautionary Principle.* Washington, D.C.: Island Press.

Rawls, John. 1999. *A Theory of Justice,* rev. ed. Cambridge, Mass.: Harvard University Press.

Reilly, John, Peter H. Stone, Chris E. Forest, Mort D. Webster, Henry D. Jacoby, and Ronald G. Prinn. 2001. "Climate Change: Uncertainty and Climate Change Assessments." *Science* 293 (July 20): 430–433.

Revkin, Andrew. 2001a. "Deals Break Impasse on Global Warming Treaty." *New York Times,* November 11.

———. 2001b. "Warming Threat." *New York Times,* June 12.

———. 2002. "Climate Talks Will Shift Focus from Emissions." *New York Times,* October 23.

Sagoff, Mark. 1988. *The Economy of the Earth.* Cambridge, U.K.: Cambridge University Press.

Schelling, Thomas. 1997. "The Cost of Combating Global Warming: Facing the Tradeoffs." *Foreign Affairs* 76: 8–14.

Schiermeier, Quirin. 2006. "A Sea Change," *Nature* 439 (January 19): 256–260.

Schmidtz, David. 2001. "A Place for Cost-Benefit Analysis." *Noûs* 11, supp.: 148–171.

Schultz, Peter, and James Kasting. 1997. "Optimal Reductions in CO_2 Emissions." *Energy Policy* 25: 491–500.

Sen, Amartya. 1980. "Equality of What?" In *Tanner Lectures on Human Values,* ed. S. M. McMurrin (Salt Lake City: University of Utah Press), pp. 195–220.

Shogren, Jason, and Michael Toman. 2000. "Climate Change Policy." Discussion paper 00–22, Resources for the Future, Washington, D.C., May 14–25. Available at http://www.rff.org.www.lib.ncsu.edu:2048.

Shue, Henry. 1992. "The Unavoidability of Justice." In *The International Politics of the Environment,* ed. Andrew Hurrell and Benedict Kingsbury (Oxford: Oxford University Press), pp. 373–397.

———. 1993. "Subsistence Emissions and Luxury Emissions." *Law and Policy* 15: 39–59. Also chapter 11 in this volume.

———. 1994. "After You: May Action by the Rich Be Contingent upon Action by the Poor?" *Indiana Journal of Global Legal Studies* 1: 343–366.

———. 1995a. "Avoidable Necessity: Global Warming, International Fairness and Alternative Energy." In *Theory and Practice, NOMOS XXXVII,* ed. Ian Shapiro and Judith Wagner DeCew (New York: New York University Press), pp. 239–264.

———. 1995b. "Equity in an International Agreement on Climate Change." In *Equity and Social Considerations Related to Climate Change,* ed. R. S. Odingo, A. L. Alusa, F. Mugo, J. K. Njihia, and A. Heidenreich (Nairobi: ICIPE Science Press), pp. 385–392.

———. 1996. "Environmental Change and the Varieties of Justice." In *Earthly Goods: Environmental Change and Social Justice,* ed. Fen Osler Hampson and Judith Reppy (Ithaca, N.Y.: Cornell University Press), pp. 9–29.

———. 1999a. "Bequeathing Hazards: Security Rights and Property Rights of Future Humans." In *Global Environmental Economics: Equity and the Limits to Markets,* ed. M. Dore and T. Mount (Oxford: Blackwell), pp. 38–53.

———. 1999b. "Global Environment and International Inequality." *International Affairs* 75: 531–545. Also chapter 5 in this volume.

———. 2001. "Climate." In *A Companion to Environmental Philosophy,* ed. Dale Jamieson (Oxford: Blackwell), pp. 449–459.

———. 2004. "A Legacy of Danger: The Kyoto Protocol and Future Generations." In *Globalisation and Equality,* ed. Keith Horton and Haig Patapan. (London: Routledge), pp. 165–178.

Singer, Peter. 2002. "One Atmosphere." In *One World: The Ethics of Globalization,* by Peter Singer (New Haven, Conn.: Yale University Press), chap. 2. Also chapter 10 in this volume.

Soroos, Marvin S. 1997. *The Endangered Atmosphere: Preserving a Global Commons.* Columbia: University of South Carolina Press.

Soule, Edward. 2000. "Assessing the Precautionary Principle." *Public Affairs Quarterly* 14: 309–328.

Toman, Michael. 2001. *Climate Change Economics and Policy.* Washington, D.C.: Resources for the Future.

Traxler, Martino. 2002. "Fair Chore Division for Climate Change." *Social Theory and Practice* 28: 101–134.

United Nations Environment Program. 1999. "Climate Change Information Kit." Available at http://www.unep.ch.iuc.

United Nations Framework Convention on Climate Change. 1992. "Framework Convention on Climate Change." Available at http://www.unfccc.int.

———. 2002. "Report of the Conference of the Parties on Its Seventh Session, Held at Marrakesh from 29 October to 10 November 2001." In *Action Taken by the Conference of the Parties, of the Addendum* 3, part 2.

U.S. Environmental Protection Agency. 2003. Information accessed at http://www.fueleconomy.gov.

U.S. National Academy of Sciences, Committee on the Science of Climate Change. 2001. *Climate Change Science: An Analysis of Some Key Questions.* Washington, D.C.: National Academies Press.

U.S. National Research Council, Committee on Abrupt Climate Change. 2002. *Abrupt Climate Change: Inevitable Surprises.* Washington, D.C.: National Academies Press.

Victor, David. 2001. *The Collapse of the Kyoto Protocol and the Struggle to Slow Global Warming.* Princeton, N.J.: Princeton University Press.

Waldron, Jeremy. 1990. "Who Is to Stop Polluting? Different Kinds of Free-Rider Problem." In *Ethical Guidelines for Global Bargains: Program on Ethics and Public Life* (Ithaca, N.Y.: Cornell University).

Wesley, E., and F. Peterson. 1999. "The Ethics of Burden-Sharing in the Global Greenhouse." *Journal of Agricultural and Environmental Ethics* 11: 167–196.

Wigley, Tom M., L. V. Ramaswamy, J. R. Christy, J. R. Lanzante, C. A. Mears, B. D. Santer, and C. K. Folland. 2006. *Temperature Trends in the Lower Atmosphere: Steps for Understanding and Reconciling Differences.* A Report by the U.S. Climate Change Science Program and the Subcommittee on Global Change Research, ed. Thomas R. Karl, Susan J. Hassol, Christopher D. Miller, and William L. Murray (Washington, D.C.).

Wingspread Statement. 1998. Available at http://www.gdrc.org/u-gov/precaution-3.html.

Woodward, Richard, and Richard Bishop. 1997. "How to Decide When Experts Disagree: Uncertainty-Based Choice Rules in Environmental Policy." *Land Economics* 73: 492–507.

Part II

The Nature of the Problem

The Economics of Climate Change

Nicholas Stern

2

Greenhouse gas (GHG) emissions are externalities and represent the biggest market failure the world has seen. We all produce emissions, people around the world are already suffering from past emissions, and current emissions will have potentially catastrophic impacts in the future. Thus, these emissions are not ordinary, localized externalities. Risk on a global scale is at the core of the issue. These basic features of the problem must shape the economic analysis we bring to bear; failure to do this will produce, and has produced, approaches to policy that are profoundly misleading and indeed dangerous.

The purpose of this chapter is to set out what I think is an appropriate way to examine the economics of climate change, given the unique scientific and economic challenges posed, and to suggest implications for emissions targets, policy instruments, and global action. The subject is complex and very wide-ranging. It is a subject of vital importance but one in which the economics is fairly young. A central challenge is to provide the economic tools necessary as quickly as possible, because policy decisions are both urgent and moving quickly—particularly following the United Nations Framework Convention on Climate Change (UNFCCC) meetings in Bali in December 2007. The relevant decisions can be greatly improved if we bring the best economic analyses and judgments to the table in real time.

A brief description of the scientific processes linking climate change to GHG emissions will help us to understand how they should shape the economic analysis. First, people, through their consumption and production decisions, emit GHGs. Carbon dioxide is especially important, accounting for around three-quarters of the human-generated global-warming effect; other relevant GHGs include methane, nitrous oxide, and hydrofluorocarbons (HFCs). Second, these flows accumulate into stocks of GHGs in the atmosphere. It is overall stocks of GHGs that matter and not their place of origin. The rate at which stock accumulation occurs depends on the "carbon cycle," including the earth's absorptive capabilities and other feedback effects. Third, the stock of GHGs in the atmosphere traps heat and results in global warming; how much depends on "climate sensitivity." Fourth, the process of global warming results in climate change. Fifth, climate change affects people, species, and plants in a variety of complex ways, most notably via water in some shape or form: storms, floods, droughts, sea-level rise. These changes will potentially transform the physical and human geography of the planet, affecting where and how we live our lives. Each of these five links involves considerable uncertainty. The absorption-stock accumulation, climate-sensitivity, and warming-climate change links all involve time lags.

The key issues in terms of impacts are not simply or mainly about global warming as such—they concern climate change more broadly. Understanding these changes requires specific analysis of how climate will be affected regionally. Levels and variabilities of rainfall depend on the functioning of weather and climate for the world as a whole. As discussed below, temperature increases of 1 °C to 5 °C on average for the world would involve radical and dangerous changes for the whole planet, with widely differing, often extreme, local impacts. Further, the challenge, in large measure, is one of dealing with the consequences of *change* and not only of comparing long-run equilibria. Under business as usual, over the next two centuries, we are likely to see change at a rate that is fast-forward in historical time and on a scale that the world has not seen for tens of millions of years.

This very brief and oversimplified description of the science carries key lessons for economics. The scientific evidence on the potential risks is now overwhelming, as demonstrated in the Intergovernmental Panel on Climate Change (IPCC) Fourth Assessment Report, or AR4 (IPCC 2007). I am not a climate scientist. As economists, our task is to take the science, particularly its analysis of risks, and think about its implications for policy. Only by taking the extraordinary position that the scientific evidence shows that the risks are definitely negligible should economists advocate doing nothing now. The science clearly shows that the probability and frequency of floods, storms, droughts, and so on, are likely to continue to grow with cumulative emissions and that the magnitude of some of these impacts could be catastrophic.

While an understanding of the greenhouse effect dates from the nineteenth century,[1] in the last decade, and particularly in the last few years, the science has fortunately started to give us greater guidance on some of the possible probability distributions linking emissions and stocks to possible warming and climate change, thus allowing us to bring to the table analytical tools on economic policy toward risk.

The brief description of the science above tells us that GHG emissions are an externality that is different from our usual examples in four key ways: (a) it is global in its origins and impacts; (b) some of the effects are very long-term and governed by a flow-stock process; (c) there is a great deal of uncertainty in most steps of the scientific chain; and (d) the effects are potentially very large, and many may be irreversible. Thus, it follows that the economic analysis must place at its core (i) the economics of risk and uncertainty; (ii) the links between economics and ethics (there are major potential policy tradeoffs both within and between generations), as well as notions of responsibilities and rights in relation to others and the environment; and (iii) the role of international economic policy. Further, the potential magnitude of impacts means that, for much of the analysis, we have to compare strategies that can have radically different development paths for the world. We cannot, therefore, rely only on the methods of marginal analysis. Here, I attempt to sketch briefly an analysis that brings these three parts of economics to center stage. It is rather surprising, indeed worrying, that much previous analysis of practical policy has relegated some or all of these three key pieces of economics to the sidelines.

The structure of the argument on stabilization is crucial, and we begin by setting that out before going into analytical detail. The choice of a stabilization target shapes much of the rest of policy analysis and discussion, because it carries strong implications for the permissible flow of emissions, and thus for emissions-reduction targets. The reduction targets, in turn, shape the pricing and technology policies.

Understanding the risks from different strategies is basic to an understanding of policy. Many articulated policies for risk reduction work in terms of targets, usually expressed in terms of emission flows, stabilization levels, or average temperature increases. The last of these has the advantage that it is (apparently) easier for the general public to understand. The problem is that this apparent ease conceals crucial elements that matter greatly to social and economic outcomes—it is the effects on storms, floods, droughts, and sea-level rise that are of particular importance, and a heavy focus on temperature can obscure this. Further, and crucially, temperature outcomes are highly stochastic and cannot be targeted directly. Emissions can be more easily controlled by policy. However, it is the stocks that shape the warming. Thus, there are arguments for and against each of the three dimensions. We shall opt for stock targets, on the basis that they are closest to the phenomenon that drives climate change and the most easily expressed in one number.

An alternative focus for policy is the price of GHGs rather than quantities. In a perfectly understood nonstochastic world, standard duality theory says that price and quantity tools are essentially mirror images and can be used interchangeably. However, where risk and uncertainty are important and knowledge is highly imperfect, we have to consider the relative merits of each. For the most part, we ignore the difference between risk and uncertainty here (where the latter is used strictly in the Knightian sense of unknown probabilities), but it is a very important issue (Henry 2006; Stern 2007, 38–39) and a key topic for further research.

We begin by setting out some of the major risks from climate change and argue that these risks point to the need for both stock and flow targets, guided by an assessment of the costs involved in achieving them. Long-term stabilization (or stock) targets are associated with a range of potential flow paths, although the stock target exerts a very powerful influence on their shape. The choice of a particular flow path would be influenced by the expected pattern of costs over time. The target flow paths can then be associated with a path for marginal costs of abatement, if we think of efficient policy designed to keep flows to the levels on the path, in particular by using a price for carbon set at the marginal abatement cost (MAC). Essentially, the economics of risk points to the need for stock and flow quantity targets and the economics of costs and efficiency to a price mechanism to achieve the targets.

A policy that tries to start with a price for marginal GHG damages has two major problems: (a) the price estimate is highly sensitive to ethical and structural assumptions on the future; and (b) there is a risk of major losses from higher stocks than anticipated, since the damages rise steeply with stocks, and many are irreversible.

Formal modeling of damages can supplement the argument in three ways. First, it can provide indicative estimates of overall damages to guide strategic risk analysis. Second, it can provide estimates of marginal damage costs of GHGs, for comparison with MACs. Third, and most important in my view, it can help to clarify key tradeoffs and the overall logic and key elements of an argument.

A useful analogy is the role of computable general equilibrium models (CGMs) in discussions of trade policy. These have much more robust foundations than aggregative models on the economics of climate change, yet their quantitative results are very sensitive to assumptions, and they leave out so much that is important to policy. Thus, most economists would not elevate them to the main plank of an argument on trade policy. That policy would usually be better founded on an understanding of economic theory and of economic history, together with country studies and particular studies of the context and issues in question.

However, as the *Stern Review* stressed, such analysis has very serious weaknesses and must not be taken too literally. It is generally forced to aggregate into a single good and in so doing misses a great deal of the crucial detail of impacts—on different dimensions and in different locations—which should guide risk analysis. It is forced to make assumptions about rates and structures of growth over many centuries. Further, it will be sensitive to the specification of ethical frameworks and parameters. Thus, its estimates of marginal social costs of damages provide a very weak foundation for policy. This type of modeling does have an important supplementary place in an analysis, but all too often it has been applied naively and transformed into the central plank of an argument.

Our analysis of risks and targets points to the need for aggregate GHG stabilization targets of less than 550 parts per million (ppm) carbon dioxide equivalent (CO_2e), arguably substantially less. This corresponds to cuts in global emissions flows of at least 30 percent, and probably around 50 percent, by 2050. These cuts may seem large in the context of (we hope) a growing world economy but are not ambitious in relation to the risks we run by exceeding 550ppm CO_2e. And given the avoided risks, the costs of around 1 percent of world GDP per annum (see section IB below) of achieving this stabilization should be regarded as relatively low. The carbon price required to achieve these reductions (up to, say, 2030) would be around, or in excess of, $30 per ton of CO_2.

This chapter incorporates many important elements of the *Stern Review*, published on the Web in October/November 2006 (see http://www.sternreview.org.uk, including postscript) and in book form (Stern 2007) a year ago but goes beyond it in many important ways—in relation to subsequent policy discussions, new evidence and analysis, and discussions in the economics literature.

There are four further parts to this chapter. The second part focuses on risks and how to reduce them and on costs of abatement. The third part examines formal modeling and damage assessment. The fourth part examines policy, in particular the role of different policy instruments. The final part outlines what I see as the central elements of a global deal or framework for collaborative policy and discusses how that deal can be built and sustained.

I. Stabilization of Stocks of Greenhouse Gases I: Risks and Costs

A. Risks and Targets

The relation between the stock of GHGs in the atmosphere and the resulting temperature increase is at the heart of any risk analysis. The preceding link in the chain, the way the carbon cycle governs the process relating emissions to changes in stocks, and the subsequent link, from global average temperature to regional and local climate change, are full of risk as well. But the stock-temperature relationship is the clearest way to begin, as it anchors everything else. Broadly conceived, it is about "climate sensitivity"—in terms of modeling, this is indicated by the expected eventual temperature increase from a doubling of GHG stocks.[2]

There are now a number of general circulation models (GCMs—also known as global climate models) that have been built to describe the links from emissions to climate change. The large ones work with a very large number of geographic cells, consume computer time extremely heavily, and can be run only on some of the world's biggest computers. Nevertheless, particularly if combined with appropriate linking to a large number of other machines, they can be run many times for different possible parameter choices. Such exercises yield Monte Carlo estimates of probability distributions of outcomes. A discussion of various methods and models can be found in Meinshausen 2006 and in chapter 1 of the *Stern Review*.

Figure 2.1 and table 2.1 are drawn from the models of the U.K.'s Hadley Centre. The work of the Hadley Centre was a particular focus of models for the *Stern Review* for a number of reasons. First, it is one of the world's finest climate-science groups, with a very large computing capacity. Second, it was close by, and the staff were extremely accessible and helpful. Third, its probability distributions are fairly cautious, balanced, and "middle of the road" (Meinshausen 2006); this judgment is sustained by a comparison of their results with the subsequently published AR4 (IPCC 2007).

Figure 2.1 and table 2.1 present estimated probabilities for eventual temperature increases (which take time to be established) relative to preindustrial times (around 1850), were the world to stabilize at the given concentration of GHGs in the atmosphere measured in ppm CO_2e. Figure 2.1 portrays 90 percent confidence intervals—the solid horizontal bars—for temperature increases. The lower bound (fifth percentile) is derived from the IPCC

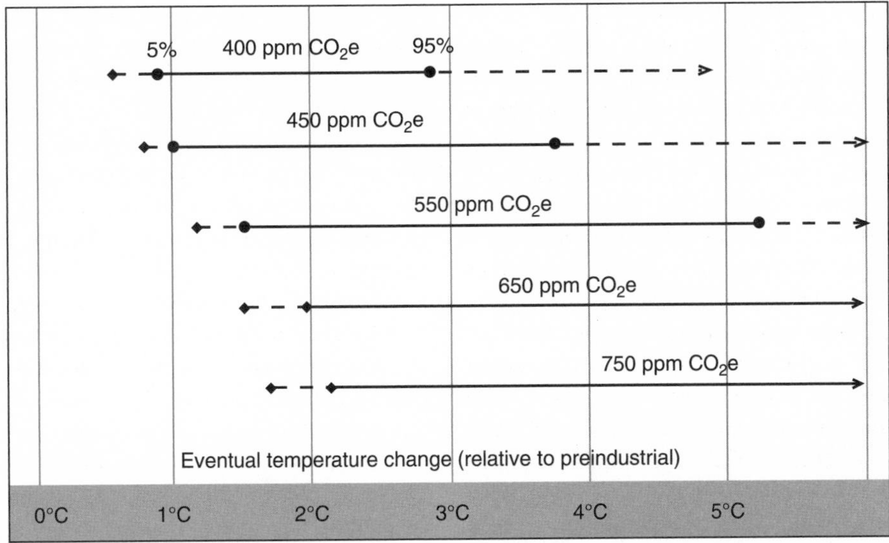

Figure 2.1. Stabilization and eventual change in temperature.
Sources: Stern 2007, p.16; Meinshausen 2006; Wigley and Raper 2001; Murphy et al. 2004.

Table 2.1.
Likelihood (%) of exceeding a temperature increase at equilibrium

Stabilization Level (in ppm CO_2e)	2°C	3°C	4°C	5°C	6°C	7°C
450	78	18	3	1	0	0
500	96	44	11	3	1	0
550	99	69	24	7	2	1
650	100	94	58	24	9	4
750	100	99	82	47	22	9

Source: Stern 2007, p. 220, with some added information.

Third Assessment Report, or TAR (Wigley and Raper 2001; IPCC 2001a, 2001b)[3] and the upper bound is from the Hadley Centre (Hadley Centre 2005; Murphy et al. 2004). The dotted bars cover the range of the 11 studies examined by Meinshausen (2006). The bar for 550 ppm CO_2e (with a 90 percent interval of 1.5 °C to 5.3 °C) approximately represents the possible range for "climate sensitivity."

Concentrations are currently around 430 ppm CO_2e (Stern 2007, p. 5—Kyoto GHGs), and are rising at around 2.5 ppm CO_2e per annum. This rate appears to be accelerating, particularly as a result of the very rapid growth of emissions in China. On fairly conservative estimates (International Energy Agency 2007), China's energy-related emissions are likely to double by 2030, taking overall emissions from 6–7 to 12–15 gigatons. There seems little doubt that, under BAU, the annual increments to stocks would average somewhere well above 3 ppm CO_2e, perhaps 4 or more, over the next century. That is likely to take us to around, or well beyond, 750 ppm CO_2e by the end of the century. If we manage to stabilize there, that would give us around a 50–50 chance of a stabilization temperature increase above 5 °C. This is a high probability of a disastrous transformation of the planet (see below).[4]

The issue is still more worrying than that of dealing with very large damages with very low probability.

Further, we should emphasize that key positive feedback from the carbon cycle—such as release of methane from the permafrost, the collapse of the Amazon, and thus the destruction of a key carbon sink, and reduction in the absorptive capacity of the oceans—has been omitted from the projected concentration increases quoted here. It is possible that stocks could become even harder to stabilize than this description suggests.

We do not really know what the world would look like at 5 °C above preindustrial times. The most recent warm period was around 3 million years ago, when the world experienced temperatures 2 °C to 3 °C higher than today (Jansen et al. 2007, p. 440). Humans (dating from around 100,000 years or so) have not experienced anything that high. Around 10,000 to 12,000 years ago, temperatures were around 5 °C lower than today, and ice sheets came down to latitudes just north of London and just south of New York. As the ice melted and sea levels rose, England separated from the continent, rerouting much of the river flow. These magnitudes of temperature changes transform the planet.

At an increase of 5 °C, most of the world's ice and snow would disappear, including major ice sheets and, probably, the snows and glaciers of the Himalayas. This would eventually lead to sea-level rises of 10 meters or more and would thoroughly disrupt the flows of the major rivers from the Himalayas, which serve countries comprising around half of the world's population. There would be severe torrents in the rainy season and dry rivers in the dry season. The world would probably lose more than half its species. Storms, floods, and droughts would probably be much more intense than they are today.

Further tipping points could be passed, which together with accentuated positive feedbacks could lead to "runaway" further temperature increase. The last time temperature was in the region of 5 °C above preindustrial times was in the Eocene period around 35 million to 55 million years ago. Swampy forests covered much of the world, and there were alligators near the North Pole. Such changes would fundamentally alter where and how different species, including humans, could live. Human life would probably become difficult or impossible in many regions that are currently heavily populated, thus necessitating large population movements, possibly or probably on a huge scale. History tells us that large movements of population often bring major conflict. And many of the changes would take place over 100 to 200 years rather than thousands or millions of years.

While there is no way that we can be precise about the magnitude of the effects associated with temperature increases of this size, it does seem reasonable to suppose that they would, in all likelihood, be disastrous. We cannot obtain plausible predictions by extrapolating from "cross-sectional" (Mendelsohn et al. 2000, p. 557) comparisons of regions with current temperature differences of around 5 °C—comparisons between, say, Massachusetts and Florida miss the point. Nor, given the nonlinearities involved, can we extrapolate from lower temperature increases (say, 2 °C) concerning which there is more evidence. Most people contemplating 5 °C increases and upward would surely attach a very substantial weight on keeping the probability of such outcomes down.

From this perspective, an examination of table 2.1 suggests that 550 ppm CO_2e is an upper limit to the stabilization levels that should be contemplated. This level is nevertheless rather dangerous, with a 7 percent probability of being above 5 °C and a 24 percent probability of being above 4 °C. The move to 650 ppm CO_2e gives a leap in probability of being above 4 °C to 58 percent and of being above 5 °C to 24 percent. Further, we should remember that the Hadley Centre probabilities are moderately conservative—one highly computationally intensive Monte Carlo estimate of climate sensitivity found a 4.2 percent probability of temperatures exceeding 8 °C (Stainforth et al. 2005). A concentration in the region of 550 ppm CO_2e is clearly itself a fairly dangerous place to be, and the danger posed by even higher concentrations looks unambiguously unacceptable. For this reason, I find it remarkable that some economists continue to argue that stabilization levels around 650 ppm CO_2e or even higher are preferable to 550 ppm or even optimal (Nordhaus 2007a, p. 166; Mendelsohn 2007, p. 95). It is important to be clear that the "climate policy ramp" (Nordhaus 2007b, p. 687) advocated by some economists involves a real possibility of devastating climatic changes.

In thinking about targets for stabilization, we have to think about more than the eventual stocks. We must also consider where we start, costs of stabilization, and possibilities of reversal, or backing out, if we subsequently find ourselves in or approaching very dangerous territory. The costs of stabilization depend strongly on where we start. Starting at 430 ppm CO_2e, stabilizing at 550 ppm CO_2e or below would likely cost around 1 percent of world GDP with good policy and timely decision making (see section IB); for stabilization at 450 ppm CO_2e, it might cost three or four times as much (possibly more). With bad policy, costs could be still higher. Note that the comparison of costs between 450 ppm and 550 ppm CO_2e illustrates the cost of delay[5]—waiting for 30 years before strong action would take us to around 530 ppm CO_2e, from which point the cost of stabilizing at 550 ppm CO_2e would likely be similar to stabilizing at 450 ppm CO_2e starting from now. Under most reasonable assumptions on growth and discounting, a flow of 1 percent of GDP for 50 to 100 years starting now would be seen as much less costly than a flow for a similar period of 4 percent or so of GDP, starting 30 years later.

It can be argued that at some future point, we might be able to turn to geoengineering, for example, firing particles into the atmosphere to keep out solar energy, analogous to the effect of major volcanic eruptions in the past. There are, however, substantial dangers associated with initiating other effects we do not understand. We might well be replacing one severe risk with another; however, extreme circumstances could require an extreme response. And there are difficult issues of global governance—would it be right for just one country, or group of countries, to do this? It seems much more sensible, at acceptable cost, to avoid getting into this position.

The above is basically the risk-management economics of climate change. For an expenditure of around 1 percent (between −1 percent and 3 percent) of world GDP (see section IB), we could keep concentration levels well below 550 ppm CO_2e and ideally below 500 ppm CO_2e. While leaving the world vulnerable, this would avoid the reckless risks implied by the higher stabilization concentrations (e.g., 650 ppm CO_2e) advocated by some economists. Thinking about the information basis for this argument also points to caution. If (as is unlikely) the risks of high concentrations turn out to be low and we have taken action, we would still have purchased a cleaner, more biodiverse, and more attractive world, at modest cost. If our actions are weak and the central scientific estimates are correct, we will be in very dangerous circumstances, from which it may be impossible, or very costly, to recover.

B. Costs of Abatement and Prices of GHGs

To this point, our discussion of targets has focused on those for the stabilization of stocks. We must now ask about implications for emissions paths and how much, with good policy, they would cost. We have already anticipated part of the broad answer—around 1 percent of world GDP per annum to get below 550 ppm CO_2e—but we must look at the argument in a little more detail.

Figure 2.2 illustrates possible paths for stabilization at 550 ppm CO_2e (thin line), 500 ppm CO_2e (dotted), and 450 ppm CO_2e (dot-dashed); the solid line is BAU. There are many paths for stabilization at a given level—see, for example, Stern 2007, p. 226—but all of them are a similar shape to those shown (if a path peaks later, it has to fall faster). And if the carbon cycle weakens, the cuts would have to be larger to achieve stabilization at a given level (see Stern 2007, p. 222). Broadly speaking, however, a path stabilizing at 550 ppm CO_2e or below will have to show emissions peaking in the next 20 years. For lower stabilization levels, the peak will have to be sooner. The magnitudes of the implied reductions between 2000 and 2050 are around 30 percent for 550 ppm CO_2e, 50 percent for 500 ppm CO_2e, and 70 percent for 450 ppm CO_2e. Cuts relative to BAU are indicated in the figure.

Figure 2.3 shows that, to achieve these cuts in emissions, it will be necessary to take action across the board and not in just two or three sectors such as power and transport. For the world as a whole, energy emissions represent around two-thirds of the total, nonenergy around one-third. Land-use change, mainly deforestation and degradation of forests, accounts for nearly 20 percent of the total. Given that the world economy is likely to be perhaps three times bigger in mid-century than it is now, absolute cuts of around 50 percent would require cuts of 80 to 85 percent in emissions per unit

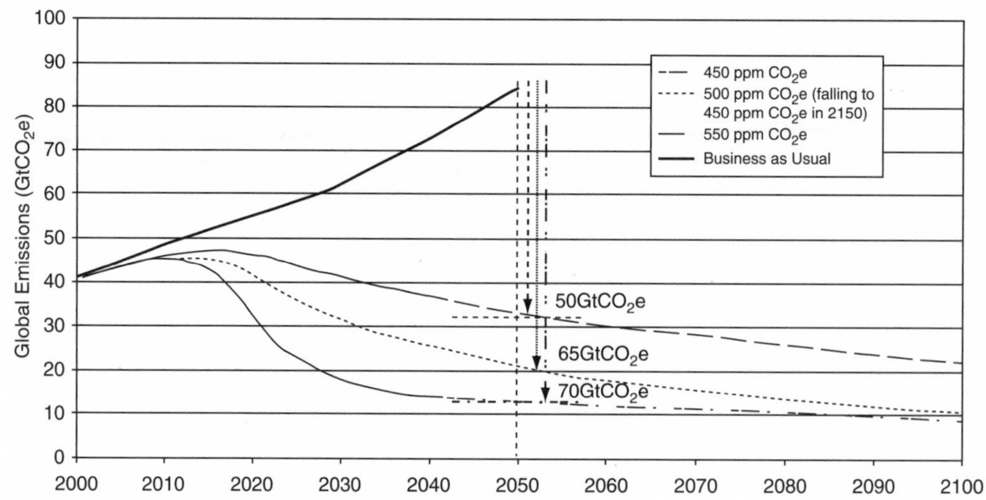

Figure 2.2. BAU and stabilization trajectories for 450–550 ppm Co_2e.
Source: Stern 2007, p. 233.

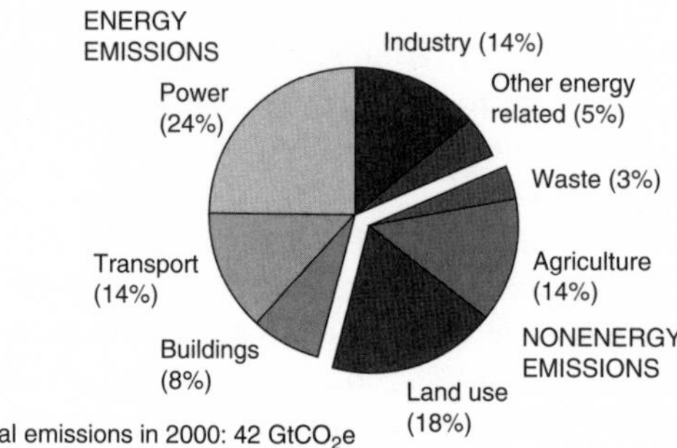

Figure 2.3. Reducing emissions requires action across many sectors.
Source: Stern 2007, p. 196.

of output. Further, since emissions from some sectors (in particular agriculture) will be difficult to cut back to anything like this extent, and since richer countries should make much bigger proportional reductions than poor countries (see section IV), richer countries will need to have close-to-zero emissions in power (electricity) and transport by 2050. Close-to-zero emissions in power are indeed possible, and this would enable close-to-zero emissions for much of transport. This would, however, require radical changes to the source and use of energy, including much greater energy efficiency. Achieving the necessary reductions would also require an end to deforestation. The totality of such reductions would, however, not result in a radical change in way of life to the extent of that brought by electricity, rail, automobiles, or the Internet.

On the path for stabilization, there would be different options for cutting emissions that would be more prominent at different times. In the earlier periods, there would be greater scope for energy efficiency and halting deforestation, and with technical progress, there will be, and already are, strong roles for different technologies in power and transport.

Various different options for abatement were discussed in chapter 9 of the *Stern Review*.[6] McKinsey has recently carried out

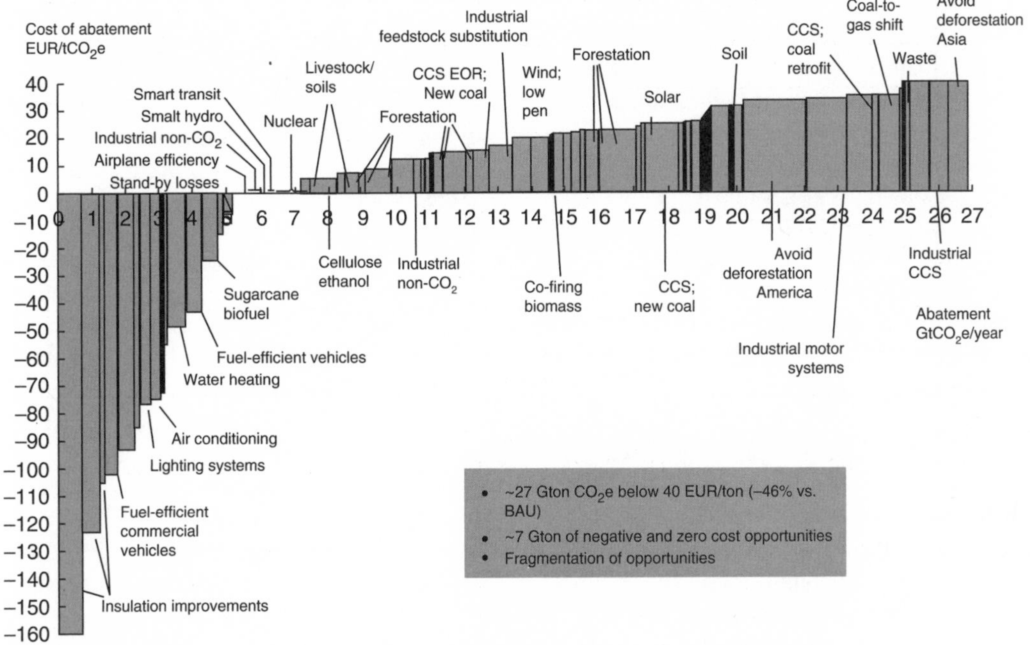

Figure 2.4. McKinsey bottom-up approach to abatement costs.

Source: Enkvist et al. 2007, p. 38.

a more detailed study (Enkvist, Nauclér, and Rosander 2007)—see figure 2.4. There are several important lessons from this type of curve. First, there are many options for reducing emissions that have negative cost; that is, they save money. Second, there is a whole range of options, and each should be explored in detail—for example, the costs associated with combating deforestation in the McKinsey curve are, in my view, far too high.[7] Third, the emissions savings from any one option will depend on what it replaces. Fourth, given the broad range of options, policy is very important—bad policy will lead to the uptake of more expensive options. Fifth, technical progress will be important and should be promoted so that the range of options is widened and costs are reduced. Finally, and of special importance, starting now in a strong way and with clear signals will allow more time for planned choices, discovery of options, and exploration of the renewal periods and timings for equipment. This is the measured, lower-cost approach. Going more slowly and then moving in haste when and if the science is confirmed still more strongly is likely to be the expensive option.

Very impotant for policy, this type of figure gives us an understanding of where carbon prices (or GHG prices more generally) should be. By 2030, cuts at the world level would have to be on the order of 20 Gt CO_2e (see figure 2.2) for stabilization at 550 ppm CO_2e. This suggests a CO_2 price of around €30 per ton.[8]

A fairly clear idea of where the carbon price should be from the point of view of necessary abatement is of great help both to policy makers and to investors. It also provides the opportunity to check against estimates of the marginal social cost of carbon (SCC) reflecting the future damage of an incremental emission. The levels quoted here for the MAC are consistent with ranges for the SCC indicated in the *Stern Review* along an abatement path for 550 ppm CO_2e stabilization.

However, the SCC is very slippery numerically, since it is so sensitive to assumptions about model structure, including future emission paths, carbon cycles, climate sensitivity, future technologies, *and* ethical approaches to valuation over the centuries to come. The SCC at time t is the expectation of the integral[9] over τ from t onward of:

the marginal social utility of consumption at τ (embodying ethical values and a particular path)

X the impact on consumption at τ of all relevant preceding temperature changes (and resultant climate change)

X the impact on a relevant temperature increase of increases in preceding carbon stocks

X the impact on all relevant stocks of an increase in carbon emissions at t, where "impact" in the above is to be interpreted as a partial derivative.

Given this sensitivity, it is remarkable how carelessly the SCC is often quoted—it is quite common, for example, for people to quote an SCC without even referring to a reference emissions path, to say nothing of all the other relevant assumptions that matter greatly.

Thus, the SCC is a very weak foundation for policy. The target approach and the calculation of the associated MAC is more attractive from the point of view both of policy toward risk and of clarity of conclusions. It is also important, however, to check prices derived from the MAC against SCC calculations and to keep policy under revision, as further information and discovery arrive. Some notion of the SCC is also useful in examining the emissions savings from, for example, transport programs or policies. If the MAC and SCC were thought to be in very different ballparks for an anticipated path, it would suggest strongly that policy revision is necessary.

Prices for abatement should be at a level that supports carbon capture and storage (CCS) for coal. Without CCS for coal, it will be difficult (and more costly) to achieve the necessary cuts, given that many countries will rely heavily on coal for power generation for the next 30 or 40 years (IEA 2006, 2007). China and India (Expert Committee on Integrated Energy Policy 2006), for example, will be using coal for around 80 percent of their electricity plants for the next 30 years or so—for the simple reasons that coal is cheap and available within their own borders, that they are familiar with the technologies, and that the plants can be erected quickly. Speed is of great importance for them, as the costs of electricity outages are very high.

The *Stern Review* also looked at top-down macro modeling of costs of emissions reductions (see also Barker, Qureshi, and Köhler 2006). Both the bottom-up and the top-down studies produced numbers in similar ranges—around 1 percent of world GDP. There is, of course, considerable uncertainty. Bad policy or delayed decisions could give higher numbers. Stronger technical progress could give lower numbers. Assumptions about substitutability between different goods and options matter, too. Since the *Stern Review* was published, there have been a number of new studies, both bottom-up and top-down. Significant examples of the former are those from McKinsey (Enkvist, Nauclér, and Rosander 2007) and the IEA (2007), both of which indicated costs either in the region we suggested or somewhat lower. Similar conclusions are drawn in the AR4 (IPCC 2007).

It is very important to recognize that costs of 1 percent of GDP do not necessarily slow medium- or long-term growth rates. They are like a one-off 1 percent increase in prices from "doing energy" in different ways. Further, there is a real possibility that incentives to discovery could generate a Schumpeterian burst of growth—on such possibilities, see recent work by Philippe Aghion (2007). The scale of markets for new technologies will be very large (IEA 2006); see also Fankhauser, Sehlleier, and Stern (2007) for an assessment of investment and employment opportunities, which are likely to be positive.[10]

Finally, reducing GHGs can bring strong benefits elsewhere. Cleaner energy can provide greater energy security and energy access. It can give reductions in local air pollution. Cleaner transport policies can increase life expectancy. Combating deforestation can protect watersheds, sustain biodiversity, and promote local livelihoods. Taking these associated benefits into account would reduce cost estimates further.

In summary, looking back after a year, we would suggest that subsequent evidence and analysis have confirmed the range of our cost estimates for stabilization or indicated that they may be on the high side. Good policy and timely decision making are, however, crucial to keeping costs down. And we would emphasize that taking a clear view now of a stabilization goal allows for a measured and careful adjustment, allowing for the replacement cycles of capital goods. To wait and see, or to rely on a "climate policy ramp," risks not only excessive and dangerous levels of stocks but also much more costly abatement if, as is likely, there is a subsequent realization that the response has been delayed and inadequate.

II. Stabilization of Stocks of Greenhouse Gases II: Modeling and Evaluation of Damages

A. Introduction

The previous section looked directly at the risks from GHGs and at the costs of action to reduce emissions and thus risks. This is the kind of judgment that people take when considering various forms of insurance, or design of buildings or infrastructure, or new medical treatments. They try to be as clear as possible on consequences and costs, bearing in mind that both are stochastic and that risk is of the essence, while also being aware that it will often be difficult to put a price or money values on consequences and risks.

It is also informative, however, to try to produce, using aggregate models, quantitative estimates of avoided damages in order to compare with costs. For climate change, that quantification may be possible for some dimensions, for some locations, for some time periods, and for some ranges of temperature increases.[11] However, the avoidance of planet-transforming change by keeping down risks of 5 °C and above is at the heart of the argument here, and it is extremely difficult to provide plausible aggregate numbers for the effects and overall damages of temperatures so far out of experience, particularly when nonlinearities may be of great importance. Nevertheless, formal modeling is central to the tools of our trade, and the exercises do have value in bringing out the logic of some important tradeoffs.

In making valuations of consequences, we have to face very difficult analytical and ethical issues. How does one value the transformation of the planet, the consequences of radical changes in ways of life, and big movements of population and associated conflict? Our standard cost-benefit analysis (CBA) tools do not give us much guidance. I have invested a lot of effort (e.g., Drèze and Stern 1987, 1990), as have many others, in developing these tools and have some understanding of what they are and where they can be applied. They are largely marginal methods, providing tools for analysis of big changes in, say, one or two markets as a result of a program. But when we are considering major strategic decisions for the world as a whole, with huge dynamic uncertainties and feedbacks, the potential contribution of an approach to decision making based on marginal methods is very limited. Rational decision making has to go back to the first principles from which the marginal methods of CBA are derived. This is not at all to use a different theory. On the contrary, it is to maintain the theory and to avoid a gross misapplication of the special (i.e., marginal) case.

The centrality of nonmarginal changes and of risk means either using the risk-analysis approach of section IA or using aggregate modeling with a social-welfare function to compare consequences. Both have their role, but for the reasons given, I would see the former as the main plank of the argument. The latter has a valuable supplementary role, which we now investigate.

In setting out a social-welfare function to evaluate damages and costs, the valuation of consequences on different dimensions—social, health, conflict, and so on—will be extremely difficult. I do not go into these issues. I focus on one issue that has, understandably, received considerable attention in discussion of the *Stern Review*: how to value benefits accruing to different people at different times. There are unavoidable ethical issues. They are the subject of section IIB. In section IIC, we examine some of the challenges, results, and sensitivities of formal modeling and comment on new evidence and discussions concerning the *Stern Review*'s damage estimates after one year.

B. Ethics—

Discounting

Much of the discussion of ethics in relation to the *Stern Review* has been focused on discounting. Sometimes, simplistic approaches to discounting conceal or obscure the underlying structural and ethical logic by shoehorning the issues into a simple discount rate specified entirely externally to the problem. However, careful use of theory and concepts is crucial. Some have argued that "the discount rate of the Stern model" is too low in relation to market rates of return. This argument has generally been thoroughly confused for a whole set of reasons. It arises from inappropriate application of a marginal method to a strongly nonmarginal context, failure to apply modern public economics, ignorance of the multigood nature of this problem, and, in some cases, ignorance of the difference between a social discount rate and a pure time discount rate. Given this pervasive confusion about the basic theory of discounting, it seems worthwhile to clarify briefly the logic of discounting as applied to climate change and relate it to some simple empirical data.

Let us start with the definition of a discount rate in policy evaluations. It is simply the proportionate rate of fall of the value of the numeraire used in the policy evaluation. In the simple case, with aggregate consumption as the numeraire, we have a social discount factor, or SDF, $\lambda(t)$, which measures the social value of a unit of consumption at time t relative to a unit at time zero. The social discount rate, or SDR, is then $-\lambda/\lambda$.

A number of general conclusions follow immediately from these basic definitions. First, the SDF and the SDR depend on a given reference path for future growth in consumption and will be different for different paths. Second, the discount rate will vary over time. Third, with uncertainty, there will be a different discount rate for each possible sequence of outcomes. Fourth, there will be a different discount rate for different choices of numeraire. In imperfect economies, the social value of a unit of private consumption may be different from the social value of a unit of private investment, which may be different from the social value of a unit of public investment. And the rates of changes of these values may be different, too.

A further key element for understanding discount rates is the notion of optimality of investments and decisions. For each capital good, if resources can be allocated without constraint between consuming the good in question and its use in accumulation, we have, for that good, the result that the social rate of return on investment (the marginal productivity of this type of good at shadow prices), the SRI, should be equal to the SDR in terms of that good (i.e., with that good as numeraire). This is intuitively clear and in optimal growth theory is a standard first-order condition. But where there are constraints on this optimization, as there usually will be in imperfect economies, this condition that the SRI equal the SDR is not generally applicable. Drèze and Stern (1987, 1990), for example, show how opportunity costs, and thus shadow prices and shadow rates of return, depend on which alternative use a unit of resource comes from. Further, in such economies, it will not generally be true that the private rate of return on investment (PRI) will be equal to the SRI. And similarly, private discount rates (PDRs) can diverge from SDRs. Such divergences can arise from all forms of market imperfections, including externalities. In this case, we have the additional complication that key players, future generations, are not directly represented. Thus, in the general case:

$$\text{PDR} \neq \text{SDR} \neq \text{SRI} \neq \text{PRI}$$

Before looking into discount rates along a given path, we should remind ourselves that the most basic mistake here is to use a marginal concept (discount rates) around a current path for strategic choices and comparisons among paths. Policy on climate change means choosing among paths with very different growth patterns for a whole collection of capital goods, including those relating to natural endowments. Thus, it is simply wrong to look at rates as currently observed, or in historical terms, which refer to existing paths. A choice among paths means also choosing the implied set of discount rates associated with the paths (Stern 2007, 27–31; for more on this issue, see Hepburn 2006). This is simply another way of expressing the old idea that the shadow prices or marginal values depend on where you are. It is absolutely fundamental here for this very nonmarginal set of choices to recognize that the social discount rates are endogenous, not exogenous. They are determined by ethical values, which have to be discussed explicitly, and by the paths that result from climate change and investment choices.

Let us suppose, however, that we go past this problem and look at discount rates around a given path, or path of choice. What can we learn from observed rates in markets? Rates at which households can borrow and lend, usually for periods of no longer than three or four decades, give a reading on their private discount rates or PDRs (assuming, they equate their discount rate with their market rate, with some appropriate treatment of uncertainty). But as this borrowing and lending take place through private decisions made by individuals acting in a market, this does not necessarily answer the relevant question in the context of climate-change decisions by a society—namely, how do we, acting together, evaluate our responsibilities to future generations over very long periods?

Rates of return on investment generally reflect private rates of return narrowly measured. They take no account of externalities, which are of the essence for this discussion. Thus, even if we think we can observe some private rates of discount for some households, and some private rates of return for some firms, we do not have a reading on the concept at issue here, the social discount rates for the key goods. Thus, observations on the PRIs and PDRs have only limited usefulness. And note that the problems that prevent the equalities in this chain, such as missing markets, unrepresented consumers, imperfect information, uncertainty, production, and consumption externalities, are all absolutely central for policy toward the problem of climate change. We come back again to a basic conclusion: the notions of ethics, with the choice of paths, together determine endogenously the discount rates. There is no market-determined rate that we can read off to sidestep an ethical discussion.

It must surely, then, be clear that it is a serious mistake to argue that the SDR should be anchored by importing one of the many private rates of return on the markets (or a rate from government manuals or a rate from outside empirical studies). Yet it is a mistake that many in the literature have made. Nordhaus (2007b, p. 690) and Weitzman (2007b), for example, substitute a market investment return of 6 percent for the SDR, thus producing a relatively high 6 percent rate of discount on future consumption. This mistakenly equates the PRI to the SRI and the SRI to the SDR. Such an approach is entirely inappropriate given the type of nonmarginal choices at issue and the risk structure of the problem and in light of developments in modern public economics, which encompasses social cost-benefit analysis and which takes account of many imperfections in the economy, including unrepresented consumers, imperfect information, the absence of first-best taxes, and so on.

If, despite these difficulties, we nevertheless insist on looking to markets for a benchmark rate of discount, what do we find? In the United Kingdom and the United States, we find (relatively) "riskless," indexed lending rates on government bonds centered around 1.5 percent over very long periods. For private very long-run rates of return on equities, we find rates centered around 6 or 7 percent (Mehra and Prescott 2003, p. 892; Arrow et al. 2004, p. 156; Kochugovindan and Nilsson 2007a, p. 64; 2007b, p. 71). Given that it is social discount rates that are at issue, and also that actions to reduce carbon are likely to be financed via the diversion of resources from consumption (via pricing) rather than from investment, it is the long-run riskless rates associated with consumer decisions that have more relevance than those for the investment-related equities. Thus, even if one were to endorse the approach of importing a discount rate from markets, when one uses the rate of return closer (but not equivalent) to the relevant concept—the risk-free rate—it is far from clear that one would obtain a rate of discount on future consumption as high as the 6 percent advocated by Nordhaus (2007b, p. 690).

Weitzman (2007c) has produced an interesting insight into the difference between the riskless rate and equity returns in terms of perceived high weights in the downside tail of equity returns—the implication being that the perceived equivalent return on equities, allowing for risk, is close to the lower riskless rates. In this context, Weitzman (2007a, 2007b), has also suggested encapsulating risk and uncertainty in some contexts into discount rates. In my view, however, it is far more transparent to treat risk directly through the approach to social welfare under uncertainty than to squash it into a single parameter that tries to reduce the problem to one of certainty.

Suppose, however, that we persisted with the argument that it is better to invest at 6 to 7 percent and then spend money on overcoming the problems of climate change later rather than spending money now on these problems. The multigood nature of the problem, together with the irreversibilities from GHG accumulation and climate change, tell us that we would be making an additional mistake. The price of environmental goods will likely have gone up very sharply, so that our returns from the standard types of investment will buy us much less in reducing environmental damage than resources allocated now (see also section I on the costs of delay).[12] This reflects the result that if environmental services are declining as stocks of the environment are depleted, then the SDR with that good as numeraire will be negative. On this, see the interesting work by Hoel and Sterner (2007), Sterner and Persson (2007), and Guesnerie (2004), and also Stern 2007, p. 60). Environmental services are also likely to be income elastic, which will further reduce the implied SDR.

Finally, we underline an unhappily common mistake—namely, confusing the pure time discount rate (PTDR) with the SDR. With a very simple single-good structure and consumption at time t having social value $u(c)e^{-\delta t}$, we have the SDF, λ, as $'(c)e^{-\delta t}$.[13] Its proportionate rate of fall (the SDR) is $\eta(\dot{c}/c) + \delta$, where η is the elasticity of the social marginal utility of consumption with respect to consumption.[14] Often, η is taken to be a constant. In this very simple case, we can now see the difference between the SDR and the PTDR. The PTDR is the rate of fall of the value of a unit of consumption, *simply because it is in the future*, quite separately from the levels of consumption enjoyed at the time. Here, the PTDR is δ. For example, with $\delta = 0$, $\eta = 1.5$, and $\dot{c}/c = 2.5$ percent, we have a social discount rate of 3.75 percent, in excess of the U.K. government's test discount rate (Her Majesty's Treasury 2003), notwithstanding a PTDR of zero. It is η and the growth rate that capture the idea that we should discount the consumption of future generations on the basis that they are likely to be richer than ourselves. This reason for discounting is, and should be, part of most models, including those of the *Stern Review*. We shall show in the next subsection that the cost, in terms of climate changes, of weak or delayed action in the formal models is much greater than that of timely and stronger action, in terms of abatement expenditure, over a range of parameter values for η.

A δ of 2 percent (3 percent)—as endorsed by many commentators such as Nordhaus (2007b) and Weitzman (2007b)—implies that the utility of a person born in 1995 (1985) would be "worth" (have a social weight) roughly half that of a person born in 1960. This type of discrimination seems very hard to justify as an ethical proposition and would be unappealing to many. Indeed, the ethical proposition that δ should be very small or zero has appealed to a long line of illustrious economists, including Frank P. Ramsey (1928, p. 543), Arthur Cecil Pigou (1932, pp. 24–25), Roy F. Harrod (1948, pp. 37–40), Robert M. Solow (1974, 9), James A. Mirrlees (Mirrlees and Stern 1972), and Amartya Sen (Anand and Sen 2000). I have heard only one ethical argument for positive δ (Beckerman and Hepburn 2007; Dietz, Hepburn, and Stern 2008) that has some traction—namely, a temporal interpretation of the idea that one will have stronger fellow feelings for those closer to one (such as family or clan) relative to those more distant. This is often explained in terms of functionality for survival of groups. However, this type of reasoning from evolutionary biology does not have much relevance when we are thinking about the survival of the planet as a whole.

For these reasons, the *Stern Review* followed the tradition established by the economists cited above, adopting and arguing strongly for a δ that exceeds zero only in order to account for the possibility of some exogenous event that would render future welfare calculations irrelevant—the exogenous extinction of humanity (for discussion of this interpretation of δ, see, e.g., Pearce and Ulph 1995 and Newbery 1992). On this basis, the *Review* adopted a δ of 0.1 percent (although even this value for δ appears to be quite large in relation to this interpretation, implying a probability of exogenous extinction of around 10 percent in 100 years). For a project or program, the probability of exogenous extinction could be substantially higher, and this is reflected in some cost-benefit manuals or approaches; in our case, however, we are considering humanity as a whole.

My overall assessment of the discussion of discounting in the context of climate change is that it is disappointing. All too often, it has failed to come to grips with the basic concepts, with the key nonmarginal and uncertainty elements at the core of the issue, and with the theories of social cost-benefit analysis and modern public economics of the last 30 or 40 years.

Distributional Judgments

Having seen the implausibility of importing a discount rate from outside the model to sidestep ethical judgments, let us turn to the ethics relating to the distribution of consumption or income, at least in its very narrow form of η within the narrower cases (as in the models that follow) where the social objective is the expectation of the integral of $\Sigma_i\, u(c_i)e^{-\delta t}$ (Stern 2007, pp. 50–54).[15] Thinking about η is, of course, thinking about value judgments—it is a prescriptive and not a descriptive exercise. But that does not mean that η is arbitrary; we can, and should, ask about "thought experiments" and observations that might inform a choice of η. In so doing, we must remember that η plays three roles, guiding (a) intratemporal distribution, (b) intertemporal distribution, and (c) attitudes to risks. We look at the relevance of empirical data for each of the three in turn.

Intratemporal Distribution. Let us begin with a thought experiment concerning direct consumption transfers in a very simple context. If A has k times the consumption of B, the social value of a unit of consumption to B is k^η times that to A for constant η. For example, for $k = 5$ and $\eta = 2$, the relative value is 25, and a transfer from A to B would be socially worthwhile even if up to 96 percent were lost along the way (the so-called leaky bucket—Okun 1975). While I might not regard that position as unacceptable, to take just one example, it appears inconsistent with many attitudes to transfers. In this sense, many would consider an η of 2 to be very egalitarian. With $\eta = 1$, the 96 percent in the example above becomes 80 percent because the unit to B is worth five times that to A. Some might regard even this position as rather egalitarian.

Value judgments are, of course, precisely that, and there will be many different positions. They will inevitably be important in this context—they must be discussed explicitly, and the implications of different values should be examined. Examples follow of what we find when we turn to empirical evidence and try to obtain implied values (the "inverse optimum" approach). Empirical evidence can inform, but not settle, discussions about value judgments. For further exploration, see Dietz, Hepburn, and Stern 2008. In using such evidence, we must constantly bear in mind two key issues. First, we must ask about the relevance of individual decisions for the societal decisions about the problem at hand—here social decisions by the world community now, bearing in mind consequences for future generations. And, second, if we infer values from decisions, we must ask whether we have modeled well the decision processes, the objectives, and the perceived structure of the problem as seen by the decision maker.

Atkinson and Brandolini (2007) have produced an interesting set of examples on empirical income distributions and actual transfer schemes in relation to welfare weights.[16] They conclude that constancy of η across a range of increases is difficult to "square with" the way many transfer schemes occur in practice; in addition, there are many examples where policies appear inconsistent with η greater than one. For example, given the current income distribution in the United States, an η of two would imply that a redistribution from the fifth-richest decile to the second-poorest decile would be welfare-improving even if only 7 percent of the transfer reached the recipient; for a transfer from the richest decile to the second-poorest, virtually any redistribution would be welfare-improving regardless of loss along the way, so long as the recipient received some benefit (Atkinson and Brandolini 2007, p. 14). Of course, interpretation of actual intratemporal tax and transfer schemes will depend on many assumptions about the structure of incentives[17] and policymaking procedures. Perhaps people think that tax-transfer disincentives are very strong and they oppose transfers for these reasons. Or notions of rights and duties may influence them. The upshot is that empirical estimates of implied welfare weights can give a wide range of η, including η below one and even as little as zero.

It is striking that there are some, such as Nordhaus (2007b) and Weitzman (2007b), who appear to argue for high η (equal to 2 or 3) in intertemporal analysis yet do not bring out how this is potentially inconsistent with standard cost-benefit analysis treatments of intragenerational distribution (which effectively assume $\eta = 0$) or with some intratemporal tax and transfer policies.

Intertemporal Distribution. In discussions of h in an intertemporal framework, there has been much focus on implied saving rates. Some (Dasgupta 2007, p. 6; Nordhaus 2007b, pp. 694–696), following arguments in Arrow 1995, pp. 12–17), have criticized the relatively high weight placed by the Stern Review on the consumption of future generations (whether via h or d) by arguing that the Review's parameter choices can, in certain scenarios, imply implausibly high optimal savings rates. As is clearly explained in the Review (Stern 2007, p. 54), with d = 0, output proportional to capital, and no technical progress, the optimal savings rate is 1/h. With h close to one, this would lead to very high optimal savings rates. At the same time, the Review also states clearly (Stern 2007, p. 54) that this result is highly dependent on model assumptions.

Brad DeLong, in a short blog entry (DeLong 2006), points out this flaw in the Dasgupta-Nordhaus position and argues that technical progress would greatly reduce the optimal savings rate. Mirrlees and Stern (1972) presented a more fully developed argument. Using a standard one-good, infinite-horizon Ramsey growth model, constant returns to scale, and a Cobb-Douglas production function, they showed that under one specification—with constant population, a competitive share of capital equal to 0.375, and 3 percent exogenous technological progress—the optimal consumption path for $\eta = 2$ and $\delta = 0$ involves a savings rate, s, between 0.19 and 0.29 (or 0.23 if constrained to a constant s). This is far below the 0.5 that would be optimal with $\eta = 2$ and $\delta = 0$ in the simpler case of output proportional to capital and no technical progress.

Just as with intragenerational values, the approach of the "inverse optimum" or implied social values does not take us very far in this context. We cannot really interpret actual saving decisions as revealing the collective view of how society acting together should see its responsibilities to the future in terms of distributional values—too much depends on assumptions about how decisions are made in a society and on how the participants perceive the workings of the future economy. Observed aggregate savings rates are sums of individual decisions, each taken from a narrow perspective. This is not the same thing as a society trying to work out responsible and ethical collective action—the crucial issue for climate change.

Attitudes to Risks. "Guidance" on η from analyses of risk and uncertainty is even less informative. We can interpret η as the parameter of relative risk aversion in the context of an expected utility model of individual behavior. However, the expected utility model is unreliable as a description of attitudes to risk.

Further, we see a whole range of behavior, from the acceptance of "unfair risks" in gambling (similar to $\eta < 0$) to extreme risk aversion in insurance (very high η). And even if behavior were somewhat more "rational" in the narrow sense of conforming to the expected utility hypothesis, it would still be unclear how sound a basis it would be for the specification of a prescriptive value for use in this context.

From this very brief discussion of empirical information, which might help us to think about η in a prescriptive context, our conclusion is that there is very little to guide us.[18] Again, we are pushed back to the standard moral philosopher's approach of trying to think through simple examples, that is, the thought experiment. It has the great virtue of facing the issues directly—it is transparent and clear.

What do we conclude about ethics and discounting in this context when we clear the various confusions out of the way? The answer is fairly simple. First, we must address the ethics directly. There is no simple market information from intertemporal choices or otherwise that can give us the answers. Second, if we express the problem in standard welfare economic terms, portraying the objective as an expectation of an integral of social utility, we cannot use marginal approximations to changes in welfare, since we are comparing strategies that yield very different paths. Third, within this framework, we may focus the discussion on elasticities of marginal social utility η and pure time discount rates δ, but in so doing we must recognize the ethical narrowness of this approach. Fourth, direct ethical discussion of η and δ suggests a broad range for η, although the consequences for simple transfers suggest that many would regard η in excess of 2 as unacceptably egalitarian; on the other hand, there appears to be little in the way of ethical arguments to support δ much above zero. Fifth, within a marginal analysis framework, the relevant concept for discounting here is the SDR. In the narrow η–δ context, with η of 1 to 2, very low δ, and growth at 1.5 to 2.5 percent,[19] we find an SDR of 1.5 to 5.0 percent, which is close to ranges for long-run consumer real borrowing rates and (at least in the U.K.) government discount rates for program evaluations.

C. Formal Modeling

Aggregate models have been popular in the economics of climate change. They attempt to integrate the science of climate change, as expressed, for example, via GCMs, with economic modeling and are termed integrated assessment models, or IAMs.

As I have argued, it is very hard to believe that models where radically different paths have to be compared, where time periods of hundreds of years must be considered, where risk and uncertainty are of the essence, and where many crucial economic, social, and scientific features are poorly understood can be used as the main quantitative plank in a policy argument. Thus, IAMs, while imposing some discipline on some aspects of the argument, risk either confusing the issues or throwing out crucial features of the problem.

A related but different point is their use, when modeling of costs of abatement is integrated with modeling of damages from emissions, as vehicles for optimization analysis. In this respect, they are still less credible. Those of us schooled in the optimal tax and optimal growth analysis of the 1960s, 1970s, and 1980s learned just how sensitive model results can be to simple structural assumptions, such as the form of preferences, production, or technical progress, even before parameter values are introduced (Atkinson and Stiglitz 1976, 1980; Deaton and Stern 1986).

The models portrayed here should be seen as helpful supplements exploring some serious logical and modeling issues related to the estimation of damages from BAU and their comparison with alternative paths. We shall see, not surprisingly, that the key assumptions influencing damage estimates concern risk and ethics. It is surprising, however, that these two issues did not occupy until recently the absolutely central position that the logic of the analysis demands. The result is that—given the recent evidence on emissions, carbon cycles, and climate-change sensitivity—most of the studies prior to a year or two ago grossly underestimated damages from BAU.

The PAGE[20] model was chosen for the work of the *Stern Review* first, because, in contrast with a large majority of preceding work, it places risk and uncertainty at center stage. It provides for a Monte Carlo analysis of explicit distributions of a large number of parameter values. Second, Chris Hope, its originator, chose the parameters and their distributions to straddle a range of climate models, IAMs, and economic models in the literature. Third, Hope kindly made the model available and was very generous with his advice. The model was described extensively in chapter 6 of the *Stern Review* as well as by Hope (2006a, 2006b) and Dietz et al. (2007a, 2007b).

Key assumptions on the form of the models and of the parameters in these models may be grouped into two broad headings: the *structural* elements that shape the estimated consequences of different kinds of emissions strategies and the *ethical* elements that shape the evaluations of different outcomes. Of the structural elements in this approach, four are crucial: the emission flows; the functioning of the carbon-cycle linking flows to stocks; the climate sensitivity linking stocks to temperature; and the damages from temperature, via climate change. Of the ethical elements, the following are crucial: the type of ethical values considered (including the role of rights and obligations); the type of outcomes introduced into evaluation functions (including separate goods or services, such as environment, health, and standard elements of consumption); the functional forms used to capture evaluations; and the parameters within those functional forms, including those covering intra- and intergenerational values. The ethical discussion should not be shoehorned into a narrow focus on just one or two parameters such as η and δ; the ethical issues and their interactions with a model structure designed to reflect a range of uncertainties are much broader and deeper.

Stern Review Damages and Sensitivity

The *Stern Review* base case had damages from BAU relative to no climate change of around 10 percent of consumption per annum measured in terms of the balanced growth equivalent, or BGE (see Mirrlees and Stern 1972). Here, the BGE for any given path is calculated from the expected social-utility integral of that path by asking what initial consumption level, growing at a given growth rate and without uncertainty, would give this expected social-utility integral? The difference between the BGEs with and without climate change can be thought of as the premium, in terms of a percentage of annual consumption, that society might be willing to pay to do away with the risk and uncertainties associated with dangerous climate change. Essentially, the BAU provides a calibration in terms of consumption (useful since "expected integrated utils" are hard to interpret) for the expected utility integral: it summarizes an average over time, space, and possible outcomes.

Table 2.2 presents some of the results of the PAGE model. The parameter η was discussed in section IIB and is the elasticity of the social marginal utility of consumption where the integrand for expected social utility is the sum over i of $N_i u(C_i/N_i)e^{-\delta t}$, and where C_i and N_i are consumption and population in region i. In the model, γ is the exponent of a power function linking temperature T to damage through the function AT^{γ} (Stern 2007, p. 660; the damages vary by region). Table 2.2 provides BGE differences (in percent) across paths without and with climate change, with a 5 to 95 percent confidence interval in brackets. We think of increases in γ as capturing increases in the structural risks[21] and of increases in η as capturing increases in aversion to inequality and risk.

Intuitively, we can think of γ as combining both the relation between temperature and damages and the distribution of temperatures arising from a certain emission path. These are, of course, distinct effects, but both an increase in γ and a broader distribution for the temperature (in particular more weight in the upper tail, either from a weakening in the carbon cycle or from higher climate sensitivity) has the effect of producing a higher probability of large damages. The effects are treated separately in the *Review* (chapter 6 and the technical annex to postscript), where many more sensitivity results are given. These two processes (damages and temperature distributions) can and

Table 2.2.
Sensitivity of total cost of climate change to key model assumptions.

Damage function exponent (γ)	Consumption elasticity of social marginal utility (η)		
	1	1.5	2
2	10.4 (2.2–22.8)	6.0 (1.7–14.1)	3.3 (0.9–7.8)
2.5	16.5 (3.2–37.8)	10.0 (2.3–24.5)	5.2 (1.1–13.2)
3	33.3 (4.5–73.0)	29.3 (3.0–57.2)	29.1 (1.7–35.1)

Source: Dietz et al. 2007b.

should be modeled separately, but here we keep the discussion and presentation as simple as possible.

While we shall discuss results in terms of the sensitivity of estimated damages with and without climate change, we must emphasize that stabilization at 550 ppm CO_2e removes around 90 percent[22] of the damages (Stern 2007, p. 333), so that we are essentially comparing two strategies, namely BAU and stability below 550 ppm CO_2e. *A key broad lesson from this type of modeling is that the costs of stabilizing below 550 ppm CO_2e are generally far lower than the costs of the damages from climate change that would thereby be avoided.* While the measurement of estimated damages may vary, this key lesson is robust to parameter changes.

In this type of modeling, results are highly sensitive to assumptions on both structural risks and ethics, suggesting that great care should be exercised in choosing the key parameters. We can illustrate the importance of these two issues in terms of both computations in the model and of general results. Replacing all random variables in the PAGE model by their modes brings down the central case of damages from BAU from 10 to 11 percent to 3 to 4 percent.[23] Thus, it is wrong to argue, as Dasgupta (2007) and Nordhaus (2007b) have, that the chapter 6 results of the *Stern Review* arise solely from assumptions related to ethics, in particular the use of $\eta = 1$ and, at least in the view of Nordhaus, a low δ. *Both* risks *and* ethics are crucial to any serious assessment of policy toward climate change and, in particular, assessment of damages from BAU.

A formal result is provided in Box 2.1, which shows that for any given set of structural risks and a utility function, pure time discounting (a key element in the ethics) can be set so that the estimated damages are as small as we please. Further, for any given pure time discounting, risks and utility can be set such that damages are as big as we please.

Recently, in a series of papers (Weitzman 2007a, 2007b), Weitzman has argued that when we consider how the various different probability distributions (particularly of climate sensitivity) that might arise in different models can or should be combined, there is a convincing case for strong weights in the tails of overall temperature and damage distributions. These can lead to divergent (i.e., infinite) estimates of expected damages. His arguments are powerful and persuasive, underlining strongly the crucial role of risk in this story and raising questions on the use of the expected utility approach.

It is interesting to note that divergence of integrals can occur in three ways in this expected utility integral: first, via uncertainty, as Weitzman emphasizes; second, via intragenerational distribution (for example, this can occur for the Pareto distribution of income; Kleiber and Kotz 2003, pp. 59–106); and third, via integration over time. Indeed, for $\eta = 1$ and $\delta = 0$, with positive growth the time integral is on the borderline of convergence (Stern 2007, 58). Thus, for $\eta = 1$ and $\delta = 0.1$ percent, the

Box 2.1. Role for both risk and ethics

- Write expected utility integral as $\int_0^\infty g(t)f(t)dt$, where $g(t) = E[\hat{u}(c)]$, and $\hat{u}$ is the welfare difference without and with climate change; $f(t)$ is the pure time discount factor. $g(t)$ will depend on model structure, policies/path and shape of $u(c)$. It is possible that $g(t)$ is infinite for some finite T (see Weitzman 2007a).
- For any given $f(t)$ we can construct $g(t)$ so that there are infinite losses from climate change, i.e. $\int_0^\infty g(t)f(t)dt = \infty$. An example is $g(t) \equiv 1/f(t)$.
- For any given $g(t)$ we can construct $f(t)$ so that $\int_0^\infty g(t)f(t)dt < \varepsilon$ for any $\varepsilon > 0$, i.e., there are arbitrarily small losses from climate change. An example is $f(t) \equiv (1/g(t))e^{-\delta t}$ with $\delta > 1/\varepsilon$.
- Clearly, both ethical values and risk play key roles

bulk of the changes (in terms of the expected utility integral)—more than 90 percent—occur after 2200. For η = 2, the proportion is around 10 percent and for η = 1.5 around 30 percent. For some (e.g., Cline 2007), this is an argument for η higher than one, and I have some sympathy with this view.

Claude Henry (2006; Stern 2007, pp. 38–39) has argued that our lack of knowledge about which of the probability distributions to use for temperature and damages is an example of Knightian uncertainty, and he shows, using recent mathematics on how the von Neumann–Morgenstern axioms might be modified, how strong weights are likely to be (or should be) attached to the worst outcomes. We might see his approach, together with that of Weitzman, as a mathematical embodiment of the precautionary principle.

Other forms of sensitivity are summarized only briefly here—see the *Stern Review* for more details. We comment on some specifics of a weakening carbon cycle on the structural side and pure time preference, intragenerational issues, and a narrower view of dimensions of damage on the ethical side. The *Stern Review* had a "base climate scenario" (Stern 2007, p. 175) which ruled out a weakening carbon cycle and included only very moderate positive natural feedbacks. These are known to be possibilities but are not sufficiently well understood to enable calibration for most modeling purposes. A "high-climate scenario" (Stern 2007, p. 175) introduced increased changes for the carbon cycle, covering plant and soil respiration and possible methane emissions from thawing permafrost, but these effects as modeled now look fairly small in relation to current scientific concerns. This added a 4 percent extra BGE loss from BAU relative to no climate change. We also experimented with higher climate sensitivity—see Stern 2007, p. 179—although we did not publish results. It seems now that the "high +" scenario discussed there may be of real relevance.

The PAGE model used in the review includes some damage estimates from nonmarket effects such as health. If these are removed, the base-case damage estimates drop from 10 percent to 5 percent. Unsurprisingly, results are sensitive to pure time discounting. A pure time discount rate of 1 percent implies, under the extinction view of discounting, only a 60 percent chance of the world surviving the next 50 years, which most would regard as a very pessimistic number. Nevertheless, as the *Review*'s technical annex to postscript shows, even with δ = 1 percent, damages from BAU are likely to be higher than the costs of a mitigation strategy that removes the bulk of the risks.

We did not carry out an examination in the model of intragenerational issues in any detail, but comparisons with other studies suggested that these could add around a quarter or more to loss estimates (for η around 1), in this case another 4 to 5 percent. Starting with the base case of BGE losses of 10 to 11 percent, these variations (−5 percent for a narrower view of damages, +4 percent for higher climate response, +4 or 5 percent for intragenerational issues) gave us the range of 5 to 20 percent losses per annum from BAU that has been widely quoted. These are averages in three senses: over time, over space, and over possible outcomes.

In chapter 13 of the *Review*, different methods for looking at stabilization are examined, starting with the bottom-up or risk-evaluation approach of chapters 1, 3, 4, and 5 (and section I). The discussion of the top-down damage modeling approach, used in this section and in chapter 6 of the *Review*, explains that the 10 to 11 percent BAU base-case damage costs are reduced to around 1 percent for stabilization at 550 ppm CO_2e (Stern 2007, p. 333), that is, the cost saving, from avoided damages of stabilizing at 550 ppm CO_2e is 9 to 10 percent. When compared with costs of stabilization at 550 ppm CO_2e of around 1 percent of world consumption or GDP[24] (see section IB above), this saving from action represents a very good return. Even if damages avoided are only 3 to 4 percent of world consumption or GDP, stabilizing below 550 ppm CO_2e is still a good deal. The basic statement that the costs of strong and timely action are much less than the costs of weak and delayed action is very robust. Let us underline again, however, that the *Review* gives stronger weight in terms of space and emphasis to the bottom-up risk evaluation approach than to the top-down aggregate modeling approach.

Comparison with Other Modelling of Damages.

Much of the earlier literature on climate modeling found damage results that were lower than the results in the *Stern Review*.[25] Much of this earlier work underestimates BAU emission flows (see below), suppresses or only lightly touches on risk, takes an extraordinarily low view of damages from temperature increases, and embodies very high pure time discounting with little explicit ethical discussion of why (see section IIB).

As figure 2.5 shows, Tol (2002) and Nordhaus (Nordhaus and Boyer 2000) essentially suppress uncertainty about climate sensitivity by using point estimates and not spreads.[26] There are some minor attempts to "add on" risk, but it is not given the central role demanded by the science and the economics. The range covered by PAGE is cautious on climate sensitivity, using only triangular distributions for its parameters—its *full* spread from all Monte Carlo runs is within the IPCC AR4 "*likely*" (66 percent confidence interval) range. The Meinshausen (2006) spread covers the 90 percent confidence interval for the full range of models he surveys, some of which go far higher.

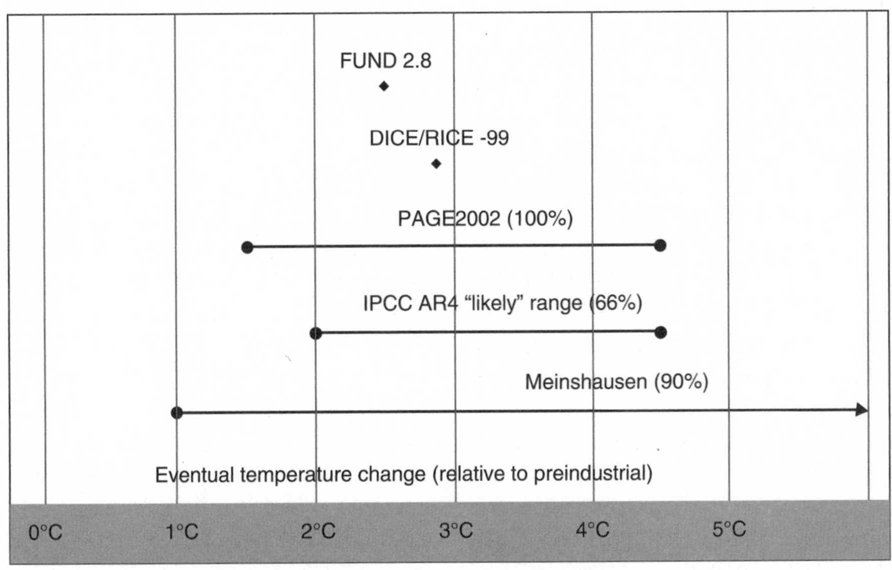

Figure 2.5. Estimates of climate sensitivity from IAMs compared with GCMs.

Source: Tol 2002; Nordhaus and Boyer 2000; Hope 2006a; 2006b; IPCC 2007; Meinshausen 2006.

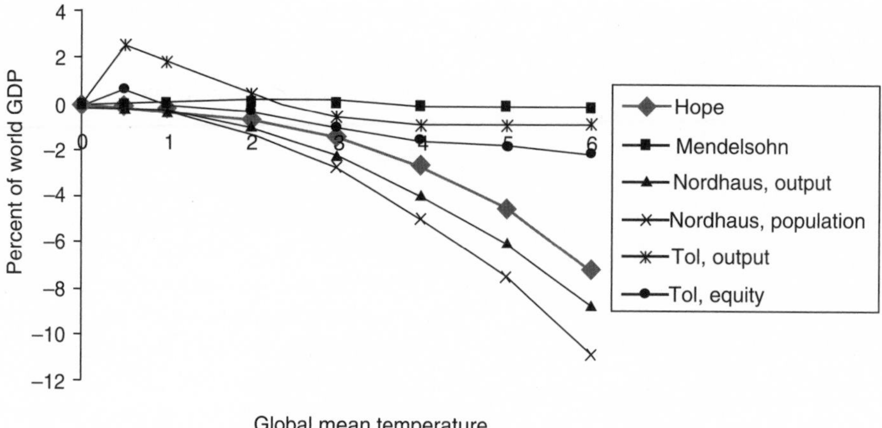

Figure 2.6. The modelled damages from climate change with increasing global temperatures.
Source: Dietz et al. 2007a.

Figure 2.6 summarizes results by Mendelsohn (Mendelsohn et al. 2000) and Tol (2002) with astonishingly low damages of 0 to 2 percent of GDP from temperature increases as high as 5 °C to 6 °C. The Nordhaus and PAGE (Hope 2006a) damages in terms of output are fairly close together, although arguably much too small in relation to the possible implications of 5 °C to 6 °C temperature increases.

These early models have given rise to a powerful and unjustified bias against strong and timely action on climate change. The question is not so much why the *Stern Review's* modeling obtained high damages under BAU as why the earlier literature made assumptions that give such low results.

D. Damages and Sensitivity, One Year On from the Review

Looking back, I think the *Review* was too cautious on all four of the key structural elements: (a) emissions growth, (b) carbon cycle, (c) climate sensitivity, and (d) damages from a given temperature.

(a) Ross Garnaut and his commission, working for the Australian government on climate change, are revisiting the emissions scenarios in the IPCC Special Report on Emissions Scenarios or SRES (IPCC 2000). In its chapter 6 model (Stern 2007, pp. 173–188), the *Stern Review* used the second highest of the four scenarios (called A2). Garnaut is now suggesting that the highest of the four, A1F1, is likely to be the best description of BAU (Garnaut 2007). Key among the reasons are the growth rates of the developing world, particularly China and India, and their continued strong emphasis on coal (Export Committee on Integrated Energy Policy 2006).

(b) The carbon cycle is likely to weaken as a result of, for example, the possible collapse of the Amazon forest at temperature increases of above 3 °C to 4 °C or the decreasing absorptive capacity of the oceans. Further, a thawing of the permafrost is likely to result in strong methane release.

(c) The climate sensitivity assumed in the *Review* is likely to be conservative (as argued in section I).

(d) The damages from given temperature increases assumed in the *Stern Review* seem very low. The *Review*'s mean damage loss (based on estimates in the economic literature) from 5 °C was around 5 percent of GDP (Stern 2007, p. 180). As argued in section I, a temperature increase of 5 °C would most likely result in massive movements of population and large-scale conflict.

Considering these structural factors together, the modeling of the *Stern Review* probably underestimated significantly the risks of high damages from BAU, perhaps by 50 percent or more if one compares the first two rows of table 2.2. Much of the earlier literature grossly underestimated the risks.

Looking at both γ and η, with the benefit of hindsight, my inclination would be to place the base case from which sensitivity analysis is undertaken farther down the diagonal of table 2.2—that is, with higher γ and higher η. As indicated in section IIB, the "weight in the far future" from $\eta = 1$ and $\delta = 0.1$ percent suggests that there is a case for raising η, although it remains true that many would see the implications of $\eta = 2$ for intragenerational distribution as very egalitarian. In a sense, moving down the ($\eta - \gamma$) diagonal is taking on board the positions of two commentators on the *Review*. Weitzman (2007a, 2007b) argued for greater emphasis on risk and uncertainty, and Dasgupta (2007) for more egalitarian values than those captured by $\eta = 1$.

In summary, one year on from the *Stern Review*, with the benefit of new scientific evidence and valuable economic discussions, my views would have been modified as follows. First, the case has been strengthened that the bottom-up, disaggregated, less formal, risk-evaluation approach is preferable to aggregate modeling in investigating the case for action. The latter is particularly weak in relation to formal optimization. Second, within aggregate modeling, we have learned still more clearly that the key issues are ethics and risks and that we have to look at them together to form a serious view on damages. Third, our own modeling probably underestimated the risks from BAU. Fourth, the reasons some earlier studies have lower damage estimates than the *Stern Review* are twofold: they badly underestimate all four of the elements just described, and in many cases, their approach to pure time discounting discriminates, unjustifiably in my view, very strongly against future generations.

III. Policy Instruments

At the heart of good policy will be a price for GHGs—this is a classic and sound approach to externalities and is crucial for an incentive structure both to reduce GHG emissions and to keep costs of abatement down. Indeed, in a world without any other imperfections, it would be a sufficient instrument for optimal policy. But it will not be enough in our world, given the risks, urgency, inertia in decision making, difficulty of providing clear and credible future price signals in an international framework, market imperfections, unrepresented consumers, and serious concerns about equity. A second plank of policy will have to embrace technology and accelerate its development. Third, policy should take account of information and transactions costs, particularly in relation to energy efficiency. Fourth, it should provide an international framework to help with combating deforestation, which is subject to a number of market failures. And fifth, policy should have a strong international focus, to promote collaboration, take account of equity, and reduce global costs.

Careful analytical investigation by economists of policies on climate change involves the whole range of the tools of our trade, including the economics of risk and uncertainty, innovation and technology, development and growth, international trade and investment, financial markets, legal issues, ethics and welfare, as well as public and environmental economics. It will no doubt require the development of further analytical methods. And it necessitates close collaboration with scientists and other social scientists.

Our focus here in this very brief discussion of policy will be on price-oriented mechanisms and on technology, but we should also note a sixth key element that is often overlooked in discussions of economic policy, namely, how preferences change as a result of public discussion. This was an integral part of John Stuart Mill's (Mill 1972 [1861], p. 262) perception of democracy and policy formation (see also the discussion in chapter 9 of Stern, Dethier, and Rogers 2005). In this context, it involves a change in public understanding of responsible behavior. Thus, people will spend time on separating out different elements of waste for recycling, or they will drive more carefully, not only because there may be a financial incentive for recycling or penalties for bad driving but also because they have a view of responsible behavior.

Pricing an externality can be done in a number of ways. First, there is carbon taxation; second, carbon trading on the basis of trade in rights to emit which are allocated or auctioned; and third, implicit pricing via regulations and standards that insist on constraints on actions or technologies that involve extra cost but imply reductions in emissions. Each of the three has different advantages and disadvantages, and all three are likely to be used. Understanding the pros and cons, where the different mechanisms can and should be used, and how to deal with problems of overlaps are all very important issues. We have the space to look briefly only at a few of the relevant considerations.

Taxes have the advantage of being implementable by individual governments without international agreement. All taxes are contentious, but those on recognized "bads" such as tobacco, alcohol, or carbon emissions may be less so than others and allow the balance of taxes to adjust away from other taxes such as income or VAT; alternative uses of revenue are possible, too, including those related to climate change. We should beware, though, of arguments about double dividends: environmental taxes have dead-weight losses in addition to their beneficial effects in addressing externalities. Taxes on GHGs would require measurement of GHGs, just as in trading, but taxes on petroleum products, coal, or other fossil fuels can act as fairly good approximations, avoiding direct emissions measurement, which can be relatively costly to small enterprises.

As discussed in section IA, where the world is perfect other than in relation to the tax in question, quantity controls and price measurements can have dual and essentially identical effects. Where there is risk, uncertainty, and imperfections in this market and in other parts of the economy, there will be price uncertainty, quantity uncertainty, or both, depending on the policies chosen and the nature of the uncertainty. Both price certainty and quantity certainty are important: firms would like clear and simple price signals for decision making; quantity overshooting on emissions is dangerous. With learning and readjustment of policy (although not so frequently as to confuse structures and issues), the difference in effects between a tax-orientated policy and a quantity/carbon-trading policy may not be so large. Given where we start, however, in my view, the danger of overshooting emissions targets is of great significance.

Tradable quotas, the second method of establishing a price for GHGs, have the advantage of providing greater certainty about quantities of emissions than taxes. The European Union Emissions Trading Scheme (EUETS) has shown that a big part of the economy can be covered (currently around one half of European emissions) with relatively low administrative burdens by focusing on major emitting industries, such as power.

By starting with allocations that are not paid for and moving to auctions, trading can build acceptance by industry because it allows for a less dramatic adjustment. Free allocations based on historical emissions do have important problems, however: they are likely to slow adjustment since immediate profit pressures are lower; they can give competitive advantages to incumbent firms that may succeed in getting large quota allocations, thus reducing competition and promoting rent seeking; and they forgo public revenue. Thus, moving to auctioning over time has strong advantages and should be a clear and transparent policy.

An aspect of quotas and trading that is crucial is their potential role in international efficiency and collaboration. Developing countries (see next section) have a strong and understandable sense of injustice. They see rich countries having first relied on fossil fuels for their development and thus being largely responsible for the existing stocks of GHGs, then telling them to find another, and possibly more costly, route to development. They feel least responsible for the position we are in, yet they will be hit earliest and hardest.

International trading provides for lower costs, from the usual arguments about international trade, and provides an incentive for poor countries to participate. These arguments on cost and collaboration are central to my view that there should be a very substantial focus on carbon trading in the policy of rich countries, with openness to international trade, backed by strong rich-country targets for reductions, in order to maintain prices at levels that will give incentives both for reduction at home and purchase abroad. Rich and poor country targets will be discussed in the next section.

Price volatility is sometimes said to be a problem with quotas and trading, and the EUETS is cited as an example. But that scheme provided some basic simple lessons that have been learned. In its first stage (2005–2007), giving away too many quotas collapsed the price. Quotas have been allocated with greater rigor and stringency in the second phase (2008–2012), and the price for that phase is currently above €20 per ton, already approaching the type of range indicated as necessary. Volatility can be reduced by (a) clarity, (b) firmness of quotas, and (c) broader and deeper markets—greater trading across sectors, periods, and countries. Particular measures for dealing with volatility should be analyzed in relation to, or after, these broader, more market-friendly approaches. And care should be taken not to restrict international trade as a result; for example, differences in caps on prices in different regions might, because of attempts to arbitrage where prices are different but fixed, make open trade difficult or impossible.

Further, difficulties arise in trading with countries that are not taking strong measures, price-based or otherwise, against climate change. There is, in principle, a case for levying appropriate border taxes on goods from countries that do not otherwise embody a carbon price. A system analogous to the operation of the border procedures for VAT could be envisaged. My own view is that this should be a last resort. There are many searching for arguments on protection that might climb on the bandwagon. The best way forward is to build international collaboration with a positive and constructive approach.

Regulation and standards can give greater certainty to industry. This can accelerate responses and allow the exploitation of economies of scale, lead-free petrol and catalytic converters are probably good examples. Misguided regulation, on the other hand, could reduce emissions in very costly ways. Again, urgency points to a role for regulation/standards, and careful economic analysis can keep costs down. In thinking about these costs, however, we should remark that there are a number of examples in the history of the motor industry where innovations on safety or pollution were resisted by industry on cost grounds, only for compliance costs to turn out to be much lower than manufacturer predictions; for Environmental Protection Agency vehicle-emissions-control programs, industry stakeholders predicted price changes to consumers that exceeded actual changes by ratios ranging from 2:1 to 6:1 (Anderson and Sherwood 2002).

While taxation, trading, and regulation will all have roles to play, it is important to think carefully about how they might interact. For example, if taxation and carbon trading overlap, there are likely to be problems in establishing a clear and uniform price for carbon, leading to confused signals and inefficiency. And strong regulatory targets such as renewables percentages could, without care, result in low demand on carbon markets.

Our discussion of technology will be very brief, but in my view, policy in this area will be of great importance—we cannot simply leave the correction of externalities to carbon markets or taxation. There is a standard argument on knowledge and technology that sees ideas and experience as having positive externalities. Figure 2.7 shows that experience is indeed important in the electricity industry—it seems that in a number of "less mature" technologies, costs can fall quite sharply with cumulative experience. Further, the rate of fall is different for different technologies. This tells us that public support for deployment—such as feed-in tariffs, which may be different for different technologies—has a strong foundation. Care

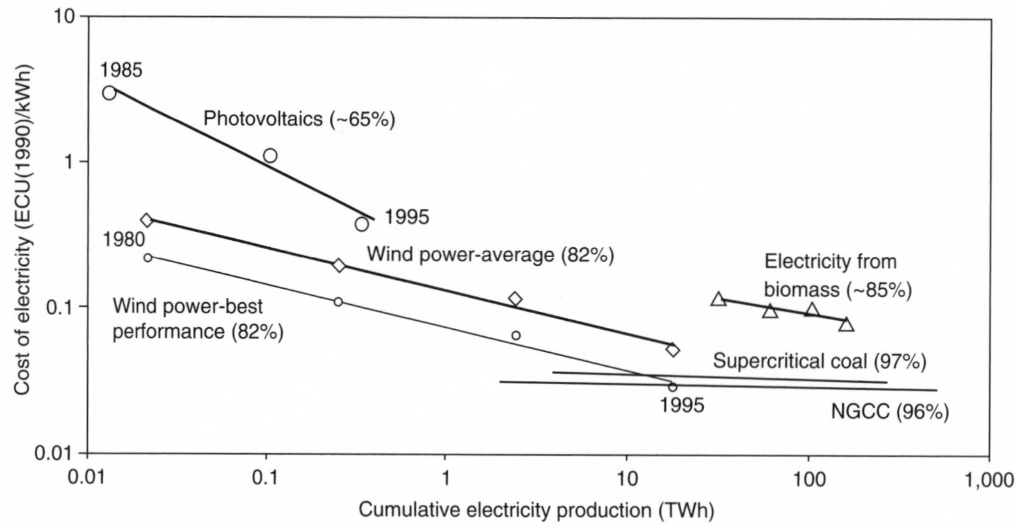

Figure 2.7. Cost of electricity for different technology.

Source: Stern 2007, 254.

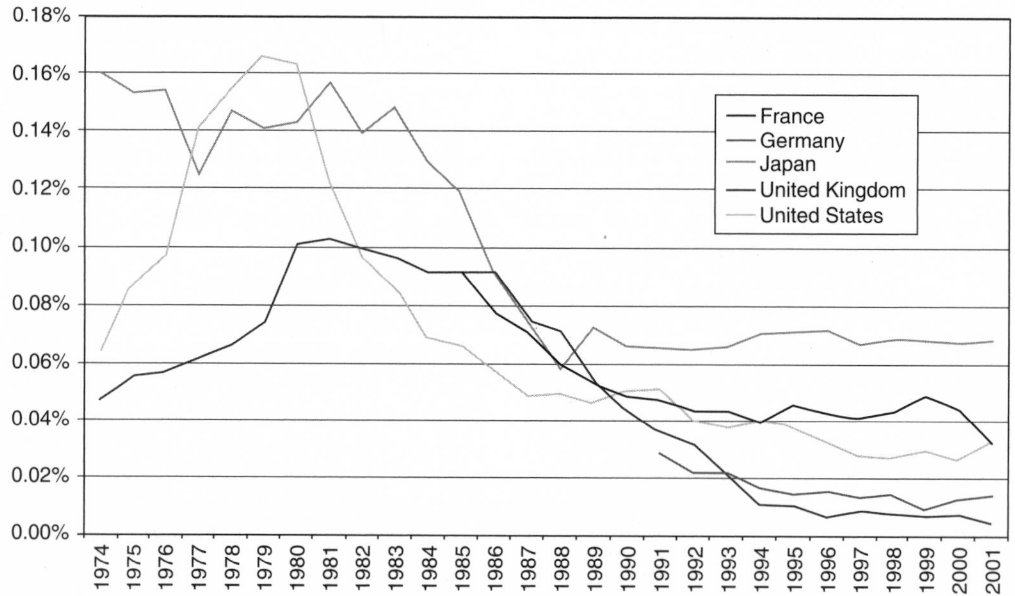

Figure 2.8. Public energy R&D investments as a share of GDP.

Source: Stern 2007, p. 401.

with applying such incentives is necessary to avoid the dangers of bureaucrats trying to pick private-sector technological "winners."

Research and development in basic technologies also require public support. It is remarkable how much public support for R&D in energy has fallen since the early 1980s (see figure 2.8). Part of this was probably a result of low energy prices,[27] but nevertheless, the now-recognized urgency of developing low-carbon technologies requires a strong reversal of this trend. Private-and public-sector R&D on energy have moved closely together, and this is an area where public-private partnership to enhance both private and social returns, and to cover different risks, will be crucial. Fortunately, the last few years have seen a number of exciting and promising developments, such as in materials and technologies, other than silicon, for photovoltaics.

The international aspects of technology are crucial, too. We all gain from reduced emissions if others adopt cleaner technologies quickly. Thus, a balance of private return to innovation, for example through patents, and rapid sharing must be found. This should be part of a global deal or framework to which we now turn.

IV. A Global Deal

Climate change is global in its origins and in its impacts. An effective response must therefore be organized globally and must involve international understanding and collaboration. Collaboration, if it is to be established and sustained, must be underpinned by a shared appreciation that the methods adopted are *effective* (on the scale required), *efficient* (they keep costs down), and *equitable* (responsibilities and costs are allocated in ways that take account of wealth, ability, and historical responsibility). The incentive structures must be such that solutions are incentive-compatible. And country-by-country political support must be built, as this is what will sustain policies over time.

Public support for action will be founded not only on recognition of the magnitude of the problem but also on the realization that it is possible to construct collaborative policies that are effective, efficient, and equitable. It is a great responsibility of economists to help design those policies. And they must do so urgently—the international discussion is moving quickly, and key decisions will be taken over the next few years.

The following is my own attempt to describe the outline of a possible global deal based on the preceding analysis and on my own intensive experience over the last two years of involvement in public discussion, taking account of the recent UNFCCC meetings at Bali. Let us begin with overall reductions targets and the allocation of responsibilities across countries. Our earlier discussion of trading, technologies, and deforestation will then allow us to see quickly the broad structure of a global deal. Let us be clear at the outset that this should not be seen in the overly formal way of a WTO discussion, founded in legal structures, with compliance driven by sanctions, and where no one is bound until the full deal is agreed. This is much more a framework in which each country, or group of countries, can assess its own responsibilities and targets with some knowledge of where the rest of the world is going and how it can interact.

On targets—a key element of *effectiveness*, or action on an appropriate scale—we should be clear how far the international discussion has already moved. The G8–G5 summit chaired by Germany in Heiligendamm in June 2007 declared a world target of 50 percent reductions by 2050. As sometimes happens in international communiqués, not all details (such as base date and levels of agreement among attendees) were clear, but it was a significant marker nonetheless. And it is broadly consistent with the type of stabilization range, around 500 ppm CO_2e, for example, discussed in section I. In what follows, unless otherwise stated, emissions reductions will be measured from 1990, covering all GHGs (in the six-gas Kyoto sense) and emissions sources. The Heiligendamm 50 percent target is for the world as a whole, and it is generally agreed (see below) that, in the spirit of the Kyoto language of "common but differentiated treatment," the richer countries should take responsibility for reductions bigger than the average. In what follows, we shall think of rich-country reductions as including those discharged by purchases on international markets.

At Bali in December 2007, three countries, Costa Rica, New Zealand, and Norway, declared targets of 100 percent reductions by

2050, "going carbon-neutral." The latter two are highly likely to need international purchases to get there. Note, too, that reductions of more than 100 percent are possible—many in developing countries would regard targets for rich countries above 100 percent as appropriate, given past history—and that such reductions that would almost inevitably involve international purchase.

California has a target of 80 percent reductions by 2050. France has its "Facteur Quatre," dividing by 4, or 75 percent reductions, by 2050 (Stern 2007, p. 516). The United Kingdom has a 60 percent target, but Prime Minister Gordon Brown indicated in November 2007 that this could be raised to 80 percent (Brown 2007). Australia, under the government elected at the end of November 2007, has now signed Kyoto and has a target of 60 percent (Rudd 2007); 80 percent is under consideration after the Garnaut review is published next summer.

Targets for 2050 seem far away, but the long lifetime of many investments means that early decisions are needed to reach them. Intermediate targets are also being set. At the European Spring Council, 20 to 30 percent targets were set for 2020; Germany has set 40 percent targets by 2020. The European Council also set targets for renewables and CCS for 2020 and beyond, but it is the overall emissions targets and their achievement that are crucial. How they are achieved country by country will vary and must take account of economic as well as environmental, social, and political considerations. At Bali, many were pressing for rich countries to accept 25 to 40 percent cuts by 2020. That is indeed in the right range for rich-country cuts of 80 percent by 2050 and is now at least an initial 2020 benchmark. Overall, in discussions of global and rich-country targets, ranges consistent with the criteria of effectiveness and equity are now the basic benchmarks, and many key commitments have been made. Delivery on targets at reasonable cost—essentially *efficiency*—is, of course, crucial and a challenge. Policies that could support this constituted the subject of section III and should be at the heart of a global deal.

Let us investigate equity in a little more detail. The history of flows and their relation to future stabilization targets should, in my view, be central to a discussion of equity. All too often, equity is seen solely or largely in terms of the relative level of future flows (for example, per capita convergence by 2050). A few numbers and a little basic arithmetic will help to understand the issues. Currently, global emission flows are around 40 to 45 Gt CO_2e, With a world population of around 6 billion, that means average global per capita emissions are around 7 tons. Given that the world population in 2050 will be around 9 billion, in order to achieve 50 percent reductions (i.e., an aggregate flow of around 20 Gt CO_2e) by then, per capita emissions will have to be 2.0 to 2.5 tons. And since around 8 billion of these people will be in currently poor countries, those countries will have to be in that range,[28] even if emissions in currently rich countries were to fall to zero. It is clear from this basic arithmetic that any effective global deal must have the currently poor countries at its center.

From the point of view of equity, the numbers are stark. The currently rich countries are responsible for around 70 percent of the existing stock and are continuing to contribute substantially more to stock increases than developing countries. The United States, Canada, and Australia each emit more than 20 tons of CO_2e (i.e., from all GHGs) per capita, Europe and Japan more than 10 tons, China more than 5 tons, India around 2 tons, and most of sub-Saharan Africa much less than 1 ton. Recent per capita CO_2 emissions (i.e., omitting other GHGs) for some countries are illustrated in figure 2.9.

In the lower part of this graph are three big, fast-growing developing countries. China is growing especially quickly. Even with fairly conservative estimates, it is likely that, under BAU, China will reach current European per capita emissions levels within 20 to 25 years. With its very large population over this time, China under BAU will emit cumulatively more than the United States and Europe combined over the last 100 years. That is one indication of the urgency of finding a global response quickly.

But let us keep focused on equity. With 80 percent reductions by 2050, Europe and Japan would be around the required two-ton global

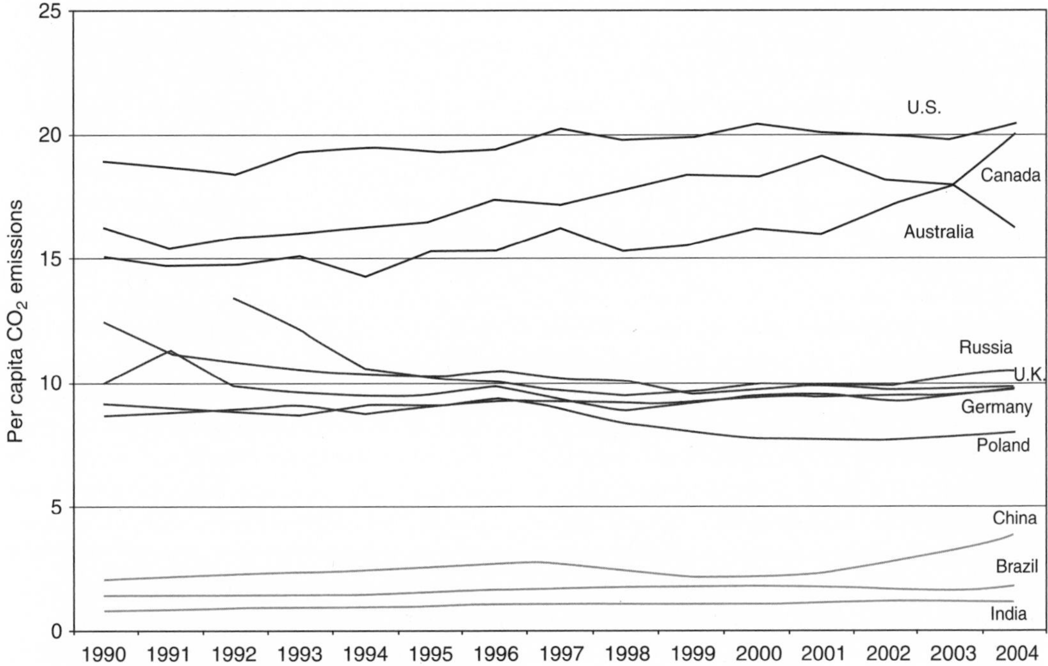

Figure 2.9. Per capita CO_2 emissions (in metric tons).

Source: CDIAC 2007.

average level. An 80 percent reduction by the United States, Australia, and Canada by 2050 would leave them around four tons, twice the required average level. Thus, a 50 percent overall reduction and an 80 percent rich-country reduction would still leave average rich-country flows above the world average in 2050.

Turning to stocks, let us think about the path from some initial level to a stock stabilization (to be specific, suppose that level is 550 ppm CO_2e) and about who consumes what along the way. We can think of the initial level as 280 ppm CO_2e, corresponding to preindustrial times (around 1850), or we could start 20 years ago (around 390 ppm CO_2e), when the problems of climate change began to receive strong policy attention, or we could start now (around 430 ppm CO_2e). One perspective on equity would be to see the difference between 280 ppm CO_2e and 550 ppm CO_2e as a reservoir sized 270 ppm CO_2e, which the world will get close to exhausting over the 200 years between 1850 and 2050. If we start the clock in the late 1980s or now, it would be a reservoir sized around 140 ppm CO_2e or 120 ppm CO_2e, respectively.

From this perspective, equalizing the per capita flows of emissions—or the size of the glass drawn per person per year from the reservoir—by 2050, shortly before it is dry, is a very weak notion of equity. It takes no account of all the guzzling that took place by the better-off over the preceding 50 to 200 years (depending on when we start the clock). There is a very big difference between a stock and a flow notion of equity. An 80 percent reduction of flows by rich countries by 2050, in the context of a 50 percent reduction overall, is not a target for which rich countries should congratulate themselves warmly as demonstrating a splendidly powerful commitment to equity. And the contract-and-converge argument for some common flow level, or for using such a level as the eventual basis of trading, on the asserted grounds that there are "equal rights to emit or pollute," does not seem to me to have special claim on our attention.[29] Rather, the target of equalizing by 2050 (allowing for trade) may be seen as a fairly pragmatic one, on which it might be possible to get agreement, and one that, while only weakly equitable, is a lot less inequitable than some other possibilities, such as less stringent targets for rich countries.

If we take any particular good, it will generally be true that rich people consume more than poor people. That is simply an expression of their being richer. In the case of the reservoir, or the "contents of the atmosphere," it is hard to think of an argument for why rich people should have more of this shared resource than poor people. They are not exchanging their labor for somebody else's, and they are not consuming the proceeds of their own land or some natural resource that lies beneath it. I do not have any special "correct" answer to the challenge of understanding equity here, but it is a challenge we cannot avoid discussing. Any global deal will have to involve some implicit or explicit understanding over the sharing of this "reservoir."

The key elements of the global deal have, with one exception, now been raised and discussed. Let me express the deal or framework in terms of two groups of three headings, the first covering targets and trade and the second covering programs for which public funding is likely to be required. This set of six policies or programs is the international part of a deal. The domestic policies will vary across countries, using different combinations of policy instruments and technologies as discussed in section III. The six elements of a global deal are expressed in bullet point form in tables 2.3 and 2.4.

The first element of the first group covers the targets. The global target was explained and justified in section I and the distribution of targets above in this section. The second, the importance of emissions trading, was emphasized in section III: the justification for a major focus on GHG trading in policy lies in its promotion of both efficiency and collaboration. Unless financing flows for the extra costs of reducing emissions are available to poor countries, they are extremely unlikely to join the effort on the scale and pace required. They feel the inequities of the situation and phenomena acutely. Just when, they argue, they are beginning to overcome poverty, in part by rapid growth, they should not be asked to slow down. Financing, together with technology demonstration and transfer, will be needed to convince them that moving to a low-carbon growth path is not the same thing as moving to a low-growth path.

The third element refers to the short- and medium-term approaches to trading between rich and poor countries. The current system, the clean development mechanism (CDM), was established by Kyoto and operates at the level

Table 2.3.
Key elements of a global deal—targets and trade.

- Confirm Heiligendamm **50%** cuts in world emissions by 2050 with rich-country cuts at least **75%**.
- Rich-country reductions and trading schemes designed to be **open to trade with other countries**, including developing countries.
- **Supply side from developing countries** simplified to allow much bigger markets for emissions reductions: "carbon flows" to rise to \$50 billion–\$100 billion per annum by 2030. Role of sectoral or technological benchmarking in "one-sided" trading to give reformed and much bigger CDM market.

Table 2.4.
Key elements of a global deal—funding.

- Strong initiatives, with public funding, on **deforestation** to prepare for inclusion in trading. For \$10 billion to \$15 billion per annum, could have a program that might halve deforestation. Importance of global action and involvement of IFIs.
- Demonstration and sharing of **technologies**: e.g., \$5 billion per annum. commitment to feed-in tariffs for CCS coal would lead to 30+ new commercial-size plants in the next 7 to 8 years.
- Rich countries to deliver on Monterrey and Gleneagles commitments on **ODA** in context of extra costs of development arising from climate change: potential extra cost of development with climate change upward of \$80 billion per annum.

of a project in a poor country (a so-called non-Annex 1 country in the Kyoto Protocol). If a firm in a rich country (an Annex 1 country) is part of a trading scheme (such as the EUETS) that recognizes the CDM, then that firm can buy an emissions reduction achieved by the project, subject to the project using technologies or approaches from an admissible list. The amount of the notional reduction comes from comparing the project with a counterfactual—what the entity doing the project might otherwise have done. Approval of a project goes through the poor-country authorities and a special institutional structure, currently in Bonn. The system is slow, cumbersome, and very "micro."

Trading on the scale required to reach the type of targets discussed (see table 2.3) requires a much simpler, "wholesale" system.[30] At the same time, to get agreement with poor countries, it will have to continue to be "one-sided," as in the CDM; that is, you can gain from innovation but are not penalized for BAU. Wholesale measures can include technological benchmarks such as employing CCS (currently excluded from CDM) or sectoral benchmarks such as getting below a certain amount of CO_2 per ton of cement. As one-sided trading measures, the benchmarks could be set ambitiously.

After these trading mechanisms have been in place (with associated technology sharing) for a while, developing countries will be able to have confidence that a trading system can work on an appropriate scale. Then it would be reasonable to ask them to accept targets consistent with overall global goals in the context of a strong set of goals by rich countries. If we look for targets from poor countries now, the only ones that would be accepted would be far too loose and would knock the bottom out of international trading, collapsing the price. And in the future, these loose targets would be likely to form a baseline for subsequent discussion. That is why a staged approach is essential if currently poor countries are to accept participation in responsible global stabilization so that by 2050, their emissions average around 2 tons per capita. Recall that this is a half or a third of China's *current* level. It is very unlikely to be possible to find financial flows on the scale required to incentivize appropriate action from the public sector of rich countries. Witness the difficulty in getting resources for Overseas Development Assistance (ODA), which will be strained still further by the challenge of adaptation (see below). The trading system provides for private flows.

The public funding requirements are grouped in three elements in table 2.4. Each would require a paper in itself for appropriate treatment, and we can give only headlines. Deforestation accounts for up to 20 percent of current emissions; the numbers are not easy to specify precisely—probably 5 to 8 Gt CO_2e per annum. These flows could be roughly halved, in my view, for around $5 per ton of CO_2, taking into account opportunity costs of land and the institutional, administrative, and enforcement measures necessary. Some have estimated higher costs (e.g., Enkvist, Nauclér, and Rosander 2007), but there appear to be large amounts of "initial" reductions available at lower costs, particularly if programs are large-scale and coordinated across countries (for further discussion, see Myers 2007, Nepstad et al. 2007; Anger and Sathaye 2007). This would help to avoid reduced deforestation in country A, simply displacing activity and thus increasing deforestation in country B. Public-sector flows can be combined with private-sector flows as avoided deforestation is brought into the carbon-trading process so that all countries are given incentives. Indeed, one of the responsibilities of the publicly funded program would be to work toward trading.

The second element in this second group, the demonstration and sharing of technologies, is urgent; financial resources must be made available and institutional arrangements designed. This is an important area for economic research. One problem of particular urgency, for reasons described above, is the demonstration of CCS for coal. There are no current plants using CCS for coal-fired generation on a commercial scale. From 2015 or 2020 on, the world will need most of its new coal-fired electricity-generation plants to be operating with CCS if it is to have any chance of realizing its targets. If CCS cannot work on the necessary scale, then we need to know soon and follow alternative strategies. At present, however, it does look promising. There is geological work to be done to identify storage capacity and careful legal and regulatory work to be done to allocate risk and responsibility. Geology and coal vary greatly across the world, and many demonstrations of commercial-scale plants are necessary. Feed-in subsidies, worldwide, of around $5 billion per annum could support 30+ such plants over the next seven to eight years and cover a broad range of examples.[31]

There should also be support for many other technologies. We do not know what the most efficient clean technologies will be in the future, and the answers are likely to vary with location. CCS is emphasized here simply because we can be fairly confident that BAU will involve a great deal of coal for electricity over the next 20 to 30 years. Perhaps it will be a medium-term technology and be replaced by others over the longer term.

Finally, in the global deal, I would emphasize an element that has not been discussed here and that will be of great importance. Even with very responsible policies, the world is likely to see an additional 1 °C to 2 °C of warming over and above the 0.8 °C it has already experienced. Adaptation will be necessary worldwide and will be particularly difficult for poor countries. The United Nations Development Program (2007, p. 15) estimated additional costs for developing countries of around $85 billion per annum by 2015. And they will presumably rise after that.

Such extra financing will be hard to find. It may be compared with the $150 billion to $200 billion per annum extra that would arise if the Organization for Economic Cooperation and Development (OECD) countries moved to 0.7 percent GDP in ODA by 2015, as many of them have promised. The ODA promises of the 2002 UN International Conference on Financing for Development in Monterrey, Mexico, in connection with the Millennium Development Goals, and of the 2005 U.K.-chaired G8 Gleneagles summit on Africa, and preceding EU commitments, were powerfully argued and justified at the time. They took little account of climate change. If that aspect is added, as it should be given the magnitude of the challenge, and combined with the historical responsibilities for stocks of GHGs and the implied consequences for poor countries, then the argument for 0.7 percent, in my view, becomes overwhelming. The *Stern Review* left the argument at that point, although a case could have been made for increasing the ODA targets.

The framework I have now described does, in my view, meet the criteria of effectiveness (it is on the right scale); efficiency (It relies heavily on markets and market-orientated innovation); and equity (it does at least give some specificity to the "common but differentiated responsibility" already accepted internationally). It builds on existing commitments and some aspects of the current discussions in international forums. It is also designed to give some realistic opportunity for the major developing countries to become strongly involved, as they must if serious targets are to be agreed on and achieved.

It is a framework that could allow all countries to move quickly along what they see to be a responsible path. What is very striking here is how broadly basic understanding has already been established. Country by country, we see targets being erected and measures being set by individual countries recognizing their own responsibilities as they see international agreement being built. People seem to understand the arguments for action and collaboration on climate change much more readily than they do for international trade. But I do not want to pretend that the problems and necessary actions are universally recognized and accepted. Scientific agreement seems broad and deep, but we cannot yet say that about economic policy or about economists. This is a time for exchange of ideas and intensive discussion. Economic policy is much too important here to be left to noneconomists.

It is intensive public discussion that will, in my view, be the ultimate enforcement mechanism. For example, in November 2007, we saw an Australian prime minister thrown out of office in part because of his perceived weakness on this issue. It is remarkable that when elections come around, politicians recognize strong public interest and demand for action. And it has become a unifying and defining issue in the structures of Europe. It has not moved at the same pace in all countries, but we are also seeing strong changes in perception in the key countries of the United States, China, and India.

Beyond discussion, there are some promising movements in world and individual country policy. The UNFCCC 13th Conference of the Parties, COP 13, in Bali in December 2007 was a major step forward, with all countries involved broadly (but not universally) recognizing the need for overall 50 percent cuts by 2050 and 25 to 40 percent cuts by rich countries by 2020 (although only the phrase "deep cuts" was agreed on). There was progress on international action on deforestation. But it was the launch of negotiations only; it was not an agreement on a shared global framework.

The discussion of that global framework will move forward strongly over the next few years. It is vital that economics and economists be more strongly involved, particularly if the criteria of efficiency and equity are to play their proper role. It is the analytical application of these two criteria to practical policy problems that is at the heart of public economics. The challenge of climate change is especially difficult because it covers so much of the economy, is so long-term, is so full of risk and uncertainty, is so demanding internationally, and is so urgent because of the problem itself and the pace of public discussion and decision making. It is also a long-term problem for analysis. We will be learning all the time, and policy will be made and reformed over coming decades.

It is dangerous, in my view, for us as economists to seem to advocate weak policy and procrastination and delay under the banner of "more research to do" or "let's wait and see." The former argument is always true, but we have the urgent challenge of giving good advice now, based on what we currently understand. And the latter, in my view, is misguided—waiting will take us into territory that we can now see is probably very dangerous and from which it will be very difficult to reverse. Acting now will give us, at fairly modest cost, a cleaner world and environment, even if, as seems very improbable, the vast majority of climate scientists have got it wrong. If we conclude that whatever the merits of the argument, it is all too difficult to make and implement policy, then we should at least be clear about the great magnitude of the risks of moving to concentrations of 650 ppm CO_2e or more, which are the likely consequences of no, weak, or delayed action.

It is hard to imagine a more important and fascinating problem for research. It will involve all of our skills and more, and it will require collaboration across disciplines. This is a time and a subject for economists to prove their worth.

Acknowledgments

I am very grateful for the advice and comments of Claire Abeillé, Dennis Anderson, Alex Bowen, Sebastian Catovsky, Peter Diamond, Simon Dietz, Ottmar Edenhofer, Sam Fankhauser, Graham Floater, SuLin Garbett, Ross Garnaut, Roger Guesnerie, Geoffrey Heal, Daniel Hawellek, Claude Henry, Cameron Hepburn, Paul Joskow, Jean-Pierre Landau, James Mirrlees, Ernesto Moniz, Steven Pacala, Nicola Patmore, Vicky Pope, Laura Ralston, Mattia Romani, John Schellnhuber, Matthew Skellern, Robert Socolow, Martin Weitzman, Dimitri Zenghelis, and all of those who worked on and guided the Stern Review team. The views expressed here are mine and do not necessarily reflect the judgments or positions of those who kindly provided advice, or of the London School of Economics, or of the U.K. government, for which I was working while leading the *Stern Review on the Economics of Climate Change*. This is dedicated to my close friend, distinctive and distinguished economist, and fine man, Andrew Glyn, who died on December 22, 2007, and whose funeral took place in Oxford, U.K., on the same day as the Ely Lecture, January 4, 2008.

Notes

1. Joseph Fourier recognized in the 1820s (Fourier 1827) that the atmosphere was trapping heat; three decades later. John Tyndall (1861) identified the types of gases responsible for the trapping; and at the end of the century, Svante Arrhenius (1896) gave calculations of the possible effects of doubling GHGs.

2. Climate modelers tend to define "doubling" in relation to preindustrial times. The relationship from stock to temperature increase is approximately logarithmic, so that doubling from other stock levels would be likely to yield a similar increase.

3. The TAR was without probabilities but Wigley and Raper produced distributions based on it. The Stern Review blended the TAR and Hadley because the former was based on international discussion, but the latter was more recent. The Stern Review used lower climate sensitivities than Hadley, although the IPCC's more recent AR4 (IPCC 2007) is closer to those used by Hadley.

4. To avoid excessive length of discussion, we focus on 5 °C, because it is an extremely dangerous increase and because its probability of occurrence under BAU is far from small. In a full analysis, one could and should look at the full range of possible concentrations and associated probability distributions for temperature increases.

5. There would be some negatives (more inflexible equipment in place) and some positives (more technical knowledge).

6. Illustrative MAC curves were provided in (Stern 2007, pp. 243, 249).

7. Erin C. Myers (2007, pp. 9–12) reviews the literature and highlights the outlier status of the McKinsey deforestation estimate; see also the discussion in section IV.

8. This is not the place to speculate about euro-dollar exchange rates over two or three decades.

9. This sketch of the calculation assumes the simple objective of the maximization of the integral of expected utility.

10. These assessments refer to the potential shifts of the demand side of labor markets—outcomes depend, of course, on market structures.

11. See, for example, chapters 1, 3, 4, and 5 of the *Stern Review*.

12. The issue is still more complex in this context, as delays in action result in environmental damage along the way, as well as increasing the cost of achieving a given stabilization level. On balance, the extra intertemporal complexity is likely to strengthen this chapter's argument in this case.

13. The SDF is the marginal utility of consumption at time t (and we normalize the SDF to one for $t = 0$). If we consider a changing population $N(t)$ and replace $u(c)$ by $Nu(c)$ where c is C/N and C is total consumption at time t, the partial derivative with respect to C is $u'(c)$.

14. Unfortunately, some, including Nordhaus (2007b) and Weitzman (2007b), have been tempted to think that a value for the PTDR can be "backed out" from this expression by equating the SDR with some market rate of return. For example, with a market investment return of 6 percent, consumption growth of 2 percent, and $\eta = 2$, one "infers" that $\delta = 2$ percent. Thus, the fallacy that the SDR can be anchored by some market rate of return leads to a second fallacy, namely, that society's PTDR can be "revealed" from market behavior (instead of requiring explicit specification on ethical grounds).

15. The summation is across individuals existing at time t and c_i is the consumption of individual i.

16. The welfare weight on an individual with consumption c is taken here as the social marginal utility of consumption at that level. To keep things simple, we assume that this depends only on the individual's consumption and not on his or her preceding consumption or the consumption of others.

17. See, for example, Stern (1976), who shows how sensitive tax calculations are to assumptions about substitutability between goods and leisure.

18. More than 30 years ago, I examined all three of these methods with no particularly strong conclusions, other than that the results covered a broad range (Stern 1977).

19. In section IIC, we consider η in this range. Higher growth rates are not examined in detail. The modeling would have to take account of a changed path of emissions with earlier damages. With risk distributions appropriate to current knowledge, our preliminary findings suggest that estimated damages from climate change are likely to be well above the cost of action to drastically reduce those risks.

20. PAGE 2002, Policy Analysis of the Greenhouse Effect 2002 Integrated Assessment Model. See Hope 2006a, 2006b.

21. To keep things simple, the results in the table have γ fixed—that is, nonstochastic. The Monte Carlo probabilities are therefore generated by the variations in the many other parameters. In the postscript to the *Review* (Stern 2007. pp. 658–671),

stochastic γ is presented. The base case of γ fixed and equal to 2 in table corresponds closely to the base case for stochastic γ in chapter 6 of the *Review* and the technical annex to postscript.

22. Measured in terms of BGE.

23. See Dietz et al. (2007a, c). The drop from replacing all random variables by their means is smaller but still substantial.

24. Over time, 1 percent of consumption and 1 percent of GDP are broadly equivalent.

25. A valuable review can be found in Heal 2007.

26. Their models (FUND for Tol and DICE/RICE for Nordhaus) can, however, be used for Monte Carlo studies.

27. Extensive privatization has probably played a role as well. For example, the U.K.'s nationalized National Coal Board and Central Electricity Generation Board had R&D departments of international distinction.

28. In this context, I am referring to absolute emissions originating in the country rather than who pays.

29. Asserting equal rights to pollute or emit seems to me to have a very shady ethical grounding. Emissions deeply damage and sometimes kill others. Do we have a "right" to do so?

30. This scale is derived from preliminary calculations using a trading model at the U.K. Department of Environment, Food, and Rural Affairs.

31. My own calculations using, for example, the McKinsey cost curve and working with power stations of a few hundred megawatts. I am grateful to Dennis Anderson for his advice.

References

Aghion, Philippe. 2007. "Environment and Endogenous Technical Change." Unpublished.

Anand, Sudhir, and Amartya Sen. 2000. "Human Development and Economic Sustainability." *World Development* 28.12: 2029–2049.

Anderson, John F., and Todd Sherwood. 2002. "Comparison of EPA and Other Estimates of Mobile Source Rule Costs to Actual Price Changes." Society of Automotive Engineers Technical Paper 2002–01–1980.

Anger, Niels, and Jayant Sathaye. 2007. "Reducing Deforestation and Trading Emissions: Economic Implications for the Post-Kyoto Carbon Market." Unpublished.

Arrhenius, Svante. 1896. "On the Influence of Carbonic Acid in the Air upon the Temperature of the Ground." *Philosophical Magazine* 41.4: 237–276.

Arrow, Kenneth J. 1995. "Intergenerational Equity and the Rate of Discount in Long-Term Social Investment." Stanford University Department of Economics Working Paper 97–005.

Arrow, Kenneth, Partha Dasgupta, Lawrence Goulder, Gretchen Daily, Paul Ehrlich, Geoffrey Heal, Simon Levin, et al. 2004. "Are We Consuming Too Much?" *Journal of Economic Perspectives* 18.3: 147–172.

Atkinson, Anthony B., and Andrea Brandolini. 2007. "On Analysing the World Distribution of Income." Unpublished.

Atkinson, Anthony B., and Joseph Stiglitz. 1976. "The Design of Tax Structure: Direct versus Indirect Taxation." *Journal of Public Economics* 6.1–2: 55–75.

———. 1980. *Lectures on Public Economics.* London: McGraw-Hill.

Barker, Terry, Mahvash S. Qureshi, and Jonathan Köhler. 2006. "The Costs of Greenhouse Gas Mitigation with Induced Technological Change: A Meta-Analysis of Estimates in the Literature." Tyndall Centre for Climate Change Research Working Paper 89.

Beckerman, Wilfred, and Cameron Hepburn. 2007. "Ethics of the Discount Rate in the Stern Review on the Economics of Climate Change." *World Economics* 8.1: 187–210.

Brown, Gordon. 2007. "Speech on Climate Change." World Wildlife Fund, London, November 19, 2007. Available at http://www.number10.gov.uk/output/Page13791.asp.

Carbon Dioxide Information Analysis Center. 2007. U.S. Department of Energy. Available at http://cdiac.ornl.gov.

Cline, William. 2007. "Comments on the Stern Review." In *Yale Symposium on the Stern Review* (New Haven, Conn.: Yale Center for the Study of Globalization), pp. 78–86. Available at www.ycsg.yale.edu/climate/stern.html.

Dasgupta, Partha. 2007. "Commentary: The Stern Review's Economics of Climate Change." *National Institute Economic Review* 199.1: 4–7.

Deaton, Angus S., and Nicholas Stern. 1986. "Optimally Uniform Commodity Taxes, Taste Differences and Lump-Sum Grants." *Economics Letters* 20.3: 263–266.

DeLong, Brad. 2006. "Partha Dasgupta Makes a Mistake in His Critique of the Stern Review." Available at http://delong.typepad.com/sdj/2006/11/partha_dasgupta.html.

Dietz, Simon, Cameron Hepburn, and Nicholas Stern. 2008. "Economics, Ethics, and Climate Change." Unpublished.

Dietz, Simon, Chris Hope, Nicholas Stern, and Dimitri Zenghelis. 2007a. "Reflections on the Stern Review (1): A Robust Case for Strong Action to Reduce the Risks of Climate Change." *World Economics* 8.1: 121–168.

Dietz, Simon, Dennis Anderson, Nicholas Stern, Chris Taylor, and Dimitri Zenghelis. 2007b. "Right for the Right Reasons: A Final Rejoinder on the Stern Review." *World Economics* 8.2: 229–258.

Dietz, Simon, Chris Hope, and Nicola Patmore. 2007c. "Some Economics of 'Dangerous' Climate Change: Reflections on the Stern Review." *Global Environmental Change* 17.3–4: 311–25.

Drèze, Jean, and Nicholas Stern. 1987. "The Theory of Cost-Benefit Analysis." In *Handbook of Public Economics*, vol. 2, ed. Alan J. Auerbach and Martin S. Feldstein (Amsterdam: Elsevier), pp. 909–989.

———. 1990. "Policy Reform, Shadow Prices, and Market Prices." *Journal of Public Economics* 42.1: 1–45.

Enkvist, Per-Anders, Tomas Nauclér, and Jerker Rosander. 2007. "A Cost Curve for Greenhouse Gas Reduction." *McKinsey Quarterly* 1: 35–45.

Expert Committee on Integrated Energy Policy. 2006. *Integrated Energy Policy: Report of the Expert Committee.* Government of India Planning Commission. New Delhi.

Fankhauser, Samuel, Friedel Sehlleier, and Nicholas Stern. 2007. "Climate Change, Innovation, and Jobs." Unpublished.

Fourier, Joseph. 1827. *"Mémoire sur les températures du globe terrestre et des espaces planétaires." Mémoires de l'Académie Royale des Sciences* 7: 569–604.

Garnaut, Ross. 2007. "Will Climate Change Bring an End to the Platinum Age?" Paper presented at the inaugural S. T. Lee Lecture on Asia and the Pacific, Australian National University, Camberra, November 29.

Guesnerie, Roger. 2004. *"Calcul économique et développement durable." Revue Économique* 55.3: 363–382.

Hadley Centre for Climate Prediction and Research. 2005. "Stabilising Climate to Avoid Dangerous Climate Change—A Summary of Relevant Research at the Hadloy Centre." Met Office. Available at http://www.metoffice.gov.uk/research/hadleycentre/pubs/brochures.

Harrod, R. F. 1948. *Towards a Dynamic Economics: Some Recent Developments of Economic Theory and Their Application to Policy.* London: MacMillan.

Heal, Geoffrey. 2007. "Climate Change Economics: A Meta-Review and Some Suggestions." Unpublished.

Henry, Claude. 2006. "Decision-Making under Scientific, Political and Economic Uncertainty." *Laboratoire d'Econométrie de l'Ecole Polytechnique Chair Développement Durable Cahier* DDX: 6–12.

Hepburn, Cameron. 2006. "Discounting Climate Change Damages: Working Note for the Stern Review." Available at http://www.economics.ox.ac.uk/members/cameron.hepburn.

Her Majesty's Treasury. 2003. *Green Book: Appraisal and Evaluation in Central Government.* London: Stationery Office.

Hoel, Michael, and Thomas Sterner. 2007. "Discounting and Relative Prices." *Climatic Change*: 84.3–4: 265–280.

Hope, Christopher. 2006a. "The Marginal Impacts of CO_2, CH_4 and SF_6 Emissions." *Climate Policy*, 6.5: 537–544.

———. 2006b. "The Marginal Impact of CO_2 from PAGE2002: An Integrated Assessment Model Incorporating the IPCC's Five Reasons for Concern." *Integrated Assessment Journal* 6.1: 19–56.

IEA. 2006. *World Energy Outlook 2006.* Paris: International Energy Agency.

———. 2007. *World Energy Outlook 2007.* Paris: International Energy Agency.

IPCC. 2000. *Emissions Scenarios: A Special Report of Working Group III of the Intergovernmental Panel on Climate Change.* Cambridge: Cambridge University Press.

Intergovernmental Panel on Climate Change. 2001a. *Climate Change 2001: Synthesis Report.* Cambridge, UK: Cambridge University Press.

Intergovernmental Panel on Climate Change. 2001b. *Climate Change 2001: Synthesis Report.* Cambridge, UK: Cambridge University Press.

———. 2007. *Climate Change 2007: The Physical Science Basis.* Cambridge, U.K.: Cambridge University Press.

Jansen, E., J. Overpeck, K. R. Briffa, J.-C. Duplessy, F. Joos, V. Masson-Delmotte, D. Olago, et al. 2007. "Palaeoclimate." In *Climate Change 2007: The Physical Science Basis*, ed. S. Solomon,

D. Qin, M. Manning, Z. Chen, M. Marquis, K. B. Avery, M. Tignor, and H. L. Miller. (Cambridge, U.K.: Cambridge University Press), pp. 433–498.

Kleiber, Christian, and Samuel Kotz. 2003. *Statistical Size Distributions in Economics and Actuarial Sciences*. Hoboken, N.J.: John Wiley.

Kochugovindan, Sree, and Roland Nilsson. 2007a. "UK Asset Returns since 1899." In *Equity Gilt Study 2007* (London: Barclays Capital), pp. 64–70.

———. 2007b. "US Asset Returns since 1925." In *Equity Gilt Study 2007* (London: Barclays Capital), pp. 71–76.

Mehra, Rajnish, and Edward C. Prescott. 2003. "The Equity Premium in Retrospect." In *Handbook of the Economics of Finance*, vol. 1B, ed. George M. Constantinides, Milton Harris, and René Stulz (Amsterdam: Elsevier), pp. 887–936.

Meinshausen, Malte. 2006. "What Does a 2 °C Target Mean for Greenhouse Gas Concentrations? A Brief Analysis Based on Multi-Gas Emission Pathways and Several Climate Sensitivity Uncertainty Estimates." In *Avoiding Dangerous Climate Change*, ed. Hans Joachim Schellnhuber, Wolfgang Cramer, Nebojsa Nakicenovic, Tom Wigley, and Gary Yohe (Cambridge, U.K.: Cambridge University Press), pp. 265–279.

Mendelsohn, Robert. 2007. "Comments on the Stern Review." In *Yale Symposium on the Stern Review* (New Haven, Conn.: Yale Center for the Study of Globalization), pp. 93–103. Available at www.ycsg.yale.edu/climate/stern.html.

Mendelsohn, Robert, Wendy Morrison, Michael Schlesinger, and Natalia Andronova. 2000. "Country-Specific Market Impacts of Climate Change." *Climatic Change* 45.3–4: 553–569.

Mill, John Stuart. 1972. "Considerations of Representative Government." In *Utilitarianism, On Liberty and Considerations on Representative Government*, ed. H. B. Acton. London: J. M. Dent. (Originally Published in 1861).

Mirrlees, James A., and Nicholas Stern. 1972. "Fairly Good Plans." *Journal of Economic Theory* 4.2: 268–288.

Murphy, James M., David M. H. Sexton, David N. Barnett, Gareth S. Jones, Mark J. Webb, Matthew Collins, and David A. Stainforth. 2004. "Quantification of Modelling Uncertainties in a Large Ensemble of Climate Change Simulations." *Nature* 430 7001: 768–772.

Myers, Erin C. 2007. "Policies to Reduce Emissions from Deforestation and Degradation (REDD) in Tropical Forests: An Examination of the Issues Facing the Incorporation of REDD into Market-Based Climate Policies." Resources for the Future Discussion Paper 07–50.

Nepstad, Daniel, Britaldo Soares-Filho, Frank Merry, Paulo Moutinho, Hermann Oliveira Rodrigues, Maria Bowman, Steve Schwartzman, Oriana Almeida, and Sergio Rivero. 2007. "The Costs and Benefits of Reducing Carbon Emissions from Deforestation and Forest Degradation in the Brazilian Amazon." Paper presented at the United Nations Framework Convention on Climate Change Conference of the Parties, Bali, December 3–14, 2007.

Newbery, David. 1992. "Long Term Discount Rates for the Forest Enterprise." Unpublished.

Nordhaus, William D. 2007a. "The Challenge of Global Warming: Economic Models and Environmental Policy." Available at http://nordhaus.econ.yale. edu/dice_mss_072407_all.pdf.

———. 2007b. "A Review of the Stern Review on the Economics of Climate Change." *Journal of Economic Literature* 45.3: 686–702.

Nordhaus, William D., and Joseph G. Boyer. 2000. *Warming the World: Economic Models of Global Warming*. Cambridge, Mass.: MIT Press.

Okun, Arthur M. 1975. *Equality and Efficiency: The Big Tradeoff*. Washington, D.C.: Brookings Institution Press.

Pearce, David, and David Ulph. 1995. "A Social Discount Rate for the United Kingdom." Centre for Social and Economic Research on the Global Environment Global Environmental Change Working Paper GEC–1995–01.

Pigou, Arthur C. 1932. *The Economics of Welfare*, 4th ed. London: Macmillan.

Ramsey, F. P. 1928. "A Mathematical Theory of Saving." *Economic Journal* 38.152: 543–559.

Rudd, Kevin. 2007. "Ratifying the Kyoto Protocol." Media release, prime minister of Australia, Canberra, December 3, 2007. Available at http://www.pm.gov.au/media/Release/2007/media_release_0003.cfm.

Solow, Robert M. 1974. "The Economics of Resources or the Resources of Economics." *American Economic Review* 64.2: 1–14.

Stainforth, D., T. Aina, C. Christensen, M. Collins, N. Faull, D. J. Frame, J. A. Kettleborough, et al. 2005. "Uncertainty in Predictions of the Climate Response to Rising Levels of Greenhouse Gases." *Nature* 433 (7024): 403–406.

Stern, Nicholas. 1976. "On the Specification of Models of Optimum Income Taxation." *Journal of Public Economics* 6.1–2: 123–162.

———. 1977. "The Marginal Valuation of Income." In *Studies in Modern Economic Analysis: The Proceedings of the Association of University Teachers of Economics, Edinburgh 1976*, ed. M. J. Artis and A. R. Nobay (Oxford: Basil Blackwell), pp. 209–254.

———. 2007. *The Economics of Climate Change: The Stern Review*. Cambridge. U.K.: Cambridge University Press.

Stern, Nicholas, Jean-Jacques Dethier, and F. Halsey Rogers. 2005. *Growth and Empowerment: Making Development Happen*. Cambridge, Mass.: MIT Press.

Sterner, Thomas, and U. Martin Persson. 2007. "An Even Sterner Review: Introducing Relative Prices into the Discounting Debate." Resources for the Future Discussion Paper 07–37.

Tol, Richard S. J. 2002. "Estimates of the Damage Costs of Climate Change, Part II: Dynamic Estimates." *Environmental and Resource Economics* 21.2: 135–160.

Tyndall, John. 1861. "On the Absorption and Radiation of Heat by Gases and Vapours." *Philosophical Magazine* 22: 169–194, 273–285.

United Nations Development Programme. 2007. *Human Development Report 2007/2008 Fighting Climate Change: Human Solidarity in a Divided World*. New York: Palgrave Macmillan.

Weitzman, Martin L. 2007a. "On Modeling and Interpreting the Economics of Catastrophic Climate Change." Unpublished.

———. 2007b. "A Review of the Stern Review on the Economics of Climate Change." *Journal of Economic Literature* 45.3: 703–724.

———. 2007c. "Subjective Expectations and Asset-Return Puzzles." *American Economic Review* 97.4: 1102–1130.

Wigley, T. M. L., and S. C. B. Raper. 2001. "Interpretation of High Projections for Global-Mean Warming." *Science* 293 (5529): 451–454.

Ethics, Public Policy, and Global Warming

Dale Jamieson

There has been speculation about the possibility of anthropogenic global warming since at least the late nineteenth century (Arrhenius 1896, 1908). At times the prospect of such a warming has been welcomed, for it has been thought that it would increase agricultural productivity and delay the onset of the next Ice Age (Callendar 1938). Other times, and more recently, the prospect of global warming has been the stuff of "doomsday narratives," as various writers have focused on the possibility of widespread drought, flood, famine, and economic and political dislocations that might result from a "greenhouse warming"-induced climate change (Flavin 1989).

Although high-level meetings have been convened to discuss the greenhouse effect since at least 1963 (see Conservation Foundation 1963), the emergence of a rough, international consensus about the likelihood and extent of anthropogenic global warming began with a National Academy Report in 1983 (National Academy of Sciences/National Research Council 1983) and meetings in Villach, Austria, and Bellagio, Italy, in 1985 (World Climate Program 1985) and in Toronto, Canada, in 1988 (Conference Statement 1988). The most recent influential statement of the consensus holds that although there are uncertainties, a doubling of atmospheric carbon dioxide from its preindustrial baseline is likely to lead to a 2.5°C increase in the earth's mean surface temperature by the middle of the twenty-first century (IPCC 1990). (Interestingly, this estimate is within the range predicted by Arrhenius 1896.) This increase is expected to have a profound impact on climate and therefore on plants, animals, and human activities of all kinds. Moreover, there is no reason to suppose that without policy interventions, atmospheric carbon dioxide will stabilize at twice preindustrial levels. According to the IPCC (1990), we would need immediate 60 percent reductions in net emissions in order to stabilize at a carbon dioxide doubling by the end of the twenty-first century. Since these reductions are very unlikely to occur, we may well see increases of 4°C by the end of the twenty-first century.

The emerging consensus about climate change was brought home to the American public on June 23, 1988, a sweltering day in

Washington, D.C., in the middle of a severe national drought, when James Hansen testified to the U.S. Senate Committee on Energy and Natural Resources that it was 99 percent probable that global warming had begun. Hansen's testimony was front-page news in the *New York Times* and was extensively covered by other media as well. By the end of the summer of 1988, the greenhouse effect had become an important public issue. According to a June 1989 Gallup poll, 35 percent of the American public worried "a great deal" about the greenhouse effect, while 28 percent worried about it "a fair amount" (Gallup Organization 1989).

Beginning in 1989, there was a media "backlash" against the "hawkish" views of Hansen and others (for the typology of "hawks," "doves," and "owls," see Glantz 1988). In 1989, the *Washington Post* (February 8), the *Wall Street Journal* (April 10), and the *New York Times* (December 13) all published major articles expressing skepticism about the predictions of global warming or minimizing its potential impacts. These themes were picked up by other media, including such mass-circulation periodicals as *Reader's Digest* (February 1990). In its December 1989 issue, *Forbes* published a hard-hitting cover story titled "The Global Warming Panic" and later took out a full-page ad in the *New York Times* (February 7, 1990) congratulating itself for its courage in confronting the "doom-and-gloomers."

The Bush administration seems to have been influenced by this backlash. The April 1990 White House conference on global warming concluded with a ringing call for more research, disappointing several European countries that were hoping for concerted action. In July at the Houston Economic Summit, the Bush administration reiterated its position, warning against precipitous action. In a series of meetings in 1992, convened as part of the IPCC process, the American government has stood virtually alone in opposing specific targets and timetables for stabilizing carbon dioxide emissions. The Bush administration has continually emphasized the scientific uncertainties involved in forecasts of global warming and also expressed concern about the economic impacts of carbon dioxide stabilization policies.

It is a fact that there are a number of different hypotheses about the future development of the global climate and its impact on human and other biological activities; and several of these are dramatically at variance with the consensus. For example, Budyko (1988) and Idso (1989) think that global warming is good for us, and Ephron (1988) argues that the injection of greenhouse gases will trigger a new Ice Age. Others, influenced by the "Gaia Hypothesis" (see Lovelock 1988), believe that there are self-regulating planetary mechanisms that may preserve climate stability even in the face of anthropogenic forcings of greenhouse gases.

Although there are some outlying views, most of the differences of opinion within the scientific community are differences of emphasis rather than differences of kind. Rather than highlighting the degree of certainty that attaches to predictions of global warming, as does Schneider (1989), for example, some emphasize the degree of uncertainty that attaches to such predictions (for example, Abelson 1990).

However, in my view, the most important force driving the backlash is not concerns about the weakness of the science but the realization that slowing global warming or responding to its effects may involve large economic costs and redistributions, as well as radical revisions in lifestyle. Various interest groups argue that they are already doing enough in response to global warming, while some economists have begun to express doubt about whether it is worth trying to prevent substantial warming (*New York Times*, November 11, 1989; White House Council of Economic Advisors 1990). What seems to be emerging as the dominant view among economists is that chlorofluorocarbons (CFCs) should be eliminated, but emissions of carbon dioxide or other trace gases should be reduced only slightly, if at all (see Nordhaus 1990; Darmstadter 1991).

There are many uncertainties concerning anthropogenic climate change, yet we cannot wait until all the facts are in before we respond. All the facts may never be in. New knowledge may resolve old uncertainties, but it may bring with it new uncertainties. And it is an important dimension of this problem that our insults to the biosphere outrun our ability to understand them. We may suffer the worst effects of the greenhouse before we can prove to everyone's satisfaction that they will occur (Jamieson 1991).

The most important point I wish to make, however, is that the problem we face is not a purely scientific problem that can be solved by the accumulation of scientific information. Science has alerted us to a problem, but the problem also concerns our values. It is about how we ought to live and how humans should relate to one another and to the rest of nature. These are problems of ethics and politics as well as problems of science.

In the first section I examine the "management" approach to assessing the impacts of, and our responses to, climate change. I argue that this approach cannot succeed, for it does not have the resources to answer the most fundamental questions that we face. In the second section I explain why the problem of anthropogenic global change is to a great extent an ethical problem and why our conventional value system is not adequate for addressing it. Finally, I draw some conclusions.

Why Management Approaches Must Fail

From the perspective of conventional policy studies, the possibility of anthropogenic climate change and its attendant consequences are problems to be "managed." Management techniques mainly are drawn from neoclassical economic theory and are directed toward manipulating behavior by controlling economic incentives through taxes, regulations, and subsidies.

In recent years economic vocabularies and ways of reasoning have dominated the discussion of social issues. Participants in the public dialogue have internalized the neoclassical economic perspective to such an extent that its assumptions and biases have become almost invisible. It is only a mild exaggeration to say that in recent years debates over policies have largely become debates over economics.

The Environmental Protection Agency's draft report *Policy Options for Stabilizing Global Climate* (U.S. Environmental Protection Agency 1989) is a good example. Despite its title, only one of nine chapters is specifically devoted to policy options, and in that chapter only "internalizing the cost of climate change risks" and "regulations and standards" are considered. For many people questions of regulation are not distinct from questions about internalizing costs. According to one influential view, the role of regulations and standards is precisely to internalize costs, thus (to echo a parody of our forefathers) "creating a more perfect market." For people with this view, political questions about regulation are really disguised economic questions (for discussion, see Sagoff 1988).

It would be both wrong and foolish to deny the importance of economic information. Such information is important when making policy decisions, for some policies or programs that would otherwise appear to be attractive may be economically prohibitive. Or in some cases there may be alternative policies that would achieve the same ends and also conserve resources.

However, these days it is common for people to make more grandiose claims on behalf of economics. As philosophers and clergymen have become increasingly modest and reluctant to tell people what to do, economists have become bolder. Some economists or their champions believe not only that economics provides important information for making policy decisions but that it provides the most important information. Some even appear to believe that economics provides the only relevant information. According to this view, when faced with a policy decision, what we need to do is assess the benefits and costs or various alternatives. The alternative that maximizes the benefits less the costs is the one we should prefer. This alternative is "efficient," and choosing it is "rational."

Unfortunately, too often we lose sight of the fact that economic efficiency is only one value, and it may not be the most important one. Consider, for example, the idea of imposing a carbon tax as one policy response to the prospect of global warming (Moomaw

[1988] 1989). What we think of this proposal may depend to some extent on how it affects other concerns that are important to us. Equity is sometimes mentioned as one other such concern, but most of us have very little idea about what equity means or exactly what role it should play in policy considerations.

One reason for the hegemony of economic analysis and prescriptions is that many people have come to think that neoclassical economics provides the only social theory that accurately represents human motivation. According to the neoclassical paradigm, welfare can be defined in terms of preference satisfaction, and preferences are defined in terms of choice behavior. From this, many (illicitly) infer that the perception of self-interest is the only motivator for human beings. This view suggests the following "management technique": if you want people to do something give them a carrot; if you want them to desist, give them a stick. (For the view that self-interest is the "soul of modern economic man," see Myers 1983.)

Many times the claim that people do what they believe is in their interests is understood in such a way as to be circular, therefore unfalsifiable and trivial. We know that something is perceived as being in a person's interests because the person pursues it; and if the person pursues it, then we know that the person must perceive it as being in his or her interests. On the other hand, if we take it as an empirical claim that people always do what they believe is in their interests, it appears to be false. If we look around the world, we see people risking or even sacrificing their own interests in attempts to overthrow oppressive governments or to realize ideals to which they are committed. Each year more people die in wars fighting for some perceived collective good than die in criminal attempts to further their own individual interests. It is implausible to suppose that the behavior (much less the motivations) of a revolutionary, a radical environmentalist, or a friend or lover can be revealed by a benefit-cost analysis (even one that appeals to the "selfish gene").

It seems plain that people are motivated by a broad range of concerns, including concern for family and friends and religious, moral, and political ideals. And it seems just as plain that people sometimes sacrifice their own interests for what they regard to be a greater, sometimes impersonal, good. (Increasingly, these facts are being appreciated in the social science literature; see, for example, Mansbridge 1990, Opp 1989, and Scitovsky 1976.)

People often act in ways that are contrary to what we might predict on narrowly economic grounds, and moreover, they sometimes believe that it would be wrong or inappropriate even to take economic considerations into account. Many people would say that choosing spouses, lovers, friends, or religious or political commitments on economic grounds is simply wrong. People who behave in this way are often seen as manipulative, not to be trusted, without character or virtue. One way of understanding some environmentalists is to see them as wanting us to think about nature in the way that many of us think of friends and lovers—to see nature not as a resource to be exploited but as a partner with whom to share our lives.

What I have been suggesting in this section is that it is not always rational to make decisions solely on narrowly economic grounds. Although economic efficiency may be a value, there are other values as well, and in many areas of life, values other than economic efficiency should take precedence. I have also suggested that people's motivational patterns are complex and that exploiting people's perceptions of self-interest may not be the only way to move them. This amounts to a general critique of viewing all social issues as management problems to be solved by the application of received economic techniques.

There is a further reason for why economic considerations should take a back seat in our thinking about global climate change: there is no way to assess accurately all the possible impacts and to assign economic values to alternative courses of action. A greenhouse warming, if it occurs, will have impacts that are so broad, diverse, and uncertain that conventional economic analysis is practically useless. (Our inability to perform reliably the economic calculations also counts against the "insurance" view favored by many "hawks," but that is another story.)

Consider first the uncertainty of the potential impacts. Some uncertainties about the global effects of loading the atmosphere with carbon dioxide and other greenhouse gases have already been noted. But even if the consensus is correct that global mean surface temperatures will increase 1.4 °C to 4.0 °C sometime in the next century because of a doubling of atmospheric carbon dioxide, there is still great uncertainty about the impact of this warming on regional climate. One thing is certain: the impacts will not be homogeneous. Some areas will become warmer, some will probably become colder, and overall variability is likely to increase. Precipitation patterns will also change, and there is much less confidence in the projections about precipitation than in those about temperature. These uncertainties about the regional effects make estimates of the economic consequences of climate change radically uncertain.

There is also another source of uncertainty regarding these estimates. In general, predicting human behavior is difficult, as recent events in central and eastern Europe have demonstrated. These difficulties are especially acute in the case that we are considering because climate change, if it occurs, will affect a wide range of social, economic, and political activities. Changes in these sectors will affect emissions of "greenhouse gases," which will in turn affect climate, and around we go again (Jamieson 1990). Climate change is itself uncertain, and its human effects are even more radically so. It is for reasons such as these that in general, the area of environment and energy has been full of surprises.

A second reason for why the benefits and costs of the impacts of global climate change cannot reliably be assessed concerns the breadth of the impacts. Global climate change will affect all regions of the globe. About many of these regions—those in which most of the world's population live—we know very little. Some of these regions do not even have monetarized economies. It is ludicrous to suppose that we could assess the economic impacts of global climate change when we have such little understanding of the global economy in the first place. Nordhaus (1990), for example, implausibly extrapolates the sectorial analysis of the American economy to the world economy for the purposes of his study.

Finally, consider the diversity of the potential impacts. Global climate change will affect agriculture, fishing, forestry, and tourism. It will affect "unmanaged" ecosystems and patterns of urbanization. International trade and relations will be affected. Some nations and sectors may benefit at the expense of others. There will be complex interactions among these effects. For this reason we cannot reliably aggregate the effects by evaluating each impact and combining them by simple addition. But since the interactions are so complex, we have no idea what the proper mathematical function would be for aggregating them (if the idea of aggregation even makes sense in this context). It is difficult enough to assess the economic benefits and costs of small-scale, local activities. It is almost unimaginable to suppose that we could aggregate the diverse impacts of global climate change in such a way as to dictate policy responses.

In response to skeptical arguments like the one that I have given, it is sometimes admitted that our present ability to provide reliable economic analyses is limited, but then it is asserted that any analysis is better than none. I think that this is incorrect and that one way to see this is by considering an example.

Imagine a century ago a government doing an economic analysis in order to decide whether to build its national transportation system around the private automobile. No one could have imagined the secondary effects: the attendant roads, the loss of life, the effects on wildlife, on communities; the impact on air quality, noise, travel time, and quality of life. Given our inability to reliably predict and evaluate the effects of even small-scale technology (e.g., the artificial heart; see Jamieson 1988), the idea that we could predict the impact of global climate change reliably enough to permit meaningful economic analysis seems fatuous indeed.

When our ignorance is so extreme, it is a leap of faith to say that some analysis is better than none. A bad analysis can be so wrong that it can lead us to do bad things, outrageous things—things that are much worse than what we would have done had we not tried to assess the costs and benefits at all (this may be the wisdom in the old adage that "a little knowledge can be a dangerous thing").

What I have been arguing is that the idea of managing global climate change is a dangerous conceit. The tools of economic evaluation are not up to the task. However, the most fundamental reason for why management approaches are doomed to failure is that the questions they can answer are not the ones that are most important and profound. The problems posed by anthropogenic global climate change are ethical as well as economic and scientific. I will explain this claim in the next section.

Ethics and Global Change

Since the end of World War II, humans have attained a kind of power that is unprecedented in history. While in the past entire peoples could be destroyed, now all people are vulnerable. While once particular human societies had the power to upset the natural processes that made their lives and cultures possible, now people have the power to alter the fundamental global conditions that permitted human life to evolve and that continue to sustain it. While our species dances with the devil, the rest of nature is held hostage. Even if we step back from the precipice, it will be too late for many or even perhaps most of the plant and animal life with which we share the planet (Borza and Jamieson 1990). Even if global climate can be stabilized, the future may be one without wild nature (McKibben 1989). Humans will live in a humanized world with a few domestic plants and animals that can survive or thrive on their relationships with humans.

The questions that such possibilities pose are fundamental questions of morality. They concern how we ought to live, what kinds of societies we want, and how we should relate to nature and other forms of life. Seen from this perspective, it is not surprising that economics cannot tell us everything we want to know about how we should respond to global warming and global change. Economics may be able to tell us how to reach our goals efficiently, but it cannot tell us what our goals should be or even whether we should be concerned to reach them efficiently.

It is a striking fact about modern intellectual life that we often seek to evade the value dimensions of fundamental social questions. Social scientists tend to eschew explicit talk about values, and this is part of the reason we have so little understanding of how value change occurs in individuals and societies. Policy professionals are also often reluctant to talk about values. Many think that rational reflection on values and value change is impossible, unnecessary, impractical, or dangerous. Others see it as a professional, political, or bureaucratic threat (Amy 1984). Generally, in the political process, value language tends to function as code words for policies and attitudes that cannot be discussed directly.

A system of values, in the sense in which I will use this notion, specifies permissions, norms, duties, and obligations; it assigns blame, praise, and responsibility; and it provides an account of what is valuable and what is not. A system of values provides a standard for assessing our behavior and that of others. Perhaps indirectly it also provides a measure of the acceptability of government action and regulation.

Values are more objective than mere preferences (Andrews and Waits 1978). A value has force for a range of people who are similarly situated. A preference may have force only for the individual whose preference it is. Whether or not someone should have a particular value depends on reasons and arguments. We can rationally discuss values, while preferences may be rooted simply in desire, without supporting reasons.

A system of values may govern someone's behavior without these values being fully explicit. They may figure in people's motivations and in their attempts to justify or criticize their own actions or those of others. Yet it may require a theorist or a therapist to make these values explicit.

In this respect a system of values may be like an iceberg—most of what is important may be submerged and invisible even to the person whose values they are. Because values are often opaque to the person who holds them, there can be inconsistencies and incoherencies in a system of values. Indeed much debate and dialogue about values involves attempts to resolve inconsistencies and incoherencies in one direction or another.

A system of values is generally a cultural construction rather than an individual one (Weiskel 1990). It makes sense to speak of contemporary American values, or those of eighteenth-century England or tenth-century India. Our individual differences tend to occur around the edges of our value system. The vast areas of agreement often seem invisible because they are presupposed or assumed without argument.

I believe that our dominant value system is inadequate and inappropriate for guiding our thinking about global environmental problems, such as those entailed by climate changes caused by human activity. This value system, as it impinges on the environment, can be thought of as a relatively recent construction, coincident with the rise of capitalism and modern science, and expressed in the writings of such philosophers as Francis Bacon ([1620] 1870), John Locke ([1690] 1952), and Bernard Mandeville ([1714] 1970; see also Hirschman 1977). It evolved in low-population-density and low-technology societies, with seemingly unlimited access to land and other resources. This value system is reflected in attitudes toward population, consumption, technology, and social justice, as well as toward the environment.

The feature of this value system that I will discuss is its conception of responsibility. Our current value system presupposes that harms and their causes are individual, that they can readily be identified, and that they are local in space and time. It is these aspects of our conception of responsibility on which I want to focus.

Consider an example of the sort of case with which our value system deals best. Jones breaks into Smith's house and steals Smith's television set. Jones's intent is clear: she wants Smith's TV set. Smith suffers a clear harm; he is made worse off by having lost the television set. Jones is responsible for Smith's loss, for she was the cause of the harm and no one else was involved.

What we have in this case is a clear, self-contained story about Smith's loss. We know how to identify the harms and how to assign responsibility. We respond to this breech of our norms by punishing Jones in order to prevent her from doing it again and to deter others from such acts, or we require compensation from Jones so that Smith may be restored to his former position.

It is my contention that this paradigm collapses when we try to apply it to global environmental problems, such as those associated with human-induced global climate change. It is for this reason that we are often left feeling confused about how to think about these problems.

There are three important dimensions along which global environmental problems such as those involved with climate change vary from the paradigm: apparently innocent acts can have devastating consequences, causes and harms may be diffuse, and causes and harms may be remote in space and time. (Other important dimensions may concern nonlinear causation, threshold effects, and the relative unimportance of political boundaries, but I cannot discuss these here; see Lee 1989.)

Consider an example. Some projections suggest that one effect of greenhouse warming may be to shift the southern hemisphere cyclone belt to the south. If this occurs the frequency of cyclones in Sydney, Australia, will increase enormously, resulting in great death and destruction. The causes of this death and destruction will be diffuse. There is no one whom we can identify as the cause of destruction in the way in which we can identify Jones as the cause of Smith's loss. Instead of a single cause, millions of people will have made tiny, almost imperceptible causal contributions—by driving cars, cutting trees, using electricity, and so on. They will have made these contributions in the course of their daily lives performing apparently "innocent" acts, without intending to bring about this harm. Moreover, most of these people will be geographically remote from Sydney, Australia. (Many of them will have no idea where Sydney, Australia, is.) Further, some people who are harmed will be remote in time from those who have harmed them. Sydney may suffer in the twenty-first century in part because of people's behavior in the nineteenth and twentieth centuries. Many small people doing small things over a long period of time together will cause unimaginable harms.

Despite the fact that serious, clearly identifiable harms will have occurred because of human agency, conventional morality would have trouble finding anyone to blame. For no one intended the bad outcome or brought it about or even was able to foresee it.

Today we face the possibility that the global environment may be destroyed, yet no one will be responsible. This is a new problem. It takes a great many people and a high level of consumption and production to change the earth's climate. It could not have been done in low-density, low-technology societies. Nor could it have been done in societies like ours until recently. London could be polluted by its inhabitants in the eighteenth century, but its reach was limited. Today no part of the planet is safe. Unless we develop new values and conceptions of responsibility, we will have enormous difficulty in motivating people to respond to this problem.

Some may think that discussion about new values is idealistic. Human nature cannot be changed, it is sometimes said. But as anyone who takes anthropology or history seriously knows, our current values are at least in part historically constructed, rooted in the conditions of life in which they developed. What we need are new values that reflect the interconnectedness of life on a dense, high-technology planet.

Others may think that a search for new values is excessively individualistic and that what is needed are collective and institutional solutions. This overlooks the fact that our values permeate our institutions and practices. Reforming our values is part of constructing new moral, political, and legal concepts.

One of the most important benefits of viewing global environmental problems as moral problems is that this brings them into the domain of dialogue, discussion, and participation. Rather than being management problems that governments or experts can solve for us, when seen as ethical problems, they become problems for all of us to address, both as political actors and as everyday moral agents.

In this chapter I cannot hope to say what new values are needed or to provide a recipe for how to bring them about. Values are collectively created rather than individually dictated, and the dominance of economic models has meant that the study of values and value change has been neglected (but see Wolfe 1989; Reich 1988). However, I do have one positive suggestion: we should focus more on character and less on calculating probable outcomes. Focusing on outcomes has made us cynical calculators and has institutionalized hypocrisy. We can each reason: since my contribution is small, outcomes are likely to be determined by the behavior of others. Reasoning in this way, we can each justify driving cars while advocating bicycles or using fireplaces while favoring regulations against them. In such a climate we do not condemn or even find it surprising that Congress exempts itself from civil rights laws. Even David Brower, the "archdruid" of the environmental movement, owns two cars, four color televisions, two video cameras, three video recorders, and a dozen tape recorders, and he justifies this by saying that "it will help him in his work to save the Earth" (*San Diego Union*, April 1, 1990).

Calculating probable outcomes leads to unraveling the patterns of collective behavior that are needed in order to respond successfully to many of the global environmental problems that we face. When we "economize" our behavior in the way that is required for calculating, we systematically neglect the subtle and indirect effects of our actions, and for this reason we see individual action as inefficacious. For social change to occur, it is important that there be people of integrity and character who act on the basis of principles and ideals.

The content of our principles and ideals is, of course, important. Principles and ideals can be eccentric or even demented. In my opinion, in order to address such problems as global climate change, we need to nurture and give new content to some old virtues such as humility, courage, and moderation and perhaps develop such new virtues as those of simplicity and conservatism. But whatever the best candidates are for 21st-century virtues, what is important to recognize is the importance and centrality of the virtues in bringing about value change.

Conclusion

Science has alerted us to the impact of humankind on the planet, one another, and all life. This dramatically confronts us with questions about who we are, our relations to nature, and what we are willing to sacrifice for various possible futures. We should confront this as a fundamental challenge to our values and not treat it as if it were simply another technical problem to be managed.

Some who seek quick fixes may find this concern with values frustrating. A moral argument will not change the world overnight. Collective moral change is fundamentally cooperative rather than coercive. No one will fall over, mortally wounded, in the face of an argument. Yet if there is to be meaningful change that makes a difference over the long term, it must be both collective and thoroughgoing. Developing a deeper understanding of who we are, as well as how our best conceptions of ourselves can guide change, is the fundamental issue that we face.

Acknowledgments

This material was discussed with an audience at the 1989 AAAS meetings in New Orleans, at the conference "Global Warming and the Future: Science Policy and Society" at Michigan Technological University, and with the philosophy departments at the University of Redlands and the University of California at Riverside. I have benefited greatly from each of these discussions. Michael H. Glantz (National Center for Atmospheric Research) commented helpfully on an earlier draft, and Karen Borza (Pennsylvania State University) contributed to the development of this chapter in many ways. I have also been helped by the comments of two anonymous referees. A preliminary version of this essay appeared in *Problemi di Bioetica*. I gratefully acknowledge the support of the Ethics and Values Studies Program of the National Science Foundation for making this research possible.

References

Abelson, Philip. 1990. "Uncertainties about Global Warming." *Science* 247 (March 30): 1529.

Amy, Douglas R. 1984. "Why Policy Analysis and Ethics Are Incompatible." *Journal of Policy Analysis and Management* 3: 573–591.

Andrews, Richard, and Mary Jo Waits. 1978. *Environmental Values in Public Decisions: A Research Agenda*. Ann Arbor: University of Michigan, School of Natural Resources.

Arrhenius, S. 1896. "On the Influence of Carbonic Acid in the Air upon the Temperature of the Ground." *Philosophical Magazine* 41: 237.

———. 1908. *Worlds in the Making*. New York: Harper.

Bacon, F. [1620] 1870. *Works*, ed. James Spedding, Robert Leslie Ellis, and Douglas Devon Heath. London: Longmans Green.

Borza, K., and D. Jamieson. 1990. *Global Change and Biodiversity Loss: Some Impediments to Response*. Boulder: University of Colorado, Center for Space and Geoscience Policy.

Budyko, M. I. 1988. "Anthropogenic Climate Change." Paper presented at the World Congress on Climate and Development, Hamburg, Germany.

Callendar, G. S. 1938. "The Artificial Production of Carbon Dioxide and Its Influence on Temperature." *Quarterly Journal of the Royal Meteorological Society* 64: 223–240.

Conference Statement. 1988. "The Changing Atmosphere: Implications for Global Security." Toronto, June 27–30.

Conservation Foundation. 1963. *Implications of Rising Carbon Dioxide Content of the*

Atmosphere. New York: Conservation Foundation.

Darmstadter, Joel. 1991. *The Economic cost of CO_2 Mitigation: A Review of Estimates for Selected World Regions.* Washington, D.C.: Resources for the Future.

Ephron, L. 1988. *The End: The Imminent Ice Age and How We Can Stop It.* Berkeley, Calif.: Celestial Arts.

Flavin, C. 1989. *Slowing Global Warming: A Worldwide Strategy.* Washington, D.C.: Worldwatch Institute.

Gallup Organization. 1989. *The Gallup Report 285: Concern about the Environment.* Washington, D.C.: Gallup Organization.

Glantz, M. 1988. "Politics and the Air around Us: International Policy Action on Atmospheric Pollution by Trace Gases." In *Societal Responses to Regional Climate Change: Forecasting by Analogy*, ed. M. Glantz (Boulder, Colo.: Westview), pp. 41–72.

Hirschman, Albert, 1977. *The Passions and the Interests.* Princeton, N.J.: Princeton University Press.

Idso, Sherwood B. 1989. *Carbon Dioxide and Global Change: The Earth in Transition.* Tempe, Ariz.: IBR Press.

IPCC. 1990. *Policymakers' Summary: Working Group III.* Geneva: World Meteorological Association and United Nations Environment Program.

Jamieson, Dale. 1988. "The Artificial Heart: Reevaluating the Investment." In *Organ Substitution Technology*, ed. D. Mathieu (Boulder, Colo.: Westview), pp. 277–296.

———. 1990. "Managing the Future: Public Policy, Scientific Uncertainty, and Global Warming." In *Upstream/Downstream: New Essays in Environmental Ethics*, ed. D. Scherer (Philadelphia: Temple University Press), pp. 67–89.

———. 1991. "The Epistemology of Climate Change: Some Morals for Managers." *Society and Natural Resources* 4: 319–329.

Lee, Keekok. 1989. *Social Philosophy and Ecological Scarcity.* New York: Routledge.

Locke, John. [1690] 1952. *The Second Treatise of Government.* Indianapolis: Bobbs-Merrill.

Lovelock, J. E. 1988. *The Ages of Gaia: A Biography of Our Living Earth.* New York: Norton.

Mandeville, B. [1714] 1970. *The Fable of the Bees*, trans. P. Harth. Hammersmith, U.K.: Penguin.

Mansbridge, Jane, ed. 1990. *Beyond Self-interest.* Chicago: University of Chicago Press.

McKibben, W. 1989. *The End of Nature.* New York: Knopf.

Moomaw, William R. [1988] 1989. "Near-term Congressional Options for Responding to Global Climate Change." Reprinted in *The Challenge of Global Warming*, ed. Dean Edwin Abrahamson (Washington, D.C.: Island), pp. 305–326.

Myers, Milton. L. 1983. *The Soul of Modern Economic Man.* Chicago: University of Chicago Press.

National Academy of Sciences/National Research Council 1983. *Changing Climate.* Washington, D.C.: National Academy Press.

Nordhaus, W. 1990. "To Slow or Not to Slow: The Economics of the Greenhouse Effect." Paper presented at the American Association for the Advancement of Science, New Orleans.

Opp, Karl-Dieter. 1989. *The Rationality of Political Protest.* Boulder, Colo.: Westview.

Reich, Robert, ed. 1988. *The Power of Public Ideas.* Cambridge, Mass.: Harvard University Press.

Sagoff, Mark. 1988. *The Economy of the Earth.* New York: Cambridge University Press.

Schneider, Stephen H. 1989. *Global Warming: Are We Entering the Greenhouse Century?* San Francisco: Sierra Club Books.

Scitovsky, Tibor. 1976. *The Joyless Economy: An Inquiry into Human Satisfaction and Consumer Dissatisfaction.* New York: Oxford University Press.

U.S. Environmental Protection Agency. 1989. *Policy Options for Stabilizing Global Climate, Draft Report to Congress*, ed. D. Lashof and D. A. Tirpak. Washington, D.C.: U.S. GPO.

Weiskel, Timothy. 1990. "Cultural Values and Their Environmental Implications: An Essay on Knowledge, Belief and Global Survival." Paper presented at the American Association for the Advancement of Science, New Orleans.

White House Council of Economic Advisors. 1990. *The Economic Report of the President.* Washington, D.C.: Executive Office of the President, Publication Services.

Wolfe, Alan. 1989. *Whose Keeper? Social Science and Moral Obligation.* Berkeley: University of California Press.

World Climate Program. 1985. *Report of the International Conference on the Assessment of the Role of Carbon Dioxide and of Other Greenhouse Gases in Climate Variations and Associated Impacts: Report on an International Conference Held at Villach, Austria, 9–15 October 1985.* Geneva: World Meteorological Organization.

4

A Perfect Moral Storm

Climate Change, Intergenerational Ethics, and the Problem of Corruption

Stephen M. Gardiner

"There's a quiet clamor for hypocrisy and deception; and pragmatic politicians respond with ... schemes that seem to promise something for nothing. Please, spare us the truth."

—ROBERT J. SAMUELSON (2005, P. 41)

The most authoritative scientific report on climate change begins by saying:

> Natural, technical, and social sciences can provide essential information and evidence needed for decisions on what constitutes "dangerous anthropogenic interference with the climate system." At the same time, *such decisions are value judgments.*[1]

There are good grounds for this statement. Climate change is a complex problem raising issues across and between a large number of disciplines, including the physical and life sciences, political science, economics, and psychology, to name just a few. But without wishing for a moment to marginalize the contributions of these disciplines, ethics does seem to play a fundamental role.

At the most general level, the reason is that we cannot get very far in discussing why climate change is a problem without invoking ethical considerations. If we do not think that our own actions are open to moral assessment, or that various interests (our own, those of our kin and country, those of distant people, future people, animals, and nature) matter, then it is hard to see why climate change (or much else) poses a problem. But once we see this, then we appear to need some account of moral responsibility, morally important interests, and what to do about both. And this puts us squarely in the domain of ethics.

At a more practical level, ethical questions are fundamental to the main policy decisions that must be made, such as where to set a global ceiling for greenhouse-gas emissions and how to distribute the emissions allowed by such a ceiling. For example, where the global ceiling is set depends on how the interests of the current generation are weighed against those of future generations, and how emissions are distributed under the global cap depends in part on various beliefs about the appropriate role of energy consumption in people's lives, the importance of historical responsibility for the problem, and the current needs and future aspirations of particular societies.

The relevance of ethics to substantive climate policy thus seems clear. But this is not the topic that I wish to take up here.[2] Instead,

I want to discuss a further, and to some extent more basic, way in which ethical reflection sheds light on our present predicament. This has nothing much to do with the substance of a defensible climate regime; instead, it concerns the process of making climate policy.

My thesis is this. The peculiar features of the climate-change problem pose substantial obstacles to our ability to make the hard choices necessary to address it. Climate change is a perfect moral storm. One consequence of this is that even if the difficult ethical questions could be answered, we might still find it difficult to act. For the storm makes us extremely vulnerable to moral corruption.[3]

Let us say that a perfect storm is an event constituted by an unusual convergence of independently harmful factors where this convergence is likely to result in substantial, and possibly catastrophic, negative outcomes. The term *the perfect storm* seems to have become prominent in popular culture through Sebastian Junger's book of that name and the associated Hollywood film.[4] Junger's tale is based on the true story of the *Andrea Gail*, a fishing vessel caught at sea during a convergence of three particularly bad storms.[5] The sense of the analogy is that climate change appears to be a perfect moral storm because it involves the convergence of a number of factors that threaten our ability to behave ethically.

As climate change is a complex phenomenon, I cannot hope to identify all of the ways in which its features cause problems for ethical behavior. Instead, I will identify three especially salient problems—analogous to the three storms that hit the *Andrea Gail*—that converge in the climate-change case. These three "storms" arise in the global, intergenerational, and theoretical dimensions, and I will argue that their interaction helps to exacerbate and obscure a lurking problem of moral corruption that may be of greater practical importance than any of them.

I. The Global Storm

The first two storms arise out of three important characteristics of the climate-change problem:

- Dispersion of causes and effects
- Fragmentation of agency
- Institutional inadequacy

Since these characteristics manifest themselves in two especially salient dimensions—the spatial and the temporal—it is useful to identify two distinct but mutually reinforcing components of the climate-change problem. I shall call the first the global storm. This corresponds to the dominant understanding of the climate-change problem, and it emerges from a predominantly spatial interpretation of the three characteristics.

Let us begin with the dispersion of causes and effects. Climate change is a truly global phenomenon. Emissions of greenhouse gases from any geographical location on the earth's surface enter the atmosphere and then play a role in affecting climate globally. Hence, the impact of any particular emission of greenhouse gases is not realized solely at its source, either individual or geographical; rather, impacts are dispersed to other actors and regions of the earth. Such spatial dispersion has been widely discussed.

Next comes the fragmentation of agency. Climate change is caused not by a single agent but by a vast number of individuals and institutions not unified by a comprehensive structure of agency. This is important because it poses a challenge to humanity's ability to respond.

In the spatial dimension, this feature is usually understood as arising out of the shape of the current international system, as constituted by states. Then the problem is that, given that there is not only no world government but also no less centralized system of global governance (or at least no effective one), it is very difficult to coordinate an effective response to global climate change.[6]

This general argument is usually given more bite through the invocation of a certain familiar theoretical model.[7] For the international situation is usually understood in game-theoretic terms as a prisoner's dilemma, or what Garrett Hardin calls a "Tragedy of the Commons."[8] For the sake of ease of exposition, let us describe the prisoner's dilemma scenario in terms of a paradigm case, that of overpollution.[9] Suppose that a number of distinct agents are trying to decide whether or not to engage in a polluting activity and that their situation is characterized by the following two claims:

> (PD1) It is *collectively rational* to cooperate and restrict overall pollution: each agent prefers the outcome produced by everyone restricting his or her individual pollution over the outcome produced by no one doing so.
>
> (PD2) It is *individually rational* not to restrict one's own pollution: when each agent has the power to decide whether or not to restrict his or her pollution, each (rationally) prefers not to do so, whatever the others do.

Agents in such a situation find themselves in a paradoxical position. On the one hand, given (PD1), they understand that it would be better for everyone if every agent cooperated, but on the other hand, given (PD2), they also know that they should all choose to defect. This is paradoxical because it implies that if individual agents act rationally in terms of their own interests, then they collectively undermine those interests.[10]

A tragedy of the commons is essentially a prisoner's dilemma involving a common resource. This has become the standard analytical model for understanding regional and global environmental problems in general, and climate change is no exception. Typically, the reasoning goes as follows. Imagine climate change as an international problem, and conceive of the relevant parties as individual countries, which represent the interests of their countries in perpetuity. Then (PD1) and (PD2) appear to hold. On the one hand, no one wants serious climate change. Hence, each country prefers the outcome produced by everyone restricting individual emissions over the outcome produced by no one doing so, and so it is collectively rational to cooperate and restrict global emissions. But on the other hand, each country prefers to free-ride on the actions of others. Hence, when each country has the power to decide whether or not to restrict its emissions, each prefers not to do so, whatever the others do.

From this perspective, it appears that climate change is a normal tragedy of the commons. Still, there is a sense in which this turns out to be encouraging news; for, in the real world, commons problems are often resolvable under certain circumstances, and climate change seems to fill these desiderata.[11] In particular, it is widely said that parties facing a commons problem can resolve it if they benefit from a wider context of interaction; this appears to be the case with climate change, since countries interact with one another on a number of broader issues, such as trade and security.

This brings us to the third characteristic of the climate-change problem: institutional inadequacy. There is wide agreement that the appropriate means for resolving commons problems under the favorable conditions just mentioned is for the parties to agree to change the existing incentive structure through the introduction of a system of enforceable sanctions. (Hardin calls this "mutual coercion, mutually agreed upon.") This transforms the decision situation by foreclosing the option of free-riding, so that the collectively rational action also becomes individually rational. Theoretically, then, matters seem simple; but in practice, things are different. The need for enforceable sanctions poses a challenge at the global level because of the limits of our current, largely national, institutions and the lack of an effective system of global governance. In essence, addressing climate change appears to require global regulation of greenhouse-gas emissions, where this includes establishing a reliable enforcement mechanism; but the current global system—or lack of it—makes this difficult, if not impossible.

The implication of this familiar analysis, then, is that the main thing needed to solve the global-warming problem is an effective system of global governance (at least for this issue). And there is a sense in which this is still good news. For, in principle at least, it should be possible to motivate countries to establish such a regime, since they ought to recognize that it is in their best interests to eliminate the possibility of free-riding and so make genuine cooperation the rational strategy at the individual as well as collective level.

Unfortunately, however, this is not the end of the story. There are other features of the climate-change case that make the necessary global agreement more difficult and so exacerbate the basic global storm.[12] Prominent among these is scientific uncertainty about the precise magnitude and distribution of effects, particularly at the national level.[13] One reason for this is that the lack of trustworthy data about the costs and benefits of climate change at the national level casts doubt on the truth of (PD1). Perhaps, some nations wonder, we might be better off with climate change than without it. More importantly, some countries might wonder whether they will at least be relatively better off than other countries and so might get away with paying less to avoid the associated costs.[14] Such factors complicate the game-theoretic situation, making agreement more difficult.

In other contexts, the problem of scientific uncertainty might not be so serious. But a second characteristic of the climate-change problem exacerbates matters in this setting. The source of climate change is located deep in the infrastructure of current human civilizations; hence, attempts to combat it may have substantial ramifications for human social life. Climate change is caused by human emissions of greenhouse gases, primarily carbon dioxide. Such emissions are brought about by the burning of fossil fuels for energy. But it is this energy that supports existing economies. Hence, given that halting climate change will require deep cuts in projected global emissions over time, we can expect that such action will have profound effects on the basic economic organization of the developed countries and on the aspirations of the developing countries.

This has several salient implications. First, it suggests that those with vested interests in the continuation of the current system—for example, many of those with substantial political and economic power—will resist such action. Second, unless ready substitutes are found, real mitigation can be expected to have profound impacts on how humans live and how human societies evolve. Action on climate change is therefore likely to raise serious, and perhaps uncomfortable, questions about who we are and what we want to be. Third, this suggests a status quo bias in the face of uncertainty. Contemplating change is often uncomfortable; contemplating basic change may be unnerving, even distressing. Since the social ramifications of action appear to be large, perspicuous, and concrete, but those of inaction appear uncertain, elusive, and indeterminate, it is easy to see why uncertainty might exacerbate social inertia.[15]

The third feature of the climate-change problem that exacerbates the basic global storm is that of skewed vulnerabilities. The climate-change problem interacts in some unfortunate ways with the present global power structure. For one thing, the responsibility for historical and current emissions lies predominantly with the richer, more powerful nations, and the poor nations are badly situated to hold them accountable. For another, the limited evidence on regional impacts suggests that the poorer nations are most vulnerable to the worst impacts of climate change.[16] Finally, action on climate change creates a moral risk for the developed nations. It embodies a recognition that there are international norms of ethics and responsibility and reinforces the idea that international cooperation on issues involving such norms is both possible and necessary. Hence attention to other moral defects of the current global system, such as global poverty, human-rights violations, and so on.[17]

II. The Intergenerational Storm

We can now return to the three characteristics of the climate-change problem identified earlier:

- Dispersion of causes and effects
- Fragmentation of agency
- Institutional inadequacy

The global storm emerges from a spatial reading of these characteristics; but I would argue that another, even more serious problem arises when we see them from a temporal perspective. I shall call this the intergenerational storm.

Consider first the dispersion of causes and effects. Human-induced climate change is a severely lagged phenomenon. This is partly because some of the basic mechanisms set in motion by the greenhouse effect—such as sea-level rise—take a very long time to be fully realized. But it is also because by far the most important greenhouse gas emitted by human beings is carbon dioxide, and once emitted, molecules of carbon dioxide can spend a surprisingly long time in the atmosphere.[18]

Let us dwell for a moment on this second factor. The IPCC says that the average time spent by a molecule of carbon dioxide in the atmosphere is in the region of 5 to 200 years. This estimate is long enough to create a serious lagging effect; nevertheless, it obscures the fact that a significant percentage of carbon dioxide molecules remain in the atmosphere for much longer periods of time, thousands and tens of thousands of years. For instance, in a recent paper, David Archer says:

> The carbon cycle of the biosphere will take a long time to completely neutralize and sequester anthropogenic CO_2. We show a wide range of model forecasts of this effect. For the best-guess cases…we expect that 17–33% of the fossil fuel carbon will still reside in the atmosphere 1 kyr from now, decreasing to 10–15% at 10 kyr, and 7% at 100 kyr. The mean lifetime of fossil fuel CO_2 is about 30–35 kyr.[19]

This is a fact, he says, that has not yet "reached general public awareness."[20] Hence he suggests that "a better shorthand for public discussion [than the IPCC estimate] might be that CO_2 sticks around for hundreds of years, plus 25% that sticks around for ever."[21]

The fact that carbon dioxide is a long-lived greenhouse gas has at least three important implications. The first is that climate change is a *resilient* phenomenon. Given that it currently does not seem practical to remove large quantities of carbon dioxide from the atmosphere or to moderate its climatic effects, the upward trend in atmospheric concentration is not easily reversible. Hence, a goal of stabilizing and then reducing carbon dioxide concentrations requires advance planning. Second, climate-change impacts are *seriously backloaded*. The climate change that the earth is currently experiencing is primarily the result of emissions from some time in the past, rather than current emissions. As an illustration, it is widely accepted that by 2000, we had already committed ourselves to a rise of at least 0.5°C and perhaps more than 1°C over the then-observed rise of 0.6°C.[22] Third, backloading implies that the full, cumulative effects of our current emissions will not be realized for some time in the future. So, climate change is a *substantially deferred* phenomenon.

Temporal dispersion creates a number of problems. First, as is widely noted, the resilience of climate change implies that delays in action have serious repercussions for our ability to manage the problem. Second, backloading implies that climate change poses serious epistemic difficulties, especially for normal political actors. For one thing, backloading makes it hard to grasp the connection between causes and effects, and this may undermine the motivation to act;[23] for another, it implies that by the time we realize that things are bad, we will already be committed to much more change, so it undermines the ability to respond. Third, the deferral effect calls into question the ability of standard institutions to deal with the problem. For one thing, democratic political institutions have relatively short time horizons—the next election cycle, a politician's political career—and it is doubtful whether such institutions have the wherewithal to deal with substantially deferred impacts. Even more seriously, substantial deferral is likely to undermine the will to act. This is because there is an incentive problem: the bad effects of current emissions are likely to fall, or fall disproportionately, on future generations, whereas the benefits of emissions accrue largely to the present.[24]

These last two points already raise the specter of institutional inadequacy. But to appreciate this problem fully, we must first say something about the temporal fragmentation of agency. There is some reason to think that the temporal fragmentation of agency might be worse than the spatial fragmentation even considered in isolation. There is a sense in which temporal fragmentation is more intractable than spatial fragmentation: in principle, spatially fragmented agents may actually become unified and able really to act as a single agent; but temporally fragmented agents cannot actually become unified and may at best only act *as if* they were a single agent.

As interesting as such questions are, they need not detain us here. Temporal fragmentation in the context of the kind of temporal dispersion that characterizes climate change is clearly much worse than the associated spatial fragmentation. The presence of backloading and deferral together brings on a new collective-action problem that adds to the tragedy of the commons caused by the global storm and thereby makes matters much worse.

The problem emerges when one relaxes the assumption that countries can be relied upon adequately to represent the interests of both their present and future citizens. Suppose that this is not true. Suppose instead that countries are biased toward the interests of the current generation. Then, since the benefits of carbon dioxide emission are felt primarily by the present generation, in the form of cheap energy, whereas the costs, in the form of the risk of severe and perhaps catastrophic climate change, are substantially deferred to future generations, climate change might provide an instance of a severe intergenerational collective-action problem. Moreover, this problem will be iterated. Each new generation will face the same incentive structure as soon as it gains the power to decide whether or not to act.[25]

The nature of the intergenerational problem is easiest to see if we compare it to the traditional prisoner's dilemma. Suppose we consider a pure version of the intergenerational problem, where the generations do not overlap.[26] (Call this the pure intergenerational problem, or PIP.) In that case, the problem can be (roughly) characterized as follows:[27]

> (PIP1) It is *collectively rational* for most generations to cooperate: (almost) every generation prefers the outcome produced by everyone restricting pollution over the outcome produced by everyone overpolluting.
>
> (PIP2) It is *individually rational* for all generations not to cooperate: when each generation has the power to decide whether or not it will overpollute, each generation (rationally) prefers to overpollute, whatever the others do.

The PIP is worse than the prisoner's Dilemma in two main respects. First, its two constituent claims are worse. On the one hand, (PIP1) is worse than (PD1) because the first generation is not included. This means not only that one generation is not motivated to accept the collectively rational outcome but also that the problem becomes iterated. Since subsequent generations have no reason to comply if their predecessors do not, noncompliance by the first generation has a domino effect that undermines the collective project. On the other hand, (PIP2) is worse than (PD2) because the reason for it is deeper. Both of these claims hold because the parties lack access to mechanisms (such as enforceable sanctions) that would make defection irrational. But whereas in normal prisoner's dilemma cases this obstacle is largely practical and can be resolved by creating appropriate institutions, in the PIP it arises because the parties do not coexist and therefore seem unable to influence each other's behavior through the creation of appropriate coercive institutions.

This problem of interaction produces the second respect in which the PIP is worse than the prisoner's dilemma. This is that the PIP is more difficult to resolve, because the standard solutions to the prisoner's dilemma are unavailable: one cannot appeal to a wider context of mutually beneficial interaction or to the usual notions of reciprocity.

The upshot of all this is that in the case of climate change, the intergenerational analysis will be less optimistic about solutions than the tragedy-of-the-commons analysis. It implies that current populations may not be motivated to establish a fully adequate global regime, since, given the temporal dispersion of effects—and especially backloading and deferral—such a regime is probably not in *their* interests. This is a large moral problem, especially since, in my view, the intergenerational problem dominates the tragedy-of-the-commons aspect in climate change.

The PIP is bad enough considered in isolation. But in the context of climate change, it is also subject to morally relevant multiplier effects. First, climate change is not a static phenomenon. In failing to act appropriately, the current generation does not simply pass an existing problem along to future people; rather, it adds to it, making the problem worse. For one thing, it increases the costs of coping with climate change: failing to act now increases the magnitude of future climate change and its effects. For another, it increases mitigation costs: failing to act now makes it more difficult to change because it allows additional investment in fossil-fuel-based infrastructure in developed and especially less developed countries. Hence, inaction raises transition costs, making future change harder than change now. Finally, and perhaps most importantly, the current generation does not add to the problem in a linear way. Rather, it rapidly accelerates the problem, since global emissions are increasing at a substantial rate; for example, total carbon dioxide emissions have increased more than four-fold in the last 50 years (see figure 4.1). Moreover, the current growth rate is about 2 percent per year.[28] Although 2 percent might not seem like much, the effects of compounding make it significant, even in the near term: "continued growth of CO_2 emissions at 2% per year would yield a 22% increase of emission rate in 10 years and a 35% increase in 15 years."[29]

Second, insufficient action may make some generations suffer unnecessarily. Suppose that at this point in time, climate change seriously affects the prospects of generations A, B, and C. Suppose, then, that if generation A refuses to act, the effect will continue for longer, harming generations D and E. This may make generation A's inaction worse in a significant respect. In addition to failing to aid generations B and C (and probably also increasing the magnitude of harm inflicted on them), generation A now harms generations D and E, which otherwise would be spared. On some views, this might count as especially egregious, since it might be said that it violates a fundamental moral principle of "Do no harm."[30]

Third, generation A's inaction may create situations where *tragic choices* must be made. One way in which a generation may

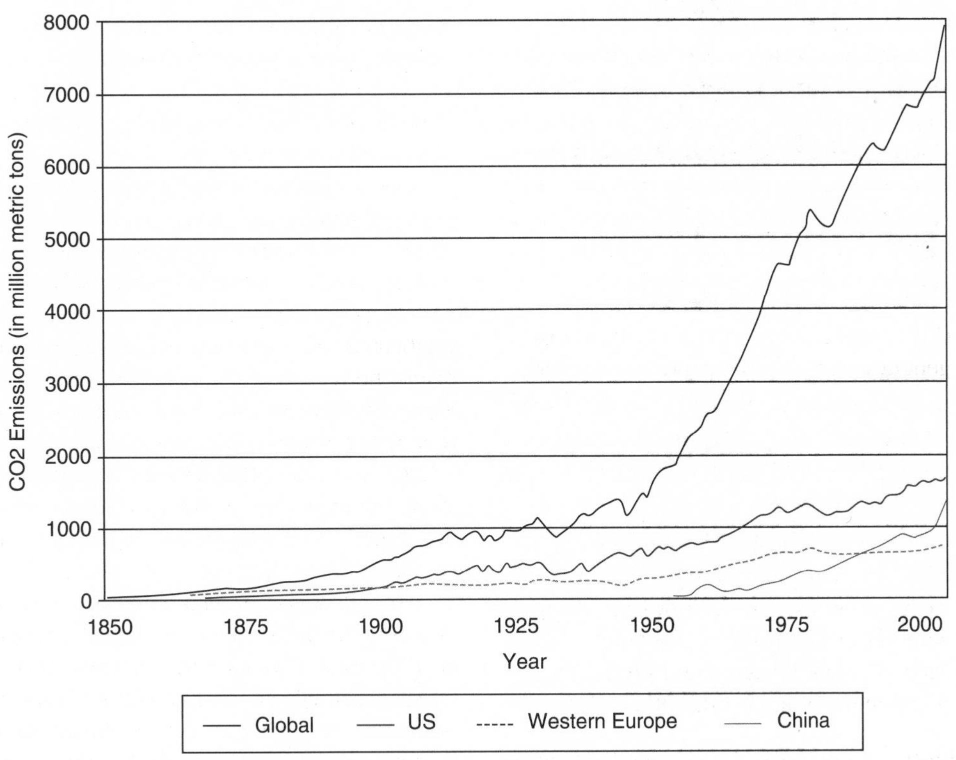

Figure 4.1. Country/region fossil-fuel CO_2 annual emissions.

act badly is if it puts in place a set of future circumstances that make it morally required for its successors (and perhaps even itself) to make other generations suffer either unnecessarily or at least more than would otherwise be the case. For example, suppose that generation A could and should act now in order to limit climate change such that generation D would be kept below some crucial climate threshold, but delay would mean that it would pass that threshold.[31] If passing the threshold imposes severe costs on generation D, then its situation may be so dire that it is forced to take action that will harm generation F—such as emitting even more greenhouse gases—that it would otherwise not need to consider. What I have in mind is this. Under some circumstances, actions that harm innocent others may be morally permissible on grounds of self-defense, and such circumstances may arise in the climate-change case.[32] Hence the claim is that if there is a self-defense exception on the prohibition on harming innocent others, one way in which generation A might behave badly is by creating a situation such that generation D is forced to call on the self-defense exception and so inflict extra suffering on generation F.[33] Moreover, like the basic PIP, this problem can become iterated: perhaps generation F must call on the self-defense exception, too, and so inflict harm on generation H, and so on.

III. The Theoretical Storm

The final storm I want to mention is constituted by our current theoretical ineptitude. We are extremely ill equipped to deal with many problems characteristic of the long-term future. Even our best theories face basic and often severe difficulties addressing basic issues such as scientific uncertainty, intergenerational equity, contingent persons, nonhuman animals, and nature. But climate change involves all of these matters and more.[34]

I do not want to discuss any of these difficulties in any detail here. Instead, I want to point out how, when they converge with one another and with the global and intergenerational storms, they encourage a new and distinct problem for ethical action on climate change: moral corruption.

IV. Moral Corruption

Corruption of the kind I have in mind can be facilitated in a number of ways. Consider the following examples of possible strategies:

- Distraction
- Complacency
- Unreasonable doubt
- Selective attention
- Delusion
- Pandering
- False witness
- Hypocrisy

The mere listing of these strategies is probably enough to make the main point here, and I suspect that close observers of the political debate about climate change will recognize many of these mechanisms as being in play. Still, I would like to make a particular point about selective attention.

Since climate change involves a complex convergence of problems, it is easy to engage in *manipulative or self-deceptive behavior* by applying one's attention selectively, to only some of the considerations that make the situation difficult. At the level of practical politics, such strategies are all too familiar. For example, many political actors emphasize considerations that appear to make inaction excusable, or even desirable (such as uncertainty or simple economic calculations with high discount rates), and action more difficult and contentious (such the basic lifestyles issue) at the expense of those that seem to impose a clearer and more immediate burden (such as scientific consensus and the pure intergenerational problem).

But selective-attention strategies might also manifest themselves more generally. And this prompts a very unpleasant thought: perhaps there is a problem of corruption in the theoretical, as well as the practical, debate. For example, it is possible that the prominence of the global storm model is not independent of the existence of the intergenerational storm but, rather, is encouraged by it. After all, the current generation may find it highly advantageous to focus on the global storm. For one thing, such a focus tends to draw attention toward various issues of global politics and scientific uncertainty, which seem to problematize action, and away from issues of intergenerational ethics, which tend to demand it. Thus, an emphasis on the global storm at the expense of the other problems may *facilitate* a strategy of procrastination and delay. For another thing, since it assumes that the relevant actors are nation-states that represent the interests of their citizens in perpetuity, the global-storm analysis has the effect of assuming away the intergenerational aspect of the climate-change problem.[35] Thus, an undue emphasis on it may obscure much of what is at stake in making climate policy, and in a way that may benefit present people.[36]

In conclusion, the presence of the problem of moral corruption reveals another sense in which climate change may be a perfect moral storm. Its complexity may turn out to be *perfectly convenient* for us, the current generation, and indeed for each successor generation as it comes to occupy our position. For one thing, it provides each generation with the cover under which it can seem to be taking the issue seriously—by negotiating weak and largely substanceless global accords, for example, and then heralding them as great achievements[37]—when really it is simply exploiting its temporal position. For another thing, all of this can occur without the exploitative generation actually having to acknowledge that this is what it is doing. By avoiding overtly selfish behavior, an earlier generation can take advantage of the future without the unpleasantness of admitting it—either to others or, perhaps more important, to itself.

Acknowledgments

This chapter was originally written for presentation to an interdisciplinary workshop on "Values in Nature" at Princeton University, the proceedings of which appeared in *Environmental Values*. I thank the Center for Human Values at Princeton and the University of Washington for research support in the form of a Laurance S. Rockefeller fellowship. I also thank audiences at Iowa State University, Lewis and Clark College, the University of Washington, the Western Political Science Association, and the Pacific Division of the American Philosophical Association. For comments, I am particularly grateful to Chrisoula Andreou, Kristen Hessler, Jay Odenbaugh, John Meyer, Darrel Moellendorf, Peter Singer, Harlan Wilson, Clark Wolf, and two anonymous reviewers. I am especially indebted to Dale Jamieson.

Notes

1. Intergovernmental Panel on Climate Change (IPCC) 2001a, p. 2; emphasis added.

2. For more on such issues, see Gardiner 2004a.

3. One might wonder why, despite the widespread agreement that climate change involves important ethical questions, there is relatively little overt discussion of them. The answer to this question is no doubt complex. But my thesis might constitute part of the answer.

4. Junger 1997.

5. This definition is my own. The term *perfect storm* is in wide usage. However, it is difficult to find definitions of it. An online dictionary of slang offers the following: "When three events, usually beyond one's control, converge and create a large inconvenience for an individual. Each event represents one of the storms that collided on the *Andrea Gail* in the book/movie titled the perfect storm" (Urbandictionary.com).

6. An anonymous reviewer objects that this is a "very American take on the matter," since "the rest of the world" (a) "is less sure that there is an utter absence of effective global governance," (b) "might argue that were it not for recent U.S. resistance, a centralized system of governance might be said to at least be in the early stages of evolution," and (c) accepts "Kyoto as a reasonable first step toward global governance on climate change." Much might be said about this, but here I can make only three

quick points. First, suppose that (a), (b), and (c) are all true. Even so, their truth does not seem sufficient to undermine the global storm; the claims are just too weak. Second, if there is a system of effective governance, then the current weakness of the international response to climate change becomes more, rather than less, surprising, and this bolsters one of my main claims in this chapter, which is that other factors need to be taken into account. Third, elsewhere I have criticized Kyoto as too weak (Gardiner 2004b). Others have criticized me for being too pessimistic here, invoking the "first step" defense (e.g., DeSombre 2004). My response is to say that it is the critics who are the pessimists: they believe that Kyoto was the best that humanity could achieve at the time; I am more optimistic about our capabilities.

7. The appropriateness of this model even to the spatial dimension requires some further specific, but usually undefended, background assumptions about the precise nature of the dispersion of effects and fragmentation of agency. But I shall pass over that issue here.

8. Hardin 1968. I discuss this in more detail in previous work, especially Gardiner 2001.

9. Nothing depends on the case being of this form. For a fuller characterization, see Gardiner 2001.

10. Some will complain that such game-theoretical analyses are misguided in general and in any case are irrelevant to the *ethics* of international affairs, since they focus on self-interested motivation. Although a full discussion is not possible here, a couple of quick responses might be helpful. First, I believe that often the best way to make progress in solving a given ethical problem is to get clear on what the problem actually is. Game-theoretic analyses are sometimes helpful here (consider their popularity in the actual literature on environmental issues in general and climate change in particular). Second, my analysis need not assume that actual human individuals, states, or generations are exclusively self-interested or that their interests are exclusively economic. (In fact, I would reject such claims.) Instead, it can proceed on a much more limited set of assumptions. Suppose, for example, that the following were the case: first, the actual, unreflective *consumption behavior* of most agents is dominated by their *perceived self-interest*; second, this is often seen in rather narrow terms; and third, it is such behavior that drives much of the energy use in the industrialized countries and so much of the problem of climate change. If such claims are reasonable, then modeling the dynamics of the global-warming problem in terms of a simplifying assumption of self-interest is not seriously misleading. For the role of that assumption is simply to suggest (a) that *if nothing is done to prevent it*, unreflective consumption behavior will dominate individual, state, and generational behavior; (b) that this is likely to lead to tragedy; and so (c) that some kind of regulation of normal consumption patterns (whether individual, governmental, market-based, or of some other form) is necessary in order to avoid a moral disaster.

11. This implies that in the real world, commons problems do not, strictly speaking, satisfy all of the conditions of the prisoner's dilemma paradigm. For relevant discussion, see Shepski 2006 and Ostrom 1990.

12. There is one fortunate convergence. Several writers have emphasized that the major ethical arguments all point in the same direction: that the developed countries should bear most of the costs of the transition—including those accruing to developing countries—at least in the early stages of mitigation and adaptation. See, for example, Singer 2002 and Shue 1999.

13. Rado Dimitrov argues that we must distinguish among different kinds of uncertainty when we investigate the effects of scientific uncertainty on international regime building and that it is uncertainties about national impacts that undermine regime formation. See Dimitrov 2003.

14. This consideration appears to play a role in U.S. deliberation about climate change, where it is often asserted that the U.S. faces lower marginal costs from climate change than other countries. See, for example, Mendelsohn 2001, Nitze 1994, and, by contrast, National Assessment Synthesis Team 2000.

15. Much more might be said here. I discuss some of the psychological aspects of political inertia and the role they play independently of scientific uncertainty in Gardiner 2009b.

16. This is so both because a greater proportion of their economies are in climate-sensitive sectors and because—being poor—they are worse placed to deal with those impacts. See IPCC 2001b, pp. 8, 16.

17. Of course, it does not help that the climate-change problem arises in an unfortunate geopolitical setting. Current international relations occur against a backdrop of distraction, mistrust, and severe inequalities of power. The dominant global actor and lone superpower, the United States, refuses to address climate change and is in

any case distracted by the threat of global terrorism. Moreover, the international community, including many of America's historical allies, distrusts its motives, its actions, and especially its uses of moral rhetoric, so there is global discord. This unfortunate state of affairs is especially problematic in relation to the developing nations, whose cooperation must be secured if the climate-change problem is to be addressed. One issue is the credibility of the developed nations' commitment to solving the climate-change problem. (See below.) Another is the North's focus on mitigation to the exclusion of adaptation issues. A third concern is the South's fear of an "abate and switch" strategy on the part of the North. Note that considered in isolation, these factors do not seem sufficient to explain political inertia. After all, the climate-change problem originally became prominent during the 1990s, a decade with a much more promising geopolitical environment.

18. For more on both claims, see IPCC 2001a, pp. 16–17.

19. Archer 2005a, p. 5. One "kyr" is 1,000 years.

20. Archer 2005b.

21. Ibid.; a similar remark occurs in Archer 2005a, p. 5.

22. Wigley 2005; Meehl et al. 2005; Wetherald et al. 2001.

23. This is exacerbated by the fact that the climate is an inherently chaotic system in any case and that there is no control against which its performance might be compared.

24. The possibility of nonlinear effects, such as in abrupt climate change, complicates this point, but I do not think it undermines it. See Gardiner 2009b.

25. Elsewhere, I have argued that it is this background fact that most readily explains the weakness of the Kyoto deal. See Gardiner 2004b.

26. Generational overlap complicates the picture in some ways, but I do not think that it resolves the basic problem. See Gardiner 2003, 2009a.

27. These matters are discussed in more detail in Gardiner 2003, from which the following description is drawn.

28. Hansen and Soto 2004; Hansen 2006; graph based on Marland et al. 2005.

29. Hansen 2006, p. 9.

30. I owe this suggestion to Henry Shue.

31. See O'Neill and Oppenheimer 2002.

32. Traxler 2002, p. 107.

33. For a related case, see Shue 2005, pp. 275–276.

34. For some discussion of the problems faced by cost-benefit analysis in particular, see Broome 1992, Spash 2002, and Gardiner 2004a, 2006.

35. In particular, it conceives of the problem as one that self-interested motivation alone should be able to solve and where failure will result in self-inflicted harm. But the intergenerational analysis makes clear that these claims are not true: current actions will largely harm (innocent) future people, and this suggests that motivations that are not generation-relative must be called upon to protect them.

36. In particular, once one identifies the intergenerational storm, it becomes clear that any given generation confronts two versions of the tragedy of the commons. The first version assumes that nations represent the interests of their citizens in perpetuity and so is genuinely cross-generational, but the second assumes that nations predominantly represent the interests of their current citizens and so is merely intragenerational. The problem is, then, that the collectively rational solutions to these two commons problems may be—and very likely are—different. (For example, in the case of climate change, it is probable that the intragenerational problem calls for much less mitigation of greenhouse-gas emissions than the cross-generational problem.) So, we cannot take the fact that a particular generation is motivated to and engages in resolving one (the intragenerational tragedy) as evidence that they are interested in solving the other (the cross-generational version). See Gardiner 2004b.

37. Gardiner 2004b.

References

Archer, David. 2005a. "Fate of Fossil Fuel CO_2 in Geologic Time." *Journal of Geophysical Research* 110: C09S05. DOI: 10.1029/2004JC002625.

———. 2005b. "How Long Will Global Warming Last?" Available at http://www.realclimate.org/index.php/archives/2005/03/how-long-will-global-warming-last/#more-134.

Broome, John. 1992. *Counting the Cost of Global Warming*. Isle of Harris, U.K.: White Horse Press.

DeSombre, Elizabeth. 2004. "Global Warming: More Common than Tragic." *Ethics and International Affairs* 18: 41–46.

Dimitrov, R. 2003. "Knowledge, Power and Interests in Environmental Regime Formation." *International Studies Quarterly* 47: 123–150.

Gardiner, Stephen M. 2001. "The Real Tragedy of the Commons." *Philosophy and Public Affairs* 30: 387–416.

———. 2003. "The Pure Intergenerational Problem." *Monist* 86: 481–500.

———. 2004a. "Ethics and Global Climate Change." *Ethics* 114: 555–600. Also chapter 1 in this volume.

———. 2004b. "The Global Warming Tragedy and the Dangerous Illusion of the Kyoto Protocol." *Ethics and International Affairs* 18: 23–39.

———. 2006. "Protecting Future Generations." In *Handbook of Intergenerational Justice,* ed. Jörg Tremmel (Cheltenham, U.K.: Edgar Elgar), pp. 148–169.

———. 2009a. "A Contract on Future Generations?" In *Theories of Intergenerational Justice,* ed. Axel Gosseries and Lukas Meyer (Oxford: Oxford University Press), pp. 77–119.

———. 2009b. "Saved by Disaster? Abrupt Climate Change, Political Inertia, and the Possibility of an Intergenerational Arms Race." *Journal of Social Philosophy* 40(2): 140–162. Special Issue on Global Environmental Issues, edited by Tim Hayward and Carol Gould.

Hansen, James. 2006. "Can We Still Avoid Dangerous Human-made Climate Change?" Talk presented at the New School University, New York, February.

Hansen, James, and Makiko Sato. 2004. "Greenhouse Gas Growth Rates." *Proceedings of the National Academy of Sciences* 101.46: 16109–16114.

Hardin, Garrett. 1968. "Tragedy of the Commons." *Science* 162: 1243–1248.

Intergovernmental Panel on Climate Change (IPCC). 2001a. *Climate Change 2001: Synthesis Report.* Cambridge, U.K.: Cambridge University Press. Available at http.//www.ipcc.ch.

———. 2001b. "Summary for Policymakers." *Climate Change 2001: Impacts, Adaptation, and Vulnerability.* Cambridge, U.K.: Cambridge University Press. Available at http.//www.ipcc.ch.

Junger, Sebastian. 1997. *A Perfect Storm: A True Story of Men Against the Sea.* New York: Norton.

Marland, G., T. Boden, and R. J. Andreas. 2005. "Global CO_2 Emissions from Fossil-Fuel Burning, Cement Manufacture, and Gas Flaring: 1751–2002." Carbon Dioxide Information Analysis Center, U.S. Department of Energy. Available at: http://cdiac.ornl.gov/trends/emis/glo.htm.

Meehl, Gerald, Warren M. Washington, William D. Collins, Julie M. Arblaster, Aixue Hu, Lawrence E. Buja, Warren G. Strand, and Haiyan Teng. 2005. "How Much More Global Warming and Sea Level Rise?" *Science* 307: 1769–1772.

Mendelsohn, Robert O. 2001. *Global Warming and the American Economy.* London: Edward Elgar.

National Assessment Synthesis Team. 2000. *Climate Change Impacts on the United States: The Potential Consequences of Climate Variability and Change.* Cambridge, U.K.: Cambridge University Press. Available at www.usgcrp.gov/usgcrp/nacc/default.htm.

Nitze, W. A. 1994. "A Failure of Presidential Leadership." In *Negotiating Climate Change: The Inside Story of the Rio Convention,* ed. Irving Mintzer and J. Amber Leonard (Cambridge, U.K.: Cambridge University Press), pp. 189–190.

O'Neill, Brian C., and Oppenheimer, Michael. 2002. "Dangerous Climate Impacts and the Kyoto Protocol." *Science* 296: 1971–1972.

Ostrom, Elinor. 1990. *Governing the Commons.* New York: Cambridge University Press.

Samuelson, Robert J. 2005. "Lots of Gain and No Pain!" *Newsweek* (February 21): 41.

Shepski, Lee. 2006. "Prisoner's Dilemma: The Hard Problem." Paper presented at the Pacific Division of the American Philosophical Association, San Francisco, March.

Shue, Henry. 1999. "Global Environment and International Inequality." *International Affairs* 75: 531–545. Also chapter 5 in this volume.

———. 2005. "Responsibility of Future Generations and the Technological Transition." In *Perspectives on Climate Change: Science, Economics, Politics, Ethics,* ed. Walter Sinnott-Armstrong and Richard Howarth (Amsterdam: Elsevier), pp. 265–284.

Singer, Peter. 2002. "One Atmosphere." In *One World: The Ethics of Globalization,* by Peter Singer (New Haven, Conn.: Yale University Press), chap. 2.

Spash, Clive L. 2002. *Greenhouse Economics: Value and Ethics.* London: Routledge.

Traxler, Martino. 2002. "Fair Chore Division for Climate Change." *Social Theory and Practice* 28: 101–134.

Wetherald, Richard T., Ronald J. Stouffer, and Keith W. Dixon 2001. "Committed Warming and Its Implications for Climate Change." *Geophysical Research Letters* 28.8: 1535–1538.

Wigley, T. M. L. 2005. "The Climate Change Commitment." *Science* 307: 1766–1769.

Part III

Global Justice and Future Generations

5

Global Environment and International Inequality

Henry Shue

My aim is to establish that three common-sense principles of fairness, none of them dependent upon controversial philosophical theories of justice, give rise to the same conclusion about the allocation of the costs of protecting the environment.

Poor states and rich states have long dealt with each other primarily upon unequal terms. The imposition of unequal terms has been relatively easy for the rich states because they have rarely needed to ask for the voluntary cooperation of the less powerful poor states. Now the rich countries have realized that their own industrial activity has been destroying the ozone in the earth's atmosphere and has been making far and away the greatest contribution to global warming. They would like the poor states to avoid adopting the same form of industrialization by which they themselves became rich. It is increasingly clear that if poor states pursue their own economic development with the same disregard for the natural environment and the economic welfare of other states that rich states displayed in the past during their development, everyone will continue to suffer the effects of environmental destruction. Consequently, it is at least conceivable that rich states might now be willing to consider dealing cooperatively on equitable terms with poor states in a manner that gives due weight to both the economic development of poor states and the preservation of the natural environment.

If we are to have any hope of pursuing equitable cooperation, we must try to arrive at a consensus about what equity means. And we need to define equity not as a vague abstraction but concretely and specifically in the context of both development of the economy in poor states and preservation of the environment everywhere.

Fundamental Fairness and Acceptable Inequality

What diplomats and lawyers call equity incorporates important aspects of what ordinary people everywhere call fairness. The concept of fairness is neither Eastern nor Western, Northern nor Southern, but universal.[1] People everywhere understand what it means to ask whether an arrangement is fair or biased toward some parties over other parties. If you own the land but I supply the labor, or you own the seed but I own the ox, or you are old but I am young, or you are female but I am male, or you have an education and I do not, or you worked long and hard but I was lazy—in situation after situation it makes perfectly good sense to ask whether a particular division of something among two or more parties is fair to all the parties, in light of this or that difference between them. All people understand the question, even where they have been taught not to ask it. What would be fair? Or, as the lawyers and diplomats would put it, which arrangement would be equitable?

Naturally, it is also possible to ask other kinds of questions about the same arrangements. One can always ask economic questions, for instance, in addition to ethical questions concerning equity. Would it increase total output if, say, women were paid less and men were paid more? Would it be more efficient? Sometimes the most efficient arrangement happens also to be fair to all parties, but often it is unfair. Then a choice has to be made between efficiency and fairness. Before it is possible to discuss such choices, however, we need to know the meaning of equity. What are the standards of equity, and how do they matter?

Complete egalitarianism—the belief that all good things ought to be shared equally among all people—can be a powerfully attractive view, and it is much more difficult to argue against than many of its opponents seem to think. I shall, nevertheless, assume here that complete egalitarianism is unacceptable. If it were the appropriate view to adopt, our inquiry into equity could end now. The answer to the question "what is an equitable arrangement?" would always be the same: an equal distribution. Only equality would ever provide equity.

While I do assume that it may be equitable for some good things to be distributed unequally, I also assume that other things must be kept equal—most important, dignity and respect. It is part of the current international consensus that every person is entitled to equal dignity and equal respect. In traditional societies in both hemispheres, even the equality of dignity and respect was denied in theory as well as practice. Now, although principles of equality are still widely violated in practice, inequality of dignity and of respect have relatively few public advocates even among those who practice them. If it is equitable for some other human goods to be distributed unequally, but it is not equitable for dignity or respect to be unequal, the central questions become "Which inequalities in which other human goods are compatible with equal human dignity and equal human respect?" and "Which inequalities in other goods ought to be eliminated, reduced, or prevented from being increased?"

When one is beginning from an existing inequality, like the current inequality in wealth between North and South, three critical kinds of justification are: justifications of unequal burdens intended to reduce or eliminate the existing inequality by removing an unfair advantage of those at the top, justifications of unequal burdens intended to prevent the existing inequality from becoming worse through any infliction of an unfair additional disadvantage upon those at the bottom, and justifications of a guaranteed minimum intended to prevent the existing inequality from becoming worse through any infliction of an unfair additional disadvantage upon those at the bottom. The second justification for unequal burdens and the justification for a guaranted minimum are the same: two different mechanisms are being used to achieve fundamentally the same purpose. I shall look at these two forms of justification for unequal burdens and then at the justification for a guaranteed minimum.

Unequal Burdens

Greater Contribution to the Problem

All over the world parents teach their children to clean up their own mess. This simple rule makes good sense from the point of view of incentive: if one learns that one will not be allowed to get away with simply walking away from whatever messes one creates, one is given a strong negative incentive against making messes in the first place. Whoever makes the mess presumably does so in the process of pursuing some benefit—for a child, the benefit may simply be the pleasure of playing with the objects that constitute the mess. If one learns that whoever reaps the benefit of making the mess must also be the one who pays the cost of cleaning up the mess, one learns at the very least not to make messes with costs that are greater than their benefits.

Economists have glorified this simple rule as the "Internalization of externalities." If the basis for the price of a product does not incorporate the costs of cleaning up the mess made in the process of producing the product, the costs are being externalized, that is, dumped upon other parties. Incorporating into the basis of the price of the product the costs that had been coercively socialized is called internalizing an externality.

At least as important as the consideration of incentives, however, is the consideration of fairness or equity. If whoever makes a mess receives the benefits and does not pay the costs, not only does he have no incentive to avoid making as many messes as he likes, but he is also unfair to whoever does pay the costs. He is inflicting costs upon other people, contrary to their interests and, presumably, without their consent. By making himself better off in ways that make others worse off, he is creating an expanding inequality.

Once such an inequality has been created unilaterally by someone's imposing costs upon other people, we are justified in reversing the inequality by imposing extra burdens upon the producer of the inequality. There are two separate points here. First, we are justified in assigning additional burdens to the party who has been inflicting costs upon us. Second, the minimum extent of the compensatory burden we are justified in assigning is enough to correct the inequality previously unilaterally imposed. The purpose of the extra burden is to restore an equality that was disrupted unilaterally and arbitrarily (or to reduce an inequality that was enlarged unilaterally and arbitrarily). In order to accomplish that purpose, the extra burden assigned must be at least equal to the unfair advantage previously taken. This yields us our first principle of equity:

> When a party has in the past taken an unfair advantage of others by imposing costs upon them without their consent, those who have been unilaterally put at a disadvantage are entitled to demand that in the future the offending party shoulder burdens that are unequal at least to the extent of the unfair advantage previously taken, in order to restore equality.[2]

In the area of development and the environment, the clearest cases that fall under this first principle of equity are the partial destruction of the ozone layer and the initiation of global warming by the process of industrialization that has enriched the North but not the South. Unilateral initiatives by the so-called developed countries (DCs) have made them rich, while leaving the less developed countries (LDCs) poor. In the process the industrial activities and accompanying lifestyles of the DCs have inflicted major global damage upon the earth's atmosphere. Both kinds of damage are harmful to those who did not benefit from Northern industrialization as well as to those who did. Those societies whose activities have damaged the atmosphere ought, according to the first principle of equity, to bear sufficiently unequal burdens henceforth to correct the inequality that they have imposed. In this case, everyone is bearing costs—because the damage was universal—but the benefits have been overwhelmingly skewed toward those who have become rich in the process.

This principle of equity should be distinguished from the considerably weaker—because entirely forward-looking—"polluter pays principle" (PPP), which requires only that all future costs of pollution (in production or consumption) be henceforth internalized into prices. Even the OECD formally adopted the PPP in 1974, to govern relations among rich states.[3]

Spokespeople for the rich countries make at least three kinds of counter arguments to this first principle of equity. These are:

1. The LDCs have also benefited, it is said, from the enrichment of the DCs. Usually it is conceded that the industrial countries have benefited more than the nonindustrialized. Yet it is maintained that, for example, medicines and technologies made possible by the lifestyles of the rich countries have also reached the poor countries, bringing benefits that the poor countries could not have produced as soon for themselves.

Quite a bit of breath and ink have been spent in arguments over how much LDCs have benefited from the technologies and other advances made by the DCs, compared to the benefits enjoyed by the DCs themselves. Yet this dispute does not need to be settled in order to decide questions of equity. Whatever benefits LDCs have received, they have mostly been charged for. No doubt some improvements have been widespread. Yet, except for a relative trickle of aid, all transfers have been charged to the recipients, who have in fact been left with an enormous burden of debt, much of it incurred precisely in the effort to purchase the good things produced by industrialization.

Overall, poor countries have been charged for any benefits that they have received by someone in the rich countries, evening that account. Much greater additional benefits have gone to the rich countries themselves, including a major contribution to the very process of their becoming so much richer than the poor countries. Meanwhile, the environmental damage caused by the process has been incurred by everyone. The rich countries have profited to the extent of the excess of the benefits gained by them over the costs incurred by everyone through environmental damage done by them, and ought in future to bear extra burdens in dealing with the damage they have done.

2. Whatever environmental damage has been done, it is said, was unintentional. Now we know all sorts of things about CFCs and the ozone layer, and about carbon dioxide and the greenhouse effect, that no one dreamed of when CFCs were created or when industrialization fed with fossil fuels began. People cannot be held responsible, it is maintained, for harmful effects that they could not have foreseen. The philosopher Immanuel Kant is often quoted in the West for having said, "Ought presupposes can"—it can be true that one ought to have done something only if one actually could have done it. Therefore, it is allegedly not fair to hold people responsible for effects they could not have avoided because the effects could not have been predicted.

This objection rests upon a confusion between punishment and responsibility. It is not fair to punish someone for producing effects that could not have been avoided, but it is common to hold people responsible for effects that were unforeseen and unavoidable.

We noted earlier that in order to be justifiable, an inequality in something between two or more parties must be compatible with an equality of dignity and respect between the parties. If there were an inequality between two groups of people such that members of the first group could create problems and then expect members of the second group to deal with the problems, that inequality would be incompatible with equal respect and equal dignity. For the members of the second group would in fact be functioning as servants for the first group. If I said to you, "I broke it, but I want you to clean it up", then I would be your master and you would be my servant. If I thought that you should do my bidding, I could hardly respect you as my equal.

It is true, then, that the owners of many coal-burning factories could not possibly have known the bad effects of the carbon dioxide they were releasing into the atmosphere, and therefore could not possibly have intended to contribute to harming it. It would, therefore, be unfair to punish them—by, for example, demanding that they pay double or triple damages. It is not in the least unfair, however, simply to hold them responsible for the damage that they have in fact done. This naturally leads to the third objection.

3. Even if it is fair to hold a person responsible for damage done unintentionally, it will be said, it is not fair to hold the person responsible for damage he did not do himself. It would not be fair, for example, to hold a grandson responsible for damage done by his grandfather. Yet it is claimed this is exactly what is being done when the current generation is held respon-

sible for carbon dioxide emissions produced in the nineteenth century. Perhaps Europeans living today are responsible for atmosphere-damaging gases emitted today, but it is not fair to hold people responsible for deeds done long before they were born.

This objection appeals to a reasonable principle, namely that one person ought not to be held responsible for what is done by another person who is completely unrelated. "Completely unrelated" is, however, a critical portion of the principle. To assume that the facts about the industrial North's contribution to global warming straightforwardly fall under this principle is to assume that they are considerably simpler than they actually are.

First, and undeniably, the industrial states' contributions to global warming have continued unabated long since it became impossible to plead ignorance. It would have been conceivable that as soon as evidence began to accumulate that industrial activity was having a dangerous environmental effect, the industrial states would have adopted a conservative or even cautious policy of cutting back greenhouse-gas emissions or at least slowing their rate of increase. For the most part this has not happened.

Second, today's generation in the industrial states is far from completely unrelated to the earlier generations going back all the way to the beginning of the industrial revolution. What is the difference between being born in 1975 in Belgium and being born in 1975 in Bangladesh? Clearly, one of the most fundamental differences is that the Belgian infant is born into an industrial society and the Bangladeshi infant is not. Even the medical setting for the birth itself, not to mention the level of prenatal care available to the expectant mother, is almost certainly vastly more favorable for the Belgian than for the Bangladeshi. Childhood nutrition, educational opportunities, and lifelong standards of living are likely to differ enormously because of the difference between an industrialized and a nonindustrialized economy. In such respects current generations are, and future generations probably will be, continuing beneficiaries of earlier industrial activity.

Nothing is wrong with the principle invoked in the third objection. It is indeed not fair to hold someone responsible for what has been done by someone else. Yet that principle is largely irrelevant to the case at hand, because one generation of a rich industrial society is not unrelated to other generations past and future. All are participants in enduring economic structures. Benefits and costs, and rights and responsibilities, carry across generations.

We turn now to a second, quite different kind of justification of the same mechanism of assigning unequal burdens. This first justification has rested in part upon the unfairness of the existing inequality. The second justification neither assumes nor argues that the initial inequality is unfair.

Greater Ability to Pay

The second principle of equity is widely accepted as a requirement of simple fairness. It states:

> Among a number of parties, all of whom are bound to contribute to some common endeavor, the parties who have the most resources normally should contribute the most to the endeavor.

This principle of paying in accordance with ability to pay, if stated strictly, would specify what is often called a progressive rate of payment: insofar as a party's assets are greater, the rate at which the party should contribute to the enterprise in question also becomes greater. The progressivity can be strictly proportional—those with double the base amount of assets contribute at twice the rate at which those with the base amount contribute, those with triple the base amount of assets contribute at three times the rate at which those with the base amount contribute, and so on. More typically, the progressivity is not strictly proportional—the more a party has, the higher the rate at which it is expected to contribute, but the rate does not increase in strict proportion to increases in assets.

The general principle itself is sufficiently fundamental that it is not necessary, and perhaps not possible, to justify it by deriving it from considerations that are more fundamental still.

Nevertheless, it is possible to explain its appeal to some extent more fully. The basic appeal of payment in accordance with ability to pay as a principle of fairness is easiest to see by contrast with a flat rate of contribution, that is, the same rate of contribution by every party irrespective of different parties' differing assets. At first thought, the same rate for everyone seems obviously the fairest imaginable arrangement. What could possibly be fairer, one is initially inclined to think, than absolutely equal treatment for everyone? Surely, it seems, if everyone pays an equal rate, everyone is treated the same and therefore fairly? This, however, is an exceedingly abstract approach, which pays no attention at all to the actual concrete circumstances of the contributing parties. In addition, it focuses exclusively upon the contribution process and ignores the position in which, as a result of the process, the parties end up. Contribution according to ability to pay is much more sensitive both to concrete circumstance and to final outcome.

Suppose that Party A has 90 units of something, Party B has 30 units, and Party C has 9 units. In order to accomplish their missions, it is proposed that everyone should contribute at a flat rate of one-third. This may seem fair in that everyone is treated equally: the same rate is applied to everyone, regardless of circumstances. When it is considered that A's contribution will be 30 and B's will be 10, while C's will be only 3, the flat rate may appear more than fair to C, who contributes only one-tenth as much as A does. However, suppose that these units represent $100 per year in income and that where C lives it is possible to survive on $750 per year but on no less. If C must contribute 3 units—$300—he will fall below the minimum for survival. While the flat rate of one-third would require A to contribute far more ($3,000) than C, and B to contribute considerably more ($1,000) than C, both A (with $6,000 left) and B (with $2,000 left) would remain safely above subsistence level. A and B can afford to contribute at the rate of one-third because they are left with more than enough, while C is unable to contribute at that rate and survive.

While flat rates appear misleadingly fair in the abstract, they do so largely because they look at only the first part of the story and ignore how things turn out in the end. The great strength of progressive rates, by contrast, is that they tend to accommodate final outcomes and take account of whether the contributors can in fact afford their respective contributions.

A single objection is usually raised against progressive rates of contribution: disincentive effects. If those who have more are going to lose what they have at a greater rate than those who have less, the incentive to come to have more in the first place will, it is said, be much less than it would have been with a flat rate of contribution. Why should I take more risks, display more imagination, or expend more effort in order to gain more resources if the result will only be that whenever something must be paid for, I will have to contribute not merely a larger absolute amount (which would happen even with a flat rate) but a larger percentage? I might as well not be productive if much of anything extra I produce will be taken away from me, leaving me little better off than those who produced far less.

Three points need to be noticed regarding this objection. First, of course, being fair and providing incentives are two different matters, and there is certainly no guarantee in the abstract that whatever arrangement would provide the greatest incentives would also be fair.

Second, concerns about incentives often arise when it is assumed that maximum production and limitless growth are the best goals. It is increasingly clear that many current forms of production and growth are unsustainable and that the last thing we should do is to give people self-interested reasons to consume as many resources as they can, even where the resources are consumed productively. These issues cannot be settled in the abstract, either, but it is certainly an open question—and one that should be asked very seriously—whether in a particular situation it is desirable to stimulate people by means of incentives to maximum production. Sometimes it is desirable, and sometimes it is not. This is an issue about ends.

Third, there is a question about means. Assuming that it had been demonstrated that the best goal to have in a specific set of circumstances involved stimulating more production of something, one would then have to ask: how much incentive is needed to stimulate that much production? Those who are preoccupied with incentives often speculate groundlessly that unlimited incentives are virtually always required. Certainly, it is true that it is generally necessary to provide some additional incentive in order to stimulate additional production. Some people are altruistic and are therefore sometimes willing to contribute more to the welfare of others even if they do not thereby improve their own welfare. It would be completely unrealistic, however, to try to operate an economy on the assumption that people generally would produce more irrespective of whether doing so was in their own interest—they need instead to be provided with some incentive. However, some incentive does not mean unlimited incentive.

It is certainly not necessary to offer unlimited incentives in order to stimulate (limited) additional production by some people (and not others). Whether people respond or not depends upon individual personalities and individual circumstances. It is a factual matter, not something to be decreed in the abstract, how much incentive is enough: for these people in these circumstances to produce this much more, how much incentive is enough? What is clearly mistaken is the frequent assumption that nothing less than the maximum incentive is ever enough.

In conclusion, insofar as the objection based on disincentive effects is intended to be a decisive refutation of the second principle of equity, the objection fails. It is not always a mistake to offer less than the maximum possible incentive, even when the goal of thereby increasing production has itself been justified. There is no evidence that anything less than the maximum is even generally a mistake. Psychological effects must be determined case by case.

On the other hand, the objection based on disincentive effects may be intended—much more modestly—simply as a warning that one of the possible costs of restraining inequalities by means of progressive rates of contribution, in the effort of being fair, may (or may not) be a reduction in incentive effects. As a caution rather than a (failed) refutation, the objection points to one sensible consideration that needs to be taken into account when specifying which variation upon the general second principle of equity is the best version to adopt in a specific case. One would have to consider how much greater the incentive effect would be if the rate of contribution were less progressive, in light of how unfair the results of a less progressive rate would be.

This conclusion that disincentive effects deserve to be considered, although they are not always decisive, partly explains why the second principle of equity is stated not as an absolute but as a general principle. It says: "the parties who have the most resources *normally* should contribute the most"—not always but normally. One reason the rate of contribution might not be progressive, or might not be as progressive as possible, is the potential disincentive effects of more progressive rates. It would need to be shown case by case that an important goal was served by having some incentive and that the goal in question would not be served by the weaker incentive compatible with a more progressive rate of contribution.

We have so far examined two quite different kinds of justifications of unequal burdens: to reduce or eliminate an existing inequality by removing an unfair advantage of those at the top and to prevent the existing inequality from becoming worse through any infliction of an unfair additional disadvantage upon those at the bottom. The first justification rests in part upon explaining why the initial inequality is unfair and ought to be removed or reduced. The second justification applies irrespective of whether the initial inequality is fair. Now we turn to a different mechanism that—much more directly—serves the second purpose of avoiding making those who are already the worst-off yet worse off.

Guaranteed Minimum

We noted earlier that issues of equity or fairness can arise only if there is something that must be divided among different parties. The existence of the following circumstances can be taken as grounds for thinking that certain parties have a legitimate claim to some of the available resources: (a) the aggregate total of resources is sufficient for all parties to have more than enough; (b) some parties do in fact have more than enough, some of them much more than enough; and (c) other parties have less than enough. American philosopher Thomas Nagel has called such circumstances radical inequality.[4] Such an inequality is radical in part because the total of available resources is so great that there is no need to reduce the best-off people to anywhere near the minimum level in order to bring the worst-off people up to the minimum: the existing degree of inequality is utterly unnecessary and easily reduced, in light of the total resources already at hand. In other words, one could preserve considerable inequality—in order, for instance, to provide incentives, if incentives were needed for some important purpose—while arranging for those with less than enough to have at least enough.

Enough for what? The answer could of course be given in considerable detail, and some of the details would be controversial (and some, although not all, would vary across societies). The basic idea, however, is of enough for a decent chance for a reasonably healthy and active life of more or less normal length, barring tragic accidents and interventions. "Enough" means the essentials for at least a bit more than mere physical survival—for at least a distinctively human, if modest, life. For example, having enough means owning not merely clothing adequate for substantial protection against the elements but clothing adequate in appearance to avoid embarrassment, by local standards, when being seen in public, as Adam Smith noted.

In a situation of radical inequality—a situation with the three features outlined above—fairness demands that those people with less than enough for a decent human life be provided with enough. This yields the third principle of equity, which states:

> When some people have less than enough for a decent human life, other people have far more than enough, and the total resources available are so great that everyone could have at least enough without preventing some people from still retaining considerably more than others have, it is unfair not to guarantee everyone at least an adequate minimum.[5]

Clearly, provisions to guarantee an adequate minimum can be of many different kinds, and, concerning many of the choices, equity has little or nothing to say. The arrangements to provide the minimum can be local, regional, national, international, or, more likely, some complex mixture of all, with secondary arrangements at one level providing a backstop for primary arrangements at another level.[6] Similarly, particular arrangements might assign initial responsibility for maintaining the minimum to families or other intimate groups, to larger voluntary associations like religious groups, or to a state bureau. Consideration of equity might have no implications for many of the choices about arrangements, and some of the choices might vary among societies, provided the minimum was in fact guaranteed.

Children, it is worth emphasizing, are the main beneficiaries of this principle of equity. When a family drops below the minimum required to maintain all its members, the children are the most vulnerable. Even if the adults choose to allocate their own share of an insufficient supply to the children, it is still quite likely that the children will have less resistance to disease and less resilience in general. And of course, not all adults will sacrifice their own share to their children. Or, in quite a few cultures, adults will sacrifice on behalf of male children but not on behalf of female children. All in all, when essentials are scarce, the proportion of children dying is far greater than their proportion in the population, which in poorer countries is already high—in quite a few poor countries, more than half the population is under the age of 15.

One of the most common objections to this third principle of equity flows precisely from this point about the survival of children. It is what might be called the overpopulation objection. I consider this objection to be ethically outrageous and factually groundless, as explained elsewhere.[7]

The other most common objection is that while it may be only fair for each society to have a guaranteed minimum for its own members, it is not fair to expect members of one society to help to maintain a guarantee of a minimum for members of another society.[8] This objection sometimes rests on the assumption that state borders—national political boundaries—have so much moral significance that citizens of one state cannot be morally required, even by considerations of elemental fairness, to concern themselves with the welfare of citizens of a different political jurisdiction. A variation on this theme is the contention that across state political boundaries, moral mandates can only be negative requirements not to harm and cannot be positive requirements to help. I am unconvinced that, in general, state political borders and national citizenship are markers of such extraordinary and overriding moral significance. Whatever may be the case in general, this second objection is especially unpersuasive if raised on behalf of citizens of the industrialized wealthy states in the context of international cooperation to deal with environmental problems primarily caused by their own states and of greatest concern in the medium term to those states.

To help to maintain a guarantee of a minimum could mean either of two things: a weaker requirement (a) not to interfere with others' ability to maintain a minimum for themselves or a stronger requirement (b) to provide assistance to others in maintaining a minimum for themselves. If everyone has a general obligation, even towards strangers in other states and societies, not to inflict harm on other persons, the weaker requirement would follow, provided only that interfering with people's ability to maintain a minimum for themselves counted as a serious harm, as it certainly would seem to. Accordingly, persons with no other bonds to each other would still be obliged not to hinder the others' efforts to provide a minimum for themselves.

One could not, for example, demand as one of the terms of an agreement that someone make sacrifices that would leave the person without necessities. This means that any agreement to cooperate made between people having more than enough and people not having enough cannot justifiably require those who start out without enough to make any sacrifices. Those who lack essentials will still have to agree to act cooperatively, if there is in fact to be cooperation, but they should not bear the costs of even their own cooperation. Because a demand that those lacking essentials should make a sacrifice would harm them, making such a demand is unfair.

That (a), the weaker requirement, holds seems perfectly clear. When, if ever, would (b), the stronger requirement to provide assistance to others in maintaining a minimum for themselves, hold? Consider the case at hand. Wealthy states, which are wealthy in large part because they are operating industrial processes, ask the poor states, which are poor in large part because they have not industrialized, to cooperate in controlling the bad effects of these same industrial processes, like the destruction of atmospheric ozone and the creation of global warming. Assume that the citizens of the wealthy states have no general obligation, which holds prior to and independently of any agreement to work together on environmental problems, to contribute to the provision of a guaranteed minimum for the citizens of the poor states. The citizens of the poor states certainly have no general obligation, which holds prior to and independently of any agreement, to assist the wealthy states in dealing with the environmental problems that the wealthy states' own industrial processes are producing. It may ultimately be in the interest of the poor states to see ozone depletion and global warming stopped, but in the medium term the citizens of the poor states have far more urgent and serious problems—like lack of food, lack of clean drinking water, and lack of jobs to provide minimal support for themselves and their families. If the wealthy states say to the poor states, in effect, "Our most urgent request of you is that you act in ways that will avoid worsening the ozone depletion and global warming that we have started," the poor states could reasonably respond, "Our most urgent request of you is assistance in guaranteeing the fulfilment of the essential needs of our citizens."

In other words, if the wealthy have no general obligation to help the poor, the poor certainly have no general obligation to help the wealthy. If this assumed absence of general obligations means that matters are to be determined by national interest rather than international obligation, then surely the poor states are as fully at liberty to specify their own top priority as the wealthy states are. The poor states are under no general prior obligation to be helpful to the wealthy states in dealing with whatever happens to be the top priority of the wealthy states. This is all the more so as long as the wealthy states remain content to watch hundreds of thousands of children die each year in the poor states for lack of material necessities, which the total resources in the world could remedy many times over. If the wealthy states are content to allow radical inequalities to persist and worsen, it is difficult to see why the poor states should divert their attention from their own worst problems in order to help out with problems that for them are far less immediate and deadly. It is as if I am starving to death, and you want me to agree to stop searching for food and instead to help repair a leak in the roof of your house without your promising me any food. Why should I turn my attention away from my own more severe problem to your less severe one, when I have no guarantee that if I help you with your problem, you will help me with mine? If any arrangement would ever be unfair, that one would.

Radical human inequalities cannot be tolerated and ought to be eliminated, irrespective of whether their elimination involves the movement of resources across national political boundaries. Resources move across national boundaries all the time for all sorts of reasons. I have not argued here for this judgment about radical inequality, however.[9] The conclusion for which I have provided a rationale is even more compelling: when radical inequalities exist, it is unfair for people in states with far more than enough to expect people in states with less than enough to turn their attention away from their own problems in order to cooperate with the much better-off in solving their problems (and all the more unfair—in light of the first principle of equity—when the problems that concern the much better-off were created by the much better-off themselves in the very process of becoming as well off as they are). The least that those below the minimum can reasonably demand in reciprocity for their attention to the problems that concern the best-off is that their own most vital problems be attended to: that they be guaranteed means of fulfilling their minimum needs. Any lesser guarantee is too little to be fair, which is to say that any international agreement that attempts to leave radical inequality across national states untouched while asking effort from the worst-off to assist the best-off is grossly unfair.

Overview

I have emphasized that the reasons for the second and third principles of equity are fundamentally the same, namely, avoiding making those who are already the worst-off yet worse off. The second principle serves this end by requiring that when contributions must be made, they should be made more heavily by the better-off, irrespective of whether the existing inequality is justifiable. The third principle serves this end by requiring that no contributions be made by those below the minimum unless they are guaranteed ways to bring themselves up at least to the minimum, which assumes that radical inequalities are unjustified. Together, the second and third principles require that if any contributions to a common effort are to be expected of people whose minimum needs have not been guaranteed so far, guarantees must be provided; and the guarantees must be provided most heavily by the best-off.

The reason for the first principle was different from the reason for the second principle, in that the reason for the first rests on the assumption that an existing inequality is already unjustified. The reason for the third principle rests on the same assumption. The first and third principles apply, however, to inequalities that are, respectively, unjustified for different kinds of reasons. Inequalities to which the first principle applies are unjustified because of how they arose, namely, some people have been benefiting unfairly by dumping the costs of their own advances upon other people. Inequalities to which the third principle applies are unjustified independently of how they arose and simply because they are radical, that is, so extreme in circumstances in which it would be very easy to make them less extreme.

What stands out is that in spite of the different content of these three principles of equity, and in spite of the different kinds of grounds upon which they rest, they all converge upon the same practical conclusion: whatever needs to be done by wealthy industrialized states or by poor nonindustrialized states about global environmental problems like ozone destruction and global warming, the costs should initially be borne by the wealthy industrialized states.

Notes

1. Or so I believe. I would be intensely interested in any evidence of a culture that seems to lack a concept of fairness, as distinguished from evidence about two cultures whose specific conceptions of fairness differ in some respects.

2. A preliminary presentation of these principles at New York University Law School has been helpfully commented upon in Thomas M. Franck, *Fairness in International Law and Institutions* (Oxford: Clarendon Press, 1997), pp. 390–391.

3. OECD Council, November 14, 1974 (Paris: OECD, 1974), p. 223.

4. See Thomas Nagel, "Poverty and Food: Why Charity Is Not Enough," in Peter G. Brown and Henry Shue, eds., *Food Policy: The Responsibility of the United States in the Life and Death Choices* (New York: Free Press, 1977), pp. 54–62. In an important recent and synthetic discussion Thomas W. Pogge has suggested adding two further features to the characterization of a radical inequality, as well as a different view about its moral status—see Thomas W. Pogge, "A Global Resources Dividend," in David A. Crocker and Toby Linden, eds. *Ethics of Consumption: The Good Life, Justice and Global Stewardship* (Lanham, Md., and Oxford: Rowman & Littlefield, 1998), pp. 501–36. On radical inequality, see pp. 502–503.

5. This third principle of equity is closely related to what I called the argument from vital interests in Henry Shue, "The Unavoidability of Justice," in Andrew Hurrell and Benedict Kingsbury, eds., *The International Politics of the Environment* (Oxford: Oxford University Press, 1992), pp. 373–397. It is the satisfaction of vital interests that constitutes the minimum everyone needs to have guaranteed. In the formulation here the connection with limits on inequality is made explicit.

6. On the importance of backstop arrangements, or the allocation of default duties, see "Afterword," in Henry Shue, *Basic Rights: Subsistence, Affluence, and US Foreign Policy*, 2d ed. (Princeton, N.J.: Princeton University Press, 1996).

7. Shue, *Basic Rights*, chap. 4.

8. This objection has recently been provided with a powerful and sophisticated Kantian formulation that deserves much more attention than space here allows—see Richard W. Miller, "Cosmopolitan Respect and Patriotic Concern", *Philosophy & Public Affairs* 27.3 (Summer 1998): 202–224.

9. And for the argument to the contrary, see Miller, "Cosmopolitan Respect and Patriotic Concern."

Energy Policy and the Further Future

The Identity Problem

Derek Parfit

I have assumed that our acts may have good or bad effects in the further future.[1] Let us now examine this assumption. Consider first:

> *The Nuclear Technician:* Some technician lazily chooses not to check some tank in which nuclear wastes are buried. As a result there is a catastrophe two centuries later. Leaked radiation kills and injures thousands of people.

We can plausibly assume that, whether or not this technician checks this tank, the same particular people would be born during the next two centuries. If he had chosen to check the tank, these same people would have later lived and escaped the catastrophe.

Is it morally relevant that the people whom this technician harms do not yet exist when he makes his choice? I have assumed here that it is not. If we know that some choice either may or will harm future people, this is an objection to this choice even if the people harmed do not yet exist. (I am to blame if I leave a man-trap on my land, which ten years later maims a five-year-old child.)

Consider next:

> *The Risky Policy:* Suppose that, as a community, we have a choice between two energy policies. Both would be completely safe for at least two centuries, but one would have certain risks for the further future. If we choose the Risky Policy, the standard of living would be somewhat higher over the next two centuries. We do choose this policy. As a result there is a similar catastrophe two centuries later, which kills and injures thousands of people.

Unlike the Nuclear Technician's choice, our choice between these policies affects who will be later born. This is not obvious but is on reflection clear.

Our identity in fact depends partly on when we are conceived. This is so on both the main views about this subject. Consider some particular person, such as yourself. You are the *n*th child of your mother, and you were conceived at time *t*. According to one view, you could not have grown from a different pair of cells. If your mother had conceived her *n*th child some months earlier or later, that child would *in fact* have grown from a different pair of cells, and so would not have been you.

According to the other main view, you could have grown from different cells, or even had different parents. This would have happened if your actual parents had not conceived a child when they in fact conceived you, and some other couple had conceived an extra child who was sufficiently *like* you, or whose life turned out to be sufficiently like yours. On this other view, that child would have been you. (Suppose that Plato's actual parents never had children and that some other ancient Greek couple had a child who wrote *The Republic, The Last Days of Socrates*, and so on. On this other view, this child would have been Plato.) Those who take this other view, while believing that you *could* have grown from a different pair of cells, would admit that this would not *in fact* have happened. On both views, it is in fact true that, if your mother had conceived her *n*th child in a different month, that child would not have been you, and *you* would never have existed.

It may help to shift to this example. A 14-year-old girl decides to have a child. We try to change her mind. We first try to persuade her that if she has a child now, that will be worse for her. She says that even if it will be, that is her affair. We then claim that if she has a child now; that will be worse for her child. If she waits until she is grown up, she will be a better mother and will be able to give her child a better start in life.

Suppose that this 14-year-old rejects our advice. She has a child now and gives him a poor start in life. Was our claim correct? Would it have been better for him if she had taken our advice? If she had, *he* would never have been born. So her decision was worse for him only if it is against his interests to have been born. Even if this makes sense, it would be true only if his life was so wretched as to be worse than nothing. Assume that this is not so. We must then admit that our claim was false. We may still believe that this girl should have waited. That would have been better for her, and the different child she would have had later would have received a better start in life. But we cannot claim that, in having *this* child, what she did was worse for *him*.

Return now to the choice between our two energy policies. If we choose the Risky Policy, the standard of living will be slightly higher over the next two centuries. This effect implies another. It is not true that whichever policy we choose, the same particular people will exist two centuries later. Given the effects of two such policies on the details of our lives, it would increasingly over time be true that people married different people. More simply, even in the same marriages, the children would increasingly be conceived at different times. (Thus the British Miners' Strike of 1974, which caused television to close down an hour early and thereby affected the timing of thousands of conceptions.) As we have seen, children conceived at different times would in fact be different children. So the proportion of those later born who would owe their existence to our choice would, like ripples in a pool, steadily grow. We can plausibly assume that after two centuries, there would no one living who would have been born whichever policy we chose. (It may help to think of this example: how many of us could truly claim, "Even if railways had never been invented, I would still have been born"?)

In my imagined case, we choose the Risky Policy. As a result, two centuries later, thousands of people are killed and injured. But if we had chosen the alternative Safe Policy, these particular people would never have existed. Different people would have existed in their place. Is our choice of the Risky Policy worse for anyone?

We can first ask, "Could a life be so bad—so diseased and deprived—that it would not be worth living? Could a life be even worse than this? Could it be worse than nothing, or as we might say worth *not* living" We need not answer this question. We can suppose that whether or not lives could be worth not living, this would not be true of the lives of the people killed in the catastrophe. These people's lives would be well worth living. And we can suppose the same of those who mourn for those killed and those whom the catastrophe disables. (Perhaps for some of those who suffer most, the rest of their lives would be worth not living. But this would not be true of their lives as a whole.)

We can next ask, "If we cause someone to exist, who will have a life worth living, do we thereby benefit this person?" This is a difficult question. Call it the question whether *causing to exist can benefit*. Since the question is so difficult, I shall discuss the implications of both answers.

Because we chose the Risky Policy, thousands of people are later killed or injured or bereaved. But if we had chosen the Safe Policy, these particular people would never have existed. Suppose we do *not* believe that causing to exist can benefit. We should ask, "If particular people live lives that are on the whole well worth living, even though they are struck by some catastrophe, is this worse for these people than if they had never existed?" Our answer must be "no." If we believe that causing to exist *can* benefit, we can say more. Since the people struck by the catastrophe live lives that are well worth living and would never have existed if we had chosen the Safe Policy, our choice of the Risky Policy is not only not worse for these people, it *benefits* them.

Let us now compare our two examples. The Nuclear Technician chooses not to check some tank. We choose the Risky Policy. Both of these choices predictably cause catastrophes, which harm thousands of people. These predictable effects both seem bad, providing at least some moral objection to these choices. In the case of the technician, the objection is obvious. His choice is worse for the people who are later harmed. But this is not true of our choice of the Risky Policy. Moreover, when we understand this case, we know that this is not true. We know that even though our choice may cause such a catastrophe, it will not be worse for anyone who ever lives.

Does this make a moral difference? There are three views. It might make all the difference, or some difference, or no difference. There might be no objection to our choice, or some objection, or the objection may be just as strong.

Some claim:

> *Wrongs Require Victims:* Our choice cannot be wrong if we know that it will be worse for no one.

This claim implies that there is no objection to our choice. We may find it hard to deny this claim or to accept this implication.

I deny that wrongs require victims. If we know that we may cause such a catastrophe, I am sure that there is at least some moral objection to our choice. I am inclined to believe that the objection is just as strong as it would have been if, as in the case of the Nuclear Technician, our choice would be worse for future people. If this is so, it is morally irrelevant that our choice will be worse for no one. This may have important theoretical implications.

Before we pursue the question, it will help to introduce two more examples. We must continue to assume that some people can be worse off than others, in morally significant ways, and by more or less. But we need not assume that these comparisons could be even in principle precise. There may be only rough or partial comparability. By "worse off" we need not mean "less happy." We could be thinking, more narrowly, of the standard of living or, more broadly, of the quality of life. Since it is the vaguer, I shall use the phrase "the quality of life." And I shall extend the ordinary use of the phrase "worth living." If one of two groups of people would have a lower quality of life, I shall call their lives to this extent "less worth living." Here is another example:

> *Depletion*: Suppose that, as a community, we must choose whether to deplete or conserve certain kinds of resources. If we choose Depletion, the quality of life over the next two centuries would be slightly higher than it would have been if we had chosen Conservation, but it may later be much lower. Life at this much lower level would, however, still be well worth living. The effects might be shown as in figure 6.1.

This case raises the same problem. If we choose Depletion rather than Conservation, this will lower the quality of life more than two centuries from now. But the particular people

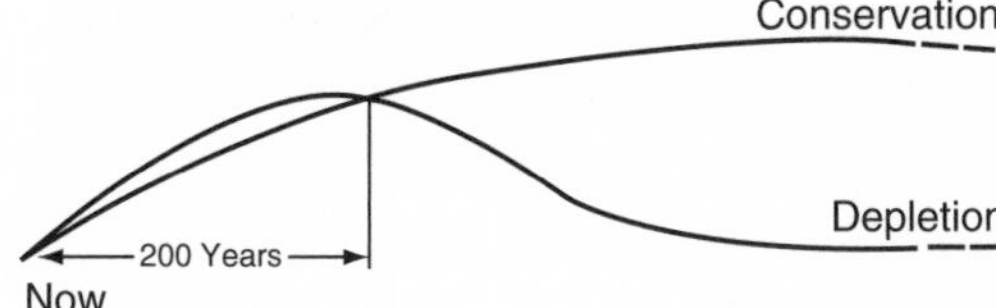

Figure 6.1. Effects of choice on future standard of Living.

who will then be living would never have existed if instead we had chosen Conservation. So our choice of Depletion is not worse for any of these people. But our choice will cause these people to be worse off than the different people who, if we had chosen Conservation, would have later lived. This seems a bad effect and an objection to our choice, even though it will be worse for no one.

Would the effect be *worse*, having greater moral weight, if it *was* worse for people? One test of our intuitions may be this. We may remember a time when we were concerned about effects on future generations but had overlooked my point about personal identity. We may have thought that a policy like Depletion would be against the interests of future people. When we saw that this was false, did we become less concerned about effects on future generations?

I myself did not. But it may help to introduce a different example. Suppose there are two rare conditions X and Y, which cannot be detected without special tests. If a pregnant woman has condition X, this will give to the child she is carrying a certain handicap. A simple treatment would prevent this effect. If a woman has condition Y when she becomes pregnant, this will give to the child she conceives the same particular handicap. Condition Y cannot be treated but always disappears within two months. Suppose next that we have planned two medical programs, but there are funds for only one, so one must be canceled. In the first program, millions of women would be tested during pregnancy. Those found to have condition X would be treated. In the second program, millions of women would be tested when they intend to try to become pregnant. Those found to have condition Y would be warned to postpone conception for at least two months. We are able to predict that these two programs would achieve results in as many cases. If there is Pregnancy Testing, 1,000 children a year would be born normal rather than handicapped. If there is Preconception Testing, there would each year be born 1,000 normal children, rather than 1,000 different handicapped children. Would these two programs be equally worthwhile?

Let us note carefully what the difference is. As a result of either program, 1,000 couples a year would have a normal rather than a handicapped child. These would be different couples, on the two programs. But since the numbers would be the same, the effects on parents and on other people would be morally equivalent. The only difference lies in the effects on the children. Note next that in judging these effects, we need have no view about the moral status of a fetus. We can suppose that it would take a year before either kind of testing could begin. When we choose between the two programs, none of the children has yet been conceived. And all of the children will become adults. So we are considering effects not on present fetuses but on future people. Assume next that the handicap in question, though it is not trivial, is not so severe as to make life doubtfully worth living. Even if it can be against our interests to have been born, this would not be true of those born with this handicap.

Since we cannot afford both programs, which should we cancel? Under one description, both would have the same effects. Suppose that conditions X and Y are the only causes of this handicap. The incidence is now 2,000 a year. Either program would halve the incidence; the rate would drop to 1,000 a year. The difference is this. If we decide to cancel Pregnancy Testing, those who are later born handicapped would be able to claim, "But for your decision, I would have been normal." Our decision will be worse for all these people. If instead we decide to cancel Preconception Testing, there will later be just as many people who are born with this handicap. But none of these could truly claim, "But for your decision, I would have been normal." But for our decision, they would never have existed; their parents would have later had different children. Since their lives, though handicapped, are still worth living, our decision will not be worse for any of these people.

Does this make a moral difference? Or are the two programs equally worthwhile? Is all that matters morally how many future lives will be normal rather than handicapped? Or does it also matter whether these lives would be lived by the very same people?

I am inclined to judge these programs equally worthwhile. If Preconception Testing would achieve results in a few more cases, I would judge it the better program. This matches my reactions to the questions asked above about our choice of the Risky Policy or of Depletion. There, too, I think it would be bad if there would later be a catastrophe, killing and injuring thousands of people, and bad if there would later be a lower quality of life. And I think that it would not be *worse* if the people who later live would themselves have existed if we had chosen the Safe Policy or Conservation. The bad effects would not be worse if they had been, in this way, worse for any particular people.

Let us review the argument so far. If we choose the Risky Policy or Depletion, this may later cause a predictable catastrophe or a decline in the quality of life. We naturally assume that these would be bad effects, which provide some objection to these two choices. Many think the objection is that our choices will be worse for future people. We have seen that this is false. But does this make a moral difference? There are three possible answers. It might make all the difference, or some difference, or no difference at all. When we see that our choice will be worse for no one, we may decide that there is no objection to this choice, or that there is less objection, or that the objection is just as strong.

I incline to the third answer. And I give this answer in the case of the medical programs. But I know some people who do not share my intuitions. How can we resolve this disagreement? Is there some familiar principle to which we can appeal?

Return to the choice of the Risky Policy, which may cause a catastrophe, harming thousands of people. It may seem irrelevant here that our choice will not be worse for these future people. Can we not deserve blame for causing harm to others, even when our act is not worse for them? Suppose that I choose to drive when drunk and in the resulting crash cause you to lose a leg. One year later, war breaks out. If you had not lost this leg, you would have been conscripted and been killed. So my drunken driving saves your life. But I am still morally to blame.

This case reminds us that in assigning blame, we must consider not actual but predictable effects. I knew that my drunken driving might injure others, but I could not know that it would in fact save your life. This distinction might apply to the choice between our two policies. We know that our choice of the Risky Policy may impose harm on future people. Suppose next that we have overlooked the point about personal identity. We mistakenly believe that whichever policy we choose, the very same people will later live. We may therefore believe that, if we choose the Risky Policy, this may be worse for future people. If we believe this, our choice can be criticized. We can deserve blame for doing what we *believe* may be worse for others. This criticism stands even if our belief is false—just as I am as much to blame even if my drunken driving will in fact save your life.

Now suppose, however, that we have seen the point about personal identity. We realize that if we choose the Risky Policy, our choice will *not* be worse for those people whom it later harms. Note that this is not a lucky guess. It is not like predicting that if I cause you to lose a leg, that will later save you from death in the trenches. We know that if we choose the Risky Policy, this may impose harms on several future people. But we also know that if we had chosen the Safe Policy, those particular people would never have been born. Since their lives will be worth living, we *know* that our choice will not be worse for them.

If we know this, we cannot be compared to a drunken driver. So how should we be criticized? Can we deserve blame for causing others to be harmed, even when we know that our act will not be worse for them? Suppose we know that the harm we cause will be fully compensated by some benefit. For us to be sure of this, the benefit must clearly outweigh the harm. Consider a surgeon who saves you from blindness, at the cost of giving you a facial scar. In scarring you, this surgeon does you harm. But he knows that his act is not worse for you. Is this enough to justify his decision? Not quite. He must not be infringing your autonomy. But this does not require that you give consent. Suppose that you are unconscious, so that he is forced to choose without consulting you. If he decides to operate, he would here deserve no blame. Though he scars your face, his act is justified. It is enough for him to know that his act will not be worse for you.

If we choose the Risky Policy, this may cause harm to many people. Since these will be future people, whom we cannot now consult, we are not infringing their autonomy. And we know that our choice will not be worse for them. Have we shown that, in the same way, the objection has been met?

The case of the surgeon shows only that the objection might be met. The choice of the Risky Policy has two special features. Why is the surgeon's act not worse for you? Because it gives you a compensating benefit. Though he scars your face, he saves you from going blind. Why is our choice of the Risky Policy not worse for those future people? Because they will owe their existence to this choice. Is this a compensating benefit? This is a difficult question. But suppose that we answer "no." Suppose we believe that to receive life, even a life worth living, is not to be benefited.[2] There is then a special reason for why, if we choose the Risky Policy, this will not be worse for the people who will later live.

Here is the second special feature. If we had chosen the Safe Policy, different people would have later lived. Let us first set aside this feature. Let us consider only the people who, given our actual choice, will in fact later live. These will be the only actual people whom our choice affects. Should the objection to our choice appeal to the effects on these people? Because of our choice, they will later suffer certain harms. This seems to provide an objection. But they owe their existence to this same choice. Does this remove the objection?

Consider a second case involving a 14-year-old girl. If this second girl has a child now, she will give him a poor start in life. But suppose she knows that because she has some illness, she will become sterile within the next year. Unless she has a child now, she can never have a child. Suppose that this girl chooses to have a child. Can she be criticized? She gives her child a poor start in life. But she could not have given *him* a better start in life, and his life will still be worth living. The effects on him do not seem to provide an objection. Suppose that she could also reasonably assume that if she has this child, this would not be worse for other people. It would then seem that there is no objection to this girl's choice—not even one that is overridden by her right to have a child.

Now return to our earlier case of a 14-year-old girl. Like the second girl, the first girl knows that if she has a child now, she will give him a poor start in life. But she could wait for several years and have another child, who would have a better start in life. She decides not to wait and has a child now. If we consider the effects only on her actual child, they are just like those of the second girl's choice. But the first girl's choice surely can be criticized. The two choices differ not in their effects on the actual children but in the alternatives. How could the second girl avoid having a child to whom she would give a poor start in life? Only by never having a child. That is why her choice seemed not to be open to criticism. She could reasonably assume that her choice would not be worse either for her actual child or for other people. In her case, that seems all we need to know. The first girl's choice has the same effects on her actual child and on others. But *this* girl could have waited and given some later child a better start in life. This is the objection to her choice. Her actual child is worse off than some later child would have been.

Return new to the choice between our two social policies. Suppose that we have chosen the Risky Policy. As a result, those who later live suffer certain harms. Is this enough to make our choice open to criticism? I suggest not. Those who later live are like the actual children of the two girls. They owe their existence to our choice, so its effects are not worse for them. The objection must appeal to the alternative.

This restores the second feature that we set aside above. When we chose the Risky Policy, we imposed certain harms on our remote descendants. Were we like the second girl, whose only alternative was to have no descendants? If so, we could not be criticized. But this is not the right comparison. In choosing the Risky Policy, we were like the first girl. If we had chosen the Safe Policy, we would have had different descendants, who would not have suffered such harms.

The objection to our choice cannot appeal only to effects on those people who will later live. It must mention possible effects on the people who, if we had chosen otherwise, would have later lived. The objection must appeal to a claim like this:

> (A) It is bad if those who live are worse off than those who might have lived.

We must claim that this is bad even though it will be worse for no one.

(A) is not a familiar principle. So we have not solved the problem that we reached above. Let us remember what that was. If we choose the Risky Policy or Depletion, this may later cause a catastrophe or a decline in the quality of life. These seemed bad effects. Many writers claim that in causing such effects, we would be acting against the interests of future people. Given the point about personal identity, this is not true. But I was inclined to think that this made no moral difference. The objection to these two choices seemed to me just as strong. Several people do not share my intuitions. Some believe that the objections must be weaker. Others believe that they disappear. On their view, our choice cannot be morally criticized if we know that it will be worse for no one. They believe that, as moral agents, we need only be concerned with the effects of our acts on all of the people who are ever actual. We need not consider people who are merely possible—those who never do live but merely might have lived. On this view, the point about identity makes a great moral difference. The effects of our two choices, the predictable catastrophe and the decline in the quality of life, can be morally totally ignored.

We hoped to resolve this disagreement by appeal to a familiar principle. I suggest now that this cannot be done. To criticize our choice, we must appeal to a claim like (A). And we have yet to explain why (A) should have any weight. To those who reject (A), we do not yet have an adequate reply.

To explain (A) and decide its weight, we would need to go deep into moral theory. And we would need to consider cases where in the different outcomes of our acts or policies, different numbers of people would exist. This is much too large a task to be attempted here.

I shall therefore end with a practical question. When we are discussing social policies, should we ignore the point about personal identity? Should we allow ourselves to say that a choice like that of the Risky Policy or of Depletion might be against the interests of people in the further future? This is not true. Should we pretend that it is? Should we let other people go on thinking that it is?

If you share my intuitions, this seems permissible. We can then use such claims as a convenient form of shorthand. Though the claims are false, we believe that this makes no moral difference. So the claims are not seriously misleading.

Suppose instead that you do not share my intuitions. You believe that if our choice of Depletion would be worse for no one, this must make a moral difference. It would then be dishonest to conceal the point about identity. But this is what, with your intuitions, I would be tempted to do. I would not *want* people to conclude that we can be less concerned about the more remote effects of our social policies. So I would be tempted to suppress the argument for this conclusion.

Theoretical Footnote

How might the attempt to justify claim (A) take us far into moral theory? Here are some brief remarks. Consider any choice between two outcomes. Figure 6.2 shows that there are three kinds of choice. These can be distinguished if

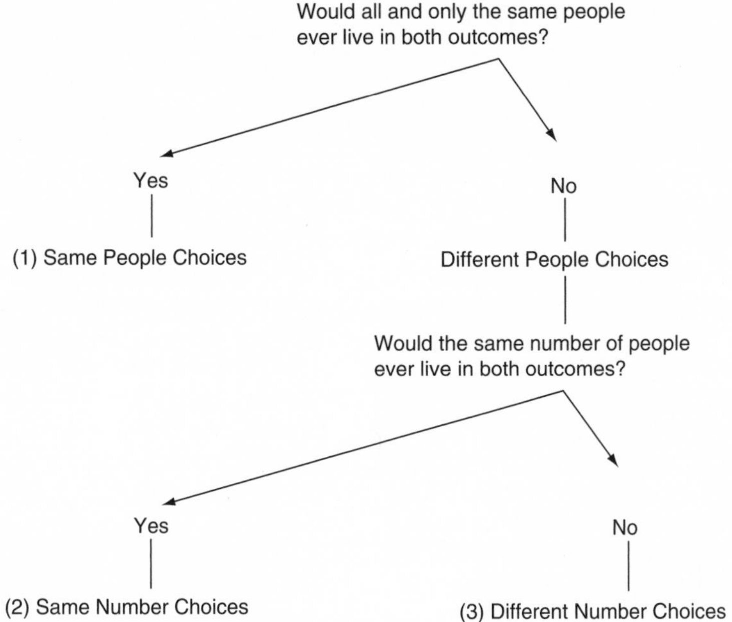

Figure 6.2. Effects of choice between two outcomes.

we ask two questions: "Would all and only the same people ever live in both outcomes?" "Would the same number of people ever live in both outcomes?"

Of these three types of choice, it is the first and third that are important. Most of our moral thinking concerns Same People Choices, where there is a given group of people whom our acts may affect. We seldom consider Different Number Choices. Those who do have found them puzzling. What this essay has discussed are the second group, Same Number Choices. These are much less puzzling than Different Number Choices. But they are not common. Once we have moved outside Same People Choices—once we are considering acts that would cause different people to exist—it is seldom true that in all of the relevant outcomes the very same numbers would exist.

According to claim (A), it is bad if those who live are worse off than those who might have lived. This claim applies straightforwardly only to Same Number Choices. Can we extend (A) to cover Different Number Choices? One extension would be the so-called Average View. On this view, it would be worse for there to be more people if the average person would be worse off. The Average View, though popular, can be shown to be implausible.[3] But this does not cast doubt on (A). What it shows is that (A) should not be thought to cover Different Number Choices. We should restate (A) to make this explicit. But (A) *can* be made to cover Same People Choices. Our restatement might be this:

> (B) If the same number of lives would be lived either way, it would be bad if people are worse off than people might have been.

The two occurrences of "people" here may refer to *different* people. That is how (B) can cover Same Number Choices. But it can also cover Same People Choices. (B) here implies that it is bad if people are worse off than *they* might have been.

Now consider a more familiar principle. This appeals to the interests of those whom our acts affect. One statement might be this:

> *The Person-Affecting Principle*, or *PAP*: It is bad if people are affected for the worse.

What is the relation between (B) and the PAP?[4] In Same People Choices, these claims coincide. If people are worse off than they might have been, they are affected for the worse. So it will make no difference whether we appeal to (B) or to the PAP.[5]

The two claims diverge only in Same Number Choices. These are what my essay has discussed. Suppose that you share my intuitions, thinking that the point about identity makes no moral difference. You then believe that in Same Number Choices we should appeal to (B) *rather than* the PAP. If we choose Depletion, this will lower the quality of life in the further future. According to (B), this is a bad effect. When we see the point about identity, we see that this effect will be worse for no one. So it is not bad according to the PAP. If we believe that the effect is just as bad, we will here have no use for the PAP. Similar remarks apply to the choice between the two medical programs. If we believe these two programs to be equally worthwhile, we shall again appeal to (B). We shall have no use for the PAP. It draws a moral distinction where, in our view, no distinction should be drawn. It is thus like the claim that it is wrong to enslave whites.

To draw these remarks together, in Same People Choices, (B) and the PAP coincide. In Same Number Choices, we accept (B) rather than the PAP. So wherever the claims diverge, we prefer (B).

There remain the Different Number Choices. Since we have restricted (B), we shall need some wider claim to cover these. Call this claim (X). I am not sure what (X) should be. But if you have shared my intuitions, we can expect this. We shall have no further use for (B). It will be implied by (X).[6] So we can expect (X) to inherit (B)'s relations to the PAP. Wherever the claims diverge, we will prefer (X). In Same People Choices, (X) will imply the PAP. It will here make no difference to which we appeal. These are the cases with which most moral thinking is concerned. This explains the reputation of the PAP. This part of morality, the part concerned with human welfare, is usually thought of in person-affecting terms. We appeal to the interests of those whom our acts affect. Even after we have found (X), we may continue to use the PAP in most cases. But it will be only a convenient form of shorthand. In some cases, (X) and the PAP will diverge. And we will here appeal to (X) rather than the PAP. We will here believe that if an effect is bad according to (X), it makes no moral difference whether it is also worse for any particular people. The PAP draws a distinction where, in our view, no distinction should be drawn. We may thus conclude that this part of morality, the part concerned with human welfare, cannot be explained in person-affecting terms. Its fundamental principle will not be concerned with whether acts will be good or bad for those people whom they affect. If this is so, many moral theories need to be revised.[7]

Notes

1. The first third of this section is adapted from my "Future Generations: Further Problems." *Philosophy & Public Affairs* 11.2 (Spring 1982).

2. Thus we might say, "We are benefited only if the alternative would not have been worse for us. If we had never existed, this would not have been worse for us." These and similar arguments I claim not to be decisive in my "Future Generations." Even if it can be in our interests to have been conceived, most of my later claims would still stand.

3. See my "Future Generations," section IX, and Jefferson McMahan's "Problems of Population Theory" in *Ethics* 92.1 (October 1981).

4. On the assumption that it cannot be in or against our interests to have been conceived. If we drop this assumption, some of the following claims need to be revised. Again, see my "Future Generations."

5. Does the equivalence go the other way? If people are affected for the worse, does this make them worse off? There is at least one exception: when they are killed. (B) should be revised to cover such exceptions. Only this ensures that in Same People Choices, (B) and the PAP always coincide.

6. Consider the best-known candidates for the role of (X): the Average and Total Views. In their hedonistic forms, the Average View calls for the greatest net sum of happiness per life lived: the Total View simply calls for the greatest

total net sum of happiness. When applied to population policy, these two views lie at opposite extremes. But when applied to Same Number Choices, both imply the hedonistic form of (B). This suggests that, whatever (X) should be, it, too, will imply (B). The difference between the candidates for (X) will be confined to Different Number Choices. This would be like the fact that only in Same Number Choices does (B) diverge from the PAP. I discuss these points more fully in my book *Reasons and Persons* (New York: Oxford University Press, 1984).

7. We can expect that we will also change our view about certain common cases (one example might be abortion). But most of our moral thinking would be unchanged. Many significant relations hold only between particular people. These include, for instance, promising, friendship, and (if we are politicians) representation. My remarks do not apply to these special relations or to the obligations they produce. My remarks apply only to our general obligations to benefit and not to harm. Since they apply only to these obligations, and they make a difference only when we can affect who will later live, my conclusion may seem overstated. But consider a (grandiose) analogy. In ordinary cases, we can accept Newton's Laws. But not in all cases. And we now believe a different theory.

7

Cosmopolitan Justice, Responsibility, and Global Climate Change

Simon Caney

It's exciting to have a real crisis on your hands when you have spent half your political life dealing with humdrum things like the environment.[1]

The world's climate is undergoing dramatic and rapid changes. Most notably, the earth has been becoming markedly warmer, and its weather has, in addition to this, become increasingly unpredictable. These changes have had, and continue to have, important consequences for human life. In this chapter, I wish to examine what is the fairest way of dealing with the burdens created by global climate change. Who should bear the burdens? Should it be those who caused the problem? Should it be those best able to deal with the problem? Or should it be someone else? I defend a distinctive cosmopolitan theory of justice, criticize a key principle of international environmental law, and, moreover, challenge the "common but differentiated responsibility" approach that is affirmed in current international environmental law.

Before considering different answers to the question of who should pay for the costs of global climate change, it is essential to be aware of both the distinct kind of theoretical challenge that global climate change raises and also the effects that climate change is having on people's lives. Section 1 thus introduces some preliminary methodological observations on normative theorizing about global climate change. In addition, it outlines some basic background scientific claims about the impacts of climate change. Section 2 examines one common way of thinking about the duty to bear the burdens caused by climate change, namely the doctrine that those who have caused the problem are responsible for bearing the burden. It argues that this doctrine, while in many ways appealing, is more problematic than might first appear and is also incomplete in a number of different ways (sections 3 through 8). In particular, it needs to be grounded in a more general theory of justice and rights. The chapter then presents an interest-based account of global environmental rights, and from this derives four principles that determine who should bear the burdens of global climate change (section 9). This account is then compared and contrasted with an alternative approach, namely the principle of "common but differentiated responsibilities" that is articulated in a number of international legal documents on the environment (section 10). Finally, in section 11 I observe that normative analyses of climate change tend to oscillate between individualist and collectivist principles.

1. Global Climate Change

Prior to beginning the normative analysis, it is necessary to make three preliminary points.

1. The topic of this chapter is one instance of what might be termed "global environmental justice," by which I mean the global distribution of environmental burdens and benefits. As such, it is worth making some methodological observations about the utility, or otherwise, of applying orthodox theories of distributive justice to climate change. How relevant are the normal theories of justice to this topic? Indeed, are they relevant at all? If they are relevant, in what ways, if any, do they need to be revised or adjusted? To answer this set of questions we may begin by observing that the standard analyses of distributive justice tend to focus on how income and wealth should be distributed among the current members of a state. To construct a theory of global environmental justice requires us to rethink three assumptions underpinning this normal analysis.[2]

First, distributive justice concerns itself with the distribution of burdens and benefits. Now conventional theories of distributive justice tend to focus on benefits such as wealth and income. It is important, then, to ask whether this framework can usefully be extended to include environmental burdens and benefits. In particular, we face the question of how to value the environment. Should it be valued because of its impact on what Rawls terms "primary goods," by which Rawls means goods such as income, wealth, liberties, opportunities, and the social bases of self-respect?[3] Or should it be valued because of its effects on what Sen and Nussbaum call "capabilities", where this refers to a person's ability to achieve certain "functionings"?[4] Here we should be alive to the distinctive aspects of the environment that might mean that its importance (for a theory of justice) cannot be captured by the orthodox liberal discourse of resources, welfare, capabilities, and so on.[5]

Second, whereas conventional theories of distributive justice concern themselves with the distribution of burdens and benefits within a *state*, the issues surrounding climate change require us to examine the *global* distribution of burdens and benefits. An appropriate analysis needs, then, to address whether the kinds of principle that should be adopted at the domestic level should also be adopted at the global level. Perhaps the two are relevantly analogous, in which case the principles that should be implemented at home should also be implemented abroad. Perhaps, however, they are so completely different that we cannot apply principles fit for the domestic realm at the global level.[6] Either way, a theory of justice that is to be applied to global climate change must, of necessity, address the question of whether the global dimensions of the issue make a morally relevant difference.

Third, global environmental justice raises questions of intergenerational justice. This is true in two senses. First, the effects of global climate change will be felt by future people, so that an adequate theory of global environmental justice must provide guidance on what duties to future generations those living at present have. It must consider whether future people have rights and whether there should be a social discount rate.[7] It must, further, explore whether the principles that apply within a generation will necessarily apply to future generations. Do the principles that apply within one generation differ from those that apply across time into the future? Some, like John Rawls, clearly think that they do, for he holds that the "difference principle" (that the basic structure should be designed to maximize the condition of the least advantaged) should govern the distribution of resources within one generation but should not be applied intergenerationally. Another principle, that of just savings, determines the obligations persons have to future generations. According to the principle of "just savings," societies should save enough so that succeeding generations are able to live in a just society. They need not pass on any more than that and certainly need not seek to maximize the condition of the least advantaged persons who will ever live.[8] Second, and furthermore, topics such as climate change require us to explore the moral relevance of decisions taken by previous generations. For example, some of the deleterious effects of industrialization are being felt now. This prompts the question of who should be responsible for dealing with the ill effects that result from earlier generations. In short, then, a theory of justice that is to apply to global climate change must address the question of how the intergenerational dimensions of the issue make a morally relevant difference.

Drawing on these, then, we can say that an adequate theory of justice in relation to climate change must explain in what ways global climate change affects persons' entitlements, and it must do so in a way that (i) is sensitive to the particularities of the environment; (ii) explores the issues that arise from applying principles at the global rather than the domestic level; and (iii) explores the intergenerational dimensions of global climate change.[9]

2. Turning now from methodological considerations to more empirical matters, an adequate analysis of the ethical dimensions of global climate change requires an empirical account of the different ways in which climate change is affecting persons' fundamental interests (by which I mean those interests that a theory of justice should seek to protect). In what follows I shall draw heavily on the findings of the Intergovernmental Panel on Climate Change (IPCC), set up in 1988 by the United Nations Environment Programme (UNEP) and the World Meteorological Organization (WMO).[10] It has now issued three assessment reports—in 1990, 1995, and 2001. For our purposes the key report is *The Third Assessment Report* published in 2001. This includes four volumes: *Climate Change 2001: The Scientific Basis; Climate Change 2001: Impacts, Adaptation and Vulnerability; Climate Change 2001: Mitigation;* and a summary of all three reports, *Climate Change 2001: Synthesis Report.* The findings of the IPCC have, of course, been criticized by a number of people—including, for example, Bjørn Lomborg—and there have, in turn, been replies to those critics.[11] I am not qualified to enter into these debates, and so I shall report the IPCC's claims without assuming that those claims are incontestable. The IPCC reports most fully on the impacts of global climate change in its report entitled *Climate Change 2001: Impacts, Adaptation and Vulnerability*. In the latter it claims that global climate change will result, inter alia, in higher sea levels and therefore threaten coastal settlements and small island states. It will also result in higher temperatures and as a consequence will generate drought, crop failure, and heatstroke. The rise in temperature will also lead to an increased incidence of malaria and cholera. To this we should also add that global climate change will result in greater weather unpredictability. This is, of course, only the briefest of summaries.[12] A fuller account will be introduced later on.

3. Having noted various ways in which climate change has harmful effects, I would now like to clarify what I mean when I refer to "the burdens of global climate change." As is commonly recognized, there are two distinct kinds of burden imposed by recent changes to the climate—what I shall term "mitigation burdens" and "adaptation burdens".[13] "Mitigation burdens," as I am defining that term, are the costs to actors of not engaging in activities that contribute to global climate change. Those who engage in a policy of mitigation bear an opportunity cost: they forgo benefits that they could have had if they had engaged in activities which involve the emission of high levels of greenhouse gases (GHGs). To make this concrete, mitigation will involve cutting back on activities like the burning of fossil fuels, and as such, it requires either that persons cut back on their use of cars, electricity, and air flight or that they invest in other kinds of energy resource. Either way, mitigation is, of course, a cost for some.[14] The second kind of burden is what I have termed "adaptation burdens". These are the costs to persons of adopting measures that enable them and/or others to cope with the ill effects of climate change. For there are ways in which people can adapt to some of the predicted outcomes of global climate change. They might, for example, spend more on drugs

designed to minimize the spread of cholera and malaria. Or they might spend more on strengthening coastal regions against rising sea levels. These, too, should obviously count as a burden, for they require resources that could otherwise be spent on other activities.

My focus in this chapter is on the question "Who should bear the costs caused by climate change?" I shall not explore the difficult question of how much we should seek to mitigate and how much we should seek to adapt. This is, of course, a key question when determining what specific concrete policies should be implemented. It is also the subject of some controversy.[15] However, I wish to set that practical issue aside and simply focus on the more abstract question of who is morally responsible for bearing the burdens of climate change where the latter is silent on the choice between adaptation and mitigation.

2. The "Polluter Pays" Principle

Let us turn now to a normative analysis of the responsibility of addressing these problems. On whose shoulders should the responsibility rest? Who is duty-bound to bear the burdens of global climate change? One common way of thinking about harms, including both environmental and nonenvironmental harms, maintains that those who have caused a problem (such as pollution) should foot the bill. In other words, the key principle is that "the polluter should pay." This principle has considerable intuitive appeal. In everyday situations we frequently think that if someone has produced a harm (they have spilled rubbish on the streets, say), then they should rectify that situation. They as the causers are responsible for the ill effects.

The "polluter pays" principle (hereafter PPP) is also one that has been affirmed in a number of international legal agreements.[16] The Organization for Economic Cooperation and Development (OECD), for example, recommended the adoption of the PPP in Council Recommendations of May 26, 1972, and November 14, 1974.[17] In addition to this, on April 21, 2004, the European Union and Council of Ministers passed a directive affirming PPP.[18] The principle has also been recommended by the Commission on Global Governance.[19] In addition to this, a number of academic commentators on the subject have applied this principle to the costs of global climate change. Henry Shue, for example, has drawn on the principle that those who have caused pollution should clear it up and has argued vigorously that members of industrialized countries have caused global climate change and hence they, and not members of developing countries, should bear the burdens of climate change.[20] Furthermore, others in addition to Shue have argued that this is the right way of thinking about bearing the burdens of global climate change. Eric Neumayer, for instance, argues that the costs of global warming should be determined according to "historical accountability".[21] We might further note that the IPCC has addressed the question in *Climate Change 2001: Mitigation.*[22] It sought not to recommend any one course of action, but it did cite the PPP, along with various others, as a possible principle of justice. How appropriate, then, is the PPP for determining the responsibility to bear the costs of climate change?

Let us begin our analysis by noting two clarificatory points.

First, the principle that the polluter pays usually means literally that if an individual actor, X, performs an action that causes pollution, then that actor should pay for the ill effects of that action. Let us call this the micro-version. One might, however, reconstruct the PPP to mean also that if actors X, Y, and Z perform actions that together cause pollution, then they should pay for the cost of the ensuing pollution in proportion to the amount of pollution that they have caused. Let us call this the macro-version. This says that polluters (as a class) should pay for the pollution that they (as a class) have caused. So, whereas the micro-version establishes a direct link between an agent's actions and the pollution suffered by others, the macro-version establishes an indirect link between, on the one hand, the actions of a group of people (e.g., emitting carbon dioxide) and, on the other hand, a certain level of pollution.

This distinction is relevant because the micro-version can be applied only when one can identify a specific burden that results from a specific act. It is, however, inapplicable in cases where one cannot trace specific burdens back to earlier individual acts. Now climate change clearly falls into this category. If an industrial plant releases a high level of carbon dioxide into the air, we cannot pick out specific individual costs that result from that particular actor and that particular action. The macro-version can, however, accommodate the causation of such effects. Even if one cannot say that A has caused this particular bit of global warming, one can say that this increase in global warming as a whole results from the actions of these actors. Furthermore, note that the macro-version can allow us to ascribe greater responsibilities to some. Even if it does not make sense to say that we can attribute a specifiable bit of global warming to each of them, we can still say that those who emit more carbon dioxide than others are more responsible than those others. In principle, then, if one had all the relevant knowledge about agents' GHG emissions, it would be possible to make individualistic assessments of just how much each agent owes. In the light of the above, then, we should interpret the PPP (when it is applied to the case of global warming) along the lines suggested by the macro-version.[23]

Second, to apply the "polluter pays" approach to climate change, we need to know "Who is the polluter?" What is the relevant unit of analysis? What kinds of entity are the polluters? Are they individuals, states, or some other entity? Furthermore, which of these entities plays the greatest role? Suppose that the relevant actor is, in fact, states; we then face the empirical question "Which particular states contribute the most?" Our answer to the question "Who pollutes?" is, of course, essential, if we accept the PPP, to enable us to allocate responsibilities and answer the question "Who should pay?"

Many of those who adopt a PPP approach to climate change appear to treat countries as the relevant unit. Shue, for example, makes constant reference to "countries" and "states".[24] Similarly, Neumayer refers always to the pollution caused by emitting countries, referring, for example, to "the Historical Emission Debt (HED*i*) of a country."[25] As he says, his view

> holds countries accountable for the amount of greenhouse gas emissions remaining in the atmosphere emanating from a country's historical emissions. It demands that the major emitters of the past also undertake the major emission reductions in the future as the accumulation of greenhouse gases in the atmosphere is mostly their responsibility and the absorptive capacity of nature is equally allocated to all human beings no matter when or where they live.[26]

In their view, then, the polluters are countries. But is this an appropriate analysis? Consider the following possibilities.

(a) Individuals. First, we might observe that individuals use electricity for heating, cooking, lighting, televisions, and computers, and, of course, they consume fossil fuels by driving cars and by taking airplane flights—all of which are responsible for carbon dioxide emissions. The Third Assessment Report of the IPCC, moreover, says in its prescriptions that individuals must change their energy-intensive lifestyles.[27] Should we say, then, that individuals should pay? If so, it would seem that instead of stating simply that each country should pay its share, we should ideally, and in principle, claim that each individual should pay his on her share.

(b) Economic corporations. Perhaps, however, it might be argued that the primary causes of greenhouse gas emissions are those economic corporations that consume vast amounts of fossil fuels and/or bring about deforestation. If this is so, then presumably the primary responsibility should accrue to them.[28]

(c) States. Maybe, however, the relevant unit of analysis is the state. As noted above, this is what many commentators on the subject assume. Since they think that states should either cut back on GHG emissions or devote resources to cover the costs of adaptation, they must think that states are the primary cause of global climate change.

(d) International regimes and institutions. However, it might perhaps be argued that one relevant factor is suprastate institutions and the nature of international law. One might, for example, think, like Thomas Pogge, that the "explanatory nationalism" adopted by position (c) is untenable, for it fails to recognize the extent to which we are part of a globally interdependent order and that this gives rise to events often seen as domestic in nature.[29] Drawing on this, might one argue that the causes of pollution are not accurately seen as "countries" or "states" but rather international institutions or the international system. Perhaps, it might be argued, existing international institutions (such as the World Trade Organization and the International Monetary Fund), by promoting economic growth, encourage countries to engage in deforestation and the high use of fossil fuels, both activities that lead to climate change.

With this taxonomy in mind, let us make three points. The first is that the likely answer to the question "Who is the polluter?" will involve reference to several different kinds of actor. The aim of the taxonomy above is not to suggest that the appropriate answer lies at one level alone. Second, we should observe that to reach the standard conclusion (namely that certain *countries* should pay), we need to show that options (a), (b), and (d) do not hold. It might, for example, be argued against (d) that international law and regimes do not have any autonomy—they are merely the creations of states, and as such, the relevant level of analysis is the actions of states. And it might be argued against (a) and (b) that it is not possible to ascertain the GHG emissions of individual persons or individual corporations. Given this, we should refer to the GHG emissions of a country as the best approximation available. Alternatively, it might be argued against (a) and (b) that the GHG of individuals or corporations is what has been permitted by the relevant state, so that the latter should be held liable. My aim, here, is not to canvass any of these options, it is simply to point out that the only way to vindicate the conclusion reached by Neumayer, Shue, and others is to establish that the relevant unit of analysis is the state and that the other options collapse into it. Of course, further empirical analysis may reveal that it is simply implausible to hold that states are the appropriate entities and we need a fine-grained analysis that traces the contributions of individuals, corporations, states, and international actors and accordingly attributes responsibilities to each of these.

Having made these two clarificatory points, let us turn now to consider some of the problems faced by the "polluter pays" approach to allocating the burdens of climate change.

3. Past Generations

One problem with applying the "polluter pays" principle to climate change is that much of the damage to the climate was caused by the policies of earlier generations. It is, for example, widely recognized that there have been high levels of carbon dioxide emissions for the last 200 years, dating back to the industrial revolution in western Europe. This poses a simple, if also difficult, problem for the "polluter pays" principle: who pays when the polluter is no longer alive? And the proposal, made by Neumayer and Shue, that the industrial economies of the first world should pay seems, on the face of it, unfair, for it does not make the actual polluters pay. Their conclusion, then, is not supported by the PPP; indeed, it violates the PPP.

This is a powerful objection. However, at least three distinct kinds of response are available to an adherent of the argument under scrutiny.

3.1. The Individualist Position

One reply is given by both Shue and Neumayer. Both raise the problem of past generations but argue that this challenge can be met. In Shue's case, his response is that the current inhabitants of a country are not "completely unrelated" to previous inhabitants, and as such, they can still bear responsibility for the actions of their ancestors. In particular, says Shue, they enjoy the benefits of the policies adopted by previous generations.[30] As he writes, "current generations are, and future generations probably will be, continuing beneficiaries of earlier industrial activity."[31] The same point is made by Neumayer, who writes:

> The fundamental counter-argument against not being held accountable for emissions undertaken by past generations is that the current developed countries readily accept the benefits from past emissions in the form of their high standard of living and should therefore not be exempted from being held accountable for the detrimental side-effects with which their living standards were achieved.[32]

Let us call this reply the "beneficiary pays" principle (BPP). Put more formally, this claims that where A has been made better off by a policy pursued by others, and the pursuit by others of that policy has contributed to the imposition of adverse effects on third parties, then A has an obligation not to pursue that policy itself (mitigation) and/or an obligation to address the harmful effects suffered by the third parties (adaptation).

So if the current inhabitants of industrialized countries have benefited from a policy of fossil-fuel consumption and that policy contributes to a process that harms others, then they are not entitled to consume fossil fuels to the same degree. Their standard of living is higher than it otherwise would have been, and they must pay a cost for that.[33]

This line of reasoning has some appeal. Two points, however, should be made about it. First, we should record that the BPP is not a revision of the "polluter pays" approach, it is an abandonment of it. It would justify imposing a burden on someone who cannot, in any conceivable sense, be said to have brought about an environmental bad but who nonetheless benefits from the policy that caused the environmental bad. In such a case that person is not a polluter but is a beneficiary. Thus, according to the PPP, they should not be allocated a duty to make a contribution to cover the environmental bad; according to the BPP, however, they should. My second point is that the application of the BPP in this instance is more problematic than it might first seem. The reason for this draws on what Derek Parfit has termed the "non-identity problem" in his seminal work *Reasons and Persons*. We need therefore to state this problem. In *Reasons and Persons* Parfit drew attention to an important feature of our moral duty to future generations. Parfit begins with the statement that who is born (which particular person) depends on exactly when their parents mated. If someone's parents had mated at a different time, then, of course, a different person would have been born. It follows from this that the policies that persons adopt at one time affect who will be born in the future. So suppose that we build factories now that have no immediate malign effects but that release poisonous fumes in 300 years. Now Parfit's point is this: the policies adopted now led to the birth of different people from those who would have been born if these policies had not been adopted. The future generations whose lives are threatened by poisonous fumes would not have been born were it not for the factory construction. So they cannot say that they were made worse off or harmed by the policy. The policy, according to Parfit, is bad but it has not made anyone worse off than they would have been if the policy had not been enacted.[34]

Now I think that a very similar point could be made against the use of the "beneficiary pays" principle by the argument under scrutiny.[35] For it claims that the policies of industrialization benefited people who are currently alive. But in the same way that using up resources did not *harm* future people, so industrialization did not make an *improvement* to the standard of living of currently existing people. We cannot say to people, "You ought to bear the burdens of climate change because without industrialization you would be much worse off than you currently are." We cannot because without industrialization the "you" to which the previous sentence refers would not exist. Industrialization has not brought advantages to these people that they would otherwise be without.[36] And since it has not, we cannot say to them, "You should pay for these because your standard of living is higher than it would have been."[37] For this reason the BPP is unable to show why members of industrialized countries should pay for the costs of the industrialization that was undertaken by previous generations.

3.2. The Collectivist Position

While the first response to the question of why later generations should pay for the industrializing policies adopted by their ancestors is a rather individualistic one, a second response to the intergenerational challenge affirms a collectivist position.[38] This approach argues that the problem we are addressing arises only if we focus on individual persons. If we focus on individuals, then making current individuals pay for pollution that stems from past generations is indeed making someone other than the polluter pay. Suppose, however, that we focus our attention on collective entities like a nation or a state (or an economic corporation). Consider a country such as Britain. It industrialized in the late eighteenth and the nineteenth centuries, thereby contributing to what would become the problem of global warming. Now if we take a collectivist approach, we might say that since Britain (the collective) emitted excessive amounts of GHGs during one period in time, then Britain (as a collective) may a hundred years later, say, be required to pay for the pollution it has caused, if it has not done so already. To make this collective unit pay *is* to make the polluter pay. So to return to the original objection, one might say that the premise of the objection (namely that the polluter is no longer alive) is incorrect.

Prior to evaluating this argument we should make three observations. First, note that although in this instance I have used an example of a nation as a collective, there is no reason to assume that it must take this form. Suppose, for instance, that there is in existence a long-standing corporation. We might argue, in a collectivist vein, that if this entity has emitted high levels of carbon dioxide in the past, then it should foot the bill now. The individual decision makers of the time might be long gone, but the corporation persists. Second, we might observe that the collectivist response is also relevant to the preceding discussion of the BPP. My objection to the use of the BPP above was that the acts that led to a higher standard of living (in this case industrialization) did not make the standard of living of currently alive persons higher than it would have been had industrialization never taken place. The collectivist perspective adds a different dimension to this, for, as Edward Page has rightly noted, the identities of nations are less changeable over time than those of individuals. Industrialization may have affected which individuals get born: because of it different people are born from those who would have been born without it: And because of this it is inaccurate to say that currently alive individuals have a higher standard of living than those same individuals would have had if industrialization had never taken place. However, the acts of industrialization did not (let us assume) bring different countries into existence from those that would otherwise have existed.[39] So to turn to the objection to the BPP: whereas we cannot say that industrialization has bestowed (net) advantages on currently existing individuals that they would otherwise be without, we *can* say that industrialization has bestowed (net) advantages on currently existing countries (such as Britain) that they would otherwise be without. The collectivist response thus enables us to defend the BPP against my Parfit-inspired objection.[40] Third, and finally, we might observe that the collectivist response coheres with the way that some political philosophers have recently argued. For example, in *The Law of Peoples*, John Rawls gave two examples that appealed to a similar kind of reasoning in order to rebut a cosmopolitan political morality. In one example Rawls asks us to compare a society that industrializes with one that eschews that path, choosing instead a more pastoral way of life. For his second example Rawls again asks us to compare two societies. One, by granting women greater reproductive autonomy, results in a more controlled population policy with fewer children being born. The other society, by contrast, does not pursue this kind of population policy. In these scenarios, concludes Rawls, self-governing peoples (liberal or decent) should take responsibility for their policies. So to take the first example,

Rawls's view is that justice does not require that the wealthy industrialized society should assist the poorer pastoral society.[41] Similar reasoning is adduced by David Miller, who argues that self-governing nations should be held accountable for their decisions.[42]

Let us now evaluate this collectivist response to the problem of past generations. It is vulnerable to two objections. First, it is not enough to draw attention to the possibility of affirming a collectivist position. We need to ascertain whether we have any reason to prefer a collectivist to an individualist approach. To vindicate the collectivist perspective, we need an argument that can show when and why it is accurate to say that a collective caused an environmental bad and hence that collective must pay. Indeed, we need an argument for why this description is better than a more individualistic one (individuals A, B, and C polluted, and so individuals A, B, and C should pay). Second, a collectivist approach is vulnerable to a troubling problem. The root problem is that it seems unfair to make individuals pay the costs generated by preceding generations. In taking a collectivist route, are we not being unfair to individuals who did not make those decisions and who might have objected violently to those decisions? Can they not reasonably complain that they were not consulted; they did not vote; they disapprove of the policies and, as such, should not be required to pay for decisions that others took? "Normally," they might add, "individuals cannot inherit debts from parents or grandparents, so why should this be any different?" For this reason, then, a collectivist response to the problems raised by the excessive GHG emissions of earlier generations is not an attractive position.[43]

3.3. A Third Response

Thus far we have examined two responses to the intergenerational objection. The first contends that people currently living in industrialized countries have benefited from pollution-causing economic growth. The second contends that the relevant causal actors are collectives that still exist today (either corporations or countries or collective units such as "the industrialized world"). A third response would be to argue that all the burdens of human-induced climate change should be paid for by existing polluters. The suggestion, then, is that current polluters should pay the costs of their pollution and that of previous generations. In this way, it might be said, the mitigation and adaptation costs of climate change are shouldered by the polluters (and not by nonpolluters). But this seems unfair: they are paying more than their due. The intuition underlying the PPP (about which more later) is that people should pay for the harm that *they* (not others) have created. It is alien to the spirit of the principle to make people pay for pollution that is not theirs. So even if the proposal does, in one sense, make polluters pay (no nonpolluters pay), it does not make sure that the costs of polluters are traced back to the particular polluters, and that is what the PPP requires.

The first objection has not challenged the claim that the polluter should pay (except for Shue's revision of the principle). Rather it has shown, first, that proponents of the PPP are not entitled to conclude that current members of industrialized states should pay for the costs of global warming. And, second, it has more generally shown that the PPP cannot say who should bear the costs of climate change caused by past generations. We might, however, raise questions about the "polluter pays" principle itself. I now want to consider several challenges to the fundamental principle.

4. Ignorance and Obligation

One doubt about the "polluter pays" principle is that it is too crude and undiscriminating in its treatment of the relevant duty bearers. What if someone did not know that performing a certain activity (such as burning fossil fuels) was harmful? And suppose, furthermore, that there was no way in which they could have known that it was harmful. In such a situation their ignorance is excusable, and it seems extremely harsh to make them pay for something that they could not have anticipated. This raises a problem for the "polluter pays" principle in general. It also has considerable relevance for the issue at hand, for it is widely accepted that many who have caused GHG emissions were unaware of the effects of their activities on the earth's atmosphere. Furthermore, their ignorance was not in any way culpable: they could not have been expected to know. This objection, note, applies in different ways to the individualist and collectivist approaches considered earlier. To the collectivist version it says that even if we can deal with past generations because the fossil fuel consumption was due to the past actions of a collective (Britain, say), this collective entity was, until the last two or three decades, excusably ignorant of the effects of fossil-fuel consumption. To the individualist version it says that even if we forget about previous generations and focus simply on those currently alive, some of those individuals responsible for high emissions levels were (excusably) unaware of their effects. The objection from ignorance, thus, has more significance for the collectivist than the individualist position. Whereas the individualistic position has to explain how we deal with the GHGs emitted by currently living persons who were in (excusable) ignorance of their effects, the collectivist position has to deal with the GHGs that were emitted by both past and present members of collectives who were in (excusable) ignorance of their effects.

To this argument we might further add that Neumayer's version of the historical approach to climate change is particularly vulnerable. For Neumayer would make current generations of a country pay for all instances where a previous generation of that country emitted more than its equal per capita entitlement.[44] But how could they be expected to know that this was the entitlement? This kind of retrospective justice would seem highly unfair.

Consider now some replies to this line of reasoning. One response to it is that this point no longer has any relevance because it has been known for a considerable period that fossil-fuel consumption and deforestation cause global climate change. This is how Peter Singer, for example, responds; for him the objection of ignorance is inapplicable for post-1990 emissions.[45] Neumayer takes the same tack, but for him the relevant cutoff point is the mid-1980s.[46]

But what of high GHG emissions that took place before 1990 (or the mid-1980s)? This first response leaves pre-1990 pollution uncovered. Individuals before that time caused carbon dioxide emissions which have contributed to global warming, and this first response cannot show that pre-1990 polluters should pay for global warming. It therefore leaves some of the burdens of global warming unaddressed. As such, it should be supplemented with an account of who should bear the burdens of climate change that result from pre-1990 GHG emissions.

Given this, let us consider a second reply. In his "Global Environment and International Inequality," Shue argues that it is not unfair to make those who have emitted high levels of GHGs bear the burden of dealing with climate change, even though at the time they were not aware of the effects of what they were doing. Shue maintains that the objection of ignorance runs together punishment for an action and being held responsible for an action. His suggestion is that it would indeed be unfair to punish someone for actions they could not have known were harmful to others. However, says Shue, it is not unfair to make them pay the costs: after all, they caused the problem.[47]

In reply, it is not clear why we should accord weight to this distinction. If one should not punish ignorant persons causing harm, why is it all right to impose financial burdens on them? More worryingly, Shue's proposal seems unfair on the potential duty bearers. As Shue himself has noted in another context, we can distinguish between the perspective of rights bearers and the perspective of duty bearers.[48] The first approach looks at matters from the point of view of rights holders and is concerned to ensure that people receive a full protection of their interests. The second approach looks at matters from the point of view of the potential duty bearers and is concerned to ensure that we do not ask too much of them. Now, utilizing this terminology, I think it is arguable that to make (excusably) ignorant harmers pay is to prioritize the interests of the beneficiaries over those of the ascribed duty bearers. It is not sensitive to the fact that the alleged duty bearers could not have been expected to know.[49] Its emphasis is wholly on the interests of the rights bearers and, as such, does not adequately accommodate the duty-bearer perspective.[50]

Neither of the two replies, then, fully undermines the objection that an unqualified "polluter pays" principle is unfair on those people who were high emitters of GHGs but who were excusably ignorant of the effects of what they were doing.[51]

5. The Impoverished

Let us turn now to another worry about taking a purely historical approach to distributing environmental responsibilities. The worry is simply that such an approach may be unfair on the impoverished. Consider, for example, a country that has in the recent past caused a great deal of pollution but that remains poor. Since it is poverty-stricken we might argue that it should not have to pay for its pollution. In this kind of situation the "polluter pays" principle appears unfair, for it asks too much of the poor.

These concerns are powerful, but we must be careful in drawing conclusions here. This argument does not establish that the "polluter pays" principle should be abandoned. Rather it suggests (if we accept the claim that countries should not be required to pay when they are extremely impoverished) that we should supplement the PPP with an additional (and competing) principle (the poor should not pay). One can, that is, take a pluralist response. In support of this conclusion, consider the following scenario. Suppose that a country that is poor creates considerable pollution. Drawing on the preceding argument, we might think that they, the polluters, should not pay. But then suppose that they suddenly become very wealthy (and, for simplicity's sake, do so for reasons absolutely unconnected with their pollution). Since they can now afford to pay for the costs of their pollution, we surely think that they should pay, and the "polluter pays" principle should now be acted upon because it can, in all fairness, be required of the polluters. Given their new-found wealth, they should compensate for the environmental bads they generated. The key point here is that the argument from poverty does not entail that a "polluter pays" approach should be abandoned. Rather it entails that we should reject a monist, or purist, approach which claims that the responsibility for addressing environmental harms should only be assigned to those who have caused them, and it argues that the PPP should be supplemented by other principles.

Another point is worth making here, namely, that the objection under consideration suggests that an adequate account of people's environmental responsibilities cannot be derived in isolation from an understanding of their "economic" rights and duties. It illustrates, that is, the case for not adopting an atomistic approach which separates the task of constructing a theory of environmental justice from a theory of economic justice and so on.[52]

6. The Egalitarian Defense

Let us turn now to the rationale often adduced in support of adopting a PPP approach to deal with the intergenerational aspects of global climate change. Those who canvass a "historical" approach to allocating the responsibilities for addressing climate change often invoke egalitarian principles of justice in support of their position. Shue, for instance, argues that current members of industrialized countries should bear the burdens of climate change on the grounds that

> Once...an inequality has been created unilaterally by someone's imposing costs upon other people, we are justified in reversing the inequality by imposing extra burdens upon the producer of the inequality. There are two separate points here. First, we are justified in assigning

> additional burdens to the party who has been inflicting costs upon us. Second, the minimum extent of the compensatory burden we are justified in assigning is enough to correct the inequality previously unilaterally imposed. The purpose of the extra burden is to restore an equality that was disrupted unilaterally and arbitrarily (or to reduce an inequality that was enlarged unilaterally and arbitrarily).[53]

In a similar vein, Neumayer argues that "historical accountability is supported by the principle of equality of opportunity."[54] And Anil Agarwal, Sunita Narain, and Anju Sharma make a similar point:

> some people have used up more than an equitable share of this global resource, and others, less. Through their own industrialization history and current lifestyles that involve very high levels of GHG emissions, industrialized countries have more than used up their share of the absorptive capacity of the atmosphere. In this regard, the global warming problem is their creation, so it is only right that they should take the initial responsibility of reducing emissions while allowing developing countries to achieve at least a basic level of development.[55]

What are we to make of these related lines of reasoning? I should like to make two points in reply. First, the egalitarian argument can work only if we take a collectivist, as opposed to an individualist, approach; second, a collectivist approach is, in this instance, implausible. Consider the first point. The egalitarian argument maintains that countries such as the United States and Britain should pay for the excessive emissions of their ancestors. So the idea is that since the United States, say, used more than its "fair" share at an earlier period in time, it must use less now to even things out. But this, of course, is taking a collectivist approach. It is claiming that since a collective entity, the United States, emitted more than its fair quota, this same collective entity should emit a reduced quota to make up. The egalitarian argument thus works if we treat communities as the relevant units of analysis. It does not, however, if we focus on the entitlements of individuals. To see this, imagine two countries that now have an identical standard of living. Now imagine that one of them, but not the other, emitted excessive amounts of greenhouse gases in the past. It is then proposed that members of the one country should, in virtue of the pollution that took place in the past, make a greater contribution to dealing with global climate change than members of the other country. The first point to note is that this policy is not mandated by a commitment to equality of opportunity. It may be true that some people in the past will have had greater opportunities than some currently living people, but that simply cannot be altered. Making their descendants have fewer opportunities will not change that. In fact, making their descendants pay for the emissions of previous generations will violate equality, because those individuals will have less than their contemporaries in other countries. So if we take an individualist position, it would be wrong to grant some individuals (those in country A) fewer opportunities than others (those in country B) simply because the people who used to live in country A emitted higher levels of GHGs.

Which position should we take—a collectivist or an individualist one? This leads to my second point. I believe that we should favor the individualist one. To see why, consider an example of two families, each with a son. Now suppose that several generations ago one of the families (family A) sent its child to a prestigious and distinguished public school (Eton College, say), whereas family B sent its equivalent child to a quite ordinary school. Now on an individualistic approach, the fact that someone's great-great-great-grandfather enjoyed more than fair opportunities does not give us any reason to give them a less than equal opportunity. But the collectivist position is committed to claiming that we should penalize the descendant. It must say that since one *family* had a greater than fair allocation of educational opportunities in the past, this must be rectified by giving it (or, rather, one of its current members) a less than equal opportunity now. But that seems just bizarre and unfair.

In short, then, the egalitarian argument for ascribing responsibilities to current members of industrialized countries is unsuccessful; it could work only if we adopted a collectivist methodology that I have argued is unfair.

7. Incompleteness

Let us turn now to two further general limitations of the "polluter pays" principle (limitations that also undermine its treatment of global climate change). The first point to be made here is that the "polluter pays" principle is incomplete, for it requires a background theory of justice and, in particular, an account of persons' entitlements. To see this, we should observe that the "polluter pays" principle maintains that if persons have exceeded their entitlements, then they should pay. Given this, to make the claim that someone should pay requires an account of what their entitlement is. In addition to this, to ascertain how much someone should pay also requires a precise account of their entitlements, for we need to know by how much they have exceeded their quota. What we really need, then, is an account of what rights, if any, people have to emit greenhouse gases. Is there no right to emit? Or is there a right to emit a certain fixed amount? In short, then, the "polluter pays" principle must be located within the context of a general theory of justice, and on its own, it is incomplete.[56]

It is worth recording here that the language used by Shue, Neumayer, and Agarwal, Narain, and Sharma illustrates the point at stake. Shue, for example, argues that those who have "taken an *unfair* advantage of others by imposing costs upon them without their consent" (my emphasis) should bear the burdens of climate change; his account thus presupposes an understanding of people's "fair" share.[57] And Neumayer's analysis is predicated on the assumption that each person has an entitlement to an equal per capita allocation of carbon dioxide emissions. He maintains that agents that have exceeded this quota therefore have a responsibility to pay extra later.[58]

Note that this last point is not an objection to the PPP. It is simply pointing out that the PPP requires supplementation.

8. Noncompliance

There is one final query that one might raise about the "polluter pays" principle (and its application to global climate change). This is that the principle is incomplete in an additional sense. It assigns primary responsibilities—the polluter bears the primary responsibility to bear the burden. Often, however, primary duty bearers fail to comply with their duties. In such circumstances we might not know who the noncompliers are. Furthermore, even if we do know who they are, we might be unable to make them comply. This prompts the question: what, if anything, should be done if primary duty bearers do not perform their duties? One option might be to leave the duties unperformed. In the case of global climate change, however, this would be reckless. In light of the havoc it wreaks on people's lives, we cannot accept a situation in which there are such widespread and enormously harmful effects on the vulnerable of the world. In the light of this, we have reason to accept a second option, one in which we assign "secondary" duty bearers. And the point here is that the PPP is simply unable to provide us with any guidance on this. Since it says only that polluters should pay, it cannot tell us who the secondary duty bearers should be when we are unable to make polluters pay. In this sense, too, it is incomplete.[59]

It may be appropriate to sum up. I have argued that the PPP approach to climate change is inadequate for a number of reasons. It cannot cope with three kinds of GHG, namely GHGs that were caused by:

(i) earlier generations (*cannot pay*);
(ii) those who are excusably ignorant (*should not be expected to pay*); and
(iii) those who do not comply with their duty not to emit excessive amounts of GHGs (*will not pay*).

Furthermore, the egalitarian argument for the historical application of the "polluter pays" principle does not work. Finally, we have seen two ways in which the historical approach is incomplete: it is silent on what should occur when people do not perform their duty, and it needs to be embedded in a theory of justice.

Two other points bear making here. First, it is interesting to return to the methodological preliminaries introduced in section 1, in particular the point that a theory of global environmental justice must be able to cope with the intergenerational dimensions of global environmental problems. For the upshot of the first objection to the PPP (the past generations objection) is that the PPP cannot easily be extended to apply in an intergenerational context. To elaborate further, it is much easier to insist that the polluter should pay when we are dealing with a single generation in which both the polluter and those affected by the pollution are contemporaries. But, as the past generations objection brings out, the principle that the polluter should pay becomes inapplicable when the pollution results from people no longer alive.

Second, although I have argued above that the "polluter pays" approach is incomplete and unable to deal with various kinds of activity that contribute to global climate change, this, of course, does not entail that it should be rejected outright. In the first instance, it rightly applies to many actors who are currently emitting excessive levels of GHGs or have, at some stage since 1990, emitted excessive amounts. So even if it should not be applied to the distant past, it can apply to the present and near past. Furthermore, even if we reject its application to the past, we may still use it for the future. That is, we can inform people of their quota and build institutions that ensure that if people exceed it, then they must make compensation.

9. Justice and Rights

Having argued that a purely "polluter pays" approach is incomplete in a number of ways, we face the question of how it should be supplemented. How should the burden of climate change be distributed?

In this section I wish to outline an alternative way of thinking about global justice and climate change, an account that avoids the weakness of a purely "polluter pays" approach. The argument begins with the assumption that:

> (P1) A person has a right to X when X is a fundamental interest that is weighty enough to generate obligations on others.

This claim draws on Joseph Raz's influential theory of rights. And it follows him in claiming that the role of rights is to protect interests that we prize greatly.[60]

The next step in the argument maintains that:

> (P2) Persons have fundamental interests in not suffering from:
> (a) drought and crop failure;
> (b) heatstroke;
> (c) infectious diseases (such as malaria, cholera, and dengue);
> (d) flooding and the destruction of homes and infrastructure;
> (e) enforced relocation; and
> (f) rapid, unpredictable, and dramatic changes to their natural, social, and economic world.

Yet, as the Third Assessment Report of the IPCC records, all the malign effects listed in (P2) will be generated by climate change. The predicted temperature increases are likely to result in drought and crop failure. They will also lead directly to more deaths through heatstroke. Furthermore, with increased temperatures there is a predicted increase in the spread of malaria, cholera, and dengue fever. In addition to this, the increased temperatures are predicted to melt ice formations and thereby contribute to a rise in sea level which will threaten coastal settlements and countries such as Bangladesh that are flat and close to sea level. As well as simply destroying buildings, homes, and infrastructure, a known effect of climate change will be to force some inhabitants of small island states and coastal settlements to relocate. Finally, we should note that the IPCC maintains that global climate change is not simply a matter of global *warming*; it will lead to high levels of unpredictable weather patterns. This jeopardizes a vital interest in stability and being able to make medium- and long-term plans.[61]

Given this, it follows that there is a strong case for the claim that:

> (C) Persons have the human right not to suffer from the disadvantages generated by global climate change.

Having adduced this argument, note that it (unlike a "polluter pays" approach) does not necessarily rest on the assumption that climate change is human-induced. Its insistence is that persons' preeminent interests be protected, and it is not, in itself, concerned with the causes of climate change. Suppose that climate change is not anthropogenic: this argument would still hold that there is a human right not to suffer from global climate change as long as humans could do something to protect people from the ill effects of climate change and as long as the duties generated are not excessively onerous. The duties that follow from this right could not, of course, be mitigation-related duties, but there could be adaptation-related duties.[62]

With this account in mind we face two questions: Who has the duty to bear the burdens of dealing with global climate change? And what are people's entitlements in terms of emitting GHGs? Let us consider the first question. Drawing on what has been argued so far, I would like to propose four different kinds of duty:

> (D1) All are under a duty not to emit greenhouse gases in excess of their quota.

> (D2) Those who exceed their quota (and/or have exceeded it since 1990) have a duty to compensate others (through mitigation or adaptation) (a revised version of the "polluter pays" principle).

But what of GHG emissions arising from (i) previous generations, (ii) excusable ignorance, and (iii) polluters who cannot be made to pay? These, we recall, were the kinds of GHG emissions that could not adequately be dealt with by a purely "polluter pays" approach. My suggestion here is that we accept the following duty:

> (D3) In the light of (i), (ii), and (iii) the most advantaged have a duty either to reduce their greenhouse-gas emissions in proportion to the harm resulting from (i), (ii), and (iii) (mitigation) or to address the ill effects of climate change resulting from (i), (ii), and (iii) (adaptation) (an "ability to pay" principle).

These first three principles are, however, inadequate. For we need also to accept that:

> (D4) In the light of (iii) the most advantaged have a duty to construct institutions that discourage future non compliance (an "ability to pay" principle).[63]

We should not take pollution as a given and then act in a reactive fashion; rather, we should be proactive and take steps to minimize the likelihood of excessive pollution. And for that reason we should accept (D_4). Let us call this the "hybrid account."[64]

The key point about this account is that it recognizes that the "polluter pays" approach needs to be supplemented, and it does so by ascribing duties to the most advantaged (an "ability to pay" approach). The most advantaged can perform the roles attributed to them, and, moreover, it is reasonable to ask them (rather than the needy) to bear this burden since they can bear such burdens more easily. It is true that they may not have caused the problem, but this does not mean that they have no duty to help solve this problem. Peter Singer's well-known example of a child drowning in a puddle brings this point out nicely.[65] Suppose that one encounters a child facedown in a puddle. The fact that one did not push the child in obviously does not mean that one does not have a duty to aid the child.

It should be noted that this account of persons' duties is incomplete, for we still need to ascertain what counts as a fair quota. As we saw above, it is only with reference to the latter that we can define what counts as *unfair* levels of GHG emissions. It is not possible, in the space available, to answer the question "What is a fair quota?" but I should like to suggest that any credible answer to that question must draw on the interest-based account presented above. That is, in ascertaining the appropriate emissions levels, we need to balance persons' interests in engaging in activities that involve the emission of GHGs on the one hand with persons' interests in not suffering the harms listed in (P_2) on the other. We also need to employ a distributive principle. I have argued elsewhere that we have good reason to prioritize the interests of the global poor.[66] For this reason I would suggest here that the least advantaged have a right to emit higher GHG emissions than do the more advantaged of the world. As Shue himself argues, it is unfair to make the impoverished shoulder the burden.[67] So my account would entail that the burden of dealing with climate change should rest predominantly with the wealthy of the world, by which I mean affluent persons in the world (not affluent countries).[68]

As such, (D_1) to (D_4) may, in practice, identify as the appropriate bearers of the duty to deal with global climate change many of the same people as a "polluter pays" approach. It does so, though, for wholly different reasons. We might, speaking loosely, say that the contrast between my hybrid account and a historical approach is that a historical approach is diachronic (concerned with actions over time and who caused the problem), whereas mine has a diachronic element but is also synchronic (concerned with how much people have now and who can bear the sacrifice). It is also important to record that (D_1) to (D_4) will target different people from a purely "polluter pays" approach in a number of situations. The two accounts identify the same duty bearers only in cases where both (i) "all those who have engaged in activities which cause global climate change are wealthy" and also (ii) "all those who are wealthy have engaged in activities which cause global climate change." But these two conditions may not apply. Consider two scenarios. In the first, a unit emits high levels of GHGs but is poor and not able to contribute much to bearing the costs of climate change. In such a case the PPP would ascribe duties to them that my hybrid account would not. Consider now the second scenario: a unit develops in a clean way and becomes wealthy. If we adopt a purely "polluter pays" approach, then this unit should not accrue obligations to bear the costs of global climate change, but according to the hybrid account they would.[69] So the hybrid account and the "polluter pays" account differ in both theory and practice.

Thus far I have introduced the hybrid account and shown how it remedies defects from which the "polluter pays" approach suffers. Some, however, might object to (D1) to (D4), and to strengthen the hybrid account further, I wish to address one objection that might be pressed against it. The objection I have in mind takes issue with (D3) in particular. It runs as follows: (D3) is unfair because it requires those who are advantaged but who have complied with (D1) and (D2) to make up for the failings of those who have not complied with their duties. And, it asks, is this not unfair? Why should those who have been virtuous be required to do yet more, as (D3) would require, because some have failed to live up to their obligations?

Several comments can be made in reply. First, it should be stressed that the hybrid account explicitly seeks to address this concern by insisting, in (D4), that institutions should be designed so as to discourage noncompliance. It aims, therefore, to minimize those demands on people that stem from the noncompliance of others. Second, we might ask the critic what the alternatives are to asking the advantaged to address the climate change caused by noncompliers (as well as that stemming from past generations and excusable ignorance). One option would be to reject (D3)—and (D4)—and ask the impoverished and needy to pay, but as we have seen, this is unfair. A second option would be to let the harm to the climate that results from the excessive GHG emissions of some go unaddressed. But the problem with this is that the ill effects that this will have on other people (drought, heatstroke, crop failure, flooding) are so dire that this is unacceptable. Such a position would combine neglect (on the part of those who have exceeded their GHG quota) with indifference (on the part of those who could address the problems resulting from the high GHG emissions of others but choose not to). And if we bear in mind that those who are adversely affected by climate change are frequently poor and disadvantaged,[70] we have yet further reason to think that the advantaged have a duty to bear the burdens of climate change that arise from the noncompliance of others. If the choice is of *either* ascribing duties to the poor and needy *or* allowing serious harm to befall people (many of whom are also poor and needy) *or* ascribing duties to the most advantaged, it would seem plausible to go for that third option.[71] One final thought: there is, we can agree, an unfairness involved in asking some to compensate for the shortcomings of others. The question is, how should we best respond to this? My suggestion is that we respond best to this as suggested above, by seeking to minimize those demands and by asking the privileged to bear this extra burden. To this we can add that the virtuous *are* being ill treated but that the right reaction for them is to take this up with noncompliers (against whom they have just cause for complaint) and not to react by disregarding the legitimate interests of those who would otherwise suffer the dire effects of climate change. For these three reasons, then, (D3) can be defended against this objection.[72]

10. A Comparison of the Hybrid Account with the Concept of Common but Differentiated Responsibility

Having outlined and defended the hybrid account, I now want to compare it with a related doctrine that is commonly affirmed in international legal documents on the environment: the concept of "common but differentiated responsibility." By doing so we can gain a deeper understanding of the hybrid account, its relation to international legal treatments of climate change, and its practical implications.

The concept of common but differentiated responsibility was given expression in the 1992 Rio Declaration, and it is worth quoting Principle 7 of the Declaration. It affirms that:

> States shall cooperate in a spirit of global partnership to conserve, protect and restore the health and integrity of the Earth's ecosystem. In view of the different contributions to global environmental degradation, States have *common but differentiated* responsibilities. The developed countries acknowledge the responsibility that they bear in the international pursuit of sustainable development in view of the pressures their societies place on the global environment and of the technologies and financial resources they command.[73]

The same idea is also affirmed in Article 3(1) of the United Nations Framework Convention on Climate Change.[74] In addition to this, the concept of common but differentiated responsibility is evident in the 1997 Kyoto Protocol. For example, the Preamble stipulates that the Protocol is "*guided* by Article 3" of the United Nations Framework Convention on Climate Change (which, as we have just seen, includes a commitment to "common but differentiated responsibility") and the principle of common but differentiated responsibility is explicitly affirmed in Article 10.[75] This account of the responsibilities generated by climate change has some similarities to the hybrid account, for they both insist that duties fall on all—compare, for example, (D1) and (D2)—and yet both also insist that different demands can be made of different parties, as in (D3) and (D4). Furthermore, both accounts allow that the duties to which a party is subject depend on (i) what they have done and (ii) what they are able to do. For example, the hybrid account maintains, in (D2), that those who have exceeded their quota should make up for that. In addition, it maintains, in (D3) and (D4), that those who are able to do more should bear more onerous duties. The same two reasons for affirming "differentiated" duties are contained within the notion of common but differentiated responsibilities.[76] This can be seen in the last sentence of Principle 7 of the Rio Declaration quoted above, which claims that "The developed countries acknowledge the responsibility that they bear in the international pursuit of sustainable development in view of *the pressures their societies place on the global environment* and of *the technologies and financial resources they command*."[77]

Having noted these commonalities, it is worth stressing some key differences. First, the principle of common but differentiated responsibility refers to the responsibilities of states. This is apparent, for example, in Article 7 of the Rio Declaration quoted above. The same is also true of Article 10 of the Kyoto Protocol.[78] The hybrid account, however, does not restrict its duties to states. Furthermore, given the considerations adduced above in section 2 in particular, we have no reason to ascribe duties only to states. A second key difference is that the principle of common but differentiated responsibility tends to be interpreted in such a way that states are held accountable for the decisions of earlier generations.[79] But such a position is, I have argued, unfair on current generations, and (D1) to (D4) do not accept these kinds of historical responsibilities. Another difference is that the principle of common but differentiated responsibility, unlike the hybrid account, does not take into account what I have termed excusable ignorance. In the light of these three differences (as well as others), (D1) to (D4) would have quite different implications from the principle of common but differentiated responsibility as that principle is conventionally interpreted. In short, then, we might say that the hybrid account is one way of interpreting the general values affirmed by the principle of common but differentiated responsibility but that it departs considerably from the standard versions of that principle affirmed in international legal documents.

11. Concluding Remarks

It is time to conclude. Two points in particular are worth stressing—one methodological and the other substantive. The methodological observation takes us to an issue that has run throughout the chapter—namely, whether we should adopt an individualist methodology or a collectivist one. This issue has cropped up in three different contexts.

First, who are the polluters? If we take an individualist approach, then we will say that for some pollution (that of earlier generations) we cannot make the polluter pay, for the individual polluters are dead. If, however, we take a collectivist approach, we will say that collective A polluted in an earlier decade (or century) and hence that it should pay for the pollution now.

Second, who has benefited from the use of fossil fuels? Because of the nonidentity problem, we cannot say to the particular individuals who are alive today, "You enjoy a higher standard of living than you would have done in a world in which industrialization had not occurred." We can, however, make this claim to, and about, collectives.

Third, who is the bearer of the right to emit greenhouse gases (individuals or collectives)? The rationale given by Shue and Neumayer and by Agarwal, Narain, and Sharma for a historical approach works only if we assume that the answer to this question is "collectives." On an individualist approach, however, the rights bearers are individuals, and it is unjust to impose sacrifices on some current individuals because, and only because, of the excessive emissions of earlier inhabitants of their country.

This chapter has argued for an individualist account, but the issue requires a much fuller analysis than has been possible here.

The second point worth stressing is that if the arguments of this chapter are correct, then one common way of attributing responsibilities (the PPP) is more problematic than is recognized. In the light of this, I have suggested an alternative view, which overcomes some of these difficulties. It should be stressed that much more remains to be done. Further theoretical and empirical analysis is needed to answer the question raised earlier on about who is (causally) responsible for global climate change. Furthermore, much more is needed on the appropriate distributive criterion and on how we ascertain a fair quota. However, what I hope to have shown is that the account I have outlined provides the beginnings of an answer.

Acknowledgments

Earlier versions of this paper were presented at the Department of Politics, University of Leicester February 16 2005—the day the Kyoto Protocol came into effect); a Conference on "Global Democracy, the Nation-State and Global Ethics" held at the Centre for the Study of Globalization at the University of Aberdeen, which was funded by the Leverhulme Trust (March 18–20, 2005); the Annual Conference of the Political Studies Association (April 5–7, 2005); the symposium on "Cosmopolitism, Global Justice and International Law" organized by the *Leiden Journal of International Law* and the Grotius Centre for International Legal Studies (April 28, 2005); and, finally, as a plenary lecture at the Seventh Graduate Conference in Political Theory at the University of Warwick (May 7, 2005). I am grateful to those present for their questions and comments. I am particularly grateful to Roland Axtmann, Marcel Brus, Matthew Clayton, John Cunliffe, Lorraine Elliott, Carol Gould, James Pattison, Fabienne Peters, Roland Pierik, Thomas Pogge, and Kok-Chor Tan. Special thanks are due to Wouter Werner for his suggestions and to my commentator at the symposium at the Grotius Centre for International Legal Studies at the Hague, Peter Rijpkema, for his response to my paper. This research was conducted as part of an Arts and Humanities Research Council research project on "Global Justice and the Environment" and I thank the AHRC for its support.

Notes

1. Margaret Thatcher in 1982 during the Falklands War. Quoted in S. Barnes, "Want to Save the Planet? Then Make Me Your Not So Benevolent Dictator" *Times*, April 9, 2005.

2. See, in this context, Rawls's discussion of the "problems of extension" and in particular his discussion of the issues surrounding how "justice as fairness" is extended to deal with the international domain, future generations, duties to the environment, and nonhuman animals (as well as its extension to the ill): "The Law of Peoples," *Collected Papers*, ed. S. Freeman (1999), 70.531 (and, more generally, 531–33). See also J. Rawls, *Political Liberalism* (1993), pp. 20–21.

3. See J. Rawls, *A Theory of Justice* (1999), pp. 54–55, 78–81, 348, and E. Kelly, ed., *Justice as Fairness: A Restatement* (2001), 57–61, 168–176.

4. See M. C. Nussbaum, *Women and Human Development: The Capabilities Approach* (2000), and A. Sen, "Capability and Well-being," in M. Nussbaum and A. Sen, eds., *The Quality of Life* (1993), pp. 30–53. For an excellent analysis of several different accounts of what should be distributed and an assessment of their implications for our evaluation of global climate change, see E. Page, *Climate Change, Justice and Future Generations* (2006), chap. 3.

5. In what way might the environment be a distinctive kind of problem? I shall not explore this question fully here but note that possible answers might be that (i) some natural resources are nonrenewable and hence their consumption is irreversible; (ii) the value of some natural resources cannot adequately be captured in monetary terms; or (iii) many environmental benefits and burdens are essentially public in nature (that is, for a contiguous group of people, either all are exposed to an environmental hazard such as air pollution or none is).

6. For more on this, and for my defense of a cosmopolitan approach, see S. Caney, *Justice beyond Borders: A Global Political Theory* (2005). See also the defenses of a cosmopolitan approach in T. Pogge "Recognized and Violated by International Law: The Human Rights of the Global Poor," and K. C. Tan, "International Toleration: Rawlsian versus Cosmopolitan," both in *Leiden Journal of International Law* 18 (2005).

7. For an analysis of the latter, see D. Parfit, *Reasons and Persons* (1986), appendix F, pp. 480–486.

8. For Rawls's claim that the principles that apply across generations are distinct from those that apply within one generation and for his affirmation of a "just savings" principle, see J. Rawls, *A Theory of Justice* (1999), pp. 251–258, and Rawls, *Justice as Fairness*, pp. 159–160. We should also note that Rawls does not simply propose two different principles to govern "justice to contemporaries" and "justice to future people"; he also adopts two different methods for deriving these principles. As is well known, his derivation of the principles

of justice to govern contemporary members of a society invokes what persons seeking to advance their own primary goods would choose in the "original position." His derivation of the principles of justice to govern future generations also invokes the original position but, in its last form, stipulates that parties should choose that principle that they would want preceding generations to have honoured. (On the latter, see *Justice as Fairness*, p. 160.) So Rawls treats intergenerational justice very differently from "justice to contemporaries"—both in the method he employs and in the conclusions he reaches. (Note that Rawls's method for deriving the "just savings" principle has changed over time: see p. 160, n. 39.)

9. See, further, S. Caney, "Global Distributive Justice and the Environment," in R. Tinnevelt and G. Verschraegen, eds., *Between Cosmopolitan Ideals and State Sovereignty: Studies on Global Justice* (2006), pp. 51–63.

10. See the IPCC's Web site: http://www.ipcc.ch/about/about.htm.

11. B. Lomborg, *The Sceptical Environmentalist: Measuring the Real State of the World* (2001), chap. 24. For one critical response, see M. A. Cole, "Environmental Optimists, Environmental Pessimists and the Real State of the World—An Article Examining *The Sceptical Environmentalist: Measuring the Real State of the World* by Bjorn Lomborg", *Economic Journal* 113 (2003): 488.

12. See J. J. McCarthy, O. F. Canziani, N. A. Leary, et al., eds., *Climate Change 2001: Impacts, Adaptation and Vulnerability—Contribution of Working Group II to the Third Assessment Report of the Intergovernmental Panel on Climate Change* (2001).

13. The distinction between mitigation and adaptation comes from the IPCC So, e.g., vol. 2 of the 2001 report of the IPCC focuses on adaptation: see, in particular, B. Smit and O. Pilifosova, "Adaptation to Climate Change in the Context of Sustainable Development and Equity," in McCarthy et al., chap. 18. Vol. 3, by contrast, focuses more on mitigation: see B. Metz, O. Davidson, R. Swart, and J. Pan, eds., *Climate Change 2001; Mitigation—Contribution of Working Group III to the Third Assessment Report of the Intergovernmental Panel on Climate Change* (2001). See also Henry Shue's illuminating analysis of the different ethical questions raised by global climate change: "Subsistence Emissions and Luxury Emissions," (1993) 15 *Law and Policy* 40 and chapter 11 in this volume; "After You: May Action by the Rich Be Contingent upon Action by the Poor?" (1994) *Indiana Journal of Global I Legal Studies* 344; and "Avoidable Necessity: Global Warming, International Fairness, and Alternative Energy," in I. Shapiro and J. Wagner deCew, eds., *Theory and Practice: NOMOS XXXVII* (1995), 240.

14. The mitigation costs incurred by an actor A are not restricted to cases where A minimizes A's own GHG emissions. Consider, e.g., the "clean development mechanism" policy enunciated in Article 12 of the Kyoto Protocol (http://unfccc.in resource/docs/convkp/kpeng.html). Under this proposal certain countries (those listed in Annex I) may be given credit for cutting GHG emissions if they support the use of development projects that enable developing countries to develop in a way that does not emit high levels of GHGs. Since what they do has the effect of lowering GHG emissions and it has a cost for them (the cost of supporting clean development), then, in principle, this cost should be included under the heading of mitigation costs: they are making a sacrifice which enables there to be a reduction in GHG emissions.

15. See, e.g., Lomborg, pp. 305–318, esp. 318. Lomborg takes the highly controversial view that it would be more cost effective to focus on "adaptation" rather than "mitigation." For a contrasting view, see J. Houghton, *Global Warming: The Complete Briefing* (2004), pp. 242–321, and M. Maslin, *Global Warming: A Very Short Introduction* (2004), pp. 136–43.

16. For two excellent treatments of the role of PPP in international environmental law to which I am much indebted, see P. Birnie and A. Boyle, *International Law and the Environment* (2002), pp. 92–95, 383–385; and P. Sands, *Principles of International Environmental Law* (2003), pp. 279–285.

17. The documents for both Council Recommendations can be found in OECD, *The Polluter Pays Principle: Definition, Analysis, Implementation* (1975).

18. See Directive 2004/35/CE of the European Parliament and of the Council (passed on April 21, 2004) on environmental liability with regard to the prevention and remedying of environmental damage. The text can be found in the Official Journal of April 30, 2004 (L143), at http://europa.eu.int/eur-lex/pri/en/oj/dat/2004/l_143/l_14320040430en00560075.pdf.

19. Commission on Global Governance, *Our Global Neighbourhood* (1995), pp. 208, 212.

20. See H. Shue "Global Environment and International Inequality" (1999) 75 *International*

Affairs, 533–537. Shue writes that his argument is not equivalent to the PPP because he interprets the PPP to be a "forward-looking" principle that says that future pollution ought to be paid for by the polluter (at 534). However, I shall interpret the PPP to refer to the view that past, current, and future pollution ought to be paid for by the polluter. Shue is therefore affirming a "polluter pays" approach, given the way in which I am defining that term.

21. See E. Neumayer, "In Defence of Historical Accountability for Greenhouse Gas Emissions" (2000) 33 *Ecological Economics*, 185–192.

22. See F. L. Toth and M. Mwandosya, "Decision-Making Frameworks" in Metz et al., pp. 669 (for mention of the PPP) and 668–673 (for general discussion).

23. Here it is interesting to note that the European Union's recent directive on environmental liability (2004/35/CE) expressly rejects what I am terming the macro-version of the polluter pays principle and affirms the micro-version. See para. 13 and Art. 4 s. 5. available at http://europa.eu.int/eur-lex/pri/en/oj/dat/2004/L143/L14320040430enoo560075.pdf.

24. See, e.g., Shue, pp. 534, 545. Shue elsewhere refers to "nations (or other parties)" ("After You" p. 361) but generally assumes that the polluters/payers are nations.

25. Neumayer, p. 186.

26. Ibid.

27. See Toth and Mwandosya, pp. 637–638.

28. As noted above, Shue maintains that industrialized countries should pay. Notwithstanding this, he also refers to the actions of "the owners of many coal-burning factories" (p. 535)—a level (b) explanation.

29. For Pogge's discussion of "explanatory nationalism," see his *World Poverty and Human Rights: Cosmopolitan Responsibilities and Reforms* (2002), pp. 15, 139–144.

30. Shue, p. 536.

31. Ibid.

32. Neumayer, p. 189. I have omitted a footnote (n. 4) which appears after the word "achieved."

33. A similar position is defended by Axel Gosseries in his illuminating and interesting paper, "Historical Emissions and Free-Riding" (2004) II *Ethical Perspectives*, 1, pp. 36–60. I came across Gosseries's paper after completing this article and hope to address it more fully subsequently.

34. Parfit, chap. 16.

35. This claim about the impossibility of benefiting future people, we should note, has also been made by Thomas Schwartz. In a pioneering paper published in 1978 he presented reasoning like that given in the last paragraph to show that the policies of present generations do not benefit future generations. Schwartz's argument is directed against population policies that are justified on the grounds that they would make future people better off. His argument, though, also tells against claims that current individuals have been made better off by industrialization (and hence that they have a duty to pay for the GHG emissions that were generated by this benefit-producing industrialization). See T. Schwartz, "Obligations to Posterity," in R. I. Sikora and B. Barry, eds., *Obligations to Future Generations* (1978), pp. 3–13. I came across Schwartz's paper only when I had finished this article and was making the final revisions. My debt here is to Parfit's work.

36. Here we should note one complication to my argument. In an appendix to *Reasons and Persons* Parfit entertains the possibility that bringing someone into existence can be said to benefit them. He does not commit himself to this view, but he does think it is a potentially plausible view. To this extent there is an asymmetry in his treatment of harm to future generations (one cannot harm future people because the dangerous policies affect who is born) and his treatment of benefit to future generations (one can benefit future people by bringing them into existence). See Parfit, pp. 487–490). See, in particular, Parfit's discussion of what he terms "the two-state requirement," where this states that "We benefit someone only if we cause him to be better off than he would otherwise at that time have been" (p. 487; see further pp. 487–488). I am grateful to Edward Page for bringing this asymmetry to my attention and for a number of very helpful discussions of the issues at stake. I shall not seek to challenge Parfit's arguments to the effect that bringing people into existence may benefit them. I would, however, maintain that to sustain the disanalogous treatment of future harm and future benefit, Parfit needs to confront the possibility that the non-identity problem undermines the claim that we can benefit future people and also needs to explain why that is not correct. Without such an argument, the non-identity problem would (as I have argued in the text) appear to undermine the BPP. Note: Schwartz's account can be contrasted with Parfit's on this point for, unlike Parfit, Schwartz explicitly claims that one cannot either harm or benefit future people, Schwarty, pp. 3–4.

37. We might, of course, say, "You should pay because you are so much better off than others," but this appeals to a quite different principle and will be taken up later.

38. A collectivist account is suggested by Edward Page in an ingenious discussion of Parfit's non-identity problem. See E. Page, "Intergenerational Justice and Climate Change" (1999) 47 *Political Studies* 1, pp. 61–66. See also J. Broome, *Counting the Cost of Global Warming* (1992), pp. 34–35.) Page, however, is not addressing the argument I am making. Rather he employs a collectivist approach to rebut Parfit's non-identity problem.

39. Page, pp. 61–66.

40. See also Schwartz's discussion of this position. Schwartz briefly considers the collectivist position described in the text. He rejects it on the grounds that what matters are benefits to individuals; benefits to collectives have no moral weight. Schwartz, pp. 6–7.

41. See J. Rawls, *The Law of Peoples with "The Idea of Public Reason Revisited"* (1999), pp. 117–118.

42. D. Miller, "Justice and Global Inequality," in A. Hurrell and N. Woods, eds., *Inequality, Globalization, and World Politics* (1999), pp. 193–196.

43. See also Gosseries, pp. 41–42, on this. For a nice discussion of the way in which collectivist approaches are insufficiently sensitive to the entitlements of individuals, see K.-C. Tan, *Justice without Borders: Cosmopolitanism, Nationalism and Patriotism* (2004), pp. 73–74. More generally, see S. Caney, "Global Equality of Opportunity and the Sovereignty of States," in T. Coates, ed., *International Justice* (2000), pp. 142–143, for a discussion of the principle at issue.

44. Neumayer, p. 186.

45. P. Singer, *One World: The Ethics of Globalization* (2002), p. 34.

46. Neumayer, p. 188. Shue also makes this kind of response but does not specify a key date after which one cannot claim excusable ignorance: Shue, p. 536.

47. Shue, pp. 535–536. Neumayer makes a similar but distinct reply, arguing that the objection runs together "blame" and "accountability" Blaming those who could not have known of the effects of their actions is unjustified, but they are nonetheless accountable: see pp. 188, 189 n. 4.

48. H. Shue, *Basic Rights: Subsistence, Affluence, and U. S. Foreign Policy* (1996), pp. 164–166.

49. A similar position is defended by Gosseries. See his nuanced and persuasive treatment of what he terms "the ignorance argument," pp. 39–41.

50. This emphasis is evident in Neumayer's discussion; see p. 188.

51. We might note a third response. Someone might reply that even though there was not incontrovertible evidence prior to 1990/1985, there was reason to think that GHGs caused global climate change. And drawing on this, they might argue that pre-1990/1985 emitters had a duty to act on the precautionary principle and therefore should have eschewed activities that released high levels of GHGs. Since they did not adopt such a precautionary approach, it might be argued, they should shoulder a proportionate share of the burdens of climate change. The argument then is neither that they did know (response 1) nor that it does not matter that they did not know (response 2); it is that they should have adopted a cautious approach, and since they did not, they are culpable. How effective this reply is depends on when we think the precautionary principle should be adopted.

52. For further discussion of this methodological point, see Caney, "Global Distributive Justice."

53. Shue, "Global Environments" pp. 533–534.

54. Neumayer, p. 188.

55. A. Agarwal, S. Narain, and A. Sharma, "The Global Commons and Environmental Justice—Climate Change," in *Environmental Justice: International Discourses in Political Economy–Energy and Environmental Policy* (2002), p. 173.

56. My argument here is analogous to Rawls's discussion of "legitimate expectations," Rawls argues that we cannot define persons' entitlements (their legitimate expectations) until we have identified a valid distributive principle and ascertained what social and political framework would best fulfill that ideal, only when we have the latter can we work out what individual persons are entitled to. In the same spint, my argument is that we cannot define people's responsibilities until we have identified a valid distributive principle and seen what social and political framework realizes that ideal. See Rawls, *A Theory of Justice*, pp. 88–89, 273–277.

57. Shue, "Global Environment," 534.

58. Neumayer, "In Defense."

59. The terminology of "primary" and "secondary" duty bearers comes from Shue, *Basic Rights*, p. 59 (see also, p. 57, 171 for relevant discussion).

60. J. Raz, *The Morality of Freedom* (1986), chap. 7.

61. For a comprehensive account of the effects of climate change and empirical support for the claim that climate change causes phenomena (a) to (f), see McCarthy et al. For instance, chap. 9 on human health provides data on the links between climate change and drought (a), heatstroke (b), and malaria, dengue, and cholera (c). Chap. 6 and 17 detail the ways in which climate change results in threats to coastal zones and small island states and thereby results in (d), (e), and (f).

62. The argument sketched above could be generalized to address other environmental burdens. For an excellent analysis of the human right not to suffer from various environmental harms, the grounds supporting this right, and the correlative duties, see J. Nickel, "The Human Right to a Safe Environment Philosophical Perspectives on Its Scope and Justification," (1993) 18 *Yale Journal of International Law*, pp. 281–295. Nickel does not discuss climate change.

63. For an emphasis on constructing fair *institutions* see Shue, *Basic Rights*, pp. 17, 59–60, 159–161, 164–166, 168–169, 173–180, and C. Jones, *Global Justice: Defending Cosmopolitanism* (1999), pp. 66–72, esp. 68–69.

64. For a general discussion of the different kinds of duties generated by human rights, see S. Caney, "Global Poverty and Human Rights: The Case for Positive Duties," in T. Pogge, ed., *Freedom from Poverty as a Human Right: Who Owes What to the Very Poor?* (2006). For an account that is similar to the hybrid account, see Darrel Moellendorf's brief but perceptive discussion in *Cosmopolitan Justice* (2002), pp. 97–100. Like the hybrid account, Moellendorf's account brings together a "polluter pays" approach with an "ability to pay" approach. There are, however, several important differences: Moellendorf's view does not take into account excusable ignorance; he does not address the question of what to do if people do not comply with their duty not to emit excessive GHGs; and he does not propose a principle akin to (D_4).

65. P. Singer, "Famine, Affluence, and Morality" (1972) 1.3 *Philosophy and Public Affairs*, pp. 229–243. I am grateful to Kok-Chor Tan for advice about how to bring out the normative appeal of (D_3) and (D_4).

66. Caney *Justice beyond Borders*, chap. 4.

67. In 'Global Environment and International Inequality,' Shue defends not just the "polluter pays" principle but also an "ability to pay" principle (pp. 537–540). In addition to this, he argues that there should be a "guaranteed minimum" threshold below which people should not fall and hence that the very poor should not pay (pp. 540–544). Shue's claim is that all three principles yield the same conclusion—affluent countries are responsible for meeting the burdens of climate change (p. 545). For further discussions where Shue has argued that the wealthy should bear the mitigation and adaptation costs of climate change and that the poor be given less demanding duties see "After You: May Action by the Rich be Contingent upon Action by the Poor?," "Avoidable Necessity: Global Warming International Fairness, and Alternative Energy," pp. 250–257, and "Subsistence Emissions and Luxury Emissions," especially pp. 42–43.

68. One common principle suggested is that all persons have an equal per capita right to emit carbon dioxide. See, e.g., R. Attfield, *Environmental Ethics* (2003), pp. 179–180; P. Baer, J. Harte, B. Haya, et al., "Equity and Greenhouse Gas Responsibility," (2000) 289 *Science*, p. 2287; T. Athanasiou and P. Baer, *Dead Heat: Global Justice and Global Warming* (2002), esp. pp. 76–97; Neumayer, "In defence", pp. 185–192; and S. Bode, "Equal Emissions Per Capita over Time—A Proposal to Combine *Responsibility* and *Equity of Rights* for Post-2012 GHG Emission Entitlement Allocation," (2004) 14 *European Environment*, pp. 300–316. For the reason given in the text, however, this seems to me unfair on the poor. See, too, Shue, "Avoidable Necessity: Global Warming, International Fairness, and Alternative Energy," pp. 250–252, and S. Gardiner, "Survey Article: Ethics and Global Climate Change" (2004) 114 *Ethics*, pp. 584–585).

69. Note that even if (a) and (b) hold and the PPP and the hybrid account identify the same people; duty bearers, they may well make different demands on different people. They would converge exactly only if (a) and (b) hold, and if (c), there was a perfect positive correlation between how much GHGs persons have emitted on the one hand and how much wealth they possess on the other. Since the PPP allocates duties according to how much GHGs persons have emitted and since the hybrid account allocates duties, in part, according to how much wealth persons have, (c) is necessary to produce total convergence in their ascription of duties.

70. For further on this, see: J. B. Smith, H.J. Schellnhuber, and M. M. Q. Mirza, "Vulnerability to Climate Change and Reasons for Concern: A Synthesis," in McCarthy et al., pp. 916, 940–941, 957–958; R. S. J. Tol, T. E. Downing, O. J. Kuik, and J. B. Smith, "Distributional Aspects of Climate Change Impacts," (2004) 14.3. *Global*

Environmental Change pp. 259–272; and D. S. G. Thomas and C. Twyman, "Equity and Justice in Climate Change Adaptation Amongst Natural-Resource-Dependent Societies," (2005) 15.2 *Global Environmental Change*, pp. 115–24.

71. For a similar line of reasoning, see Moellendorf's persuasive discussion of who should deal with the ill effects caused by the GHG emissions of earlier generations. Moellendorf convincingly argues that it would be wrong to ask anyone other than the most advantaged to bear the burdens of the GHG emissions of earlier generations. See Moellendorf, p. 100. My claim is that the same reasoning shows that the advantaged should also bear the costs stemming from noncompliance.

72. As Wouter Werner has noted, the hybrid account may also encounter a problem of noncompliance. What, it might be asked, should happen if some of those designated by (D3) to deal with the problems resulting from some people's noncompliance do not themselves comply with (D3)? Three points can be made in response. First, we should recognize that even if this is a problem for the hybrid account, it does not give us reason to reject it but rather to expand on it. The PPP, for example, fares no better; indeed, it fares worse, for the hybrid account, unlike the PPP, at least addresses the issue of noncompliance. Second, if *some* of those designated to perform (D3) fail to do so, then one reply is that at least some of their shortfall should be picked up by others of those designated by (D3). The third and final point to make is that there will be limits on this. That is, one can ask only so much of those able to help, though where we draw this line will be a matter of judgment and will depend on, among other factors, how much they are able to help and at what cost to themselves. For further pertinent inquiry into persons' moral obligations when others fail to do their duty, see Parfit's pioneering discussion of "collective consequentialism" (pp. 30–31), and Liam Murphy's extended analysis of this issue in *Moral Demands in Non-ideal Theory* (2000).

73. Available at http://www.un.org/documents/ga/conf151/aconf15126-1annext.htm (emphasis added).

74. See http://unfccc.int/resource/ccsites/senegal/conven.htm.

75. For the Preamble and Art. 10 of the Kyoto Protocol, see http://unfccc.int/resource/docs/convkp/kpeng.html. My understanding of the principle of common but differentiated responsibility is indebted to Lavanya Rajamani's instructive discussion of that principle as it appears in the Rio Declaration, the UNFCCC, and the Kyoto Protocol; see her "The Principle of Common but Differentiated Responsibility and the Balance of Commitments under the Climate Regine", (2000) 9.2 *Review of European Community and International Environmental Law*, pp. 120–131. See also the useful discussions by P. G. Harris, "Common but Differentiated Responsibility: The Kyoto Protocol and United States Policy" (1999) 7.1 *New York University Environmental Law Journal*, pp. 27–48, and Sands, pp. 55–56, 285–289, 362.

76. See, on this, Rajamani, esp., and Sands, p. 286.

77. Available at http://www.un.org/documents/ga/conf151/aconf15126-1annex1.htm (emphasis added).

78. Available at http://unfccc.int/resource/docs/convkp/kpeng.html.

79. See Rajamani, pp. 121–122. Rajamani goes further than this and writes that the principle of common but differentiated responsibility requires that states that have emitted high levels of GHGs in the past have a duty specifically to pursue a policy of *mitigation* and that this duty falls on them (pp. 125, 126, 130). The hybrid account, by contrast, is not, of necessity, committed to this claim.

Deadly Delays, Saving Opportunities

Creating a More Dangerous World?

Henry Shue

Will there really be a "morning"?
Is there such a thing as "Day"?
Could I see it from the mountains
If I were as tall as they?

—EMILY DICKINSON

We now know that anthropogenic emissions of greenhouse gases (GHGs) are interfering with the planet's climate system in ways that are likely to lead to dangerous threats to human life (not to mention nonhuman life)[1] and that are likely to compromise the fundamental well-being of people who live at a later time.[2] We have not understood this for very long—for most of my life, for example, we were basically clueless about climate. Our recently acquired knowledge means that decisions about climate policy are no longer properly understood as decisions entirely about *preferences of ours* but also crucially about the *vulnerabilities of others*—not about the question "How much would we like to spend to slow climate change?" but about "How little are we in decency permitted to spend in light of the difficulties and the risks of difficulties to which we are likely otherwise to expose people, people already living and people yet to live?" For we now realize that the carbon-centered energy regime under which we live is modifying the human habitat, creating a more dangerous world for the living and for posterity. Our technologically primitive energy regime based on setting fire to fossil fuels is storing up, in the planet's radically altering atmosphere, sources of added threat for people who are vulnerable to us and cannot protect themselves against the consequences of our decisions for the circumstances in which they will have to live—most notably, whichever people inherit the worn-and-torn planet we vacate.[3]

Large Disasters, Considerable Likelihoods

As we academics love to note, matters are, of course, complicated. Let's look at a few of the complications, concentrating on some concerning risk.[4] Mostly, we are talking about risks because, although we know strikingly much more about the planetary climate system than we did a generation ago, much is still unknown and unpredictable. I will offer three comments about risk. The third comment is the crucial one and makes a strong claim about a specific type of risk, with three distinctive features. After illustrating the three features with the effects of a possible bird-flu epidemic, I then argue somewhat more fully that the three features are also jointly characteristic of the effects of climate change, with strong implications for how we should regard our recently discovered complicity in producing climate change and thereby worsening the circumstances to which whoever succeeds us will need to adapt. Then I will consider the specific implications for what it is most essential and urgent to do.

The first point to be made about risk and climate change, however, is that not everything is uncertain, for two reasons. One is simply that some threatening changes in the climate are already occurring, as practically every informed person acknowledges. I would not have the scientific knowledge to sort through very many specifics, but clearly, for example, patterns of rainfall and storm intensity have already changed somewhat, resulting in both flooding and drought.[5] The other reason is that unless virtually all human understanding about the climate were completely misguided, other changes are practically certain to occur; for example, sea level will surely rise significantly.[6] If nothing else, the volume of the water would increase from the rise in temperature that has already straightforwardly been measured. And other factors are converging on sea-level rise, such as amazingly rapid melting of Arctic and Greenland ice that both directly increases the amount of water in the ocean, when the melting ice was previously on land (i.e., was an ice sheet, not an ice shelf),[7] and indirectly warms the planet by reducing albedo through elimination of the reflectivity of the snow. Some island nations in the South Pacific are already well into the process of being submerged by rising sea levels. Nothing in my argument to follow turns on how much is already in fact happening, since I will mostly be discussing risk of future events, but it is simply factually misleading to talk as if climate change is all risk only and nothing untoward is happening yet.

Now, what about the risks, which virtually anyone will acknowledge are fortunately still most of the problem? The second point to note about risk is that it is highly significant morally whether one is choosing a risk for oneself or imposing it, conditionally or unconditionally, on others. A certain level of risk may be a reasonable one for me to choose for myself but not a reasonable one for me to impose on others. Therefore, even if the level of risk from climate change imposed on future generations were the same as the risk for us—of course, it is not remotely the same—it might still be unreasonable for us to impose it on them, even if it were not unreasonable for us to choose it for ourselves. I am free to choose to mountain-climb, but I could not reasonably propose that an experience of mountain-climbing be a requirement for graduation from university, because mountain-climbing is too dangerous to require of others generally. That we are imposing risks that others will inherit at birth is extremely important.

Risk is most often explained as the product of magnitude and probability. The magnitude is a measure of the seriousness of the loss risked, and the probability is a measure of the likelihood of that loss occurring. Some corporate and government opponents of vigorous action to slow climate change, especially coal and oil interests, have made much of alleged "uncertainty." In many cases they have purposely distorted the science and wildly exaggerated the extent of our current ignorance; Steve Vanderheiden has aptly characterized this intentional smoke blowing as "manufactured uncertainty."[8] The tobacco companies always claimed that the connections between smoking and bad health were uncertain; the coal and oil companies claim the connections between carbon combustion and bad climate are uncertain. Neither connection is uncertain. But what I want to show here is that there is a crucial kind of cases in which a considerable degree of uncertainty does not matter even if we are appropriately uncertain without having been tricked by industry and government propaganda.

I will defend the suggestion—this is the third, and chief, point about risk—that there are cases in which one can reasonably, and indeed ought to, ignore entirely questions of probability beyond a certain minimal level of likelihood. These are cases with three features: (1) *massive loss*: the magnitude of the possible losses is massive; (2) *threshold likelihood*: the likelihood of the losses is significant, even if no precise probability can be specified, because (a) the mechanism by which the losses would occur is well understood, and (b) the conditions for the functioning of the mechanism are accumulating; and (3) *non-excessive costs*: the costs of prevention are not excessive (a) in light of the magnitude of the possible losses and (b) even considering the other important demands on our resources.[9] Where these three features are all present, one ought to try urgently to make the outcome progressively more unlikely until the marginal costs of further efforts become excessive, irrespective of the outcome's precise prior probability, which may not be known in any case. We know that our actions now are opening the doors to some terrible outcomes; we ought to reclose as many of these doors as we can. The suggestion, then, is that these three features jointly constitute a sufficient set for prompt and robust action to be required.[10] When all three conditions are present, action ought to be taken urgently and vigorously. Doing nothing but calling for further research is morally irresponsible, I will now argue. Obviously, further research is also good provided that it is not a substitute for effective action.

Basically, the argument is that because the magnitude of particular losses is so serious, the only acceptable probability is as close as possible to zero, provided this reduction in likelihood can be achieved at a cost that is not inordinate. Some losses would be utterly intolerable, especially "losses" involving massive deprivations of necessities to which all people, regardless of individual identity, have rights simply as human beings. This applies to (a) some cases in which the probability is known and small but still significant and (b) some cases in which the probability cannot be calculated but can be known to be significant (because the relevant mechanism is understood and the conditions for its functioning are appearing). Only the latter would be a case of uncertainty in the technical sense, that is, an event with no calculable probability. Obviously, several aspects of this argument each need separate discussion.

I begin with a preliminary reminder about uncertainty. That something is uncertain in the technical sense, that is, has no calculable probability, in no way suggests that its objective probability, if known, would be small.[11] There is a grand illusion here: if we cannot see what the probability is, it must be small. Perhaps we assume a visual metaphor: we cannot see the probability because it is too small to see, so it must be really tiny. This inference is totally groundless. If all we know is that the probability cannot be calculated, then we do not know anything about what it is; if we do not know anything about what it is, then we do not know whether it is small or large.

However, we might have independent evidence that a likelihood is either small or large, without being able to calculate the probability. Cases of type (b) just mentioned above are such cases in which we cannot calculate a probability but know on other grounds that the likelihood is significant. The point now is that the simple fact that the probability is uncertain does not entail that it is small. Thinking so would be like thinking that if you are not sure where a city is located, the city must be small. We may simply be totally overlooking an entire dimension of a problem that will turn out to be huge. Things can be invisible for reasons other than being small. Some probabilities unknown at one time turn out later to be very large. Often the universe has major surprises for us, some very unpleasant.

Next we turn to cases illustrating the three features that I think require us to push the probability as close as we can to zero, whatever exactly it is now, given that it is significant. One example of the kind of case I have in mind is the reasons for the measures now being taken to prevent a possible bird-flu pandemic.[12] In a bird-flu pandemic, (1) the losses would be massive; (2) the likelihood of occurrence is significant even if it cannot be calculated because the mechanism of occurrence is well understood and conditions for its functioning have appeared; and (3) the costs of prevention, while far from negligible, are (extremely) moderate in light of (a) the possible losses and (b) even the other legitimate demands on resources.

First, the losses could be massive. Tens of millions of people died from the 1918 flu epidemic that helped to end World War I.[13] Now that we have enhanced globalization, including rapid movements of large numbers of people for great distances, it is entirely possible that deaths from a flu pandemic would be in the hundreds of millions of people—a modern, global plague.

And second, we understand the mechanisms by which this would happen and can see conditions favorable for the working of the mechanism arising. This second one is the "antiparanoia" requirement, designed to narrow the range of possibilities on which we need to act. By requiring a clear mechanism we avoid reacting similarly to every imaginable threat. If all the oxygen on earth burst into flame, that, too, would be a disaster, but we do not know of any way that could happen. The specification of a clear mechanism is the central contributor to our conviction that the probability is significant in spite of our not being able to calculate it.

Human flu is highly contagious, and the active bird flu has already rapidly mutated several times. Nothing naturally prevents a mutation into a form directly transmissible from human to human—it is the precise probability of this occurring that is unknown. Once the mutated flu was passing directly among humans, it would move quickly if no directly applicable vaccine had been prepared in sufficient quantity in advance of the outbreak of the pandemic. Vaccine has a production time of months using current technology; this is the problem of lead time, which is monumentally important in the case of climate change. It would take a very long time, depending on how many labs were manufacturing vaccine, to produce, say, 1 billion doses, which would still leave five out of six humans unprotected, providing only enough vaccine for a population the size of either China or India.

Meanwhile, the virus might mutate again, making the vaccine already produced until that time ineffective. So actually, the best argument for doing nothing to prevent a pandemic would be the fatalistic argument that it was impossible to stay ahead of the virus. But it is not known to be impossible—that, too, is uncertain—so I think we should try, as to some degree we are, because, third, preparing facilities for the manufacture of vaccine in large quantities, while expensive, is not prohibitively so.[14] It would be difficult to imagine a better investment of public funds than subsidizing this manufacturing capacity.

What I want to emphasize is that no precise probability of the pandemic plays any role in the argument whatsoever, primarily because the magnitude of the possible loss is so great—tens of millions or hundreds of millions of human lives. Another probability that would matter would be a virtual certainty that attempts at prevention would fail, making the funds spent on expanded production of vaccine a waste; even so, unless the costs were astronomical—on the order of a perpetual boondoggle such as the dysfunctional U.S. ballistic-missile defenses, for example—would it even begin to seem unreasonable to try. Extra manufacturing capacity for flu vaccine would cost only a tiny fraction of what is currently being wasted on misguided military systems.

Obviously, what I am next going to suggest is that a number of phenomena that could result from climate change, especially climate change allowed to build up even more momentum before anything serious is done to slow it, are like the possible flu pandemic in having the three key features: (1) the possible losses are massive; (2) while the precise probability of these losses occurring is unknown, their likelihood is significant because the mechanism by which they would occur is well understood and conditions for its functioning are falling into place, and (3) the costs of preventing these losses are not excessive—at least for now—in light of the magnitude of the possible losses, even taking into account the other important current demands on resources.

The three features must apply to each potential loss that is given weight in deciding what to do, and, of course, that each feature applies is an empirical claim that must be established with detailed scientific argument. My only hope here is to formulate a reasonable set of criteria; I lack the knowledge to make all of the various cases that the criteria are in fact satisfied. So I will merely briefly indicate the kind of empirical cases that need to be spelled out. We can count on the rapidly developing science to spell them out.

How might massive losses arise from climate change? For example, among ecosystems, agricultural systems are especially touchy.[15] Crops for humans need to be edible, which basically means they need to be just right. It cannot be too hot or too cold, too wet or too dry. If they are underripe, they cannot be eaten; if they are overripe, they cannot be eaten. If the rain comes too soon, they parch later; if the rain comes too late, they have already shriveled or will rot. Farmers already gamble on the weather. Climate change is long-term weather change. Gambling on climate change is raising the odds greatly against the already-wagering farmers, who keep us alive, when they are lucky.[16]

Generally speaking, if the weather changes faster than the crops can adapt, there is trouble, that is, shortage of food. Severe shortage in one place tends to mean higher prices in other places, if those whose own agriculture failed have enough money to import food. The famine can be exported, but it cannot be made to evaporate. As Amartya Sen demonstrated in *Poverty and Famines*, those with high incomes bid up the price of food, and those with low incomes starve.[17] So in the case of climate change, too, (1) the potential human losses could be massive. This is for many reasons, but a lethal one is disruption of food supplies, causing volatile food prices. Others include the need for massive relocations of population from low-lying shores inundated by rising sea levels.

(2) The mechanisms leading from burning fossil fuel, above all, to the climate changes are increasingly well understood.[18] Those connecting the climatic changes in turn to human misery were already well understood, since necessities as elemental as food and shelter are directly assaulted by the physical phenomena constituting climate change, such as more intense storms and atypical weather.

(3) The costs of prevention are moderate, although far from insignificant.[19] First come the "no regrets" measures that eliminate current costly energy waste and thereby improve living standards and reduce dependence on Middle Eastern dictatorships such as Saudi Arabia which are lightning rods for terrorism and entice heedless Western politicians into needless wars and bloated military budgets. Much of what we need to give up next after the economically and politically profitable, no-regrets reductions are frivolous preferences, life-shortening luxuries, and pointless indulgences. What we must give up after those depends on how long we continue to make the problem worse by continuing to derive our energy from fossil fuel before we begin to make it better by switching to alternative sources.[20] Plainly, delay will not make the necessary transition less painful—it will only shift it off us and onto others.

The Creation of a More Dangerous World

So far we have only a quick overview of the case of climate change, and now we need to look at selected aspects a little more thoroughly. The nature of what I have so far been vaguely referring to as "massive losses" can be specified more precisely. I will examine four aspects of danger, acknowledging three and setting aside the fourth.

Creating Danger

First, and most significant, failing to deal with climate change constitutes not only failing to protect future generations but inflicting adversity on them by making their circumstances more difficult and dangerous than they would have been without as much climate change, and more difficult and dangerous than circumstances are now for us.[21] If the current climate change were a naturally occurring problem, like some effects of human aging, and we did nothing to deal with it, we would leave future generations facing a problem that was only as severe when we bequeathed it as when we inherited it. We would have failed to provide protection—done nothing to make their lives less dangerous. That would be blameworthy, but what we would be guilty of would be a "sin of omission": neglecting to provide protection for subsistence rights that was ours to give if we had chosen to bother.[22]

Failing to deal with our climate change is not like that, because the current climate change is not naturally occurring. Political choices about energy policy are causing climate change. At some points in the planet's history, climate change has occurred naturally, but the climate change happening now is, as the scientists say, anthropogenic: people are causing it, by bringing about the emission of increasing amounts of greenhouse gases such as the CO_2 from the burning of fossil fuels in car engines and electricity-generating plants. Human activities are undermining the environmental conditions to which human beings have successfully adapted, making the environmental conditions for future generations more threatening for them than the present conditions are for us. "Doing nothing" about climate change in the sense of simply continuing business as usual is—far from actually doing nothing—continuing to change the environmental conditions that future generations will face for the worse. To persist in the activities that make climate change worse, and thereby make living conditions for future generations worse, is not merely to decline to provide protection. It is to inflict danger, and to inflict it on people who are vulnerable to us and to whom we are invulnerable.[23] The relationship is entirely asymmetric: they are at our mercy, but we are out of their reach. Causation runs through time in only one direction. Lucky for us.

Endangering Additional Generations

Second, failing to deal with climate change constitutes inflicting danger on additional generations that could have been spared. It is not only that future generations that are already fated to be adversely affected by the GHGs that have already been injected into the atmosphere by previous generations since the spread of the industrial revolution will face more adverse conditions of life than if we had managed to get a grip on our fossil-fuel consumption. Yet later generations, the great-great-great-grandchildren rather than the grandchildren, that might have been spared this problem if it had been solved sooner, will suffer from it. Suppose that if our generation did whatever it ought to do to stop accelerating climate change, the effects of climate change would have become manageable by some Generation L. If we do not do what we ought, and everything else remains the same, then at the very least the next generation, Generation M, will suffer from climate change. So, besides making life more treacherous for every generation from A to L, we would have inflicted completely avoidable problems on Generation M (and doubtless others), which would have been free of these problems if we had restrained our environmentally damaging activities, assuming only that tackling the problem sooner means solving it sooner.

This assumption would not be straightforwardly true if, say, some technology needed to mature before it could be successfully applied to climate change and attempts to employ it sooner would be futile. If we had reason to believe this was the situation, however, we would have no basis for merely increasing fossil-fuel consumption as usual. First, we could, instead of attempting to use the immature technology before it was ready, be seriously investing in research on improving the technology or on alternative technologies, rather than simply indulging in our own high-emissions consumption. Our investment now might allow an intermediate generation still to implement the by-then-mature technology in time to save Generation M. Second, and more important, we do not need to develop any new technologies in order simply to cease wasteful and frivolous uses of fossil fuels and to defeat shortsighted politicians who block policies that would make the wasteful pay and that would create disincentives for excessive emissions. Time passes while the problem remains untackled, so additional generations will suffer. But this is not the worst.

Creating Additional Dangers

Third, failing to deal with climate change constitutes not simply continuing to make the environment for human life more threatening but unnecessarily creating opportunities for it to become significantly more dangerous by feeding upon itself through positive feedbacks that would otherwise not have occurred—creating opportunities for the danger to escalate one or more levels. We have hardly scratched the surface of the seriousness of continued delay in facing the challenge of climate change. Climate change is dynamic. It involves many poorly understood feedbacks, negative as well as positive. It is conceivable that a continued worsening will trigger a negative feedback, such as an increase in the kinds of clouds that reflect sun waves back away from the earth, that will actually improve the situation for humans. Unknowns remain. But some of the best-understood and most likely feedbacks are positive, compounding the problem. For example, if emissions of CO_2 cause the Arctic tundra to thaw, as they appear well on the way to doing, the thawing tundra will release vast amounts of methane (CH_4), which is a far more powerful GHG per unit than CO_2 and will make climate change significantly more severe than it would have been if the tundra had not thawed.[24]

The opportunities we create for net positive feedbacks to occur may not be taken, or the positive feedback may somehow be more than canceled out by some now only more dimly foreseeable negative feedback. But it still seems wrong to create the opportunity for the positive feedback for no good reason. If I play Russian roulette with your head for my amusement as you doze and the hammer of the revolver falls on an empty chamber, I will have done you no physical harm. But I will have seriously wronged you by subjecting you to that unnecessary risk. We do no wrong when we unavoidably inflict risks on future generations, or even perhaps if we have compelling reasons for doing so where it would be avoidable. But we do wrong them if we subject them to opportunities for matters to worsen severely for no good reason except that we could not be bothered to change our comfortable habits and that the owners of the coal and oil reserves are greedy for maximum return. We can be justified in imposing a risk on others when the harm to ourselves from avoiding the risk to them would be severe—perhaps even if it would only be significant—but not when avoiding the imposition of the risk on them would cause us only mild inconvenience, or even serious but manageable difficulty, or leave us merely rich, not superrich.

The fourth aspect of danger is the most fearsome. For completeness, I need to mention it, but I will not rely on it in my argument.

Creating Desperate Dangers

Fourth, failing to deal with climate change constitutes not only unnecessarily creating opportunities for the planetary environment to become significantly worse for humans (and other living things) but unnecessarily creating opportunities for it to become catastrophically worse. It is not merely that (1) we make living conditions more dangerous for some generations that already will suffer from climate change and that (2) we make conditions dangerous for one or more generations that could have been secure from the threats of climate change and that (3) we create opportunities for the environment to degenerate severely. Worse still, (4) we could contribute to turning severe problems into literally insoluble problems. Or, of course, possibly not—this would, once again, be a question of the justifiability of avoidably imposing risks of adversity on defenseless others.

Unnecessarily imposing a risk of uncontrollable change—change that the people subject to it could neither steer nor stop—would be much like creating, for no good reason, a highly contagious fatal disease and leaving it behind without a cure for future generations to contend with.

Various mechanisms for runaway climate change are well understood and have in fact operated in the past. A runaway climate is certainly possible in the future because it has been actual in the past. A general category employed by scientists is abrupt climate change, which can be defined as "a large-scale change in the climate system that takes place over a few decades or less, persists (or is anticipated to persist) for at least a few decades, and causes substantial disruptions in human and natural systems."[25] We know, for example, that rapid warming can lead to abrupt cooling, because it did in the Younger Dryas roughly 10,000 years ago and, as we know from astoundingly informative ice cores, several times far earlier.[26] So there is no doubt that something devastating to humans could happen if climate change crosses a threshold that we can cause it to cross or prevent it from crossing.

The 2007 report from Working Group I of the IPCC, however, is skeptical about abrupt climate change in the current century: "Abrupt climate changes, such as the collapse of the West Antarctic Ice Sheet, the rapid loss of the Greenland Ice Sheet or large-scale changes of ocean circulation systems, are not considered likely to occur in the 21st century, based on currently available model results. However, the occurrence of such changes becomes increasingly more likely as the perturbation of the climate system progresses."[27] While we should, I think, take little comfort from the fact that our own century might be safe from the most extreme possibilities, if the report is correct in its judgment, the possibility of desperate danger does not, then, fully satisfy my second condition, threshold likelihood. Although we understand various mechanisms that could lead to runaway climate change, we do not yet have strong reason to believe that the conditions in which those mechanisms operate are coming together—at least, not yet. So I return to the previous point: creating additional but noncatastrophic danger by creating opportunities for positive feedbacks to cause climate change to escalate one or more levels above where it is already destined to go.

And the possibility of such severe danger, even short of desperate danger, is more than enough to concern us. The ones who need to worry about severe climate change are the most vulnerable, including children yet to be born, who may reap the whirlwind if we sow the wind. Those who will suffer most, if anyone does, will be people with absolutely no past role in causing the problem and with no other kind of responsibility for it (and other species, most with no capacity for morally responsible action but full capacity for suffering and frustration). This would put the kind of wrong done by the avoidable precipitation of severe climate change, it seems to me, in the general moral category of the infliction of damage or the risk of damage on the innocent and the defenseless. This is far worse than simply neglecting to protect rights, as wrong as that is, and is more like recklessly dropping bombs without knowing or caring whom they might hit. Can someone seriously argue that we are not morally responsible for avoiding the wreaking of such havoc?

And—feature three, once more—the human costs of preventing climate change from becoming severe could be modest, if well managed and begun promptly.[28] Much of our current GHG emissions serve worthy, even essential or admirable, goals. But substantial portions of it results from thoughtlessness, laziness, and wastefulness; and much serves purposes that are opulent, frivolous, or pointless.[29] I do not want to sound like a puritan; perhaps we are all free to engage in a certain amount of frivolity and pointless joy—at least, if we do no serious harm to others. On the other hand, much commends a life of simplicity, although I will not press that point here.[30] The main point here is that frivolous and pointless GHG emissions, far from being harmless, may be storing up threatening problems for whoever lives in future generations. There is low-emissions frivolity and high-emissions frivolity. I take no position here on low-emissions frivolity. High-emissions frivolity is another matter: it can be a serious threat to many other living things.

The overall picture, then, is that for the sake of benefits to ourselves that are, even if not forbidden, utterly insignificant, we are inflicting on whoever comes after us an unknown but substantial risk of a significantly more dangerous world—a dangerous world that would be to no minor extent our own creation: collateral damage from the primitive energy regime now fueling our lifestyle, not intended but no longer unforeseen. Even collateral damage in war is required to be proportional to the achievement of something important through a necessary action. To what present necessity would severe adversity on the part of successive generations of humans who succeed us be proportional?

Proportionality and Relativity

Judgments about proportionality—especially proportionality between incommensurable values such as qualities of human life and quantities of financial costs—cannot be precise.[31] I want to emphasize, however, the presence of two relativities in, but the absence of a third from, the proposed set of three jointly sufficient conditions for prompt and robust action. As already mentioned, the third factor, nonexcessive financial cost, is obviously not independent of the first factor, magnitude of human losses. What would be an excessive cost for preventing relatively smaller human losses might not be excessive for preventing relatively larger human losses. I take this to be the plainest of common sense. We cannot quantify very usefully, I think, but we can rank: a cost that would be excessive for preventing one additional destructive Atlantic hurricane per year might not be excessive for preventing the flooding of scores of the world's major cities by rising sea levels. Reasonable expenditure is obviously relative to the seriousness of the losses prevented.

I am tempted to say that no cost would be excessive for avoiding severe climate change that could lead to distortions of agriculture and yield additional starvation by way of global food-price fluctuations.[32] Such undercutting of the food system would be a monumental human tragedy. But the fact is that, as rich as we humans are in 2010, our financial resources are finite, so costs must, second, also be assessed in light of other legitimate current demands on resources. Right now, on the order of 18 million people are dying each year of readily remediable chronic poverty for want of relatively small sums of money and related institutional changes.[33] One could not sanely claim that unlimited sums should be devoted to blocking the possibility of future severe climate change if that entailed that one would, in consequence, refuse to spend what it would take to eliminate current severe poverty. This specific dilemma, however, is totally false: the budget for climate change does not need to be deducted from the budget for chronic poverty. It could be deducted from the budget for misguided military adventures.[34] Nevertheless, the point remains: at some level, expenditures on even the avoidance of dangerous climate change could be excessive, compared not to folly but to legitimate alternative uses. So in principle, what count as proportionate expenditures on the mitigation of climate change designed to stabilize it at a less dangerous level must be conceded to be relative not only to the losses that could occur if the expenditure is not made on prevention of climate change but also to the losses that would occur if, as is now far from being the case, the climate expenditure had to be taken away from other genuinely urgent matters. Therefore, the third condition within the sufficient set needed to be stated in a way that makes reasonable costs relative to the extent of the human losses that are the subject of the first condition and relative to other real—as opposed to politically manufactured—emergencies.[35]

What the costs do not need to be relative to, however, is a possible additional consideration: the probability of the massive losses. The second of the three features required for the set sufficient for action, threshold likelihood, has been formulated in order to deal with likelihood by relying on a threshold, not relying directly on probability. This is where I recommend diverging fundamentally from the standard manner of dealing with risk, which normally multiples the magnitude of possible losses by the probability of the losses occurring. My single most crucial claim here is that we ought not to discount huge possible losses by their probability when the likelihood of their occurrence is above some threshold level.[36] We cannot spend vast sums to prevent every catastrophe that is simply conceivable or barely possible. The likelihood must rise above a minimum threshold, as I have repeatedly emphasized. In the case of climate change, I believe this threshold is passed when (a) a mechanism and (b) emerging conditions for its working have been established. This is surely not the only basis on which the threshold can be satisfied.[37] But the essential point is that once the threshold is passed, one takes vigorous action until—

third feature—the costs of doing so become excessive. In sum, reasonable costs of action are relative to how massive the possible losses are if the expenditures are not made and to how great the losses are if the expenditures are diverted from other important uses, but reasonable costs of action are not directly relative to the probability of those losses occurring when the possible losses are massive and their likelihood is above a minimal threshold. One does not discount by the probability; one checks to see whether there is a significant likelihood, based on solid evidence, that massive losses may occur. If so, one takes preventive action.

If it is certain that one person will die, one can say that the probability of a death is 1 and the magnitude of death is 1; on the usual way of calculating risk, $1 \times 1 = 1$. If 1,000,000 people might die, but the probability is known to be 0.000001, the usual calculation of risk is: $1{,}000{,}000 \times 0.000001 = 1$. Arithmetically, the two risks are equal. I have my doubts about whether we ought to respond to a one-in-a-million chance that 1 million people will die in the same manner that we respond to the certainty that one person will die; I am inclined to think that we should do much more in the case of the possible deaths of 1 million. This, however, is familiar and contentious territory. What I am claiming here is the following: (1) if we know that 1 million people might die, and we know that the likelihood is significant, then we should take action to prevent the million deaths until the costs of those actions are clearly excessive; and (2) one way we know that the likelihood is significant is when (a) we understand the mechanism by which the deaths are likely to occur and (b) we have begun to create the conditions that lead the mechanism to function.[38]

Much more needs to be worked out before we can judge with clarity how vigorous, expensive, and urgent our efforts ought to be to reduce our chances of making our own descendants miserable. For now, however, we are in absolutely no danger of overshooting and simply need to make a serious beginning. And we have seen that our responsibilities for the climate change we are producing are of a different, more demanding, kind from the responsibilities conventionally assumed, even by those who acknowledge our responsibility. Two aspects above all are clear: (1) we are called upon not only to provide security for the members of humanity who live later but also to refrain from causing them dangers; and (2) even if the worst does not eventuate, the lesser dangers we may cause are quite sufficient to ground responsibility for robust action now.[39]

The Most Essential Precaution

What specifically should we do? Here is where the science really matters. The single most important fact about climate change will be the historic peak level of atmospheric concentration of greenhouse gases, and what is crucial to where the concentration peaks is the percentage of the carbon now safely sequestered underground in the form of coal and oil that are extracted and injected into the atmosphere as CO_2. Of course, other GHGs matter as well. However, if we burn all of the fossil fuel under the surface of the earth, the atmospheric concentration of CO_2 will quadruple.[40] Business as usual is misleadingly packaged for PR purposes as the "preservation of diversity" in energy sources.[41]

Either the carbon under the planet's surface is injected into the air through burning or not. It can be kept out of the atmosphere either by being left where it is now under the ground or the sea or by being burned only after effective carbon-sequestration techniques are developed. The opposition of interests is sharp: what is good for those who want all of the carbon extracted and burned with or without effective sequestration is bad for the climate and for the other 99.999 percent of humanity. And waiting for the price to rise until fossil fuels become noncompetitive greatly risks—as far as I can see, guarantees—that too much carbon will already have been injected into the planet's layer of GHGs before the price rises high enough to cut demand. The friends of fossil fuel—the carbon peddlers—have joined the enemies of humanity.

That is a strong statement. The grounds for it are the underlying science, the physical dynamics of climate. Climate change is driven by the atmospheric concentration of GHGs; this is what determines how much radiation is trapped on the planet. The atmospheric concentration is driven by annual emissions in excess of those compatible with the climate humans evolved in adaptation to—call that the sustainable rate of GHG emissions. Every year that the annual rate of emissions is larger than the sustainable rate, the atmospheric concentration grows. That is the stinger: every year that we fail to bring carbon emissions (and other GHG emissions) down to a sustainable level, the atmospheric concentration expands and more heat is trapped inside it. The atmospheric concentration has been expanding now for a century and a half. In recent years, it has been ballooning faster almost every year: the rate of increase is increasing.[42] "It is *very likely* that the average rates of increases in CO_2, as well as in the combined radiative forcing from CO_2, CH_4 and N_2O concentration increases, have been at least five time faster over the period from 1960 to 1999 than over any other 40-year period during the past two millennia prior to the industrial era."[43]

Even if the rate of increase were not increasing, as it is, the underlying arithmetic would be inexorable. The relation between unsustainable annual emissions and the atmospheric concentration is roughly like the relation between annual budget deficits and the national debt, or like the relation between annual population growth and size of total population at stabilization. The longer it takes a country to go down to replacement levels of fertility—the more years of growth in population—the larger the population size when the country stops growing. The more years of budget deficits, the larger the national debt is when the budget is finally balanced. In the best circumstances imaginable, the more years of unsustainable emissions, the higher the atmospheric concentration of GHGs when the concentration stabilizes—if it ever does. If we do not stop until we have pumped all of the oil and dug all of the coal, we will have the largest possible level of carbon dioxide concentrated in the atmosphere. And as far as we can tell, the larger the atmospheric concentration, the greater the disruption of the climate to which humans were adapted. At a minimum, we create a risk of greater disruption.

Matters are worse in two respects. We face political inertia and physical inertia. One cannot change the energy regime overnight because the superrich who own and distribute the fossil fuels have powerful political friends and articulate intellectual defenders.[44] Politics guarantees that high carbon emissions will continue for some time. That is bad enough. But the physical problem of lead time, analogous to the cultivation period for flu vaccine, is almost unimaginably daunting. In general, the whole planetary mechanism of atmosphere, oceans, and surface-level weather has enormous inertia overall once it is moving in a particular direction.[45] This is not the kind of dynamic process that gets reversed in a hurry. But the worst news may be specifically about CO_2, the most important GHG:

> A[n atmospheric] lifetime for CO_2 cannot be defined.... The behaviour of CO_2 is completely different from the trace gases with well-defined lifetimes. Stabilisation of CO_2 emissions at current levels would result in a continuous increase of atmospheric CO_2 over the 21st century and beyond.... In fact, only in the case of essentially complete elimination of emissions can the atmospheric concentration of CO_2 ultimately be stabilised at a constant level.... More specifically, the rate of emission of CO_2 currently greatly exceeds its rate of removal, and the slow and incomplete removal implies that small to moderate reductions in its emissions would not result in the stabilisation of CO_2 concentrations, but rather would only reduce the rate of its growth in coming decades. A 10% reduction in CO_2 emissions would be expected to reduce the growth rate by 10%, while a 30% reduction in emissions would similarly reduce the growth rate of atmospheric CO_2 concentrations by 30%.[46]

I repeat the critical finding: "only in the case of essentially complete elimination of emissions can the atmospheric concentration of CO_2 ultimately be stabilised at a constant level." It is, therefore, urgent to move aggressively now to cut CO_2 emissions sharply.

This science has strong implications for how we think about policy toward climate change. We need to ask, "What must we do now to keep the total atmospheric concentration below a dangerous level?" not "By how much would we like to reduce our emissions?" We need to focus on the target, which is lowering the risk of great danger, and reason back along the means-ends connections to what we must do now.

And the costs of lowering the risk of severe threat can be affordable if action begins soon enough. The longer we wait to start, the more it is likely to cost and the more abrupt the reductions in emissions would later have to be in order to keep the atmospheric concentration below a dangerous level.[47] How much we will need to tighten our belts depends on how rapid the transition to alternative energy is. Defenders of the carbon status quo say that to reduce emissions as much as scientists suggest would decimate the economy by depriving it of energy. But that is only if the economy continues to be dependent on fossil fuel. The economy can remain vibrant, while we avoid potential danger, as long as its energy source is not fossil fuel. The key is to move away from fossil fuel sooner, not later, before price rises force a switch. We need to get down to sustainable levels of annual GHG emissions, not when oil or—heaven help the future—coal "runs out" and not when its price rises too high, but as soon as possible, leaving as much carbon as possible in the ground, where it is harmless, or burning it only after we understand how to sequester the CO_2 for a very long time.[48]

We have been considering *imposing* risks on the vulnerable of the future. One natural objection would take the line: What do you mean, "imposing"? Will not future generations be able to make choices for themselves? Well, they will choose from the range of options we leave them.[49] Here are two vital factors they cannot choose because these will have been determined by earlier generations like us: (1) the size of the atmospheric concentration of GHGs already present (and unlikely to decline significantly during the succeeding century insofar as CO_2 is a factor) and (2) the dominant energy regime. If they inherit, say, an atmospheric concentration triple pre–industrial revolution levels and a still entrenched fossil-fuel regime like the one we labor under now—still digging that coal and pumping that oil—the people of the future are screwed. We would have been complicit in the imposition of a range of choice containing no good options in two ways: (1) we would have made the atmospheric concentration larger than it needed ever to become, and (2) we would have cooperated in the maintenance of an antiquated and corrosive fossil-fuel regime, and the high-casualty foreign policies serving it, that humans need to escape from.

We do not have a long time left to do the job, even now. If we cannot soon reverse the political inertia of favoritism toward fossil-fuel interests, the date of technological transition—that is, the date when the atmospheric concentration of GHGs ceases to increase—recedes into the future, and the level at which the atmospheric concentration finally stabilizes grows meanwhile like a planetary cancer, condemning more and more people to environmental danger and potentially undermining the ecological preconditions for sustainable human economies.[50]

If we in the present allow the continuing acceleration of a steady deterioration in the climate, the generation of today's students—or shockingly, even my own generation—could turn out to have had it as good as it gets. For the well-being and security of humans, history could be all downhill from here. Philosophers and economists used to think of the problem of intergenerational justice as the problem of the just-savings rate; the danger was that we might shortchange ourselves by saving or investing too much of our own resources for the sake of people in the future because each future generation would in any case—it was assumed—be better off than the previous one. So we needed to discount the value of benefits to people in the future. The specter of climate change means, by contrast, that we may be confronting the issue of the just-deterioration rate. How much worse off than the previous generation can we permit the next to be? And will we allow the deterioration to continue until critical thresholds for human security are passed? Economic sustainability has ecological preconditions (unless one makes the assumption of literally infinite substitutability, which is not unusual among conventional economists but fantastic nevertheless if extended to the entire environment, including climate).

One way to characterize in moral terms the choice to run a genuine risk of massive loss for those who follow us is that it would be the voluntary and knowing infliction of a grievous wrong. We would have chosen to leave open the possibility of great distress, or even disaster, when, at relatively little cost to ourselves, we could have closed off that possibility. We could have protected people in the future against threats to their well-being; instead, we would have increased the threats and left them vulnerable to threats they likely cannot handle. Yet, however appropriate this first moral characterization, most people do not respond well to being threatened with a guilty label, and there is no need to try to lay a guilt trip on our own generation.

Opportunity for a Legacy of Security

A much more positive moral characterization of the situation we now face is equally appropriate: thanks to the remarkable ingenuity of the scientists of the present day, invaluable understanding of the dynamics of the planetary climate system has been gained that places us in the position to provide vital protection to people in the future who would very likely otherwise find it impossible to protect themselves. Apart from blind technological optimism, we have no grounds for expecting that humans in the next century would have the capacity to protect themselves if we do nothing toward that purpose. But we have the capacity to leave them a legacy of security instead of a legacy of danger.[51]

The spectacular opportunity opened to us by our new understanding of the climate—most important, the realization that we must not allow much more of the carbon under the soil and the sea to be injected into the atmosphere, and certainly not all of it—is that we can protect future generations by keeping as much as possible of the remaining fossil fuel right where it is now. Bottom line: Do not leave your descendants—and more important, the descendants of much poorer people, such as most people in Africa—in avoidable danger. Instead, provide them with security. Create an energy regime that will leave as much as possible of the remaining sequestered carbon out of circulation.

We can have all we need economically, and much of what we want but do not need, while promptly moving away from burning fossil fuels to alternative energy sources. No vital interests are at stake in the choice among energy sources for those of us who do not own coal and oil. But many vital interests are at stake for those in the future whose fates are vulnerable to our choices. We can leave them social institutions that will protect them—in particular, a cleaned-up energy regime that does not vomit GHGs into the sky. An energy regime not based on fossil fuels will make the worst effects of climate change that are now increasingly likely once again nearly impossible. Let us seize the opportunity to bequeath this magnificent gift of protection against vulnerability.

Acknowledgments

The gestation of this chapter has been long and painful, and I have been assisted in wrestling with various versions of it by responsive audiences at San Diego State University, University of Washington, Cornell University, University of Oslo, University of Oxford, New York University Law School, University of Edinburgh, and the University of Tennessee. In the last round, I have benefited especially from comments by Wilfred Beckerman and Nicole Hassoun.

Notes

1. "Between a quarter and a third of the world's wildlife has been lost since 1970, according to data compiled by the Zoological Society of

London." See "Wildlife Populations 'Plummeting,'" *BBC News*, May 16, 2008. Available at http://newsvote.bbc.co.uk/mpapps/pagetools/print/news.bbc.co.uk/2/hi/uk_news/7403989. For an attempt to calculate the purely economic value of the loss of biodiversity now to be expected, see Pavan Sukhdev, *The Economics of Ecosystems & Biodiversity (TEEB): An Interim Report* (European Communities, 2008). This is being described as the *Stern Review* (see note 20 below) for biodiversity and is available online at http://ec.europa.eu/environment/nature/biodiversity/economics/pdf/teeb_report.pdf. Also see http://ec.europa.eu/environment/nature/biodiversity/economics/index_en.htm.

2. In accepting Article 2 of the Framework Convention on Climate Change in 1992, world leaders committed themselves to achieve "stabilization of greenhouse gas concentrations in the atmosphere at a level that would prevent dangerous anthropogenic interference with the climate system." From <http://unfccc.int/essential_background/convention/background/items/1349.php. So far, they have failed to stabilize concentrations at all.

3. Many philosophers are preoccupied with what is known as the nonidentity problem, which was formulated by Thomas Schwartz in two contemporaneous pieces, "Obligations to Posterity," in *Obligations to Future Generations,* ed. R. I. Sikora and Brian Barry (Philadelphia: Temple University Press, 1978), pp. 3–13; and "Welfare Judgments and Future Generations," *Theory and Decision* 11 (1979): 181–194; and by Derek Parfit, *Reasons and Persons* (New York: Oxford University Press, 1984). As far as I can see, individual nonidentity has no implications at all for what we ought to do. At most, it has some implications for how we explain our moral judgments.

4. How to think about the imposition of risk is an exceedingly difficult question. One valuable collection is Paul Slovic, ed., *The Perception of Risk* (London and Sterling, Va.: Earthscan, 2000).

5. Andrew C. Revkin, "New Climate Report Foresees Big Changes," *New York Times*, May 28, 2008; and U.S. Climate Change Science Program, *Weather and Climate Extremes in a Changing Climate*, Final Report, Synthesis and Assessment Product 3.3 (June 2008). Available at <http://www.climatescience.gov/Library/sap/sap3–3/final-report/default.htm.

6. "There is strong evidence that global sea level gradually rose in the 20th century and is currently rising at an increased rate, after a period of little change between AD 0 and AD 1900. Sea level is projected to rise at an even greater rate in this century." Nathaniel L. Bindoff, Jürgen Willebrand, Vincenzo Artale, et al., "Observations: Oceanic Climate Change and Sea Level," in *Climate Change 2007: The Physical Science Basis, Contribution of Working Group I to the Fourth Assessment Report of the Intergovernmental Panel on Climate Change,* ed. Susan Solomon, Dahe Qin, Martin Manning, et al. (Cambridge and New York: Cambridge University Press, 2007), pp. 408–414, at p. 409.

7. The chunk of Antarctic ice that collapsed in February–March 2008 was a piece of an already-floating ice shelf, the Wilkins ice shelf, and will not contribute to sea-level rise. See Associated Press, "Chunk of Antarctic Ice Collapses," *New York Times*, March 26, 2008.

8. See Steve Vanderheiden, *Atmospheric Justice: A Political Theory of Climate Change* (New York: Oxford University Press, 2008), pp. 192–202. He documents some of the intentional deceptions of U.S. citizens about the science for which the Bush-Cheney White House was notorious.

9. Obviously, the third feature is not independent of the first but proportional to it. My three features fit generally within what Neil A. Manson has called "the three part structure of the precautionary principle," consisting of a damage condition, a knowledge condition, and a remedy; see his penetrating article "Formulating the Precautionary Principle," *Environmental Ethics* 24 (2002): 263–274. My suggestion is also, I think, a variant of what Stephen M. Gardiner very fruitfully isolates as a "core precautionary principle"; see "A Core Precautionary Principle," *Journal of Political Philosophy* 14 (2006): 33–60. Whether either Manson or Gardiner would accept my specific suggestion is, of course, another matter. I will for the most part ignore the more general philosophical issues that they perceptively consider, although I do not believe I run afoul of anything they have established.

10. The suggestion is thus parallel to the suggestion of a sufficient set of conditions for states to take responsibility for the effects of their actions on affected people outside the territory they govern in Henry Shue, "Eroding Sovereignty," in *The Morality of Nationalism,* ed. Robert McKim and Jeff McMahan (New York: Oxford University Press, 1997), pp. 353–354. In both cases, the idea is that one has responsibility to protect against what might otherwise be the effects of one's own actions the people who would be vulnerable and unable

to protect themselves, in the one case across space (including national boundaries) and in the other case across time.

11. I realize that not everyone accepts that there are objective probabilities.

12. I am not hinting at any connection between climate change and bird-flu; this is a comparison.

13. Gina Kolata, *Flu: The Story of the Great Influenza Pandemic of 1918 and the Search for the Virus that Caused It* (New York: Farrar, Straus and Giroux, 1999).

14. The conduct of the U.S. occupation of Iraq currently costs $16 billion per month (the annual budget of the United Nations). Is one more likely to be killed by a terrorist based in Iraq or by bird flu?

15. Stephen H. Schneider, Serguei Semenov, Anand Patwardhan, et al., "Assessing Key Vulnerabilities and the Risk from Climate Change," in *Climate Change 2007: Impacts, Adaptation and Vulnerability, Contribution of Working Group II to the Fourth Assessment Report of the Intergovernmental Panel on Climate Change*, ed. Martin Parry, Osvaldo Canziani, Jean Palutikof, et al. (Cambridge: Cambridge University Press, 2007), pp. 779–810. And see generally Cynthia Rosenzweig and Daniel Hillel, *Climate Change and the Global Harvest: Potential Impacts of the Greenhouse Effect on Agriculture* (New York: Oxford University Press, 1998), and *Climate Variability and the Global Harvest: Impacts of El Nino and Other Oscillations on Agro-ecosystems* (New York: Oxford University Press, 2008).

16. See, for example, David Fogarty, "Warming Globe to Test Farmers' Adaptability," *International Herald Tribune*, May 5, 2008, p. 13.

17. Amartya K. Sen, *Poverty and Famines: An Essay on Entitlement and Deprivation* (New York: Oxford University Press, 1983).

18. Hervé Le Treut, Richard Somerville, Ulrich Cubasch, et al., "Historical Overview of Climate Change Science," in *Climate Change 2007: The Physical Science Basis, Contribution of Working Group I to the Fourth Assessment Report of the Intergovernmental Panel on Climate Change,* ed. Susan Solomon, Dahe Qin, Martin Manning, et al. (Cambridge and New York: Cambridge University Press, 2007), pp. 93–127. Also see Andrew C. Revkin, "Strong Action Urged to Curb Warming," *New York Times*, June 11, 2008.

19. For evidence that large reductions in greenhouse-gas emissions by the United States are possible at low cost, see Jon Creyts, Anton Derkach, Scott Nyquist, et al., *Reducing U.S. Greenhouse Gas Emissions: How Much at What Cost?* U.S. Greenhouse Gas Abatement Mapping Initiative (Chicago: McKinsey, 2007).

20. For one argument that the costs of mitigation, or abatement, will only go up, and steeply, see Nicholas Stern, *The Economics of Climate Change: The Stern Review* (Cambridge: Cambridge University Press, 2007).

21. This is a different formulation of the general thesis I advanced in "Climate," in *A Companion to Environmental Philosophy*, ed. Dale Jamieson (Malden, Mass., and Oxford: Blackwell, 2001), pp. 449–459.

22. For one argument for subsistence rights, see Henry Shue, *Basic Rights: Subsistence, Affluence, and U.S. Foreign Policy*, 2d ed. (Princeton, N.J., and Oxford.: Princeton University Press, 1996).

23. See Henry Shue, "Responsibility to Future Generations and the Technological Transition," in *Perspectives on Climate Change: Science, Economics, Politics, Ethics,* ed. Walter Sinnott-Armstrong and Richard B. Howarth (Amsterdam: Elsevier, 2005), pp. 265–283.

24. See "Permafrost Threatened by Rapid Retreat of Arctic Sea Ice, NCAR/NSIDC Study Finds," *National Snow and Ice Data Center Media Advisory*, June 10, 2008.

25. John P. McGeehin, John Barron, David M. Anderson, and David Verardo, *Abrupt Climate Change* (Washington, D.C.: U.S. Climate Change Science Program, 2008), p. 1. Also see Hans Joachim Schnellnhuber et al., eds., *Avoiding Dangerous Climate Change* (Cambridge: Cambridge University Press, 2006).

26. See Richard B. Alley, *The Two-Mile Time Machine: Ice Cores, Abrupt Climate Change, and Our Future* (Princeton, N.J.: Princeton University Press, 2000); U.S. National Academy of Sciences, National Research Council, Committee on Abrupt Climate Change, *Abrupt Climate Change: Inevitable Surprises* (Washington, D.C.: National Academy Press, 2002), pp. v, 24–36; M. Vellinga and R. A. Wood, "Global Climatic Impacts of a Collapse of the Atlantic Thermohaline Circulation," *Climatic Change* 54.3 (2002): 251–267; and Eystein Jansen, Jonathan Overpeck, Keith R. Briffa, et al., "Palaeoclimate," in *Climate Change 2007: The Physical Science Basis, Contribution of Working Group I to the Fourth Assessment Report of the Intergovernmental Panel on Climate Change,* ed. Susan Solomon, Dahe Qin, Martin Manning, et al. (Cambridge and New York: Cambridge University Press, 2007), pp. 433–497.

27. Gerald A. Meehl, Thomas F. Stocker, William D. Collins, et al., "Global Climate

Projections," in *Climate Change 2007: The Physical Science Basis, Contribution of Working Group I to the Fourth Assessment Report of the Intergovernmental Panel on Climate Change,* ed. Susan Solomon, Dahe Qin, Martin Manning, et al. (Cambridge and New York: Cambridge University Press, 2007), p. 818. They add: "Catastrophic scenarios suggesting the beginning of an ice age triggered by a shutdown of the MOC [meridional overturning circulation] are thus mere speculations" (p. 818). Compare Juliet Eilperin, "Debate on Climate Shifts to Issue of Irreparable Change: Some Experts on Global Warming Foresee 'Tipping Point' When It Is Too Late to Act," *Washington Post,* January 29, 2006, p. A1.

28. See Creyts, Derkach, Nyquist, et al., *Reducing U.S. Greenhouse Gas Emissions.*

29. A distinction between "survival emissions" and "luxury emissions" was advocated in Anil Agarwal and Sunita Narain, *Global Warming in an Unequal World: A Case of Environmental Colonialism* (New Delhi: Centre for Science and Environment, 1991), p. 5. I pursued their suggestion in Henry Shue, "Subsistence Emissions and Luxury Emissions," *Law & Policy* 15.1 (January 1993): 39–59 (chapter 11 in this volume). I did not think of these two kinds of emissions as exhaustive: many emissions are neither subsistence nor luxury. Steve Vanderheiden treats "survival emissions" and "luxury emissions" as exhaustive categories in *Atmospheric Justice*, pp. 67–73, 242–243.

30. For provocative reflections on the value of a simpler life, see Duane Elgin, *Voluntary Simplicity: Toward a Way of Life That Is Outwardly Simple, Inwardly Rich*, rev. ed. (New York: William Morrow, 1993); Wallace Kaufman, *Coming Out of the Woods: The Solitary Life of a Maverick Naturalist* (Cambridge, Mass.: Perseus, 2000); and David E. Shi, *The Simple Life: Plain Living and High Thinking in American Culture* (New York and Oxford: Oxford University Press, 1985).

31. One can adopt a common metric and then quantify, of course, as some economists do, but the choice of common metric is determined largely by convenience, and the assignments of relative value are highly arbitrary. How many dollars is a decline in the quality of nutrition in Myanmar in 2410 worth now?

32. Nicole Hassoun knows how tempted, and I am grateful to her for discussions of the issues underlying the three conditions.

33. See Thomas Pogge, "Severe Poverty as a Human Rights Violation," in *Freedom from Poverty as a Human Right,* ed. Thomas Pogge (Oxford: Oxford University Press, 2007), pp. 11–53.

34. As I write, the Bush-Cheney "war of choice"—the totally unnecessary, ill-considered, and egregiously counterproductive plunge into the invasion and occupation of Iraq—is costing, as already mentioned, $16 billion per month in operating costs. The elimination of such murderous folly would free up vast sums. One year of the budget for the Iraq occupation would take care of chronic poverty plus several years of vigorous action on climate change. For fuller consideration of some of the issues underlying the precipitate rush into war in Iraq, see Henry Shue and David Rodin, eds., *Preemption: Military Action and Moral Justification* (Oxford: Oxford University Press, 2007).

35. Bryan G. Norton makes this general point in defense of his preferred guide, the safe minimum standard (SMS): "save the resource, provided the costs of doing so are bearable"; see *Sustainability: A Philosophy of Adaptive Ecosystem Management* (Chicago: University of Chicago Press, 2005), p. 346. Similarly, he builds affordability into his statement of the precautionary principle: "take affordable steps to avoid catastrophe tomorrow" (p. 352). "Bearability" and "affordability" depend, for Norton, on which other extremely important matters also require resources.

36. Weapons of mass destruction (WMDs) in Iraq could have fit the same principle if, for example, competent UN inspectors had been finding substantial evidence of their existence. See Shue and Rodin, *Preemption.* It is clear, however, that this eagerly launched war had entirely different purposes and that the case based on WMDs was politically concocted as a pretext; see Mark Danner, *The Secret Way to War: The Downing Street Memo and the Iraq War's Buried History* (New York: New York Review Books, 2006). It is profoundly ironic, and deeply evil, that the Bush-Cheney administration, while suppressing incontrovertible evidence of climate change gathered by thousands of the world's best scientists—see Vanderheiden, *Atmospheric Justice*, chap. 6—was manufacturing phony evidence to justify a war that is consuming many times the resources needed to deal with the real problem they vigorously covered up. Bare politically motivated assertion was accepted as the justification for what has become the longest war in American history, while a powerful body of scientific support for action to mitigate climate change was simply denied. See Mark Mazzetti and Scott Shane, "Bush Overstated Evidence on Iraq, Senators Report," *New York Times*, June 6, 2008; and U.S. Senate, 110th Congress, 2nd Session, Select

Committee on Intelligence, "Report on Intelligence Activities Relating to Iraq" (June 2008). Even NASA's press releases were for years distorted by political appointees; see Andrew C. Revkin, "NASA Office Is Criticized on Climate Reports," *New York Times*, June 3, 2008.

37. I would think that one of the most productive avenues to explore is what other grounds might satisfy the second condition. The generic form of the second condition is that the likelihood of the losses specified in the first condition must be above some minimum threshold. The specific form this condition takes in the cases of bird flu and climate change is mechanism and emerging conditions for functioning of the mechanism. But the generic condition could well be instantiated by other specific forms of threshold.

38. One million deaths is, of course, simply an example. There are other kinds of "massive losses" besides deaths, such as calamitous declines in standards of living or the undermining of civilization. If no one died but everyone had to live like cavemen, that would be a massive loss.

39. A splendid survey and analysis of the literature on the moral case is Stephen M. Gardiner, "Ethics and Global Climate Change," *Ethics* 114.3 (April 2004): 555–600 (chapter 1 in this volume). I tried to establish that each of three independently persuasive arguments all led in this same general direction in Henry Shue, "Global Environment and International Inequality," *International Affairs* 75.3 (1999): 531–545, reprinted in *Environmental Ethics: What Really Matters, What Really Works*, ed. David Schmidtz and Elizabeth Willott (New York: Oxford University Press, 2001), and as chapter 5 in this volume.

40. James F. Kasting, "The Carbon Cycle, Climate, and the Long-Term Effects of Fossil Fuel Burning," *Consequences: The Nature & Implications of Environmental Change*, vol. 4, no. 1 (1998). Available at http://www.gcrio.org/CONSEQUENCES/vol4no1/carbcycle.html.

41. TV ads in 2008 successfully opposed effective action on climate change in the U.S. Senate and were sponsored by the coal and oil lobbies. ExxonMobil ads, for example, say, "We're going to need them all," meaning, "Do not cut back on oil."

42. See David Adam, "World Carbon Dioxide Levels Highest for 650,000 Years, Says US Report," *Guardian*, May 13, 2008; and U.S. Department of Commerce, National Oceanic and Atmospheric Administration, Earth System Research Laboratory, Global Monitoring Division, "Trends in Atmospheric Carbon Dioxide—Mauna Loa." Available at http://www.esrl.noaa.gov/gmd/ccgg/trends/.

43. Jansen, Overpeck, Briffa, et al., "Palaeoclimate," p. 436. Emphasis in original; "very likely" is used technically in that volume to mean a greater than 90 percent probability; see p. 23.

44. For the general picture, see Daniel Yergin, *The Prize: The Epic Quest for Oil, Money, & Power* (New York: Simon & Schuster, 1992); and Michael T. Klare, *Resource Wars: The New Landscape of Global Conflict* (New York: Henry Holt, 2001).

45. See Meehl, Stocker, Collins, et al., "Global Climate Projections," pp. 747–843.

46. Ibid., pp. 824–825.

47. Nicholas Stern, "The Economics of Climate Change," *American Economic Review* 98.2 (May 2008): 1–37 (chapter 2 in this volume).

48. It is not clear that there is any such thing as "running out," in any case. If coal becomes expensive enough, it may simply join its prettier cousin the diamond as a luxury good.

49. See the analysis of this as the "domination of posterity" in John Nolt, "Greenhouse Gas Emission and the Domination of Posterity," in *The Ethics of Global Climate Change*, ed. Denis Arnold (New York: Cambridge University Press, forthcoming).

50. For the initial development of the conception of the technological transition, see Shue, "Responsibility to Future Generations."

51. For an account of how the clean development mechanism of the Kyoto Protocol is failing to deal with the danger, see Henry Shue, "A Legacy of Danger: The *Kyoto Protocol* and Future Generations," in *Globalisation and Equality*, ed. Keith Horton and Haig Patapan (London and New York: Routledge, 2004), pp. 164–178.

9

Climate Change, Human Rights, and Moral Thresholds

Simon Caney

It is widely recognized that anthropogenic climate change will have harmful effects on many human beings and in particular on the most disadvantaged. In particular, it is projected to result in flooding, heat stress, food insecurity, drought, and increased exposure to waterborne and vector-borne diseases. Various different normative frameworks have been employed to think about climate change. Some, for example, apply cost-benefit analysis to climate change. The *Stern Review* provides a good example of this approach.[1] It proceeds by comparing the costs (and any benefits) associated with anthropogenic climate change with the costs and any benefits of a program for combating climate change. On this basis, it argues that an aggressive policy of mitigation and adaptation is justified. Whereas the costs of combating climate change, according to Stern, are quite low, the costs of "business of usual" would be considerable. Other analysts adopt a second perspective and conceive of climate change in terms of its impact on security.[2] For example, the High Representative and the European Commission to the European Council issued a statement on *Climate Change and International Security,* which argues that climate change is "a threat multiplier which exacerbates existing trends, tensions and instability."[3] It argues that climate change will contribute to the following kinds of insecurities: tensions over scarce resources; land loss and border disputes; conflicts over energy sources; conflict prompted by migration; and tensions between those whose emissions caused climate change and those who will suffer the consequences of climate change.[4] In addition to the "economic" approach and "security-based" approach, some adopt a third different perspective, according to which the natural world has intrinsic value. This ecological approach condemns human-induced climate change because it is an instance of humanity's domination and destruction of the natural world.

For all of their merits, these three perspectives omit an important consideration: the impact of climate change on persons' fundamental human rights. In this chapter, I argue that it is appropriate to analyze climate change in terms of its impact on human rights. A human-rights approach, I maintain, provides an appropriate way in which to evaluate the effects of climate change. There are historical precedents for applying human rights to evaluate environmental change. Principle 1 of the 1972 Stockholm Declaration of the United Nations Conference on the Human Environment declares that "[m]an has the fundamental right to freedom, equality and adequate conditions of life, in an environment of a quality that permits a life of dignity and well-being, and he bears a solemn responsibility to protect and improve the environment for present and future generations."[5] More recently, on November 14, 2007, a conference of AOSIS members adopted the Male Declaration on Human Dimension of Global Climate Change.[6] This invoked "the fundamental right to an environment capable of supporting human society and the full enjoyment of human rights," and it expressed concern

> that climate change has clear and immediate implications for the full enjoyment of human rights including *inter alia* the right to life, the right to take part in cultural life, the right to use and enjoy property, the right to an adequate standard of living, the right to food, and the right to the highest attainable standard of physical and mental health.[7]

The Human Rights Council of the United Nations has since passed a resolution finding that "climate change poses an immediate and far-reaching threat to people and communities around the world and has implications for the full enjoyment of human rights."[8]

I believe that this is a promising approach. In what follows, I argue that:

1. Climate change jeopardizes some key human rights.
2. A "human-rights"-centered analysis of the impacts of climate change enjoys several fundamental advantages over other dominant ways of thinking about climate change.
3. A "human-rights"-centered analysis of the impacts of climate change has far-reaching implications for our understanding of the kind of action that should be taken and who should bear the costs of combating climate change.

I. The Nature of Human Rights

It is useful to begin with an analysis of "human rights." The concept of human rights has several components. I highlight four. Human rights (i) are grounded in persons' "humanity," (ii) represent moral thresholds, (iii) respect each and every individual, and (iv) take general priority over other values. Let us consider each of these in turn.

(i) *Humanity*. Human rights refer to those rights that persons have qua human beings. There are a number of different kinds of rights. H. L. A. Hart, for example, distinguishes between "special rights" and "general rights." Special rights, in his account, are rights that persons have by virtue of some action that they and some other party have performed (e.g., they have signed a contract or one has authorized the other to do something) or by virtue of a special relationship (e.g., they have been born into one state and therefore have the rights of citizenship).[9] These special rights can be contrasted with what Hart terms general rights. These are the rights that persons have in virtue of their humanity, and not because of the nation or state into which they were born or any actions that they have performed. Hart's concept of general rights captures well the traditional understanding of human rights. They are the rights that persons possess independently of any social convention or social practice. They are grounded in a respect for persons' humanity.

(ii) *Moral thresholds*. Human rights represent moral "thresholds" below which people should not fall. They designate the most basic moral standards to which persons are entitled. This point is nicely conveyed by Henry Shue, who writes that "[b]asic rights are the morality of the depths. They specify the line beneath which no one is to be allowed to sink."[10] As such, they are only part of a complete political morality. They leave room for other moral ideals and values. To reiterate, they simply designate the most fundamental moral requirements that individuals can claim of others.

(iii) *Universal protection*. Related to this, human rights represent the entitlements of *each and every individual* to certain minimal standards of treatment, and they generate obligations on all persons to respect these basic minimum standards. Article 1 of the Universal Declaration of Human Rights (1948) captures this well. As it states, "[a]ll human beings are born free and equal in dignity and rights." A human-rights approach thus stands opposed to aggregative political moralities that simply sum the interests of all with a view to increasing the total social good. A human-rights approach insists on the protection of the entitlements of *all* individuals and condemns any tradeoffs that would leave some below the minimum moral threshold.

(iv) *Lexical priority*.[11] Finally, human rights generally take priority over moral values, such as increasing efficiency or promoting happiness.[12] They constrain the pursuit of other moral and political ideals, and if there is a clash between not violating human rights on the one hand and promoting welfare on the other, then the former should take priority.

In short, then, and combining each of the four properties above, we may say that human rights specify minimum moral thresholds to which all individuals are entitled, simply by virtue of their humanity, and which override all other moral values.[13]

Two further points bear noting about the concept of human rights. First, it is conventional to distinguish between positive and negative rights, where positive rights require others to perform certain actions and where negative rights require others simply to abstain from certain actions. To illustrate the difference, one might affirm that there is a negative right not to be tortured. This generates duties on all not to perform this kind of action. Alternatively, one might affirm a positive right, say, to education. This requires not simply that others do not deprive persons of education but also that others perform positive actions to ensure that all have access to education.[14]

Finally, it bears noting that there are a variety of different justifications of human rights. Following Thomas Nagel, I shall distinguish between "intrinsic" and "instrumental" justifications of human rights.[15] An intrinsic, or deontological, approach is grounded in the idea of respect for persons. It holds that to violate persons' human rights is to fail to show them the respect that they are owed. It does not, in Kant's phrase, treat persons as ends in themselves. Nagel himself adopts an intrinsic approach. He defends human rights on the grounds that they reflect the "value of inviolability."[16] Persons, in this view, have a certain "moral status" or standing and should not be treated as potential means to an end.[17] To view them as potentially usable in this way is to fail to recognize their inviolability. This intrinsic rationale for human rights can be contrasted with instrumental or teleological approaches. The latter justify human rights on the grounds that they enable each person to enjoy certain fundamental goods. Unlike deontological accounts, they justify human rights in terms of their consequences for people's lives and the state of affairs produced. Human rights, on this second account, are valuable because they enable people to be autonomous or to achieve a decent standard of living.[18] To give one recent example, in his important work *Justice, Legitimacy, and Self-Determination: Moral Foundations for International Law*, Allen Buchanan argues that human rights have value because they protect interests that "are constitutive of a decent life; they are necessary conditions for human flourishing."[19] A similar position is taken by Martha Nussbaum, who argues that human rights are valuable because they protect vital "capabilities" that are necessary to lead a decent life.[20] The teleological position is also defended by James Griffin in his work *On Human Rights*.[21] In what follows, I am neutral between the intrinsic and instrumental accounts.[22] Both, I suggest, will endorse the human rights I propose.

II. Climate Change and Human Rights

Having clarified the concept of human rights, I now want to turn to the linkages between anthropogenic climate change and human rights. Climate change, I argue, jeopardizes three key human rights: the human right to life, the human right to health, and the human right to subsistence. Each will be examined in turn.

Before discussing these human rights, it is worth drawing attention to one aspect of the arguments that follow. In the case of each of the human rights that I identify, I present what I take to be the least contentious and most modest formulation of the human right in question and show that even using such minimal conceptions of human rights, anthropogenic climate change violates human rights. In doing so, I am *not* rejecting other more expansive interpretations of each of these human rights. My point is that one does not need to rely on more controversial or ambitious conceptions of human rights in order to see how climate change jeopardizes human rights.[23]

§1. The right to life has been conceptualized in various ways. Controversies surround what entities hold this right (do fetuses have a right to life?) and what exceptions apply to it (consider, for example, debates concerning the justifiability of capital punishment and killing during warfare). The claim that I wish to defend does not require us, however, to take a stand on either of these controversial issues.

> *HR1, the human right to life:* Every person has a human right not to be "arbitrarily deprived of his life" (International Covenant on Civil and Political Rights, 1976, Article 6.1).

Two comments are in order here. First, note that this formulation of the right to life conceives of it simply as a negative right. As such, it does not make the more contentious claim that persons have a positive right to have their life saved from all kinds of threats. Second, HR1 makes reference to "arbitrarily" depriving people of life. The point of this wording is to allow the possibility that it might, in principle, be justifiable to deprive people of their lives. Such a loss of life would not be "arbitrary." As noted above, some might hold that capital punishment is justified and hence would reject HR1 if it claimed that all loss of life counts as human-rights violation. By insisting that only "arbitrary" loss of life counts as a rights violation (and by allowing the possibility that capital punishment can be a nonarbitrary loss of life), one avoids this controversy. This addition does not have any further implications, but it is important to present as compelling a conception of the human right to life as possible.

Once we interpret the human right to life along the lines suggested by HR1 and thereby avoid the controversies mentioned above, it is clear that it would be endorsed by both deontological and teleological approaches to human rights. If recognizing the value of inviolability entails anything, it surely entails that one does not act so as to deprive people arbitrarily of their lives. It is similarly clear (obvious, even) that from a teleological point of view, persons have a right that others do not arbitrarily deprive them of their own lives. This is a necessary condition of leading a minimally decent life.

Having identified a plausible conception of the human right to life, we see clearly that anthropogenic climate change violates this right. It does so in at least two ways. First, climate change is projected to result in an increased frequency of severe weather events, such as tornadoes, hurricanes, storm surges, and floods, and these can lead to a direct loss of life. Storm surges can have a devastating effect. R. F. Mclean and Alla Tsyban write, for example:

> Storm-surge flooding in Bangladesh has caused very high mortality in the coastal population (e.g., at least 225,000 in November 1970 and 138,000 in April 1991), with the highest mortality among the old and weak.... Land that is subject to flooding—at least 15% of the Bangladesh land area—is

> disproportionately occupied by people living a marginal existence with few options or resources for adaptation.[24]

Climate change will also produce flooding and landslides, and these can be devastating. The Fourth Assessment Report of the IPCC reports that "[i]n 1999, 30,000 died from storms followed by floods and landslides in Venezuela. In 2000/2001, 1,813 died in floods in Mozambique".[25] In addition to severe weather events, climate change will also involve heat waves, and these, too, will lead to loss of life. For example, studies have found that a five-day heat wave in Chicago in 1995 led to at least 700 extra deaths.[26] Furthermore, the heat wave of 2003 in western Europe also resulted in a considerable increase in death from respiratory, cardiovascular, and cerebrovascular problems brought on by the heat wave. A. Haines, R. S. Kovats, D. Campbell-Lendrum, and C. Corvalan report, for example:

> More than 2000 excess deaths were reported in England and Wales during the major heat wave that affected most of western Europe in 2003.... The greatest impact on mortality occurred in France, where it was estimated that 14800 excess deaths occurred during the first 3 weeks of August 2003 than would be expected for that time of year. Deaths in Paris increased by 140%.[27]

In virtue of both of these mechanisms, we may conclude that the current anthropogenic climate change violates the human right to life.[28]

§2. The effects of climate change will not be restricted to its impact on the human right to life. They also undermine the human right to health. Again, though, we need to be careful in framing this right. A canonical statement of the right to health can be found in the International Covenant on Economic, Social and Cultural Rights (ICESCR) (1976), which affirms "the right of everyone to the enjoyment of the highest attainable standard of physical and mental health" (Article 12.1). In a similar vein, the Convention on the Rights of the Child (CRC) (1990) asserts "the right of the child to the enjoyment of the highest attainable standard of health" (Article 24.1).

These maximalist conceptions of the right to health will be challenged by some. A critic might balk at the claim that all are entitled to "the highest attainable standard of physical and mental health." He or she might contend that to attain the highest possible standard of health would require diverting all resources to this single objective, and this would be implausible, given the need to resource other important rights or moral objectives.[29] In light of these possible concerns, I propose a less ambitious conception of the human right to health.

> *HR2, the human right to health:* All persons have a human right that other people do not act so as to create serious threats to their health.

This differs from the ICESCR and CRC conceptions in two related ways. First, it does not require people to maximize the health of all. Second, it does not affirm a positive right to be (maximally) healthy. It affirms only a negative right that persons do not harm the health of others. Note, however, that HR2 is, of course, presupposed by the interpretation of the human right to health found in the ICESCR. The latter also holds that persons should not act in such a way as to create an unhealthy environment; it is just that it goes much farther as well, calling for positive action to ensure the highest attainable standard of health.[30]

Again, it is, I hope, clear that both deontological and a teleological approaches would vindicate HR2. Judged from a deontological point of view, the argument for HR2 is that acting to expose others to dangerous diseases manifests a lack of respect for their status as free and equal persons. To engage in activities that create serious health hazards for others constitutes a severe failure to recognize their moral standing and their inherent dignity as persons. The teleological approach would similarly endorse HR2. The capacity to lead a decent life requires that persons are not exposed to serious threats to their health. Their capacity for agency, their ability to pursue their conception of the good, will be undermined, if not thwarted altogether, by disease and injury.

With this in mind, let us now turn our attention to the health effects of climate change. There is by now an extensive literature chronicling the severe health effects of anthropogenic climate change. The Fourth Assessment Report of the IPCC notes, for example, that anthropogenic climate change will:

- "increase the number of people suffering from . . . disease and injury from heatwaves, floods, storms, fires and droughts";
- increase the range of malaria in some places but decrease it in others;
- increase "the burden of diarrhoeal diseases";
- "increase cardio-respiratory morbidity . . . associated with ground-level ozone"; and
- "increase the number of people at risk of dengue."[31]

The IPCC reports that "[c]limate change is projected to increase the burden of diarrhoeal diseases in low-income regions by approximately 2 to 5% in 2020."[32] It adds that dengue, too, will increase dramatically, and it reports research that estimates that "in the 2080s, 5–6 billion people would be at risk of dengue as a result of climate change and population increase, compared with 3.5 billion people if the climate remained unchanged."[33] Human-induced climate change thus clearly results in a variety of different threats to the human right to health.

§3. Thus far, we have seen how anthropogenic climate change undermines two fundamental human rights. Let us turn now to the third human right that I claim is harmed by anthropogenic climate change.

> *HR3, the human right to subsistence*: All persons have a human right that other people do not act so as to deprive them of the means of subsistence.

Note that HR3 is more minimal than the human right to food affirmed in human-rights documents. Both the ICESCR and the Universal Declaration of Human Rights (1948) appear to affirm a positive right to food. For instance, the ICESCR asserts "the right of everyone to an adequate standard of living for himself and his family, including adequate food" (Article 11.1), and Article 25.1 of the Universal Declaration of Human Rights uses similar wording. Furthermore, the ICESCR also simply asserts "the fundamental right of everyone to be free from hunger" (Article 11.2). These formulations presuppose HR3 but go farther, insisting that there is also a positive right to receive aid to ensure that no one suffers from hunger no matter what the cause of that hunger.[34]

Note, further, that HR3 enjoys support from both deontological and teleological perspectives. From a deontological perspective, the claim is that to deprive others of the possibility of meeting their basic needs is to treat them without due respect. To deny others the ability to satisfy their subsistence needs fails to acknowledge their moral standing and their dignity as persons. This is especially so when, as is the case with climate change, the majority of emissions come from the advantaged, who do not need to engage in such health-endangering behavior.[35] In the teleological view, this would again endorse HR3. Food and drinkable water are necessary preconditions of the ability to act and pursue even minimal goals.

If we turn now to consider the impacts of climate change, it is clear that anthropogenic climate change violates this right. Four different mechanisms should be noted. First, temperature increases will lead to drought and thereby undermine food security. Anthony Nyong and Isabelle Niang-Diop report, for example, that "[i]n southern Africa, the area having water shortages will have increased by 29% by 2050, the countries most affected being Mozambique, Tanzania and South Africa."[36] Second, sea-level rises will involve loss of land to the sea and thus hit agriculture badly. This is especially clear in countries such as Bangladesh. Third, flooding will also lead to crop failure. Fourth, and finally, freak weather events will also destroy agriculture. The upshot of these processes is that people will be deprived of the means of subsistence. Bill Hare, for instance, reports that recent research suggests that there will be "45–55 million extra people at risk of hunger by the 2080s for 2.5°C warming, which rises to 65–75 million for a 3°C warming."[37]

§4. Thus far, we have seen that anthropogenic climate change violates three fundamental human rights. Lest this argument is misunderstood, it is important to make several additional clarificatory remarks. First, if the impacts of climate change were entirely the result of natural phenomena and were not traceable to human causes, then the preceding argument would not succeed. HR1 states that persons have a human right that *other people* do not deprive them of their lives, and so if persons lose their lives because of purely natural causes, then HR1 is intact. Similarly, HR2 states that persons have a human right that *other people* do not act so as to create serious threats to their health. And, as we have just seen, HR3 holds that all persons have a human right that *other people* do not act so as to deprive them of the means of subsistence. Climate scientists are unequivocal that the current and projected future climate change stems from human activities, and given this, the three preceding claims all hold. The threats to life, health, and subsistence that many face, and that many more shall face unless mitigation and adaptation occur, are threats that are the products of the actions of other people.[38]

Second, it is worth emphasizing and repeating the point that the aim of the preceding argument is to show how climate change undermines human rights while at the same time appealing to as uncontroversial premises as possible. For that reason, I have focused on the three rights given above and not on other, more contentious candidates, and I have also relied on what I take to be the most uncontroversial formulations of those rights. The aim is to identify absolutely fundamental human rights that can enjoy ecumenical support from a wide variety of different ethical perspectives. The rights not to be killed, not to have one's health jeopardized, and not to be deprived of the means necessary for subsistence are all, I suggest, rights that can be adopted from within a wide variety of different conceptions of the good and ethical worldviews.

Third, having noted this, it is nonetheless worth mentioning that there are other possible human-rights implications of climate change. For example, it is arguable that climate change jeopardizes a human right to development (HR4). Furthermore, one might argue that there is a human right not to be forcibly evicted (HR5) and that climate change violates this because people from coastal settlements and small island states will be forced to leave.

Fourth, it should be stressed that to say that climate change jeopardizes human rights is, of course, not to say that it may not also be criticized on a variety of other grounds. To take just one example, the stance defended here is compatible with the claim that anthropogenic climate change is objectionable because it is wrong for humanity to treat the natural world in such a hubristic fashion.[39] My claim is that the human-rights impacts of climate change are serious and should be addressed; it is not that they are the only morally relevant impacts of climate change.

III. Supplementary Considerations

In the previous section, I argued that climate change threatens the enjoyment of fundamental human rights. The case for a "human-rights"-centered analysis of the impacts of climate change can, however, be strengthened further, and I want to draw attention to the additional insights that a human-rights approach brings over cost-benefit analyses (CBAs) and security-based analyses.

A human-rights analysis enjoys three related advantages over a CBA. These all stem from the fact that the latter aggregates the costs and benefits felt by individuals and then selects the policy that maximizes the good. It has long been recognized that one implication of this kind of aggregative consequentialist approach is that it could call for outcomes in which some suffer greatly but their disutility is outweighed by enormous benefits to others. Unlike a human-rights approach, a CBA has only a partial and contingent commitment to the basic interests and entitlements of the most vulnerable. This problematic aspect of cost-benefit analysis manifests itself at several points in discussions about climate change. Consider the following three illustrations of this flaw.

§1. *Climate impacts.* One example of this kind of problem can be found in Bjørn Lomborg's book *Cool It.* Lomborg argues that although climate

risk and threatening others

change leads to loss of life from heat stress, it also leads to a much greater decrease in mortality from cold during the winter and that this good outweighs out the bad.[40] Anthropogenic climate change should, therefore, not be condemned. Indeed, other things being equal, it is morally required. To propose this, though, is to propose engaging in activities that one knows will directly kill some and harm others' health and ability to subsist. This would strike many as morally unacceptable even if it has the side effect of saving some lives. A human-rights approach, however, rules out such policies.[41]

§2. *Intergenerational equity.* A second illustration of the point in hand concerns the question of whether it is appropriate to devote resources to mitigation now for the benefit of future people. It is sometimes argued that because, and to the extent that, future generations are wealthier than current generations, it would be wrong to mitigate.[42] This, however, is not a compelling argument if it turns out that future generations are wealthier than current generations but that some in the future are deprived of the basic necessities of human life. In virtue of its aggregative nature, a cost-benefit approach is concerned only with the total amount of utility, and therefore the total wealth of current and future generations, and it is indifferent to the plight of the very severely disadvantaged if their disutility is outweighed by the utility of others. A human-rights approach, however, is not vulnerable to this charge because it establishes moral thresholds below which persons should not fall.

§3. *Risk and uncertainty.* A third illustration of the point at hand arises from the risks and uncertainties associated with climate change. Climate scientists repeatedly stress that the projections of future changes to the earth's climate are not certain and that they are characterized by both risk and uncertainty. A cost-benefit approach will respond to risks by multiplying the probability of an event with the utility/disutility of that event, thereby arriving at the expected utility. However, by doing so, it ignores a morally relevant aspect of current climate change, namely that *some* persons are imposing grave risks on *others*. It matters a great deal whether those who are taking risks are exposing just themselves to serious risks or whether they are exposing others to serious risks. In the former case, one might say that as long as the risk takers are sufficiently well informed and rational, then their choice is permissible. The second situation is, however, quite different, for some are posing a threat to the rights of *others*. A CBA cannot capture the relevance of this distinction, since its concern is simply with the aggregate level of expected utility. A human-rights approach, however, captures the importance of this distinction because it disaggregates the impacts of climate change and is concerned with ensuring that none falls beneath a certain threshold. As such, it would condemn as unjust a situation in which some (who are advantaged) expose others (who are vulnerable) to risks that threaten the latter's basic interests. Similarly, it would permit the first kind of risk taking on the grounds that persons are within their rights to expose themselves to risk. A human-rights perspective can thus deal better with the risk and uncertainty associated with climate change.

§4. If we turn now from CBA to the security-oriented approach presented in the introduction, we find a similar problem but for a different reason. This, too, will generate only a contingent and partial commitment to protecting the most vulnerable. It gives us reason to be concerned about climate change only if, because, and to the extent that it results in violent conflict.[43] It follows from this that in those cases where climate change causes death, disease, malnutrition, and starvation but in which it does not in turn lead to conflict, it is silent and would devote no resources to assist those threatened by dangerous climate change. It therefore fails to have an unconditional concern with the most disadvantaged. Its commitment to them is contingent on conditions that may not be met.

In short, then, a human-rights approach will thus protect the vulnerable, whereas a CBA fails to do so because of its aggregative character, and a security-based approach fails to do so because its concern is only with climate change that causes conflict.

IV. The Implications of a Human-Rights Approach

Having argued that climate change undermines fundamental human rights and that this way of thinking about the impacts of climate change enjoys an advantage over cost-benefit analysis, I now want to reflect on several implications of applying a human-rights approach to the impacts of climate change. First, and most obviously, a human-rights approach requires us to adopt a discriminating approach to the impacts of climate change and would therefore not take into account all of the impacts of climate change. From a purely human-rights approach, only those effects that violate rights should be taken into account.[44]

A second implication of a human-rights approach is that it requires us to reconceive the way in which one thinks about the costs involved in mitigation and adaptation. Some have argued that it would be extremely expensive to prevent dangerous climate change and hence that humanity should not do this. If, however, it is true that climate change violates human rights, then this kind of reasoning is inappropriate. Suppose that someone builds a restaurant in their garden and makes a large profit from this. Suppose, however, that this restaurant releases fumes that threaten the lives of others nearby (thereby jeopardizing their human right to life), and it also leaks pollution into the water supply (thereby violating their human right to health). Those committed to human rights will condemn this as unjust and call for the owner of the restaurant not to engage in such rights-violating behavior. If the owner protests that this would be very expensive, the appropriate reply is that this is not germane. If a person is violating human rights, then he or she should desist even if it is costly. Suppose that (as seems highly likely) the abolition of slavery was immensely costly to slave owners. It does not follow from this that slave owners should be allowed to continue in their rights-violating activity.[45] The implications for mitigation and adaptation are clear. That mitigation and adaptation would be *costly* similarly does not in itself entail that they should not be adopted. If emitting greenhouse gases results in rights violations, it should stop, and the fact that it is expensive does not tell against that claim. A human-rights approach thus requires us to reframe the issues surrounding the costs of mitigation and adaptation.

A human-rights approach to climate change has a third implication. If, as argued above, climate change violates human rights, then it follows that compensation is due to those whose rights have been violated. The conventional approach to climate change identifies only two kinds of response to climate change: mitigation and adaptation. The IPCC's Assessment Reports, for example, operate with this dualistic framework. The IPCC defines mitigation as an "*anthropogenic* intervention to reduce the anthropogenic forcing of the *climate system*; it includes strategies to reduce *greenhouse gas sources* and emissions and enhancing *greenhouse gas sinks*."[46] Adaptation is then defined as an "[a]djustment in natural or *human systems* in response to actual or expected climatic stimuli or their effects, which moderates harm or exploits beneficial opportunities."[47] Broadly put, mitigation seeks to minimize changes to the climate system, and adaptation seeks to adjust human institutions in order to cope with the changes to the climate system. This, however, is too narrow a framework, for if there is insufficient mitigation and thus changes to the climate occur, and if, further, there is insufficient adaptation, then the fundamental human rights to life, health, and subsistence will be violated. And where human rights have been violated, those who have been wronged (if they are still alive) are entitled to compensation. A human-rights approach thus generates duties of mitigation and duties of adaptation, and (given the changes to the climate that are in process and given the likely lack of adequate adaptation) it also entails duties of compensation.

It is important to stress that compensation is fundamentally different from adaptation. The point of adaptation is to prevent the changes to the natural world having a malign impact on people's vital interests and human rights. If adaptation is successfully implemented, then people's rights would be protected. The case for compensation, by contrast, arises when and because persons' rights were not protected. One might put it thus: the point of adaptation is to protect and uphold rights, and the point of compensation is to redress the fact that people's rights have been violated.

This third point draws our attention to a fourth implication of adopting a human-rights approach to climate change: that it affects the way in which one should think about inflicting harms on others and the role that compensation may play in our decision making. On one way of thinking about harms, if one imposes one cost on people but also bestows on them a benefit, then the two may cancel each other out, and the affected person has no cause for complaint. This assumes that harms and benefits are commensurable and that the shortfall represented by a harm is erased by the allocation of a benefit. A human-rights approach adopts a different approach to the imposition of harms. For if one has a human right not to suffer a certain harm, then it is wrong to violate that with a view to giving a compensatory sum to counterbalance the harm. To give an example, it is obviously impermissible for one person to assault someone else with a view to giving them a large benefit in order somehow to cancel out the harm. Similarly, one cannot destroy someone's property and then simply write a check and then think that the victim has no cause for complaint. He or she does. The point here is that if a person has a human right (and indeed, any other kind of right), then that generates a duty to respect it, and it is not acceptable to violate that duty with a view to making compensation. Of course, as was argued above, if people do in fact violate rights, then there is a case for compensation. This, however, does not give one permission to engage in rights violations, and it does not undermine the key point that a human-rights approach rejects the tradeoff between burdens and benefits that other approaches endorse.[48]

Let us turn now to a fifth corollary of a human-rights approach to climate change. A human-rights approach guides not simply our evaluation of the impacts of climate change but also the distribution of the duties to uphold the human rights threatened by climate change. It should inform who is obligated to pay for the costs of mitigation and adaptation. The central point here is that if we accept a set of fundamental human rights, then it follows that any program of combating climate change should itself also not violate these rights. Thus, any international treaty distributing emission rights and any national-level climate action plan should not jeopardize the human rights to health, life, and subsistence. In practice, this requires that the least advantaged—those whose human rights are most vulnerable—should not be required to bear the burden of combating climate change

Finally, it is worth remarking that a human-rights perspective provides a useful way of conceptualizing Article 2 of the United Nations Framework Convention on Climate Change (UNFCCC), which states that the objective of the UNFCCC is to achieve a "stabilization of greenhouse gas concentrations in the atmosphere at a level that would prevent *dangerous anthropogenic interference* with the climate system" (1992, my emphasis). What counts as a "dangerous" anthropogenic interference is clearly, in part, a normative issue. It cannot be resolved by science alone, for that can at most tell us the kinds of changes that are likely to occur. To determine when the changes are "dangerous," we need some normative principle or principles. My proposal, in this context, is that dangerous climate change should be interpreted as climate change that systematically undermines the widespread enjoyment of human rights.

V. Concluding Remarks

The important links between climate change and human rights have been neglected. In this chapter, I have sought to address this lacuna. I have defended three distinct conclusions:

1. Climate change jeopardizes human rights and in particular the human rights to life, health, and subsistence (section II).

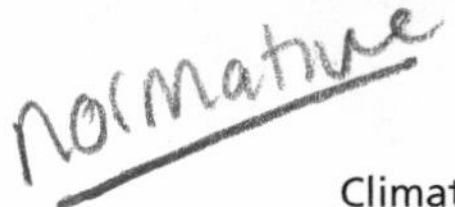

2. Analyzing the impacts of climate change in terms of its effects on human rights enjoys advantages over other ways of evaluating the impacts of climate change (section III).
3. Endorsing a human-rights framework for evaluating the impacts of climate change has implications for our understanding of who should bear the burdens of climate change and what kinds of policies are appropriate (section IV).[49]

As I noted above, I am not claiming that a human-rights approach captures *all* of the morally relevant impacts of climate change. My argument is simply that a human-rights perspective has important insights, and any account of the impacts of climate change that ignores its implications for people's enjoyment of human rights is fundamentally incomplete and inadequate.

Acknowledgments

The research for this chapter was undertaken while I held a Leverhulme Research Fellowship, and the paper was completed while I held an ESRC Climate Change Leadership Fellowship. I am grateful to both the Leverhulme Trust and the ESRC for their support. I am also grateful to Stephen Humphreys for his comments on an earlier draft.

Notes

1. Nicholas Stern, *The Economics of Climate Change: The Stern Review* (Cambridge, U.K.: Cambridge University Press, 2007).

2. It is very important to distinguish this traditional type of security-based argument, with its emphasis on violent conflict, from other conceptions of security. It should, for example, be contrasted with the concept of "human security." The latter breaks with notions of security that define it wholly in terms of the extent of violent conflict and defines it more broadly. A canonical characterization of human security can be found in the UNDP's 1994 *Human Development Report*. It is argued there that human security includes "economic security," "food security," "health security," "environmental security," "personal security," "community security," and "political security," United Nations Development Program, *Human Development Report 1994: New Dimensions of Human Security* (Oxford: Oxford University Press, 1994), chap. 2, esp. pp. 24–25. My concern here is with traditional conceptions of security. For a good application of the concept of human security to climate change, see Karen O'Brien, "Are We Missing the Point? Global Environmental Change as an Issue of Human Security," *Global Environmental Change* 16.1 (2006): 1–3.

3. The paper can be found at http://www.consilium.europa.eu/ueDocs/cms_Data/docs/pressData/en/reports/99387.pdf. The quotation is from p. 2.

4. *Climate Change and International Security,* section II.

5. From http://www.unep.org/Documents.multilingual/Default.asp?DocumentID=97&ArticleID=1503.

6. From http://www.ciel.org/Publications/Male_Declaration_Nov07.pdf.

7. From http://www.ciel.org/Publications/Male_Declaration_Nov07.pdf.

8. This was agreed at the seventh session of the Human Rights Council on March 26, 2008 (A/HRC/7/L.21/Rev.1).

9. H. L. A. Hart, "Are There Any Natural Rights?" *Philosophical Review* 64.2: 183–188. I dissent from one aspect of Hart's characterization of general rights. He ascribes general rights to all humans capable of choice, whereas I ascribe general/human rights to all humans whether or not they can exercise choice. Hart's position here follows from his commitment to the "choice" theory of rights, which he pioneered and defended in "Are There any Natural Rights?" I endorse the alternative theory of rights, which has come to be termed the "interest" theory of rights. For a canonical statement of this approach, see Joseph Raz, *The Morality of Freedom* (Oxford: Clarendon Press, 1986), chap. 7. Evaluating the debate between the choice theory and the interest theory would take us too far afield.

10. Henry Shue, *Basic Rights: Subsistence, Affluence, and U. S. Foreign Policy,* 2d ed. (Princeton, N.J.: Princeton University Press, 1996), p. 18.

11. The concept of "lexical priority" comes from John Rawls, *A Theory of Justice,* rev. ed. (Oxford: Oxford University Press, 1999), pp. 37–38. As Rawls employs this term, to say that A enjoys lexical priority over other values is to say that it is morally more urgent and may not be sacrificed to pursue any of these other values.

12. This priority may not be absolute in all circumstances. One can, of course, imagine situations where sacrificing the rights of one will save very many people. Some might then condone the sacrifice of one right in such scenarios. Three points should be made here. First, these refer to exceptional cases, and so one might say (as I do above) that human rights generally take priority. Second, even if one thinks that an individual human right may be violated, one may hold that such a violation is permissible only to honor other human rights. So even if an individual human right might be overridden, this does not entail that human rights as a category can be overridden to further some other goal. Indeed, the standard cases presented to show that human rights might be overridden always present examples in which the case for violating one human right (e.g., torturing a terrorist suspect) is that it would uphold other human rights (e.g., the right to life) of many others. Finally, though I cannot argue the point here, I agree with those who argue that even if one could conceive of a case where in principle violating one human right would protect more human rights, institutionalizing it in practice would be wrong because it would in all likelihood lead to unjustified human-rights violations. Accepting that in a hypothetical situation a right might be violated does not show that in practice institutions should be given the power to do so, simply because one might think that the relevant decision makers are fallible or might abuse the power. See Peter Jones, *Rights* (Basingstoke, U.K.: Macmillan, 1994), pp. 203–204.

13. The account I have sketched conforms to what Charles Beitz terms an "orthodox" conception of human rights. See Charles Beitz, "Human Rights and the Law of Peoples," in *The Ethics of Assistance: Morality and the Distant Needy,* ed. Deen Chatterjee (Cambridge, U.K.: Cambridge University Press, 2004), pp. 193–214. He contrasts the "orthodox" account with what he terms the "practical" account. The latter maintains that human rights should be defined in terms of the role they play in political practice. More precisely, human rights, in this view, specify the conditions under which some kind of intervention in another society is justified. Beitz raises a number of objections to the orthodox conception and proposes the practical conception as a superior alternative. For Beitz's description of the practical account, see "Human Rights and the Law of Peoples," esp. pp. 201–205, and also Charles Beitz, "Human Rights as a Common Concern," *American Political Science Review* 95.2 (2001): 269–282.

14. This is a necessarily abbreviated discussion of this distinction. For a fuller analysis, see Shue, *Basic Rights,* chap. 2; and Simon Caney, "Global Poverty and Human Rights: The Case for Positive Duties," in *Freedom from Poverty as a Human Right: Who Owes What to the Very Poor?* ed. Thomas Pogge (Oxford: Oxford University Press, 2007), pp. 275–302.

15. Thomas Nagel, "Personal Rights and Public Space," *Philosophy and Public Affairs* 24.2 (1995): 86.

16. Ibid., pp. 89–93.

17. Ibid. Nagel here is developing ideas defended by Frances Kamm and Warren Quinn (see p. 89, n. 3).

18. This does not exhaust the different approaches to grounding human rights. For a contrasting view, see that expressed by John Rawls in *The Law of Peoples with "The Idea of Public Reason Revisited"* (Cambridge, Mass.: Harvard University Press, 1999). Rawls approaches human rights in a different way. He argues that human rights perform three roles: (i) they specify an essential condition for any "decent" society, (ii) if they are honored, then any kind of intervention is illegitimate, and (iii) they constrain the extent of permissible diversity among different societies (p. 80; see also pp. 79–81). Rawls proposes a set of human rights that "liberal" and "decent" nonliberal peoples can both embrace, and he rejects an account of human rights that is predicated on a commitment to liberalism (pp. 37, 65).

19. Allen Buchanan, *Justice, Legitimacy, and Self-Determination: Moral Foundations for International Law* (Oxford: Oxford University Press, 2004), p. 127. See more generally Buchanan's excellent analysis of the nature of, and case for, human rights in *Justice, Legitimacy, and Self-Determination*, chap. 3.

20. Martha Nussbaum, "Capabilities and Human Rights," in *Global Justice and Transnational Politics: Essays on the Moral and Political Challenges of Globalization,* ed. Pablo de Greiff and Ciaran Cronin (Cambridge: Mass.: MIT Press, 2002), pp. 117–149.

21. James Griffin, *On Human Rights* (Oxford: Oxford University Press, 2008), esp. pp. 33–37, 57–82.

22. I have defended an instrumental approach in *Justice beyond Borders: A Global Political Theory* (Oxford: Oxford University Press, 2005), chap. 3.

23. My approach here is indebted to that advanced by Thomas Pogge in his pioneering work on global poverty. See Thomas Pogge, *World Poverty and Human Rights: Cosmopolitan Responsibilities and Reforms,* 2d ed. (Cambridge, U.K.: Polity, 2008). I do disagree with some aspects of Pogge's methodology. See on this, Simon Caney "Global Poverty and Human Rights."

24. R. F. Mclean and Alla Tsyban, "Coastal Zones and Marine Ecosystems," in *Climate Change 2001: Impacts, Adaptation, and Vulnerability—Contribution of Working Group II to the Third Assessment Report of the Intergovernmental Panel on Climate Change,* ed. James J. McCarthy, Osvaldo F. Canziani, Neil A. Leary, David J. Dokken, and Kasey S. White (Cambridge, U.K.: Cambridge University Press, 2001), pp. 366–367.

25. Ulisses Confalonieri and Bettina Menne, "Human Health," in *Climate Change 2007: Impacts, Adaptation and Vulnerability—Contribution of Working Group II to the Fourth Assessment Report of the Intergovernmental Panel on Climate Change,* ed. Martin Parry, Osvaldo Canziani, Jean Palutikof, Paul van der Linden, and Clair Hanson (Cambridge, U.K.: Cambridge University Press, 2007), p. 398.

26. Jonathan Patz et al., "The Potential Health Impacts of Climate Variability and Change for the United States: Executive Summary of the Report of the Health Sector of the U.S. National Assessment," *Environmental Health Perspectives* 108.4 (2000): 370.

27. A. Haines, R. S. Kovats, D. Campbell-Lendrum, and C. Corvalan "Climate Change and Human Health: Impacts, Vulnerability, and Mitigation,: *Lancet* 367 (June 24, (2006): 2103.

28. Of course, one cannot specify in advance which particular individuals will suffer, but this does not undermine the moral point that the actions in question undermine human rights. If a saboteur weakens a viaduct on which people drive to work so that after a while it will collapse under the weight of traffic, he or she violates the human rights of those who subsequently plunge to their death even if no one can predict in advance who will suffer from this fate. For instructive remarks, see Joel Feinberg, "The Rights of Animals and Unborn Generations," in *Rights, Justice, and the Bounds of Liberty: Essays in Social Philosophy* (Princeton, N.J.: Princeton University Press, 1980), pp. 181–182.

29. Such a critic, though, should take into account General Comment No. 14 (2000) on Article 12 of the ICESCR, which elaborates how this concept is to be interpreted. General Comment No. 14 can be found in Sofia Gruskin, Michael A. Grodin, George J. Annas, and Stephen P. Marks, eds., *Perspectives on Health and Human Rights* (New York and London: Routledge, 2005), pp. 473–495.

30. Note in this context that General Comment No. 14 on Article 12 of the ICESCR makes clear that the human right to health "extends to the underlying determinants of health, such as food and nutrition, housing, access to safe and potable water and adequate sanitation, safe and healthy working conditions, and a healthy environment." See paragraph 4 of General Comment No. 14. This point is reiterated in General Comment No. 14, "I. Normative Content of Article 12," paragraphs 11, 12, 15).

31. Confalonieri and Menne, "Human Health," p. 393.

32. Ibid., p. 407.

33. Ibid., p. 408.

34. HR3 is closest in formulation to Article 1.2 of the International Covenant on Civil and Political Rights (1976), which states that "In no case may a people be deprived of its own means of subsistence." HR3, though, refers to the entitlements of individuals, not those of "a people."

35. For relevant data, see "Gas Exchange: CO_2 Emissions 1990–2006," *Nature* 447.7148 (2007): 1038; and Michael R. Raupach, Gregg Marland, Philippe Ciais, Corinne Le Quéré, Josep G. Canadell, Gernot Klepper, and Christopher B. Field, "Global and Regional Drivers of Accelerating CO_2 Emissions," *Proceedings of the National Academy of Sciences of the United States of America* 104.24 (2007): esp. 10292.

36. Anthony Nyong and Isabelle Niang-Diop, "Impacts of Climate Change in the Tropics: The African Experience," in *Avoiding Dangerous Climate Change,* ed. Hans Joachim Schellnhuber, Wolfgang Cramer, Nebojsa Nakicenovic, Tom Wigley, and Gary Yohe (Cambridge, U.K.: Cambridge University Press, 2006), p. 237.

37. Bill Hare, "Relationship between Increases in Global Mean Temperature and Impacts on Ecosystems, Food Production, Water and Socio-economic Systems," in *Avoiding Dangerous Climate Change,* ed. Hans Joachim Schellnhuber, Wolfgang Cramer, Nebojsa

Nakicenovic, Tom Wigley, and Gary Yohe (Cambridge, U.K.: Cambridge University Press, 2006), p. 179.

38. The IPCC states, "It is *very likely* that anthropogenic greenhouse gas increases caused most of the observed increase in global average temperatures since the mid-20th century." Susan Solomon, Dahe Qin, and Martin Manning, "Technical Summary," in *Climate Change 2007: The Physical Science Basis—Contribution of Working Group I to the Fourth Assessment Report of the Intergovernmental Panel on Climate Change,* ed. Susan Solomon, Dahe Qin, Martin Manning, Melinda Marquis, Kristen Averyt, Melinda M. B. Tignor, Henry Leroy Miller Jr., and Zhenlin Chen (Cambridge, U.K.: Cambridge University Press, 2007), p. 60. Note that it is arguable that it would be possible for people to violate these three human rights even if climate change were not anthropogenic. Humans can violate the three human rights in two different ways. The first (and most obvious) route is for humans to emit high levels of greenhouse gases and to destroy carbon sinks, which will in turn produce high temperatures, increased precipitation, and severe weather events. The second route is for humans to design social and political institutions that leave people vulnerable to the physical impacts of climate change. Suppose that climate change were nonanthropogenic (and so route 1 was inapplicable), but politicians could implement an effective program of adaptation and design institutions that would safeguard the vital interests of people in life, health, and subsistence but chose not to do so. They could then be said to violate the human rights of others to life, health, and subsistence because they would be acting in such a way as to create threats to life, health, and subsistence.

39. This view has been defended by Dale Jamieson in "What's Wrong with Climate Change?" unpublished paper presented at conference on "Global Justice and Climate Change," Oxford, September 2007.

40. Bjørn Lomborg, *Cool It: The Skeptical Environmentalist's Guide to Global Warming* (London: Marshall Cavendish, 2007), pp. 13–18.

41. See also Edward A. Page, *Climate Change, Justice and Future Generations* (Cheltenham, U.K.: Edward Elgar, 2006), p. 34.

42. For this viewpoint, see Bjørn Lomborg, *The Skeptical Environmentalist: Measuring the Real State of the World* (Cambridge, U.K.: Cambridge University Press, 2001), p. 314; William Nordhaus, "Discounting in Economics and Climate Change," *Climatic Change* 37.2 (1997): 317; William Nordhaus, "The Question of Global Warming: An Exchange," *New York Review of Books* 55.14 (September 25, 2008): 93.

43. As stressed in note 2 above, I am concerned here only with traditional conceptions of security of the type expressed in the introduction. My arguments are not directed against "human security" and attempts to argue that climate change jeopardizes human security.

44. In general terms, this means that impacts that lead to less *preference satisfaction* or less *economic growth* do not count. In more concrete terms, this means that impacts on tourism, say, or on the insurance industry are not relevant except insofar as they bear on the realization of people's human rights. IPCC reports tend to refer to the impacts of climate change on both tourism and the insurance industry. See, for example, Tom Wilbanks and Patricia Romero Lankao, "Industry, Settlement and Society," in *Climate Change 2007: Impacts, Adaptation and Vulnerability—Contribution of Working Group II to the Fourth Assessment Report of the Intergovernmental Panel on Climate Change,* ed. Martin Parry, Osvaldo Canziani, Jean Palutikof, Paul van der Linden, and Clair Hanson (Cambridge, U.K.: Cambridge University Press, 2007), pp. 357–390. My point is not that impacts on tourism do not matter but that we need to distinguish between those impacts on the tourist industry that undermine human rights (for example, those whose livelihood depends on it) and those that do not.

45. Part of the point here is about baselines. It is true that the slave owners will be much worse compared to the status quo before abolition, but the point is that this is an illegitimate and inappropriate baseline to employ to assess what their entitlements should be.

46. For this definition, see "Appendix I: Glossary," in *Climate Change 2007: Impacts, Adaptation and Vulnerability—Contribution of Working Group II to the Fourth Assessment Report of the Intergovernmental Panel on Climate Change,* ed. Martin Parry, Osvaldo Canziani, Jean Palutikof, Paul van der Linden, and Clair Hanson (Cambridge, U.K.: Cambridge University Press, 2007), p. 878.

47. For this definition, see ibid., p. 869.

48. For illuminating discussion, see Clive L. Spash, *Greenhouse Economics: Value and Ethics* (London and New York: Routledge, 2002), pp. 231–236; and Henry Shue, "Bequeathing Hazards: Security Rights and Property Rights of Future

Humans," in *Global Environmental Economics: Equity and the Limits to Markets,* ed. Mohammed H. I. Dore and Timothy D. Mount (Malden, Mass., and Oxford: Blackwell, 1999), pp. 40–43.

49. In focusing on these links between climate change and human rights, I am not claiming that this exhausts the relevant connections between human rights and climate change. Two other connections are worth noting. First, it is arguable that persons have a human right to have an input into any decision-making process that affects their fundamental interests. On this basis, one may argue that persons have a human right to shape the political process by which decisions about mitigation and compensation are made. One might call this the human right to procedural justice. Second, it is also arguable that the extent to which people are able to adapt to dangerous climate change is a function of the extent to which their basic human rights are respected. The more their rights have been violated, the less they are able to adapt to climate change. This second theme is explored by Jon Barnett in "Human Rights and Vulnerability to Climate Change," in *Human Rights and Climate Change,* ed. Stephen Humphreys (Cambridge, U.K.: Cambridge University Press, 2009), pp. 257–271.

Part IV

Policy Responses to Climate Change

One Atmosphere

Peter Singer

The Problem

There can be no clearer illustration of the need for human beings to act globally than the issues raised by the impact of human activity on our atmosphere. That we all share the same planet came to our attention in a particularly pressing way in the 1970s when scientists discovered that the use of chlorofluorocarbons (CFCs) threatens the ozone layer shielding the surface of our planet from the full force of the sun's ultraviolet radiation. Damage to that protective shield would cause cancer rates to rise sharply and could have other effects, for example, on the growth of algae. The threat was especially acute to the world's southernmost cities, since a large hole in the ozone was found to be opening up each year over Antarctica, but in the long term, the entire ozone shield was imperiled. Once the science was accepted, concerted international action followed relatively rapidly with the signing of the Montreal Protocol in 1985. The developed countries phased out virtually all use of CFCs by 1999, and the developing countries, given a 10-year period of grace, are now moving toward the same goal.

Getting rid of CFCs has turned out to be just the curtain raiser: the main event is climate change, or global warming. Without belittling the pioneering achievement of those who brought about the Montreal Protocol, the problem was not so difficult, for CFCs can be replaced in all their uses at relatively little cost, and the solution to the problem is simply to stop producing them. Climate change is a very different matter.

The scientific evidence that human activities are changing the climate of our planet has been studied by a working group of the Intergovernmental Panel on Climate Change (IPCC), an international scientific body intended to provide policy makers with an authoritative view of climate change and its causes. The group released its *Third Assessment Report* in 2001, building on earlier reports and incorporating new evidence accumulated over the previous five years. The report is the work of 122 lead authors and 515 contributing authors, and the research on which it was based was reviewed by 337 experts. Like any scientific document it is open to criticism from other scientists, but it reflects a broad consensus of leading scientific opinion and is by far the most authoritative view at present available on what is happening to our climate.

The *Third Assessment Report* finds that our planet has shown clear signs of warming over the past century. The 1990s were the hottest decade, and 1998 the hottest year, recorded over the 140 years for which meteorological records have been kept. As 2001 drew to a close, the World Meteorological Organization announced that it would be second only to 1998 as the hottest year recorded. In fact nine of the ten hottest years during this period have occurred since 1990, and temperatures are now rising at three times the rate of the early 1900s.[1] Sea levels have risen by between 10 and 20 centimeters over the past century. Since the 1960s snow and ice cover has decreased by about 10 percent, and mountain glaciers are in retreat everywhere except near the poles. In the past three decades the El Niño effect in the southern hemisphere has become more intense, causing greater variation in rainfall. Paralleling these changes is an unprecedented increase in concentrations of carbon dioxide, methane, and nitrous oxide in the atmosphere, produced by human activities such as burning fossil fuels, the clearing of vegetation, and (in the case of methane) cattle and rice production. Not for at least 420,000 years has there been so much carbon dioxide and methane in the atmosphere.

How much of the change in climate has been produced by human activity, and how much can be explained by natural variation? The *Third Assessment Report* finds "new and stronger evidence that most of the warming observed over the last fifty years is attributable to human activities," and, more specifically, to greenhouse-gas emissions. The report also finds it "very likely" that most of the rise in sea levels over the past century is due to global warming.[2] Those of us who have no expertise in the scientific aspects of assessing climate change and its causes can scarcely disregard the views held by the overwhelming majority of those who do possess that expertise. They could be wrong—the great majority of scientists sometimes are—but in view of what is at stake, to rely on that possibility would be a risky strategy.

What will happen if we continue to emit increasing amounts of greenhouse gases and global warming continues to accelerate? The *Third Assessment Report* estimates that between 1990 and 2100, average global temperatures will rise by at least 1.4 °C and perhaps by as much as 5.8 °C.[3] Although these average figures may seem quite small—whether tomorrow is going to be 20 °C or 22 °C isn't such a big deal—even a 1 °C rise in average temperatures would be greater than any change that has occurred in a single century for the past 10,000 years. Moreover, some regional changes will be more extreme and are much more difficult to predict. Northern landmasses, especially North America and central Asia, will warm more than the oceans or coastal regions. Precipitation will increase overall, but there will be sharp regional variations, with some areas that now receive adequate rainfall becoming arid. There will also be greater year-to-year fluctuations than at present—which means that droughts and floods will increase. The Asian summer monsoon is likely to become less reliable. It is possible that the changes could be enough to reach critical tipping points at which the weather systems alter or the directions of major ocean currents, such as the Gulf Stream, change.

What will the consequences be for humans?

- As oceans become warmer, hurricanes and tropical storms that are now largely confined to the tropics will move further from the equator, hitting large urban areas that have not been built to cope with them. This is a prospect that is viewed with great concern in the insurance industry, which has already seen the cost of natural disasters rise dramatically in recent decades.[4]
- Tropical diseases will become more widespread.
- Food production will rise in some regions, especially in the high northern latitudes,

and fall in others, including sub-Saharan Africa.

- Sea levels will rise by between 9 and 88 centimeters.

Rich nations may, at considerable cost, be able to cope with these changes without enormous loss of life. They are in a better position to store food against the possibility of drought, to move people away from flooded areas, to fight the spread of disease-carrying insects, and to build seawalls to keep out the rising seas. Poor nations will not be able to do so much. Bangladesh, the world's most densely populated large country, has the world's largest system of deltas and mudflats, where mighty rivers like the Ganges and the Brahmaputra reach the sea. The soil in these areas is fertile, but the hazards of living on such low-lying land are great. In 1991 a cyclone hit the coast of Bangladesh, coinciding with high tides that left 10 million people homeless and killed 139,000. Most of these people were living on mudflats in the deltas. People continue to live there in large numbers because they have nowhere else to go. But if sea levels continue to rise, many peasant farmers will have no land left. As many as 70 million people could be affected in Bangladesh, and a similar number in China. Millions more Egyptian farmers on the Nile delta also stand to lose their land. On a smaller scale, Pacific island nations that consist of low-lying atolls face even more drastic losses. Kiribati, placed just to the west of the International Date Line, was the first nation to enter the new millennium. Ironically, it may also be the first to leave it, disappearing beneath the waves. High tides are already causing erosion and polluting fragile sources of fresh water, and some uninhabited islands have been submerged.

Global warming would lead to an increase in summer deaths due to heat stress, but these would be offset by a reduced death toll from winter cold. Much more significant than either of these effects, however, would be the spread of tropical diseases, including diseases carried by insects that need warmth to survive. The *Third Assessment Report* considers several attempts to model the spread of diseases like malaria and dengue but finds that the research methodology is, at this stage, inadequate to provide good estimates of the numbers likely to be affected.[5]

If the Asian monsoon becomes less reliable, hundreds of millions of peasant farmers in India and other countries will go hungry in the years in which the monsoon brings less rain than normal. They have no other way of obtaining the water needed for growing their crops. In general, less reliable rainfall patterns will cause immense hardship among the large proportion of the world's population who must grow their own food if they want to eat.

The consequences for nonhuman animals and for biodiversity will also be severe. In some regions plant and animal communities will gradually move further from the equator, or to higher altitudes, following climate patterns. Elsewhere that option will not be available. Australia's unique alpine plants and animals already survive only on the country's highest alpine plains and peaks. If snow ceases to fall on their territory, they will become extinct. Coastal ecosystems will change dramatically, and warmer waters may destroy coral reefs. These predictions look ahead only as far as 2100, but even if greenhouse-gas emissions have been stabilized by that time, changes in climate will persist for hundreds, perhaps thousands, of years. A small change in average global temperatures could, over the next millennium, lead to the melting of the Greenland ice cap, which, added to the partial melting of the West Antarctic Ice Sheet, could increase sea levels by six meters.[6]

All of this forces us to think differently about our ethics. Our value system evolved in circumstances in which the atmosphere, like the oceans, seemed an unlimited resource, and responsibilities and harms were generally clear and well defined. If someone hit someone else, it was clear who had done what. Now the twin problems of the ozone hole and climate change have revealed bizarre new ways of killing people. By spraying deodorant at your armpit in your New York apartment, you could, if you use an aerosol spray propelled by CFCs, be contributing to the skin-cancer deaths, many years later, of people living in Punta Arenas, Chile. By driving your car, you could be releasing carbon dioxide that is part of a causal chain leading to lethal floods in Bangladesh.[7] How can we adjust our ethics to take account of this new situation?

Rio and Kyoto

That seemingly harmless and trivial human actions can affect people in distant countries is just beginning to make a significant difference to the sovereignty of individual nations. Under existing international law, individuals and companies can sue for damages if they are harmed by pollution coming from another country, but nations cannot take other nations to court. In January 2002, Norway announced that it would push for a binding international "polluter pays" scheme for countries. The announcement followed evidence that Britain's Sellafield nuclear power plant is emitting radioactive wastes that are reaching the Norwegian coastline. Lobsters and other shellfish in the North Sea and the Irish Sea have high levels of radioactive technetium-99.[8]

The Sellafield case has revealed a gap in environmental legislation on a global basis. Norway is seeking an international convention on environmental pollution, first at the European level and then, through the United Nations, globally. The principle is one that is difficult to argue against, but if Norway can force Britain to pay for the damage Britain's leaking nuclear plant causes to Norway's coastline, will not nations like Kiribati be able to sue America for allowing large quantities of carbon dioxide to be emitted into the atmosphere, causing rising sea levels to submerge its island homes? Although the link between rising sea levels and a nation's emissions of greenhouse gases is much more difficult to prove than the link between Britain's nuclear power plant and technetium-99 found along the Norwegian coast, it is hard to draw a clear line of principle between the two cases. Yet accepting the right of Kiribati to sue for damages for American greenhouse-gas emissions makes us "one world" in a new and far more sweeping sense than we ever were before. It gives rise to a need for concerted international action.

Climate change entered the international political arena in 1988, when the United Nations Environment Program and the World Meteorological Office jointly set up the Intergovernmental Panel on Climate Change. In 1990 the IPCC reported that the threat of climate change was real, and a global treaty was needed to deal with it. The United Nations General Assembly resolved to proceed with such a treaty. The United Nations Framework Convention on Climate Change was agreed to in 1992 and opened for signature at the Earth Summit, or, more formally, the United Nations Conference on Environment and Development, which was held in Rio de Janeiro in the same year. This "framework convention" has been accepted by 181 governments. It is, as its name suggests, no more than a framework for further action, but it calls for greenhouse gases to be stabilized at safe levels, and it says that the parties to the convention should do this "on the basis of equity and in accordance with their common but differentiated responsibilities and respective capabilities." Developed nations should "take the lead in combating climate change and the adverse effects thereof". The developed nations committed themselves to 1990 levels of emissions by the year 2000, but this commitment was not legally binding.[9] For the United States and several other countries, that was just as well, because they came nowhere near meeting it. In the United States, for example, by 2000, carbon dioxide emissions were 14 percent higher than they were in 1990. Nor was the trend improving, for the increase between 1999 and 2000 was 3.1 percent, the biggest one-year increase since the mid-1990s.[10]

The framework convention builds in what is sometimes called the "precautionary principle," calling on the parties to act to avoid the risk of serious and irreversible damage even in the absence of full scientific certainty. The convention also recognizes a "right to sustainable development," asserting that economic development is essential for addressing climate change. Accordingly, the Rio Earth Summit did not set any emissions-reduction targets for developing countries to meet.

The framework convention set up a procedure for holding "conferences of the parties" to assess progress. In 1995, this conference decided that more binding targets were needed. The result, after two years of negotiations, was the 1997 Kyoto Protocol, which set targets for 39 developed nations to limit or reduce their greenhouse-gas emissions by 2012. The limits and reductions were designed to reduce total emissions from the developed nations to a level at least 5 percent below 1990 levels. The national targets vary, however, with the European Union nations and the United States having targets of 8 percent and 7 percent, respectively, below 1990 levels and other nations, such as Australia, being allowed to go over their 1990 levels. These targets were arrived at through negotiations with government leaders, and they were not based on any general principles of fairness, nor much else that can be defended on any terms other than the need to get agreement.[11] This was necessary since under the prevailing conception of national sovereignty, countries cannot be bound to meet their targets unless they decide to sign the treaty that commits them to do so. To assist countries in reaching their targets, the Kyoto Protocol accepted the principle of "emissions trading," by which one country can buy emissions credits from another country that can reach its target with something to spare.

The Kyoto conference did not settle the details of how countries could meet their targets, for example, whether they would be allowed credits for planting forests that soak up carbon dioxide from the atmosphere, and how emissions trading was to operate. After a meeting at the Hague failed to reach agreement on these matters, they were resolved at further meetings held in Bonn and Marrakech in July and November 2001, respectively. There, 178 nations reached a historic agreement that makes it possible to put the Kyoto Protocol into effect. American officials, however, were merely watching from the sidelines. The United States was no longer a party to the agreement. Later, Prime Minister John Howard announced that Australia would follow the lead set by the United States and refuse to ratify the agreement, despite his nation having received more generous terms in the protocol than any other developed nation.

The Kyoto agreement will not solve the problem of the impact of human activity on the world's climate. It will only slow the changes that are now occurring. For that reason, some skeptics have argued that the likely results do not justify the costs of putting the agreement into effect. In an article in the *Economist*, Bjørn Lomborg writes:

> Despite the intuition that something drastic needs to be done about such a costly problem, economic analyses clearly show that it will be far more expensive to cut carbon-dioxide emissions radically than to pay the costs of adaptation to the increased temperatures.[12]

Lomborg is right to raise the question of costs. It is conceivable, for example, that the resources the world is proposing to put into reducing greenhouse-gas emissions could be better spent on increasing assistance to the world's poorest people, to help them develop economically and so cope better with climate change. But how likely is it that the rich nations would spend the money in this manner?. Their past record is not encouraging. A comparatively inefficient way of helping the poor may be better than not helping them at all.[13]

Significantly, Lomborg's highly controversial book, *The Skeptical Environmentalist*, offers a more nuanced picture than the bald statement quoted above. Lomborg himself points out that, even in a worst-case scenario in which Kyoto is implemented in an inefficient way, "there is no way that the cost will send us to the poorhouse." Indeed, he says, one could argue that whether we choose to implement the Kyoto Protocol or to go beyond it and actually stabilize greenhouse gases:

> the total cost of managing global warming *ad infinitum* would be the same as deferring the [economic] growth curve by less than a year. In other words we would have to wait until 2051 to enjoy

> the prosperity we would otherwise have enjoyed in 2050. And by that time the average citizen of the world will have become twice as wealthy as she is now.[14]

Lomborg does claim that the Kyoto Protocol will lead to a net loss of $150 billion (U.S.). This estimate assumes that there will be emissions trading within the developed nations but not among all nations of the world. It also assumes that the developing nations will remain outside the protocol—in which case the effect of the agreement will be only to delay, by a few years, the predicted changes to the climate. But if the developing nations join in once they see that the developed nations are serious about tackling their emissions, and if there is global emissions trading, then Lomborg's figures show that the Kyoto pact will bring a net benefit of $61 billion (U.S.).

These estimates all assume that Lomborg's figures are sound—a questionable assumption, for how shall we price the increased deaths from tropical diseases and flooding that global warming will bring? How much should we pay to prevent the extinction of species and entire ecosystems? Even if we could answer these questions, and agree on the figures that Lomborg uses, we would still need to consider his decision to discount all future costs at an annual rate of 5 percent. A discount rate of 5 percent means that we consider losing $100 today to be the equivalent of losing $95 in a year's time, the equivalent of losing $90.25 in two years' time, and so on. Obviously, then, losing something in, say, 40 years' time isn't going to be worth much, and it wouldn't make sense to spend a lot now to make sure that you don't lose it. To be precise, at this discount rate, it would only be worth spending $14.20 today to make sure that you don't lose $100 in 40 years' time. Since the costs of reducing greenhouse-gas emissions will come soon, whereas most of the costs of not doing anything to reduce them fall several decades into the future, this makes a huge difference to the cost-benefit equation. Assume that unchecked global warming will lead to rising sea levels, flooding valuable land in 40 years' time. With an annual discount rate of 5 percent, it is worth spending only $14.20 to prevent flooding that will permanently inundate land worth $100. Losses that will occur a century or more hence dwindle to virtually nothing. This is not because of inflation—we are talking about costs expressed in dollars already adjusted for inflation. It is simply discounting the future.

Lomborg justifies the use of a discount rate by arguing that if we invest $14.20 today, we can get a (completely safe) return of 5 percent on it, and so it will grow to $100 in 40 years. Though the use of a discount rate is a standard economic practice, the decision about which rate should be used is highly speculative, and assuming different interest rates, or even acknowledging uncertainty about interest rates, would lead to very different cost-benefit ratios.[15] There is also an ethical issue about discounting the future. True, our investments may increase in value over time, and we will become richer, but the price we are prepared to pay to save human lives, or endangered species, may go up just as much. These values are not consumer goods like TVs or dishwashers, which drop in value in proportion to our earnings. They are things like health, something that the richer we get, the more we are willing to spend to preserve. An ethical, not an economic, justification would be needed for discounting suffering and death, or the extinction of species, simply because these losses will not occur for 40 years. No such justification has been offered.

It is important to see Kyoto not as the solution to the problem of climate change but as the first step. It is reasonable to raise questions about whether the relatively minor delay in global warming that Kyoto would bring about is worth the cost. But if we see Kyoto as a necessary step for persuading the developing countries that they, too, should reduce greenhouse-gas emissions, we can see why we should support it. Kyoto provides a platform from which a more far-reaching and also more equitable agreement can be reached. Now we need to ask what that agreement would need to be like to satisfy the requirement of equity or fairness.

What Is an Equitable Distribution?

In the second of the three televised debates held during the 2000 U.S. presidential election, the candidates were asked what they would do about global warming. George W. Bush said:

> I'll tell you one thing I'm not going to do is I'm not going to let the United States carry the burden for cleaning up the world's air, like the Kyoto treaty would have done. China and India were exempted from that treaty. I think we need to be more even-handed.

There are various principles of fairness that people often use to judge what is fair or "even-handed". In political philosophy, it is common to follow Robert Nozick in distinguishing between "historical" principles and "time-slice" principles.[16] A historical principle is one that says we can't decide, merely by looking at the present situation, whether a given distribution of goods is just or unjust. We must also ask how the situation came about; we must know its history. Are the parties entitled, by an originally justifiable acquisition and a chain of legitimate transfers, to the holdings they now have? If so, the present distribution is just. If not, rectification or compensation will be needed to produce a just distribution. In contrast, a time-slice principle looks at the existing distribution at a particular moment and asks if that distribution satisfies some principles of fairness, irrespective of any preceding sequence of events. I shall look at both of these approaches in turn.

A Historical Principle: "The Polluter Fays" or "You Broke It, Now You Fix It"

Imagine that we live in a village in which everyone puts their wastes down a giant sink. No one quite knows what happens to the wastes after they go down the sink, but since they disappear and have no adverse impact on anyone, no one worries about it. Some people consume a lot, and so have a lot of waste, while others, with more limited means, have barely any, but the capacity of the sink to dispose of our wastes seems so limitless that no one worries about the difference. As long as that situation continues, it is reasonable to believe that in putting waste down the sink, we are leaving "enough and as good" for others, because no matter how much we put down it, others can also put as much as they want, without the sink overflowing. This phrase "enough and as good" comes from John Locke's justification of private property in his *Second Treatise on Civil Government*, published in 1690. In that work Locke says that "the earth and all that is therein is given to men for the support and comfort of their being." The earth and its contents "belong to mankind in common." How, then, can there be private property? Because our labor is our own, and hence when we mix our own labor with the land and its products, we make them our own. But why does mixing my labor with the common property of all humankind mean that I have gained property in what belongs to all humankind, rather than lost property in my own labor? It has this effect, Locke says, as long as the appropriation of what is held in common does not prevent there being "enough and as good left in common for others."[17]

Locke's justification of the acquisition of private property is the classic historical account of how property can be legitimately acquired, and it has served as the starting point for many more recent discussions. Its significance here is that, if it is valid and the sink is, or appears to be, of limitless capacity, it would justify allowing everyone to put what they want down the sink, even if some put much more than others down it.

Now imagine that conditions change, so that the sink's capacity to carry away our wastes is used up to the full, and there is already some unpleasant seepage that seems to be the result of the sink being used too much. This seepage causes occasional problems. When the weather is warm, it smells. A nearby waterhole where our children swim now has algal blooms that make it unusable. Several respected figures in the village warn that unless usage of the sink is cut down, all the village water supplies will be polluted. At this point, when we continue to throw our usual wastes down the sink, we are no longer leaving "enough and as good" for others, and hence our right to unchecked waste disposal becomes questionable. For the sink belongs to us all in common, and by using it without restriction now, we are depriving others of their right to use the sink in the same way without bringing about results none of us wants. We have an example of the well-known "tragedy of the commons."[18] The use of the sink is a limited resource that needs to be shared in some equitable way. But how? A problem of distributive justice has arisen.

Think of the atmosphere as a giant global sink into which we can pour our waste gases. Then once we have used up the capacity of the atmosphere to absorb our gases without harmful consequences, it becomes impossible to justify our usage of this asset by the claim that we are leaving "enough and as good" for others. The atmosphere's capacity to absorb our gases has become a finite resource on which various parties have competing claims. The problem is to allocate those claims justly.

Are there any other arguments that justify taking something that has, for all of human history, belonged to human beings in common, and turning it into private property? Locke has a further argument, arguably inconsistent with his first argument, defending the continued unequal distribution of property even when there is no longer "enough and as good" for others. Comparing the situation of American Indians, where there is no private ownership of land, and hence the land is not cultivated, with that of England, where some landowners hold vast estates and many laborers have no land at all, Locke claims that "a king of a large and fruitful territory there [i.e., in America] feeds, lodges, and is clad worse than a day labourer in England."[19] Therefore, he suggests, even the landless laborer is better off because of the private, though unequal, appropriation of the common asset, and hence should consent to it. The factual basis of Locke's comparison between English laborers and American Indians is evidently dubious, as is its failure to consider other, more equitable ways of ensuring that the land is used productively. But even if the argument worked for the landless English laborer, we cannot defend the private appropriation of the global sink in the same way. The landless laborer who no longer has the opportunity to have a share of what was formerly owned in common should not complain, Locke seems to think, because he is better off than he would have been if inegalitarian private property in land had not been recognized. The parallel argument to this in relation to the use of the global sink would be that even the world's poorest people have benefited from the increased productivity that has come from the use of the global sink by the industrialized nations. But the argument does not work, because many of the world's poorest people, whose shares of the atmosphere's capacity have been appropriated by the industrialized nations, are not able to partake in the benefits of this increased productivity in the industrialized nations—they cannot afford to buy its products—and if rising sea levels inundate their farmlands, or cyclones destroy their homes, they will be much worse off than they would otherwise have been.

Apart from John Locke, the thinker most often quoted in justifying the right of the rich to their wealth is probably Adam Smith. Smith argued that the rich did not deprive the poor of their share of the world's wealth, because:

> The rich only select from the heap what is most precious and agreeable. They consume little more than the poor, and in spite of their natural selfishness and rapacity, though they mean only their own conveniency, though the sole end which they propose from the labours of all the thousands whom they employ, be the gratification of their own vain and insatiable desires, they divide with the poor the produce of all their improvements.[20]

How can this be? Because, Smith tells us, it is as if an "invisible hand" brings about a distribution of the necessaries of life that is "nearly the same" as it would have been if the world had been divided up equally among all its inhabitants. By that Smith means that in order to obtain what they want, the rich spread their wealth throughout the economy. But while Smith knew that the rich could be selfish and rapacious, he did not imagine that the rich could, far from consuming "little more" than the poor, consume many times as much of a scarce resource as the poor do.

The average American, by driving a car, eating a diet rich in the products of industrialized farming, keeping cool in summer and warm in winter, and consuming products at a hitherto unknown rate, uses more than 15 times as much of the global atmospheric sink as the average Indian. Thus Americans, along with Australians, Canadians, and to a lesser degree Europeans, effectively deprive those living in poor countries of the opportunity to develop along the lines that the rich ones themselves have taken. If the poor were to behave as the rich now do, global warming would accelerate and almost certainly bring widespread catastrophe.

The putatively historical grounds for justifying private property put forward by its most philosophically significant defenders—writing at a time when capitalism was only beginning its rise to dominance over the world's economy—cannot apply to the current use of the atmosphere. Neither Locke nor Smith provides any justification for the rich having more than their fair share of the finite capacity of the global atmospheric sink. In fact, just the contrary is true. Their arguments imply that this appropriation of a resource once common to all humankind is not justifiable. And since the wealth of the developed nations is inextricably tied to their prodigious use of carbon fuels (a use that began more than 200 years ago and continues unchecked today), it is a small step from here to the conclusion that the present global distribution of wealth is the result of the wrongful expropriation by a small fraction of the world's population of a resource that belongs to all human beings.

For those whose principles of justice focus on historical processes, a wrongful expropriation is grounds for rectification or compensation. What sort of rectification or compensation should take place in this situation?

One advantage of being married to someone whose hair is a different color or length from your own is that when a clump of hair blacks the bath outlet, it's easy to tell whose hair it is. "Get your own hair out of the tub" is a fair and reasonable household rule. Can we, in the case of the atmosphere, trace back what share of responsibility for the blockage is due to which nations? It isn't as easy as looking at hair color, but a few years ago researchers measured world carbon emissions from 1950 to 1986 and found that the United States, with about 5 percent of the world's population at that time, was responsible for 30 percent of the cumulative emissions, whereas India, with 17 percent of the world's population, was responsible for less than 2 percent of the emissions.[21] It is as if, in a village of 20 people all using the same bathtub, one person had shed 30 percent of the hair blocking the drainhole and three people had shed virtually no hair at all. (A more accurate model would show that many more than three had shed virtually no hair at all. Indeed, many developing nations have per capita emissions even lower than India's.) In these circumstances, one way of deciding who pays the bill for the plumber to clear out the drain would be to divide it up proportionately to the amount of hair from each person that has built up over the period that people have been using the tub, and has caused the present blockage.

There is a counterargument to the claim that the United States is responsible for more of the problem, per head of population, than any other country. The argument is that because the United States has planted so many trees in recent decades, it has actually soaked up more carbon dioxide than it has emitted.[22] But there are many problems with this view. One is that the United States has been able to reforest only because it earlier cut down much of its great forests, thus releasing the carbon into the atmosphere. As this suggests, much depends on the time period over which the calculation is made. If the period includes the era of cutting down the forests, then the United States comes out much worse than if it starts from the time in which the forest had been cut but no reforestation had taken place. A second problem is that forest regrowth, while undoubtedly desirable, is not a long-term solution to the emissions problem but a temporary and one-shot expedient, locking up carbon only while the trees are growing. Once the forest is mature and an old tree dies and rots for every new tree that grows, the forest no longer soaks up significant amounts of carbon from the atmosphere.[23]

At present rates of emissions—even including emissions that come from changes in land use like clearing forests—contributions of the developing nations to the atmospheric stock of greenhouse gases will not equal the built-up contributions of the developed nations until about 2038. If we adjust this calculation for population—in other words, if we ask when the contributions of the developing nations per person will equal the per person contributions of the developed nations to the atmospheric stock of greenhouse gases—the answer is not for at least another century.[24]

If the developed nations had had, during the past century, per capita emissions at the level of the developing nations, we would not today be facing a problem of climate change caused by human activity, and we would have an ample window of opportunity to do something about emissions before they reached a level sufficient to cause a problem.

So, to put it in terms a child could understand, as far as the atmosphere is concerned, the developed nations broke it. If we believe that people should contribute to fixing something in proportion to their responsibility for breaking it, then the developed nations owe it to the rest of the world to fix the problem with the atmosphere.

Time-slice Principles

The historical view of fairness just outlined puts a heavy burden on the developed nations. In their defense, it might be argued that at the time when the developed nations put most of their cumulative contributions of greenhouse gases into the atmosphere, they could not know of the limits to the capacity of the atmosphere to absorb those gases. It would therefore be fairer, it may be claimed, to make a fresh start now and set standards that look to the future rather than to the past.

There can be circumstances in which we are right to wipe the slate clean and start again. A case can be made for doing so with respect to cumulative emissions that occurred before governments could reasonably be expected to know that these emissions might harm people in other countries. (Although, even here, one could argue that ignorance is no excuse and a stricter standard of liability should prevail, especially since the developed nations reaped the benefits of their early industrialization.) At least since 1990, however, when the Intergovernmental Panel on Climate Change published its first report, solid evidence about the hazards associated with emissions has existed.[25] To wipe the slate clean on what happened since 1990 seems unduly favorable to the industrialized nations that have, despite that evidence, continued to emit a disproportionate share of greenhouse gases. Nevertheless, in order to see whether there are widely held principles of justice that do not impose such stringent requirements on the developed nations as the "polluter pays" principle, let us assume that the poor nations generously overlook the past. We would then need to look for a time-slice principle to decide how much each nation should be allowed to emit.

An Equal Share for Everyone

If we begin by asking, "Why should anyone have a greater claim to part of the global atmospheric sink than any other?" then the first and simplest response is "No reason at all." In other words, everyone has the same claim to part of the atmospheric sink as everyone else. This kind of equality seems self-evidently fair, at least as a starting point for discussion, and perhaps, if no good reasons can be found for moving from it, as an end point as well.

If we take this view, then we need to ask how much carbon each country would be allowed to emit and compare that with what they are now emitting. The first question is what total level of carbon emissions is acceptable. The Kyoto Protocol aimed to achieve a level for developed nations that was 5 percent below 1990 levels. Suppose that we focus on emissions for the entire planet and aim just to stabilize carbon emissions at their present levels. Then the allocation per person conveniently works out at about one metric ton per year. This therefore becomes the basic equitable entitlement for every human being on this planet.

Now compare actual per capita emissions for some key nations. The United States currently produces more than five metric tons of carbon per person per year. Japan, Australia, and western European nations have per capita emissions that range from around 1.6 to 4.2 metric tons, with most below 3.0. In the developing world, emissions average 0.6 metric tons per capita, with China at 0.76 and India at 0.29.[26] This means that to reach an "even-handed" per capita annual emissions limit of one metric ton of carbon per person, India would be able to increase its carbon emissions to more than three times what they now are. China would be able to increase its emissions by a more modest 33 percent. The United States, on the other hand, would have to reduce its emissions to no more than one-fifth of present levels.

One objection to this approach is that allowing countries to have allocations based on the number of people they have gives them insufficient incentive to do anything about population growth. But if the global population increases, the per capita amount of carbon that each country is allocated will diminish, for the aim is to keep total carbon emissions below a given level. Therefore a nation that increases its population would be imposing additional burdens on other nations. Even nations with zero population growth would have to decrease their carbon outputs to meet the new, reduced per capita allocation.

By setting national allocations that are tied to a specified population, rather than allowing national allocations to rise with an increase in national population, we can meet this objection. We could fix the national allocation on the country's population in a given year, say 1990, or the year that the agreement comes into force. But since different countries have different proportions of young people about to reach reproductive age, this provision might produce greater hardship in those countries that have younger populations than in those that have older populations. To overcome this, the per capita allocation could be based on an estimate of a country's likely population at some given future date. For example, estimated population sizes for the next 50 years, which are already compiled by the United Nations, might be used.[27] Countries would then receive a reward in terms of an increased emissions quota per citizen if they achieved a lower population than had been expected, and a penalty in terms of a reduced emissions quota per citizen if they exceeded the population forecase—and there would be no impact on other countries.

Aiding the Worst-off

Giving everyone an equal share of a common resource like the capacity of the atmosphere to absorb our emissions is, I have argued, a fair starting point, a position that should prevail unless there are good reasons for moving from it. Are there such reasons? Some of the best-known accounts of fairness take the view that we should seek to improve the prospects of those who are worst off. Some hold that we should assist the worst-off only if their poverty is due to circumstances for which they are not responsible, like the family, or country, into which they were born or the abilities they have inherited. Others think we should help the worst-off irrespective of how they have come to be so badly off. Among the various accounts that pay special attention to the situation of the worst-off, by far the most widely discussed is that of John Rawls. Rawls holds that when we distribute goods, we can only justify giving more to those who are already well off if this will improve the position of those who are worst off. Otherwise, we should give only to those who are, in terms of resources, at the lowest level.[28] This approach allows us to depart from equality, but only when doing so helps the worst-off.

Whereas the strict egalitarian is vulnerable to the objection that equality can be achieved by "leveling down", that is, by bringing the rich down to the level of the poor without improving the position of the poor, Rawls's account is immune to this objection. For example, if allowing some entrepreneurs to become very rich will provide them with incentives to work hard and set up industries that provide employment for the worst-off, and there is no other way to provide that employment, then that inequality would be permissible.

That there are today very great differences in wealth and income among people living in different countries is glaringly obvious. It is equally evident that these differences depend largely on the fact that people are born into different circumstances, rather than because they have failed to take advantage of opportunities open to them. Hence if we were to follow Rawls's principle, in distributing the atmosphere's capacity to absorb our waste gases safely, we could only accept a distribution that improves the situation of those who, through no fault of their own, are at the bottom of the heap. We would have to reject any distribution that reduced the living standard in poor countries, at least as long as the rich countries are clearly better off than the poor countries.[29] To put this more concretely, if, to meet the limits set for the United States, taxes or other disincentives are used that go no further than providing incentives for Americans to drive more fuel-efficient cars, it would not be right to set limits on China that prevent the Chinese from driving cars at all.

In accordance with Rawls's principle, the only grounds on which one could argue against rich nations bearing *all* the costs of reducing emissions would be that to do so would make the poor nations even worse off than they would have been if the rich nations were not bearing all the costs. It is possible to interpret President Bush's announcement of his administration's policy on climate change as an attempt to make this case. Bush said that his administration was adopting a "greenhouse-gas intensity approach" which seeks to reduce the amount of greenhouse gases the United States emits per unit of economic activity. Although the target figure he mentioned—an 18 percent reduction over 10 years—sounds large, if the U.S. economy continues to grow as it has in the past, such a reduction in greenhouse-gas intensity will not prevent an *increase* in the total quantity of greenhouse gases that the United States emits. But Bush justified this by saying "economic growth is the solution, not the problem" and "the United States wants to foster economic growth in the developing world, including the world's poorest nations."[30]

Allowing nations to emit in proportion to their economic activity—in effect, in proportion to their gross domestic product—can be seen as encouraging efficiency, in the sense of leading to the lowest possible level of emissions for the amount produced. But it is also compatible with the United States continuing to emit more emissions, because it is producing more goods. That will mean that other nations must emit less, if catastrophic climate change is to be averted. Hence for Bush's "economic growth is the solution, not the problem" defense of a growth in U.S. emissions to succeed as a Rawlsian defense of continued inequality in per capita emissions, it would be necessary to show that United States production not only makes the world as a whole better off but also makes the poorest nations better off than they would otherwise be.

The major ethical flaw in this argument is that the primary beneficiaries of U.S. production are the residents of the United States itself. The vast majority of the goods and services that the United States produces—89 percent of them—are consumed in the United States.[31] Even if we focus on the relatively small fraction of goods produced in the United States that are sold abroad, U.S. residents benefit from the employment that is created, and, of course, U.S. producers receive payment for the goods they sell abroad. Many residents of other countries, especially the poorest countries, cannot afford to buy goods produced in the United States, and it isn't clear that they benefit from U.S. production.

The factual basis of the argument is also flawed: the United States does not produce more efficiently, in terms of greenhouse gas emissions, than other nations. Figures published by the U.S. Central Intelligence Agency show that the United States is well above average in the amount of emissions per head it produces in proportion to its per capita GDP (see figure 10.1). On this basis the United States, Australia, Canada, Saudi Arabia, and Russia are relatively inefficient producers, whereas developing countries like India and China join European nations like Spain, France, and Switzerland in producing a given value of goods per head for a lower-than-average per capita level of emissions.[32]

Because the efficiency argument fails, we must conclude that a principle that requires

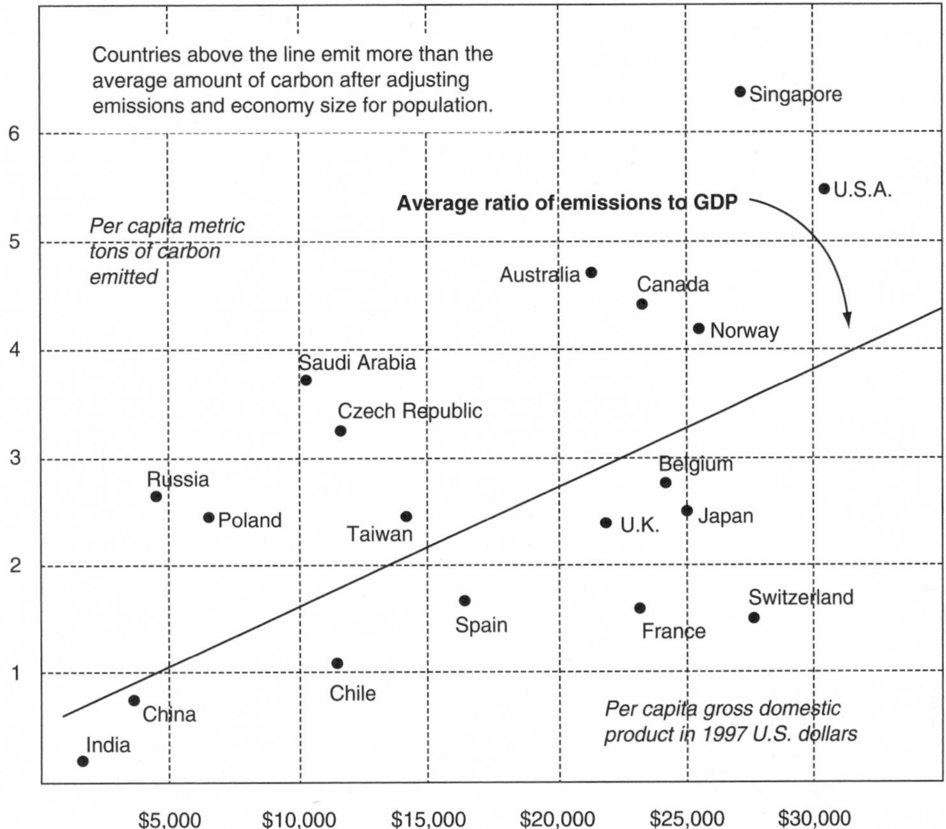

Figure 10.1. Emissions and gross domestic product.
Sources: CIA: Carbon Dioxide Information Analysis Center.

us to distribute resources so as to improve the level of the worst-off would still, given the huge resource gap between rich and poor nations, make the rich nations bear all of the costs of the required changes.

The Greatest Happiness Principle

Classical utilitarians would not support any of the principles of fairness discussed so far. They would ask what proposal would lead to the greatest net happiness for all affected—net happiness being what you have left when you deduct the suffering caused from the happiness brought about. An advocate of preference utilitarianism (a more contemporary version of utilitarianism) would instead ask what proposal would lead to the greatest net satisfaction of preferences for all concerned. But in this context, the difference between the two forms of utilitarianism is not very significant. What is much more of a problem for either of these views is to indicate how one might do such a calculation. Evidently, there are good utilitarian reasons for capping the emissions of greenhouse gases, but what way of doing it will lead to the greatest net benefits?

Perhaps it is because of the difficulty of answering such broad questions about utility that we have other principles, like the ones we have been discussing. They give you easier answers and are more likely to lead to an outcome that approximates the best consequences (or is at least as likely to do so as any calculation we could make without using those principles). The principles discussed above can be justified in utilitarian terms, although each for somewhat different reasons. To go through them in turn:

1. The principles that "the polluter pays," or more generally "you broke it, you fix it,"

provides a strong incentive to be careful about causing pollution, or breaking things. So if it is upheld as a general rule, there will be less pollution, and people will be more careful in situations where they might break something, all of which will be to the general benefit.

2. The egalitarian principle will not, in general, be what utilitarians with perfect knowledge of all the consequences of their actions would choose. Where there is no other clear criterion for allocating shares, however, it can be an ideal compromise that leads to a peaceful solution, rather than to continued fighting. Arguably, that is the best basis for defending "one person, one vote" as a rule of democracy against claims that those who have more education, or who pay more taxes, or who have served in the military, or who believe in the one true God, or who are worse off should have additional votes because of their particular attributes.[33]

3. In practice, utilitarians can often support the principle of distributing resources to those who are worst off, because when you already have a lot, giving you more does not increase your utility as much as when you have only a little. One of the 1.2 billion people in the world living on $1 per day will get much more utility out of an additional $100 than will someone living on $60,000 per year. Similarly, if we have to take $100 from someone, we will cause much less suffering if we take it from the person earning $60,000 than if we take it from the person earning $365 a year. This is known as "diminishing marginal utility." When compared with giving resources to meet someone's core needs, giving further resources "at the margin" to someone else whose core needs have already been satisfied will lead to diminished utility. Hence a utilitarian will generally favor the worst-off when it comes to distributing resources. In contrast to Rawls, however, a utilitarian does not consider this principle to be absolute. The utilitarian always seeks the greatest overall benefit, and it is only a broad rule of thumb that this will generally be obtained by adding to the stock of resources of those who have the least.

The utilitarian would also have to take into account the greater hardship that might be imposed on people living in countries that have difficulty in complying with strict emission standards because their geography or climate compels their citizens to use a greater amount of energy to achieve a given level of comfort than do people living elsewhere. Canadians, for example, could argue that it would simply not be possible to live in many parts of their country without using above average quantities of energy to keep warm. Residents of rich countries might even advance the bolder claim that, since their affluent residents have become used to traveling by car, and keeping their houses cool in warm, humid weather, they would suffer more if they have to give up their energy-intensive lifestyle than poorer people will suffer if they never get the chance to experience such comforts.

The utilitarian cannot refuse to consider such claims of hardship, even when they come from those who are already far better off than most of the world's people. As we shall see, however, these claims can be taken into account in a way that is compatible with the general conclusion to which the utilitarian view would otherwise lead: that the United States, Australia, and other rich nations should bear much more of the burden of reducing greenhouse-gas emissions than the poor nations—perhaps even the entire burden.

Fairness: A Proposal

Each of the four principles of fairness I have considered could be defended as the best one to take, or we could take some in combination. I propose—both because of its simplicity, hence its suitability as a political compromise, and because it seems likely to increase global welfare—that we support the second principle, that of equal per capita future entitlements to a share of the capacity of the atmospheric sink, tied to the current United Nations projection of population growth per country in 2050.

Some will say that this is excessively harsh on industrialized nations like the United States, which will have to cut back the most on their output of greenhouse gases. But we have now seen that the equal per capita shares principle is much more indulgent to the United States, Australia, and other developed nations than other principles for which there are strong arguments. If, for example, we combined the "polluter pays" principle with the equal-share principle, we would hold that until the excessive amounts of greenhouse gases in the atmosphere that the industrialized nations have put there have been soaked up, the emissions of industrialized nations ought to be held down to much *less* than a per capita equal share. As things stand now, even on an equal per capita share basis, for at least a century the developing nations are going to have to accept lower outputs of greenhouse gases than they would have had to if the industrialized nations had kept to an equal per capita share in the past. So by saying, "Forget about the past, let's start anew", the pure equal per capita share principle is a lot more favorable to the developed countries than a historically based principle would be.

The fact that 178 nations, including every major industrial nation in the world except the United States, have now indicated their intention to ratify the Kyoto Protocol makes the position of the United States particularly odious from an ethical perspective. Australia's position is certainly no better, for even though its total greenhouse-gas output is relatively minor, it is very high when calculated on a per capita basis—according to an Australian government report, among the highest in the world.[34] Thus Australia produces roughly the same quantity of greenhouse gases as Italy, although Italy's population is three times as large as Australia's. Moreover, Australia was offered a particularly generous deal, allowing it to increase its greenhouse-gas emissions by 8 percent over 1990 levels when other nations, on average, had to make a 5 percent cut. On top of that, further concessions granted at the Bonn meeting made it easier for Australia to meet its targets, by allowing countries to take into account carbon absorbed by increased forest plantations.

The claim that the Protocol does not require the developing nations to do their share does not stand up to scrutiny. Americans and Australians who think that even the Kyoto Protocol requires their nation to sacrifice more than it should are really demanding that the poor nations of the world commit themselves to a level that gives them, in perpetuity, lower levels of greenhouse-gas production per head of population than the rich nations have. How could that principle be justified? Alternatively, if that is not what the U.S. and Australian governments are proposing, what exactly are they proposing?

It is true that there are some circumstances in which we are justified in refusing to contribute if others are not doing their share. If we eat communally and take turns cooking, then I can justifiably feel resentment if there are some who eat but never cook or carry out equivalent tasks for the good of the entire group. But that is not the situation with climate change, in which the behavior of the industrialized nations has been more like that of a person who has left the kitchen tap running but refuses either to turn it off or to mop up the resulting flood, until you—who spilt an insignificant half-glass of water onto the floor—promise not to spill any more. Now the other industrialized nations have agreed to turn off the tap (to be strictly accurate, to restrict the flow), leaving the United States (the biggest culprit) and Australia together in their refusal to commit to reducing emissions.

Although it is true that the Kyoto Protocol does not initially bind the developing nations, it is generally understood that the developing countries will be brought into the binding section of the agreement after the industrialized nations have begun to move toward their targets. That was the procedure with the successful Montreal Protocol concerning gases that damage the ozone layer, and there is no reason to believe that it will not also happen with the Kyoto Protocol. China, by far the largest greenhouse-gas emitter of the developing nations and the only one with the potential to rival the total—not, of course, per capita—emissions of the United States in the foreseeable future, has already, even in the absence of any binding targets, achieved a substantial decline in fossil-fuel carbon dioxide emissions, thanks to improved efficiency in coal use. Emissions fell from a high of 909 million metric tons of carbon in 1996 to 848 million metric tons in 1998. Meanwhile, U.S. emissions reached an all-time high of 1,906 million metric tons of carbon in 2000, an increase of 2.5 percent over the previous year.[35]

The real objection to allocating the atmosphere's capacity to absorb greenhouse gases to nations on the basis of equal per capita shares is that it would be tremendously dislocating for the industrialized nations to reduce their emissions so much that within five, ten or fifteen years they were not producing more than their share, on a per capita basis, of some acceptable level of greenhouse gases. But fortunately there is a mechanism that, while fully compatible with the equal per capita share principle, can make this transition much easier for the industrialized nations, while at the same time producing great benefits for the developing nations. That mechanism is emissions trading.

Emissions trading works on the same simple economic principle of trade in general: if you can buy something from someone else more cheaply than you can produce it yourself, you are better off buying it than making it. In this case, what you can buy will be a transferable quota to produce greenhouse gases, allocated on the basis of an equal per capita share. A country like the United States that is already producing more gases than its share will need its full quota and then some, but a country like Russia that is below its share will have excess quota that it can sell. If the quota were not transferable, the United States would immediately have to reduce its output to about 20 percent of what it now produces, a political impossibility. In contrast, Russia would have no incentive to maintain its levels of greenhouse-gas emissions well below its allowable share. With emissions trading, Russia has an incentive to maximize the amount of quota it can sell, and the United States has, at some cost, an opportunity to acquire the quotas it needs to avoid total disruption of the economy.[36]

Although some may think that emissions trading allows the United States to avoid its burdens too easily, the point is not to punish nations with high emissions but to produce the best outcome for the atmosphere. Permitting emissions trading gives us a better hope of doing this than prohibiting emissions trading does. The Kyoto Protocol as agreed to in Bonn and Marrakesh allows emissions trading between states that have binding quotas. Thus Russia will have quota to sell, but countries such as India, Bangladesh, Mozambique, Ethiopia, and many others will not. Emissions trading would be much more effective, and have far better consequences, if all nations were given binding quotas based on their per capita share of the designated total emissions. As we saw earlier, even the environmental skeptic Bjørn Lomborg accepts that with global emissions trading, the Kyoto Protocol produces a net economic benefit. Moreover, global emissions trading would give the world's poorest nations something that the rich nations very much want. They would have, at last, something that they can trade in exchange for the resources that will help them to meet their needs. This would be, on most principles of justice or utility, a very good thing indeed. It could also end the argument about making the developing nations part of a binding agreement on emissions, because the developing nations would see that they have a great deal to gain from binding quotas.

Since global emissions trading is both possible and desirable, it also answers two objections to allocating greenhouse-gas emissions quotas on the basis of equal per capita shares. First, it answers the objection raised when discussing a utilitarian approach to these problems—that countries like Canada might suffer undue hardship if forced to limit emissions to the same per capita amount as, say, Mexico, because Canadians need to use more energy to survive their winters. But global emissions trading means that Canada would be able to buy the quota it requires from other countries that do not need their full quota. Thus the market would provide a measure of the additional burden put on the world's atmosphere by keeping one's house at a pleasant temperature when it is too cold, or too hot, outside. Citizens of rich countries could choose to pay that price and keep themselves warm, or cool, as the case may be. They would not, however, be claiming a benefit for themselves that they were not prepared to allow poor countries to have, because the poor countries would benefit by having emissions quotas to sell. The claim of undue hardship therefore does not justify allowing rich countries to have a higher per capita emissions quota than poor countries.

Second, global emissions trading answers the objection that equal per capita shares would lead to inefficient production because countries with little industrialization would be able to continue to manufacture goods even though they emit more greenhouse gases per unit of economic activity than highly industrialized nations, while the highly industrialized nations would have to cut back on their manufacturing capacity, even though they produce less emissions per unit of economic activity. But as we have seen, the present laissez-faire system allows emitters to reap economic benefits for themselves, while imposing costs on third parties that may or may not share in the benefits of the polluters' high productivity. That is neither a fair nor an efficient outcome. A well-regulated system of per capita entitlements combined with global emissions trading would, by internalizing the true costs of production, lead to a solution that is both fair and efficient.

There are two serious objections, one scientific and one ethical, to global emissions trading. The scientific objection is that we do not have the means to measure emissions accurately for all countries. Hence it would not be possible to know how much quota these countries have to sell or need to buy. This is something that needs more research, but it should not prove an insuperable obstacle in the long run. As long as estimates are fair, they do not need to be accurate to the last metric ton of carbon. The ethical objection is that while emissions trading would benefit poor countries if the governments of those countries used it for the benefit of their people, some countries are run by corrupt dictators more interested in increasing their military spending or adding to their Swiss bank accounts. Emissions trading would simply give them a new way of raising money for these purposes.

My proposed solution to the ethical objection is to refuse to recognize a corrupt dictatorial regime, interested only in self-preservation and self-enrichment, as the legitimate government of the country that has excess quota to sell. In the absence of any legitimate government that can receive payments for quota, the sale of quota could be managed by an international authority answerable to the United Nations. That authority could hold the money it receives in trust until the country has a government able to make a credible claim that the money will be used to benefit the people as a whole.

Down from the Clouds?

To cynical observers of the Washington scene, all this must seem absurdly lacking in political realism. George W. Bush's administration spurned the Kyoto Protocol, which allows the United States to continue to produce at least four times its per capita share of carbon dioxide. Since 1990, U.S. emission levels have already risen by 14 percent. The halfhearted measures for energy conservation proposed by the Bush administration will, at best, slow that trend. They will not reverse it. So what is the point of discussing proposals that are far *less* likely to be accepted by the U.S. government than the Kyoto Protocol?

The aim of this chapter is to help us to see that there is no *ethical* basis for the present distribution of the atmosphere's capacity to absorb greenhouse gases without drastic climate change. If the industrialized countries choose to retain this distribution (as the United States does), or to use it as the starting point for a new allocation of the capacity of the global sink (as the countries that accept the Kyoto Protocol do), they are standing simply on their presumed rights as sovereign nations. That claim, and the raw military power these nations wield, makes it impossible for anyone else to impose on them a more ethically defensible solution. If we, as citizens of the industrialized nations, do not understand what would be a fair solution to global warming, then we cannot understand how flagrantly self-serving the position of those opposed to signing even the Kyoto Protocol is. If, on the other hand, we can convey to our fellow citizens a sense of what would be a fair solution to the problem, then it may be possible to change the policies that are now leading the United States to block international cooperation on something that will have an impact on every being on this planet.

Let us consider the implications of this situation a little further. Today the overwhelming majority of nations in the world are united in the view that greenhouse-gas emissions should be significantly reduced, and only the United States and Australia, of all the industrialized nations, have said that they are not prepared to commit themselves to a binding treaty that will achieve this goal.

Such a situation gives impetus to the need to think about developing institutions or principles of international law that limit national sovereignty. It should be possible for people whose lands are flooded by sea-level rises due to global warming to win damages from nations that emit more than their fair share of greenhouse gases. Another possibility worth considering is sanctions. There have been several occasions on which the United Nations has used sanctions against countries that have been seen as doing something gravely wrong. Arguably, the case for sanctions against a nation that is causing harm, often fatal, to the citizens of other countries is even stronger than the case for sanctions against a country like South Africa under apartheid, since that government, iniquitous as its policies were, was not a threat to other countries. Is it inconceivable that one day a reformed and strengthened United Nations will invoke sanctions against countries that do not play their part in global measures for the protection of the environment?

Notes

1. "This Year Was the 2nd Hottest, Confirming a Trend, UN Says," *New York Times*, December 19, 2001, p. A5.

2. J. T. Houghton et al., eds., *Climate Change 2001: The Scientific Basis: Contribution of Working Group I to the Third Assessment Report of the Intergovernmental Panel on Climate*, United Nations Environment Program and Intergovernmental Panel on Climate Change, Cambridge University Press, Cambridge, U.K., 2001, Summary for Policymakers. Available at www.ipcc.ch/pub/tar/wgi/index. htm. See also *Reconciling Observations of Global Temperature Change*, Panel on Reconciling Temperature Observations, National Research Council, National Academy of Sciences, Washington, D.C., 2000. Available at www.nap.edu/books/0309068916/html. For another example of recent research indicating that anthropogenic climate change is real, see Thomas J. Crowley, "Causes of Climate Change over the Past 1000 Years," *Science*, July 14, 2000, pp. 270–277.

3. Houghton et al., *Climate Change 2001.*

4. Munich Reinsurance, one of the world's largest insurance companies, has estimated that the number of major natural disasters has risen from 16 in the 1960s to 70 in the 1990s. Cited by Christian Aid, Global Advocacy Team Policy Position Paper, *Global Warming, Unnatural Disasters and the World's Poor*, November 2000. Available at www.christianaid.org.uk/indepth/ooiiglob/globwarm.htm.

5. James McCarthy et al., eds., *Climate Change 2001: Impacts, Adaptation, and Vulnerability, Contribution of Working Group II to the Third Assessment Report of the Intergovernmental Panel on Climate Change*, United Nations Environment Program and Intergovernmental Panel on Climate Change, Cambridge University Press, Cambridge, U.K., 2001, chap. 9.7. Available at www.ipcc.ch/pub/tar/wg2/index.htm.

6. Houghton et al., *Climate Change 2001.*

7. See Dale Jamieson, "Ethics, Public Policy, and Global Warming," *Science, Technology, and Human Values* 17.2 (Spring 1992): 139–153, and "Global Responsibilities: Ethics, Public Health, and Global Environmental Change," *Indiana Journal of Global Legal Studies* 5.1 (Fall 1997): 99–119.

8. "Norway Wants Sanctions for Cross Border Polluters," Reuters News Service, February 1, 2002. Available at www.planetark.org/dailynewsstory.cfm/newsid/14316/story.htm.

9. *United Nations Framework Convention on Climate Change*, Article 4, section 2, subsections (a) and (b). Available at www.unfccc.int/resource/conv/conv.html. *Guide to the Climate Change Negotiation Process.* Available at www.unfccc.int/resource/process/components/response/respconv.html.

10. "U.S. Carbon Emissions Jump in 2000," *Los Angeles Times*, November 11, 2001, p. A36, citing figures released by the U.S. Department of Energy's Energy Information Administration on November 9, 2001.

11. Eileen Claussen and Lisa McNeilly, *The Complex Elements of Global Fairness*, Pew Center on Global Climate Change, Washington, D.C., October 29, 1998. Available at www.pewclimate.org/projects/pol_equity.cfm.

12. Bjørn Lomborg, "The Truth about the Environment," *Economist*, August 2, 2001. Available at www.economist.com/science/displayStory.cfm?Story_ID=718860&CFID=3046335&CFTOKEN=88404876.

13. For details of the very modest amounts of aid given by the rich nations to the world's poorest people, see Peter Singer, *The Life You Can Save* (Random House, New York, 2009), chaps. 3 and 7.

14. Bjorn Lomborg, *The Skeptical Environmentalist*, (Cambridge, U.K.: Cambridge University Press, 2001), p. 323.

15. See Richard Newell and William Pizer, *Discounting the Benefits of Future Climate Change Mitigation: How Much Do Uncertain Rates Increase Valuations?* Pew Center on Global Climate Change, Washington, D.C., December 2001. Available at www.pewclimate.org/projects/econ_discounting.cfm.

16. Robert Nozick, *Anarchy, State and Utopia* (New York: Basic Books, 1974), p. 153.

17. John Locke, *Second Treatise on Civil Government*, ed. C. B. Macpherson, (Indianapotis: Hacket, 1980), sec. 27, p. 19.

18. See Garrett Hardin, "The Tragedy of the Commons," *Science* 162 (1968): 1243–1248.

19. Locke, *Second Treatise*, sec. 41.

20. Adam Smith, *A Theory of the Moral Sentiments* (Amherst, N.Y.: Prometheus, 2000), IV, i. p. 10.

21. Peter Hayes and Kirk Smith, eds., *The Global Greenhouse Regime: Who Pays?* (London: Earthscan, 1993), chap. 2, table 2.4. Available at www.unu.edu/unupress/unupbooks/80836e/80836E08.htm.

22. See S. Fan, M. Gloor, J. Mahlman, S. Pacala, J. Sarmiento, T. Takahashi, and P. Tans, "A Large Terrestrial Carbon Sink in North America Implied by Atmospheric and Oceanic Carbon Dioxide Data and Models," *Science* 282 (October 16, 1998): 442–446.

23. William Schlesinger and John Lichter, "Limited Carbon Storage in Soil and Litter of Experimental Forest Plots under Increased Atmospheric CO_2," *Nature* 411 (May 24, 2001): pp. 466–469.

24. Duncan Austin, José Goldemberg, and Gwen Parker, "Contributions to Climate Change: Are Conventional Metrics Misleading the Debate?" World Resource Institute Climate Protection Initiative, Climate Notes. Available at www.igc.org/wri/cpi/notes/metrics.html.

25. The Intergovernmental Panel on Climate Change *First Assessment Report* was published in three volumes. See especially J. T. Houghton, G. J. Jenkins, and J. J. Ephraums, eds., *Scientific Assessment of Climate Change—Report of Working Group I* (Cambridge, U.K.: Cambridge University Press, 1990). For details of the other volumes, see www.ipcc.ch/pub/reports.htm.

26. See G. Marland, T. A. Boden, and R. J. Andres, *Global, Regional, and National Fossil Fuel CO_2 Emissions*, Carbon Dioxide Information Analysis Center, Oak Ridge, Tenne. Available at cdiac.esd.ornl.gov/trends/emis/top96.cap. These are 1996 figures.

27. Paul Baer et al., "Equity and Greenhouse Gas Responsibility," *Science* 289 (September 29, 2000): p. 2287; Dale Jamieson, "Climate Change and Global Environmental Justice," in P. Edwards and C. Miller, eds., *Changing the Atmosphere: Expert Knowledge and Global Environmental Governance*, (Cambridge, Mass.: MIT Press, 2001), pp. 287–307.

28 See John Rawls, *A Theory of Justice*, especially pp. 65–83. For a different way of giving priority to the worst-off, see Derek Parfit, "Equality or Priority?" Lindley Lecture, University of Kansas, November 21, 1991, reprinted in Matthew Clayton and Andrew Williams, eds., *The Ideal of Equality*, (London: Macmillan, 2000).

29. This is Rawls's "difference principle," applied without the restriction to national boundaries that are difficult to defend in terms of his own argument.

30. "President Announces Clear Skies and Global Climate Change Initiative," Office of the Press Secretary, White House, February 14, 2002. Available at www.whitehouse.gov/news/releases/2002/02/20020214-5.html. For amplification of the basis of the administration's policy, see Executive Office of the President, Council of Economic Advisers, *2002 Economic Report of the President*, (Washington, D.C.: U.S. Government Printing Office, 2002), chap. 6, pp. 244–249, available at http://w3.access.gpo.gov/eop/.

31. National Council on Economic Education, "A Case Study: United States International Trade in Goods and Services—May 2001." Available at www.econedlink.org/lessons/index.cfm?lesson=EM196.

32. Andrew Revkin, "Sliced Another Way: Per Capita Emissions," *New York Times*, June 17, 2001, section 4, p. 5.

33. For discussion of equal votes as a compromise, see my *Democracy and Disobedience*, (Oxford, Clarendon Press, 1973), pp. 30–41.

34. Ross Garnaut, *The Garnaut Climate Change Review*, Commonwealth of Australia, Canberra, 2008, chap. 7, sec. 1. Available at http://www.garnautreview.org.au/chp7.htm.

35. Energy Information Administration, *Emissions of Greenhouse Gases in the United States 2000*, DOE/EIA-0573 (2000), U.S. Department of Energy, Washington, D.C., November 2001, page vii. Available at www.eia.doe.gov/pub/oiaf/1605/cdrom/pdf/ggrpt/057300.pdf.

36. See Jae Edmonds et al., *International Emissions Trading and Global Climate Change: Impacts on the Cost of Greenhouse Gas Mitigation.* Report prepared for the Pew Center on Global Climate Change, December 1999. Available at www.pewclimate.org/projects/econ_emissions.cfm.

Subsistence Emissions and Luxury Emissions

Henry Shue

11

1. Introduction

The United Nations Framework Convention on Climate Change adopted in Rio de Janeiro at the United Nations Conference on Environment and Development (UNCED) in June 1992 establishes no dates and no dollars. No dates are specified by which emissions are to be reduced by the wealthy states, and no dollars are specified with which the wealthy states will assist the poor states to avoid an environmentally dirty development like our own. The convention is toothless because throughout the negotiations in the Intergovernmental Negotiating Committee during 1991 to 1992, the United States played the role of dentist: whenever virtually all the other states in the world (with the notable exceptions of Saudi Arabia and Kuwait) agreed to convention language with teeth, the United States insisted that the teeth be pulled out.

The Clinton administration now faces a strategic question: should the next step aim at a comprehensive treaty covering all greenhouse gases (GHGs) or at a narrower protocol covering only one, or a few, gases, for example, only fossil-fuel carbon dioxide (CO_2)? Richard Stewart and Jonathan Wiener (1992) have argued for moving directly to a comprehensive treaty, while Thomas Drennen (1993) has argued for a more focused beginning. I will suggest that Drennen is essentially correct that we should not try to go straight to a comprehensive treaty, at least not of the kind advocated by Stewart and Wiener. First I would like to develop a framework into which to set issues of equity or justice of the kind introduced by Drennen.

II. A Framework for International Justice

A. Four Kinds of Questions

It would be easier if we faced only one question about justice, but several questions are not only unavoidable individually but are entangled with one another. In addition, each question can be given not simply alternative answers but answers of different kinds. In spite of this multiplicity of possible answers to the multiplicity of inevitable and interconnected questions, I think we can lay out the issues fairly clearly and establish that commonsense principles converge to a remarkable extent upon what ought to be done, at least for the next decade or so.

Leaving aside the many important questions about justice that do not have to be raised in order to decide how to tackle threats to the global environment, we will find four questions that are deeply involved in every choice of a plan for action: (1) What is a fair allocation of the costs of preventing the global warming that is still avoidable? (2) What is a fair allocation of the costs of coping with the social consequences of the global warming that will not in fact be avoided? (3) What background allocation of wealth would allow international bargaining, about issues like (1) and (2), to be a fair process? (4) What is a fair allocation of emissions of greenhouse gases (over the long-term and during the transition to the long-term allocation)? Our leaders can confront these four questions explicitly and thoughtfully, and thereby hope to deal with them more wisely, or they can leave them implicit and unexamined and simply blunder into positions on them while thinking only about the other economic and political considerations that always come up. What leaders cannot do is evade taking actions that will in fact be just or unjust. The subject of justice will not go away. Issues of justice are inherent in the kinds of choices that must immediately be made. Fortunately, these four issues that are intertwined in practice can be separated for analysis.

1. Allocating the Costs of Prevention

Whatever sums are spent in the attempt to prevent additional warming of the climate must somehow be divided up among those who are trying to deal with the problem. The one question of justice that most people readily see is this one: Who should pay for whatever is done to keep global warming from becoming any worse than necessary?

One is tempted to say, "To keep it from becoming any worse than it is already going to be as a result of gases that are already in the air." Tragically, we will in fact continue to make it worse for some time, no matter how urgently we act. Because of the industrial revolution, the earth's atmosphere now contains far more accumulated CO_2 than it was normal for it to contain during previous centuries of human history. This is not speculation: bubbles of air from earlier centuries have been extracted from deep in the polar ice, and the CO_2 in these bubbles has been directly measured. Every day we continue to make large net additions to the total concentration of CO_2.

Several industrial nations have unilaterally committed themselves to reducing their emissions of CO_2 by the year 2000 to the level of their emissions in 1990. This may sound good, and it is obviously better than allowing emissions levels to grow in a totally uncontrolled manner, as the United States and many other industrial nations are doing. The 1990 level of emissions, however, was making a net addition to the total every day, because it was far in excess of the capacity of the planet to recycle CO_2 without raising the surface temperature of the planet. A reduction to the 1990 level of emissions means *reducing the rate* at which we are adding to the atmospheric total to a rate below the current rate of addition, but it also means *continuing to add to the total.*

Stabilizing emissions at a level as high as the 1990 level will not stabilize temperature—it will continue the pressure to drive it up. In order to stabilize temperature, emissions must be reduced to a level at which the accumulated *concentration* of CO_2 in the atmosphere is stabilized. CO_2 must not be added by human processes faster than natural processes can handle it by means that do not raise the surface temperature. Natural processes will, of course, have to "handle" whatever concentration of CO_2 we choose to produce, one way or another; some of those ways involve adjustments in parameters like surface temperature that *we* will have a hard time handling. There is, therefore, nothing magic about the 1990 level of emissions. On the contrary, at that historically unprecedented level of emissions, the atmospheric concentration would continue to expand rapidly—it merely would not expand as quickly as it will at present levels or at the higher business-as-usual future levels now to be expected.

Emissions must be stabilized at a much lower level than the 1990 level, which means that emissions must be sharply reduced. The most authoritative scientific consensus said that in order to stabilize the atmospheric concentration of CO_2, emissions would have to be reduced below 1990 levels by more than 60 percent! (Houghton, Jenkins, and Ephraums 1990, xviii, table 2). Even if this international scientific consensus somehow were a wild exaggeration and the reduction needed to be, say, a reduction of only 20 percent from 1990 levels, we would still face a major challenge. Every day that we continue to add to the growing concentration, we increase the size of the reduction from current emissions necessary to stabilize the concentration at an acceptable total.

The need to reduce emissions, not merely to stabilize them at an already historically high level, is only part of the bad news for the industrial countries. The other part is that the CO_2 emissions of most countries that contain large percentages of the human population will be rising for some time. I believe that the emissions from these poor, economically less developed countries also ought to rise insofar as this rise is necessary to provide a minimally decent standard of living for their now impoverished people. This is, of course, already a (very weak) judgment about what is fair, namely, that those living in desperate poverty ought not to be required to restrain their emissions, thereby remaining in poverty, in order that those living in luxury should not have to restrain their emissions. Anyone who cannot see that it would be unfair to require sacrifices by the desperately poor in order to help the affluent avoid sacrifices will not find anything else said in this article convincing, because I rely throughout on a common sense of elementary fairness. Any strategy of maintaining affluence for some people by keeping other people at or below subsistence is, I take it, patently unfair because so extraordinarily unequal—intolerably unequal.

Be the fairness as it may, the poor countries of the globe are in fact not voluntarily going to refrain from taking the measures necessary to create a decent standard of living for themselves in order that the affluent can avoid discomfort. For instance, the Chinese government, presiding over more than 22 percent of humanity, is not about to adopt an economic policy of no growth for the convenience of Europeans and North Americans already living much better than the vast majority of Chinese, whatever others think about fairness. Economic growth means growth in energy consumption, because economic activity uses energy. And growth in energy consumption, in the foreseeable future, means growth in CO_2 emissions.

In theory, economic growth could be fueled entirely by forms of energy that produce no greenhouse gases (solar, wind, geothermal, nuclear [fission or fusion], and hydroelectric). In practice, these forms of energy are not now economically viable (which is not to say that none of them would be if public subsidies, including government-funded research and development, were restructured). China specifically has vast domestic coal reserves, *the* dirtiest fuel of all in CO_2 emissions, and no economically viable way in the short run of switching to completely clean technologies or importing the cleaner fossil fuels, like natural gas, or even the cleaner technologies for burning its own coal, which do exist in wealthier countries. In May 1992, Chen Wang-xiang, general secretary of China's Electricity Council, said that coal-fired plants would account for 71 to 74.5 percent of the 240,000 megawatts of generating capacity planned for China by the year 2000 (Bureau of National Affairs 1992). So, until other arrangements are made and financed, China will most likely be burning vast and rapidly increasing quantities of coal with, for the most part, neither the best available coal-burning technology nor the best energy technology overall. The only alternative China actually has with its current resources is to choose to restrain its economic growth, which it will surely not do, rightly or wrongly. (I think rightly.)

Fundamentally, then, the challenge of preventing additional avoidable global warming takes this shape: How does one reduce emissions for the world as a whole while accommodating increased emissions by some parts of the world? The only possible answer is: By *reducing* the emissions by one part of the world by an amount *greater than* the increase by the other parts that are increasing their emissions.

The battle to reduce total emissions should be fought on two fronts. First, the increase in emissions by the poor nations should be held to the minimum necessary for the economic development that they are entitled to. From the point of view of the rich nations, this would serve to minimize the increase that their own reductions must exceed. Nevertheless, the rich nations must, second, also reduce their own emissions somewhat, however small the increase in emissions by the poor, if the global total of emissions is to come down while the contribution of the poor nations to that total is rising. The smaller the increase in emissions necessary for the poor nations to rise out of poverty, the smaller the reduction in emissions necessary for the rich nations—environmentally sound development by the poor is in the interest of all.

Consequently, two complementary challenges must be met—and paid for—which is where the less obvious issues of justice come in.[1] First, the economic development of the poor nations must be as "clean" as possible—maximally efficient in the specific sense of creating no unnecessary CO_2 emissions. Second, the CO_2 emissions of the wealthy nations must be reduced by more than the amount by which the emissions of the poor nations increase. The bills for both must be paid: someone must pay to make the economic development of the poor as clean as possible, and someone must pay to reduce the emissions of the wealthy. These are the two components of the first issue of justice: allocating the costs of prevention.

2. Allocating the Costs of Coping

No matter what we do for the sake of prevention from this moment forward, it is highly unlikely that all global warming can be prevented, for two reasons. First, what the atmospheric scientists call a "commitment to warming" is already in place simply because of all the additional greenhouse gases that have been thrust into the atmosphere by human activities since around 1860. Today is already the morning after. We have done whatever we have done, and now its consequences, both those we understand and those we do not understand, will play themselves out, if not this month, some later month. Temperature at the surface level—at our level—may or may not already have begun to rise. But the best theoretical understanding of what would make it rise tells us that it will sooner or later rise because of what we have already done—and are unavoidably going to continue doing in the short and medium term. Unless the theory is terribly wrong, the rise will begin sooner rather than later. In the century and a quarter between the beginnings of the industrial revolution and 1993, and especially in the half-century since World War II, the industrializing nations have pumped CO_2 into the atmosphere with galloping vigor. As of today the concentration has already ballooned.

Second, even if starting tomorrow morning everyone in the world made every exertion she could possibly be expected to make to avoid as much addition as possible to today's concentration, we would continue to add CO_2 much faster than it can be recycled without a rise in temperature for an indeterminate number of years to come. A sudden huge decline in the rate of addition to the total is not physically, not to mention economically, feasible. Further, needless to say, not everyone in the world is prepared to make every reasonable exertion. The "leadership" of the United States, the world's largest injector of CO_2 into the planet's atmosphere, will not even commit itself to cap CO_2 emissions by 2000 at 1990 levels, as easy as that would be (and as little good as it by itself would do). Consequently, even a good-faith transition to sustainable levels of CO_2 emissions would make the problem of warming worse for quite a few years before it could begin to allow it to become better. Years of fiddling while our commitment to the warming of future generations expands will make their problem considerably worse still than it already has to be.

The second issue of justice, then, is: How should the costs of coping with the unprevented human consequences of global warming be allocated? The two thoughts that immediately spring to mind are, I believe, profoundly misguided; they are, crudely put, to-each-his-own and wait-and-see. The first thought is: Let each nation that suffers negative consequences deal with "its own" problems, since this is how the world generally operates. The second is: Since we cannot be sure what negative consequences will occur, it is only reasonable to wait and see what they are before becoming embroiled in arguments about who should pay for dealing with which of them. However sensible these two strategic suggestions may seem, I believe that they are quite wrong and that this issue of paying for coping is both far more immediate and much more complex than it seems. This brief overview is not the place to pursue the arguments in any depth, but I would like to telegraph why I think these two obvious-seeming solutions need at the very least to be argued for.

To-each-his-own. Instantly adopting this solution depends upon assuming without question a highly debatable description of the nature of the problem, namely, as it was put just above, "let each nation that suffers negative consequences deal with 'its own' problems." The fateful and contentious assumption here is that whatever problems arise within one's nation's territory are *its own*, in some sense that entails that it can and ought to deal with them on its own, with (only) its own resources. This assumption depends in turn upon both of two implicit and dubious premises.

First, it is taken for granted that every nation now has all its own resources under its control. Stating the same point negatively, one can say that it is assumed that no significant proportion of any nation's own resources are physically, legally, or in any other way outside its own control. This assumes, in effect, that the international distribution of wealth is perfectly just, requiring no adjustments whatsoever across national boundaries! To put it mildly, that the world is perfectly just as it is, is not entirely clear without further discussion. Major portions of the natural resources of many of the poorer nations are under the control of multinational firms operated from elsewhere. Many Third World states are crippled by burdens of international debt contracted for them, and then wasted, by illegitimate authoritarian governments. Thus, the assumption that the international distribution of wealth is entirely as it should be is hard to swallow.

Second is an entirely independent question that is also too quickly assumed to be closed. It is taken for granted that no responsibility for problems resulting within one nation's territory could fall upon another nation or upon other actors or institutions outside the territory. Tackling this question seriously means attempting to wrestle with slippery issues about the causation of global warming and about the connection, if any, between causal responsibility and moral responsibility, issues to be discussed more fully later. Once the issues are raised, however, it is certainly not a foregone conclusion, for instance, that coastal flooding in Bangladesh (or the total submersion of, for example, the Maldives and Vanuatu) would be entirely the responsibility of, in effect, its victims and not at least partly the responsibility of those who produced, or profited from, the greenhouse gases that led to the warming that made the ocean water expand and advance inland. On quite a few readings of the widely accepted principle of "the polluter pays" those who caused the change in natural processes that resulted in the human harm would be expected to bear the costs of making the victims whole. Once again, I am not trying to settle the question here, but merely to establish that it is indeed open until the various arguments are heard and considered.

Wait-and-see. The other tactic that is supposed to be readily apparent and eminently sensible is: Stay out of messy arguments about the allocation of responsibility for potential problems until we see which problems actually arise—we can then restrict our arguments to real problems and avoid imagined ones. Unfortunately, this, too, is less commonsensical than it may sound. To see why, one must step back and look at the whole picture.

The potential costs of any initiative to deal comprehensively with global warming can be divided into two separate accounts, corresponding to two possible components of the initiative. The first component, introduced in the previous section of this chapter, is the attempted prevention of as much warming as possible, the costs of which can be thought of as falling into the prevention account. The second component, briefly sketched in this section, is the attempted correction of, or adjustment to—what I have generally called "coping with"—the damage done by the warming that for whatever reasons goes unprevented.

It may seem that if costs can be separated into prevention costs and coping costs, the two kinds of costs could then be allocated separately, and perhaps even according to unrelated principles. Indeed, the advice to wait and see about any coping problems assumes that precisely such independent handling is acceptable. It assumes in effect that prevention costs can be allocated—or that the principles according to which they will be allocated, once they are known, can be agreed upon—and prevention efforts put in motion, before the possibly unrelated principles for allocating coping costs need to be agreed upon. What is wrong with this picture of two basically independent operations is that what is either a reasonable or a fair allocation of the one set of costs may—I will argue, does—depend upon how the other set of costs is allocated. The respective principles for the two allocations must not merely not be unrelated but be complementary.

In particular, the allocation of the costs of prevention will directly affect the ability to cope later of those who abide by their agreed-upon allocation. To take an extreme case, suppose that what a nation was being asked to do for the sake of prevention could be expected to leave it much less able to cope with "its own" unprevented problems, on its own, than it would be if it refused to contribute to the prevention efforts—or refused to contribute on the specific terms proposed—and instead invested all or some of whatever it might have contributed to prevention in its own preparations for its own coping. For example, suppose that in the end more of Shanghai could be saved from the actual eventual rise in sea level due to global warming if China simply began work immediately on an elaborate and massive, Dutch-style system of sea walls, dikes, canals, and sophisticated floodgates—a kind of Great Sea Wall of China—rather than spending its severely constrained resources on, say, purification technologies for its new coal-fueled electricity-generating plants and other prevention measures. From a strictly Chinese point of view, the Great Sea Wall might be preferable even if China's refusal to contribute to the prevention efforts resulted in a higher sea level at Shanghai than would result if the Chinese did cooperate with prevention (but then did not have time or resources to build the Sea Wall fast enough or high enough).

This fact that the same resources that might be contributed to a multilateral effort at prevention might alternatively be invested in a unilateral effort at coping raises two different questions, one primarily ethical and one primarily nonethical (although these two questions are not unrelated, either). First, would it be fair to expect cooperation with a multilateral initiative on prevention, given one particular allocation of those costs, if the costs of coping are to be allocated in a specific other way (which may or may not be cooperative)? Second, would it be reasonable for a nation to agree to the one set of terms, given the other set of terms—or, most relevantly, given that the other set of terms remained unspecified? Doing your part under one set now while the other set is up for grabs later leaves you vulnerable to the possibility of the second set's being stacked against you in spite of, or because of, your cooperation with the first set. It is because the fairness and the reasonableness of any way of allocating the costs of prevention depends partly upon the way of allocating the costs of coping that it is both unfair and unreasonable to propose that binding agreement should be reached now concerning prevention, while regarding coping we should wait and see.

3. The Background Allocation of Resources and Fair Bargaining

This last point about potential vulnerability in bargaining about the coping terms, for those who have already complied with the prevention terms, is a specific instance of a general problem so fundamental that it lies beneath the surface of the more obvious questions, even though it constitutes a third issue of justice requiring explicit discussion. The outcome of bargaining among two or more parties, such as various nations, can be binding upon those parties that would have preferred a different outcome only if the bargaining situation satisfies minimal standards of fairness. An unfair process does not yield an outcome that anyone ought to feel bound to abide by if she can in fact do better. A process of bargaining about coping in which the positions of some parties were too weak precisely because they had invested so much of their resources in prevention would be unfair in the precise sense that those parties that had already benefited from the invested resources of the consequently weakened parties were exploiting that very weakness for further advantage in the terms on which coping would be handled.

In general, of course, if several parties (individuals, groups, or institutions) are in contact with each other and have conflicting preferences, they obviously would do well to talk with each other and simply work out some mutually acceptable arrangement. They do not need to have and apply a complete theory of justice before they can arrive at a limited plan of action. If parties are more or less equally situated, the method by which they should explore the terms on which different parties could agree upon a division of resources or sacrifices (or a process for allocating the resources or sacrifices) is actual direct bargaining. Other things being equal, it may be best if parties can simply work out among themselves the terms of any dealings they will have with each other.

Even lawyers, however, have the concept of an unconscionable agreement; and ordinary nonlawyers have no difficulty seeing that voluntarily entered agreements can have objectionable terms if some parties were subject, through excessive weakness, to undue influence by other parties. Parties can be unacceptably vulnerable to other parties in more than one way, naturally, but perhaps the clearest case is extreme inequality in initial positions. This means that morally acceptable bargains depend upon initial holdings that are not morally unacceptable—not, for one thing, so outrageously unequal that some parties are at the mercy of others.

Obviously this entails in turn that the recognition of acceptable bargaining presupposes knowledge of standards for fair shares, which are one kind of standard of justice. If we do not know whether the actual shares that parties currently hold are fair, we do not know whether any actual agreement they might reach would be morally unconscionable. The simple fact that they all agreed is never enough. The judgment that an outcome ought to be binding presupposes a judgment that the process that produced it was minimally fair. While this may not mean that they must have "a complete theory of justice" before they can agree upon practical plans, it does mean that they need to know the relevant criteria for minimally fair shares of holdings before they can be confident that any plan they actually work out should in any way constrain those who might have preferred different plans.

If bargaining among nations about the terms on which they will cooperate to prevent global warming is to yield any outcome that can be morally binding on the nations who do not like it, the "initial" holdings at the time of the bargaining must be fair. Similarly, the "initial" holdings at the time of the bargaining about the terms on which they will cooperate to cope with the unprevented damage from global warming depend, once again, upon minimally fair shares at that point. Holdings at the point of bargaining over the arrangements for coping will have been influenced by the terms of the cooperation on prevention. Consequently, one requirement upon the terms for prevention is that they should not result in shares that would be unfair at the time that the terms of coping are to be negotiated. The best way to prevent unfair terms of coping would appear to be to negotiate both sets of terms at the same time and to design them to be complementary and fair taken together. This would deal with all the first three issues of justice at once. First, however, one needs to know the standard of fairness by which to judge. This is the third issue of justice.

4. Allocating Emissions: Transition and Goal

The third kind of standard of justice is general but minimal, general in that it concerns all the resources and wealth that contribute to the distribution of bargaining strength and weakness and minimal in that it specifies not thoroughly fair distributions but distributions not so unfair as to undermine the bargaining process. The fourth kind of standard is neither so general nor so minimal. It is far less general because its subject is not the international distribution of all wealth and resources, but the international distribution only of greenhouse-gas emissions in particular. And rather than identifying a minimal standard, it identifies an ultimate goal: What distribution of emissions should we be trying to end up with? How should shares of the limited global total of emissions of a greenhouse gas like CO_2 be allocated among nations and among individual humans? Once the efforts at prevention of avoidable warming are complete, and once the tasks of coping with unprevented harms are dealt with, how should the scarce capacity of the globe to recycle the net emissions be divided?

So far, of course, nations and firms have behaved as if each of them had an unlimited and unshakable entitlement to discharge any amount of greenhouse gases that it was convenient to release. Everyone has simply thrust greenhouse gases into the atmosphere at will. The danger of global warming requires that a ceiling—probably a progressively declining ceiling—be placed upon total net emissions. This total must somehow be shared among the nations and individuals of the world. By what process and according to what standards should the allocation be done?

I noted above the contrast between the minimal and general third kind of standard and this fourth challenge of specifying a particular (to greenhouse emissions) final goal. I should also indicate a contrast between this fourth issue and the first two. Both of the first two issues are about the allocation of costs: Who pays for various undertakings (preventing warming and coping with unprevented warming)? The fourth issue is about the allocation of the emissions themselves: Of the total emissions of CO_2 compatible with preventing global warming, what percentage may, say, China and India use—and, more fundamentally, by what standard do we decide? Crudely put, issues one and two are about money, and issue four is about CO_2. We need separate answers to "Who pays?" and to "Who emits?" because of the distinct possibility that one nation should, for any of a number of reasons, pay so that another nation can emit more. The right answer about emissions will not simply fall out of the right answer about costs, or vice versa.[2]

We will be trying to delineate a goal, a just pattern of allocation of something scarce and valuable, namely greenhouse-gas emissions capacities. However, a transition period during which the pattern of allocation does not satisfy the ultimate standard may well be necessary because of political or economic obstacles to an immediate switch away from the status quo. For instance, current emissions of CO_2 are very nearly as unequal as they could possibly be: a few rich countries with small populations are generating the vast bulk of the emissions, while the majority of humanity, living in poor countries with large populations, produces less altogether than the rich minority. It seems reasonable to assume that, whatever exactly will be the content of the standard of justice for allocating emissions, the emissions should be divided somewhat more equally than they currently are. Especially if the total cannot be allowed to keep rising, or must even be reduced, the per capita emissions of the rich few will have to decline so that the per capita emissions of the poor majority can rise.

Nevertheless, members of the rich minority who do not care about justice will almost certainly veto any change they consider too great an infringement upon their comfort and convenience, and they may well have the power and wealth to enforce their veto. The choice at that point for people who are committed to justice might be between vainly trying to resist an almost certainly irresistible veto and temporarily acquiescing in a far-from-ideal but significant improvement over the status quo. In short, the question would be: Which compromises, if any, are ethically tolerable? To answer this question responsibly, one needs guidelines for transitions as well as ultimate goals—not, however, guidelines for transitions instead of ultimate goals but guidelines for transitions *in addition to* ultimate goals. For one central consideration in judging what is presented as a transitional move in the direction of a certain goal is the distance traveled toward the goal. The goal must have been specified in order for this assessment to be made.[3]

B. Two More Kinds of Questions

A principle of justice may specify to whom an allocation should go, from whom the allocation should come, or, most usefully, both. The distinction between the questions, from whom and to whom, would seem too obvious to be worth comment except that "theories" of justice actually tend in this regard to be only half-theories. They tend, that is, to devote almost all their attention to the question "to whom" and to fail to tackle the challenges to the firm specification of the sources for the recommended transfers. This is one legitimate complaint practical people tend to have against such "theories": "You have shown me it would be nice if so-and-so received more, but you have not told me who is to keep less for that purpose—I cannot assess your proposal until I have heard the other half."

Unfortunately, the answer to "From whom?" does not flow automatically from all answers to "To whom?" Often a given specification of the recipients of transfers leaves open a wide variety of possible allocations of the responsibility for making the transfers. For instance, if the principle governing the allocation of certain transfers were "to those who had been severely injured by the pollution from the process," the potential sources of the transfers would include those who were operating the process, the owners of the firm that authorized the process, the insurance company for the firm, the agency that was supposed to be regulating the process, society in general, only the direct beneficiaries of the process and no one else, and so on. Quite often proposals about justice are not so much wrong as too incomplete to be judged either right or wrong.

I have phrased the first four kinds of issues about justice, which arise from different aspects of the challenge of global warming, as, in effect, "From whom?" questions, precisely because this is the neglected side of the discussion of justice. What we are now noticing is simply that there is, in addition, always the question "To whom?" It is more likely that "To whom?" will have an obvious answer than it is that "From whom?" will, but it is always necessary to check. If we are discussing the costs of coping, for example, it might seem obvious that from whomsoever the transfers should come, they should go to those having the most difficulty coping. However, if the specification of the sources of the transfers is "those who caused the problem being coped with," then country A, which did in fact cause the problem in country X, might be expected to assist country X, and not country Y, even though country Y was having much more difficulty coping (but with problems that were not A's responsibility). Not much, unfortunately, is obvious, although I will try to show that a great deal is actually fairly simple, given commonsense principles of fairness.

One vital point that this abstract example of A, X, and Y illustrates is that answers to "From whom?" and answers to "To whom?" are interconnected. Once one has an answer to one question or the other, certain answers to the remaining question are inappropriate, and sometimes, another answer to the remaining question becomes the only one that really makes any sense. Often these logical connections are very helpful.

C. Two Kinds of Answers

We saw, in section A, that if one thinks hard enough about how the international community should respond to global warming, questions about justice arise unavoidably at four points:

1. Allocating the costs of prevention.
2. Allocating the costs of coping.
3. The background allocation of resources and fair bargaining.
4. Allocating emissions: transition and goal.

And in section B we have just now observed that besides these more difficult questions about identifying the bearers of responsibility who should be the sources of any necessary transfers, there is always in principle, and often in practice, a further question in each case about the appropriate recipients of any transfers.

Before attempting to sort out specific proposed answers to this array of questions, it is helpful, I think, to notice that individual principles of justice for the assignment of responsibility fall into one or the other of two general kinds, which I will call fault-based principles and no-fault principles. A well-known fault-based principle is "the polluter pays," and a widely accepted no-fault principle is "payment according to ability to pay." The principle of payment according to ability to pay is no-fault in the sense that alleged fault, putative guilt, and past misbehavior in general are all completely irrelevant to the assignment of responsibility to pay. Those with the most should pay at the highest rate, but this is not because they have done wrong in acquiring what they own, even if they have in fact done wrong. The basis for the assignment of progressive rates of contribution, which are the kind of rates that follow from the principle of payment according to ability to pay, is not how wealth was acquired but simply how much is held.

In contrast, the "polluter pays" principle is based precisely upon fault or causal responsibility. "Why should I pay for the cleanup?" "Because you created the problem that has to be cleaned up." The kind of fault invoked here need not be a moralized kind—the fault need not be construed as moral guilt so much as simply a useful barometer or symptom to be used to assign the burden of payment to the source of the need for the payment. That is, one need not, in order to rely upon this principle, believe that polluters are wicked or even unethical in some milder sense (although one can also believe they are). The rationale for relying upon "polluter pays" could, in particular, be an entirely amoral argument about incentives: the polluter should pay because this assignment of cleanup burdens creates the strongest disincentive to pollute. Even so, this would be a fault-based principle in my sense of "fault-based," which simply means that the inquiry into who should pay depends upon a factual inquiry into the origins of the problem. The moral responsibility for contributing to the solution of the problem is proportional to the causal responsibility for creating the problem. The pursuit of this proportionality can itself in turn have a moral basis (guilty parties deserve to pay) or an amoral basis (the best incentive structure makes polluters pay). The label "fault-based" has the disadvantage that it may sound as if it must have a moral basis, which it may or may not have, as well as having a moral implication about who ought to pay, which it definitely does have.

An alternative label, which avoids this possible moralistic misunderstanding of "fault-based," would have been to call this category of principles not "fault-based" but "causal" or "historical," since such principles make the assignment of responsibility for payment depend upon an accurate understanding of how the problem in question arose. This, however, has the greater disadvantage of suggesting as the natural label for what I call "no-fault" principles, "acausal," or "ahistorical" principles. That would, I think, be more misleading still because it would make the no-fault principles sound much more ethereal and oblivious to the facts than they are. "Payment according to ability to pay" does not call for an inquiry into the origins of the problem, but neither is it ahistorical or acausal. A historical analysis or a view about the dynamics of political economy might be a part of the rationale for an ability-to-pay principle, so it would be seriously misleading to label this principle "ahistorical" or "acausal" just because it does not depend upon a search for the villain in the not necessarily moralistic sense in which "fault-based" principles do depend upon identifying the villain, that is, in the sense of who produced the problem. So, I will stick with "fault-based" for principles according to which the answer to "From whom?" depends upon an inquiry into the question "By whom was this problem caused?" and to "no-fault" for principles according to which "From whom?" can be answered on grounds other than an analysis of the production of the problem.

Principles for answering the second kind of question noted in section B, "to whom transfers should be made," also fall into the general categories of fault-based and no-fault. The principle "Make the victims whole" is ultimately fault-based in that the rightful recipients of required transfers are identified as specifically those who suffered from the faulty behavior on the basis of which it will be decided from whom the transfers should come. On this principle, the transfers should come from those who caused the injury or harm and go to those who suffered the injury or harm. Indeed, one of the great advantages of fault-based principles is precisely that their cause-and-effect structure provides complementary answers to both questions: transfers go to those negatively affected, from those who negatively affected them. This specific principle, "Make the victims whole," embodies a perfectly ordinary view—and an especially clear one, since it also partly answers the third question, how much should be transferred, by indicating that the transfer should be at least enough to restore the victims to their condition prior to the infliction of the harm. The victims (to whom) are to be "made whole" (how much—minimum amount, anyway) by those who left them less than whole (from whom). This principle does not completely answer the question of "How much?" because it leaves open the option that the victims are entitled to more than enough merely to restore them to their condition *ex ante*; that is, it leaves open the possibility of additional compensation.

An ordinary example of a kind of no-fault principle for answering to whom an allocation should go is "Maintain an adequate minimum." Naturally, the level of what was claimed to be the minimum would have to be specified and defended for this to be a usably concrete version of this kind of principle. It has the general advantage of all no-fault principles, however, in that no inquiry needs to be conducted into who was in fact injured, who injured them, how much they were injured and to what extent their problems had other sources, and so forth. Transfers go to those below the minimum until they reach the minimum; then something else happens (for example, they are retrained for available jobs). Quite a bit of information is still needed to use such a no-fault principle, both to justify the original specification of the minimum level and to select those who are in fact below it. Yet this information is of different types from the information needed to apply a fault-based principle: one does not need an understanding of possibly highly complex systems of causal interactions and positive and negative feedbacks and/or lengthy chains of historical connections among potentially vast numbers of agents and multiple levels of analysis. The information needed to apply no-fault principles tends to be contemporaneous information about current functioning, which is often easier to obtain than the convincing analysis of fault needed for the use of a fault-based principle.

The evident disadvantage of a no-fault principle for specifying to whom transfers go is that it lacks the kind of naturally complementary identification of from whom the transfers should come that flows from the cause-and-effect structure of fault-based principles. In particular, it does not imply that the transfers should come from whoever caused those who are below the minimum to be below the minimum; in fact, it does not even assume that there is any clear answer or, for that matter, any meaningful question of the form "Who caused those below the minimum to be there?" The consequence of the absence of the convenient complementary answers implied by fault-based principles is that with no-fault principles the answers to the "To whom?" question and the "From whom?" question must be argued for and established separately, not by a single argument like arguments about fault. It might be, for example, that if the answer to the question "To whom?" is "Those below the minimum," the answer to the question "From whom?" may be "Those with the greatest ability to pay." The point, however, is that the argument for using "ability to pay" to answer the one question and the argument for using "maintenance of a minimum" to answer the other question have to be two separate arguments.

III. Comprehensiveness versus Justice

With this framework in mind one can return to the choice between the recommendations of Stewart and Wiener on the one hand and Drennen on the other. Stewart and Wiener make a kind of mistake that is often made by lawyers who take economics too seriously and equity not seriously enough. One of their chief arguments in favor of moving directly to a comprehensive treaty is that under a comprehensive treaty each nation could engage in what I will call homogenizing calculations of cost-effectiveness (Stewart and Wiener 1992, pp. 93–95). A major advantage of the comprehensive treaty is supposed to be that each nation could look at all uses of all GHGs and select the least-cost options. That is, one could begin the reduction of GHG emissions by eliminating the specific sources of specific gases the elimination of which would produce the smallest subtraction of economic value. The crucial feature of this approach to cost-effectiveness—the feature that leads me to call the approach "homogenizing"—is that all gas sources (every source of every gas) are thrown into the same pot. Not a single distinction is made among gas sources, not even the distinction between essential and nonessential.

Now it may initially seem strange not to embrace a thoroughgoing least-cost-first approach, but I would like to try to argue not only that hesitation is not unreasonable but, further, that equity demands qualifications on that approach. First, I would like to explain my earlier slur against some economics and expand a bit on the worry about "homogenization." For standard economic analysis everything is a preference: the epicure's wish for a little more seasoning and the starving child's wish for a little water, the collector's wish for one more painting and the homeless person's wish for privacy and warmth—all are preferences. Quantitatively, they are different because some are backed up by a greater "willingness to pay" than others, but qualitatively a preference is a preference. For a few purposes, perhaps, we might choose to treat preferences only quantitatively, in terms of willingness to pay. To choose, however, to discard all the qualitative distinctions built up during the evolution of human history is to deprive ourselves of a rich treasure of sophistication and subtlety. Some so-called preferences are vital, and some are frivolous. Some are needs, and some are mere wants (not needs). The satisfaction of some "preferences" is essential for survival, or for human decency, and the satisfaction of others is inessential for either survival or decency.

Distinctions like the one between needs and wants, or the one between the urgent and the trivial, are of course highly contested and messy, which is why we yearn for the simplicity provided by everything being a so-called preference, differing only in strength (willingness to pay). To ignore *these* distinctions, however, is to discard the most fundamental differences in kind that we understand. This is a general complaint against much mainstream economics. My specific complaint against Stewart and Wiener does not depend upon this stronger, more general one—I mention the general thesis because my specific thesis concerns a parallel form of clarity-abandoning homogenization.

To suggest simply that it is a good thing to calculate cost-effectiveness across all sources of all GHGs is to suggest that we ignore the fact that some sources are essential and even urgent for the fulfillment of vital needs and other sources are inessential or even frivolous. What if, as is surely in fact the case, some of the sources that it would cost least to eliminate are essential and reflect needs that are urgent to satisfy, while some of the sources that it would cost most to eliminate are inessential and reflect frivolous whims? What if, to be briefly concrete, the economic costs of abandoning rice paddies are less than the economic costs of increasing miles-per-gallon in luxury cars? Does it make no difference that some people need those rice paddies in order to feed their children, but no one needs a luxury car?

It would be nuts not to follow the principle of least-cost-first as long as one was dealing with matters of comparable significance. To do otherwise would be to choose a more expensive means to an end that could be reached by a less expensive means—that *is* fundamentally irrational. While the elimination of *N* thousand hectares of rice paddy might well cost less in economic terms than the tightening of corporate average fuel economy (CAFE) standards enough to produce the same reduction in GHG emissions, however, the human consequences of reducing food production and of reducing inefficient combustion are far from comparable in their effects upon the quality of life—indeed, in the case of the food, upon the very possibility of life. These are not two different means to the very same end. The ends of the two different measures could be the same in the amount of GHG emissions they eliminate, but the ends are otherwise as different as reducing vital supplies of food and making luxury a tad more costly. Consequently, to apply a homogenizing form of cost-effectiveness calculation, as if the two measures differed only in how much they each cost to produce the same reduction in emissions, is seriously to distort reality. This kind of comprehensiveness obscures distinctions that are fundamental, most notably the distinction between necessities and luxuries.

The central point about equity is that it is not equitable to ask some people to surrender necessities so that other people can retain luxuries. It would be unfair to the point of being outrageous to ask that some (poor) people spend more on better feed for their ruminants in order to reduce methane emissions so that other (affluent) people do not have to pay more for steak from less crowded feedlots in order to reduce their methane and nitrous oxide emissions, even if less crowded feedlots for fattening luxury beef for the affluent world would cost considerably more than a better quality of feed grain for maintaining the subsistence herds of the poor.

It is of course a different story if *all* incremental costs for reducing emissions, wherever incurred, are to be allocated according to ability to pay. If the beef eaters will pay for the better feed grain for the subsistence herds of the poor with *additional* funds not already owed for some other purpose like development assistance, it might, as far as the equity of the arrangements for reducing emissions goes, not be unfair to start with the least-cost measures. The least-cost measure paid for by those *most* able to pay is not at all the same as the least-cost measure paid for by those *least* able to pay. In terms of the framework laid out above, this would combine a no-fault answer to the question "From whom?" (ability to pay) with a no-fault answer to the question "To whom?" (maintenance of an adequate minimum).

If these two answers were fully justified—naturally, they require fuller argument—one or the other of two routes ought to be followed.[4] The homogenizing form of calculation of cost-effectiveness could be neutralized if it were accompanied by a firm commitment that costs are to be paid according to ability to pay and the actual establishment of mechanisms for enforcing the necessary transfers. Otherwise the costs ought to be partitioned—perhaps more than once but surely at least once—into costs that impinge upon necessities for the poor and costs that only impinge upon luxuries for the wealthy.

Drennen has suggested one type of such a partitioning, and he has based it upon two kinds of considerations: centrally upon the consideration of equity but also upon the difficulties in measuring agricultural and other biological emissions of methane and verifying any mandated reductions. (It is ironic that the United States, which held up the control of nuclear weapons for years with exaggerated worries about verification, now wants to plunge ahead with the much messier matter of methane.) Drennen has arrived at his partitioning between what should be within the scope of the treaty and what should not by combining type of gas and type of use. The gases that Drennen would have under protocol control are CO_2 and methane, presumably because they are the largest contributors to global warming that are subject to human control. (Water vapor is larger but not controllable.) The uses that he would like to see reduced are the "industrial-related" ones, not the agricultural ones. Drennen's strategy is to control nonbiological anthropogenic emissions of CO_2 and methane. This is a much more sophisticated formulation than the much-discussed one that deals only with fossil-fuel CO_2. I would nevertheless suggest that a definition of the scope different from Drennen's, based even more directly on considerations of equity, would be preferable; although I also acknowledge that practical considerations may require that Drennen's pair, non-industrial-related CO_2 and non-industrial-related methane, still be used as proxies for, and as the nearest practical approximation of, the specifications that flow directly from attention to equity.

My main doubts—and they are relatively minor—are about the division into industrial-related and non-industrial-related. This division reflects quite well, as I understand them, the difficulties of measurement and verification of methane emissions. It is much easier to calculate (and change!) leaks from natural-gas pipelines than to calculate (and change!) emissions from various varieties of rice and various species of ruminants. Precisely this division reflects less well, however, the concern about equity that Drennen and I share. Just as the methane emissions from beef feedlots are in service of the desires of the wealthy, many of the CO_2 emissions in China and India *could be*—I am not assuming that they all in fact are—in service of the needs of the poor. Some agricultural methane emissions are a luxury, and some industrial CO_2 emissions are a necessity. By the standard of equity we do not want to leave all the former uncontrolled and control all the latter.

Now insofar as we really cannot measure, or even accurately estimate, biological emissions of methane, the lack of perfect fit with equity does not for now matter. Yet I am somewhat impressed by the contention by Stewart and Wiener that we should be able to arrive at accurate enough estimates for our purposes, especially if, as I contend, our purposes ought not to include the control of subsistence emissions. For the sake of scientific understanding we must eventually be able to measure all kinds of emissions, but the measurements of the kinds that we are not going to control can be rough in the beginning.

We should not have a homogenized—undifferentiated—market in emissions allowances in which the wealthy can buy up the allowances of the poor and leave the poor unable to satisfy even their basic needs for lack of emissions allowances. Drennen's partition between industrial-related and non-industrial-related may be the best approximation we can in practice make: most agricultural emissions are probably for subsistence, and many industrial emissions are not. Better still, if it is practical, would be a finer partitioning that left the necessary industrial activities of the developing countries uncontrolled (not, of course, unmeasured) and brought the unnecessary agricultural services of the developed world, as well as their superfluous industrial activities, under the system of control.

If there is to be an international market in emissions allowances, the populations of poor regions could be allotted inalienable—unmarketable—allowances for whatever use they themselves consider best. Above the inalienable allowances, the market could work its magic, and the standard of cost-effectiveness could reign supreme. But the market for emissions allowances would not be fully comprehensive, as Stewart and Wiener recommend. The poor in the developing world would be guaranteed a certain quantity of protected emissions, which they could produce as they choose. This would allow them some measure of control over their lives rather than leaving their fates at the mercy of distant strangers.

Notes

1. Less obvious, that is, than the issue whether the poor should have to sacrifice their own economic development so that the rich can maintain all their accustomed affluence. As already indicated, if someone honestly thought this demand could be fair, we would belong to such different worlds that I do not know what I could appeal to that we might have in common.

2. For imaginative and provocative suggestions about the final allocations of emissions themselves, see Agarwal and Narain (1991).

3. A serious attempt to deal with issues about a compromise transition is in Grubb and Sebenius's chapter, "Participation, Allocation and Adaptability in International Tradeable Emission Permit Systems for Greenhouse Gas Control," in *Climate Change* (1992).

4. I have provided some relevant arguments in Shue (1992). Justice, or equity, is also discussed in several chapters in the UNCTAD publication, *Combating Global Warming: A Study on a Global System of Tradeable Carbon Emission Entitlements* (1992).

References

Agarwal, Anil, and Sunita Narain. 1991. *Global Warming in an Unequal World: A Case of Environmental Colonialism*. New Delhi: Centre for Science and Environment.

Drennen, Thomas E. 1993. "After Rio: Measuring the Effectiveness of the International Response." *Law & Policy* 15: 15–37.

Grubb, Michael, and James K. Sebenius. 1992. "Participation, Allocation and Adaptability in International Tradeable Emission Permit Systems for Greenhouse Gas Control." In *Climate Change: Designing a Tradeable Permit System* (Paris: OECD).

Houghton, J. T., G. J. Jenkins, and J. J. Ephraums. 1990. *Climate Change: The IPCC Scientific Assessment.* Report by Working Group I. New York: Cambridge University Press.

Shue, Henry. 1992. "The Unavoidability of Justice." In *The International Politics of the Environment: Actors, Interests, and Institutions*, ed. Andrew Hurrell and Benedict Kingsbury. Oxford: Clarendon Press.

Stewart, Richard B., and Jonathan B. Wiener. 1992. "The Comprehensive Approach to Global Climate Policy: Issues of Design and Practicality." *Arizona Journal of International and Comparative Law* 9: 83–113.

Bureau of National Affairs. 1992. "Top Environmental Official Welcomes Summit Aid Pledges from Developed Nations—China." *International Environment Reporter: Current Reports* 15 (July 1): 444.

United Nations Conference on Trade and Development (UNCTAD). 1992. *Combating Global Warming: A Study on a Global System of Tradeable Carbon Emission Entitlements.* New York: United Nations.

12

Greenhouse Development Rights

A Framework for Climate Protection That Is "More Fair" Than Equal Per Capita Emissions Rights

Paul Baer, with Tom Athanasiou, Sivan Kartha, and Eric Kemp-Benedict

Introduction

There is a fairly broad consensus among both the philosophers who write about climate change and the majority of the climate-policy community that efforts to reduce greenhouse-gas emissions—"mitigation" in the jargon—should not harm the ability of poor countries to grow economically and to reduce as rapidly as possible the widespread poverty their citizens suffer. Indeed, this principle of a "right to development" has been substantially embraced in the United Nations Framework Convention on Climate Change (UNFCCC) itself.[1] Yet as the evidence of the risks from climate change has continued to mount and calls have grown for more stringent mitigation targets, the need to give substance to this right has come into conflict with the evident unwillingness of already "developed" countries to pay the costs of adequately precautionary mitigation.

The long and the short of it is that almost any reasonable ethical principles lead to the conclusion that, as Henry Shue (1999) put it straightforwardly, "the costs [of mitigation] should initially be borne by the wealthy industrialized states." In the words of the UNFCCC, "the developed country Parties should take the lead in combating climate change and the adverse effects thereof," and this point is embodied in practical terms in the Kyoto Protocol itself, in which only the 40 developed "Annex I" countries have binding emissions limits. Yet particularly because of the rejection of Kyoto by the United States but also because of the weak efforts at mitigation that have taken place so far in Europe, Japan, and other industrialized countries, we find ourselves in a situation in which precaution requires that emissions be reduced extremely soon in poor countries, too, but the rich countries can't yet be said to have fulfilled their obligations to "take the lead."

The delay in taking action so far, the increasing evidence of current climate-change impacts and greater risks than previously estimated, and the speed with which we must now move all imply substantially greater costs for adequately precautionary action than were previously estimated. And as the costs of meeting a precautionary target rise, there are really only three options: the poor start to pay sooner, the rich pay even more, or the target is weakened (we leave aside, for the moment, the important distinction between rich and poor *countries* and rich and poor *people*).

This is the context within which we have been developing our "Greenhouse Development Rights" (GDRs) framework. Fundamentally, GDRs is designed to allocate the costs of extremely rapid reductions (as well as costs for adaptation) in a way that protects a "right to development" by transparently linking obligations to *capacity* (ability to pay) and *responsibility* (prior emission of greenhouse pollution). While ours is not the first framework proposal to attempt to quantify capacity and responsibility,[2] we add two innovations: the inclusion of the distribution of income *within* countries and the definition of a "development threshold" relative to the income of *individuals,* not the per capita income of countries. In this way, we eliminate the need for an arbitrary dividing line between developed and developing countries, and can straightforwardly identify a continuum along which each country's obligations are demonstrably proportional to the responsibility and capacity of its population.

In theory, this approach—on the (admittedly strong) assumption that countries implemented their domestic policies in a way that matched the progressivity of the global calculations—could address several critical problems. From the perspective of the average citizen of the industrialized countries, it could ensure that they were not paying higher "bills" for climate policy than the small but significant wealthy minority in the developing countries, and in this way help build political support for more stringent regulations. Second, by including all countries under a similar global framework, it would reduce and potentially eliminate the problem of "leakage" of pollution, production, and employment from industrialized to developing countries. And by identifying obligations for poor countries that were closely tied to the actual responsibility and capacity of the fractions of their populations who meet or exceed rich-world levels of consumption and pollution while exempting the poor majority, it could make possible the steep global emissions reductions necessary for the long-run sustainability of human development and poverty-reduction efforts.

However, in practice, the GDRs approach makes demands on both the industrialized countries and the wealthy in developing countries that are far beyond their evident current "willingness to pay"; in particular, given the failure of the Annex I countries to "take the lead," the non–Annex I countries are (as recently as the international climate negotiations in Copenhagen in December of 2009) unwilling to renegotiate the basic distinction between developed countries (with binding targets) and developing countries (without such targets). Unsurprisingly, this makes our proposal unrealistic, at least at the present time. For this reason, we describe our proposal as a "reference framework," in which we suggest *what would be fair*, if principle-based burden sharing on the basis of capacity and responsibility were taken seriously and if adequate mitigation targets were under serious consideration.

This is not to say that GDRs is only relevant in some hypothetical future; we believe that the principles and calculations we offer can usefully be incorporated, in some partial way, even in the current negotiating situation when the removal or modification of the distinction between Annex I and non–Annex I countries is not yet on the table. Nonetheless, our concern here is not with this question of immediate relevance but rather with the underlying principles and how they differ in their justification and implications from other equity-driven approaches to climate policy.

Before offering a more detailed description of the GDRs framework, we will present two substantial background arguments. First, we will put our work in the context of debates over development and global poverty and inequality, particularly the layering of inequality *within* countries on top of inequality *between* countries, and attempt to show how this raises additional political and ethical complexities. Second, we will show how our work relates to debates over the ethics of the "allocation problem"—the question of how to allocate now-scarce pollution rights among countries or, alternatively, how to allocate the costs of reducing emissions. In particular, we discuss why we explicitly reject allocation based on equal per capita emissions rights, a proposal that still finds substantial favor among both philosophers and advocates for climate justice.

In what follows, it is important to keep in mind that ethical theorizing, particularly about broad questions of social justice, is inseparable from beliefs about causality and probability in the real world. Ethicists considering climate change have—appropriately, we believe—framed their questions in terms of what might actually be feasible, not merely what is ideal. Of course, the relationship between the "ideal" and the "practical" is a dialectical one. Both will change over time, influenced by each other; individuals at any one time may agree on one and disagree on the other. The discussion here should be read in that spirit.

Poverty, Development, Inequality, and the Climate Challenge

Poverty, and often severe poverty, is widespread outside the 25 or so "rich" countries, which contain less than one-sixth of the world's population. Estimates of the number of people living on less than two dollars a day are on the order of 2 billion to 3 billion.[3] Preventable disease, high rates of infant mortality, and malnutrition are ubiquitous in the so-called developing world.

In this context, in which the focus is on the depth and breadth of poverty and the associated human deprivation, it is normal to see the world as largely being divided into rich countries and poor countries, whose interests are to significant extent in opposition. Indeed, in climate negotiations, trade negotiations, and international fora more generally, the poor countries generally negotiate as a bloc, in order to pursue perceived common interests. This has particularly been true in climate negotiations, in which, in spite of the differing interests of countries as diverse as Saudi Arabia, China, Bangladesh, and Tuvalu, the members of the G77/China group (representing almost all of the non–Annex I countries) all present a common front in their interactions with the industrialized Annex I nations.

A more nuanced view of the world would, of course, point out precisely that many of these countries that are still grouped as "developing" are hardly poor.[4] Furthermore, even truly poor countries are typically governed by rich people, whose interests arguably lie closer to those of the citizens of the rich countries than to those of their own poor citizens; similarly, the interests of rich-world elites are more aligned with poor-world elites than with the poor in the rich world (Frank 1972). In spite of these class divisions and alliances, however, countries rich and poor are divided not merely by class but also by political ideology, with the broad left-right spectrum representing a range of commitments to economic egalitarianism versus economic libertarianism ("liberalism" in the European sense), cross-cut by varying degrees of authoritarianism or democracy.

These basic facts are not generally controversial, although moral judgment about their significance certainly is. However, for a variety of reasons, scholarly discussions of the climate crisis tend to take states as unitary actors and to assume that governments represent the common interests of their citizens in international negotiations.[5] Therefore, the task of defining an acceptably fair agreement is generally seen by philosophers and policy analysts as trying to define the rights and responsibilities of states in ways that appropriately reflect those states' broad characteristics, notably in terms of average wealth (capacity) and GHG emissions levels (responsibility).

We raise this point for several reasons. First among these is the ambiguity of the generally agreed premise that the requirement to reduce global GHG emissions should not unfairly impede the right of poor countries to develop. In particular, the concept of "development" is itself notoriously ambiguous (and in fact contested), and among its common meanings are both "macroeconomic growth" and "poverty alleviation," which are related but obviously not identical.

The importance of this ambiguity lies in the following. The *moral* case for the complete or partial exemption of poor countries from mitigation obligations is that, given the continuing existence of severe and widespread poverty, anything that significantly slows down the rate at which this poverty can be alleviated requires extraordinary justification. For a variety of reasons, but primarily because of the vast inequality in per capita emissions and per capita income between rich and poor countries, asking poor countries to invest substantially in reducing GHGs prima facie fails this simple moral test.

The point was famously put by Henry Shue:

> Whatever justice may positively require, it does not permit that poor nations be told to sell their blankets [compromise their development strategies] in order that the rich nations keep their jewellery [continue their unsustainable lifestyles]. (Shue 1992, p. 397; quoted by Grubb 1995, p. 478)

Yet—leaving aside for the moment the fact that some nominally "developing" countries are plainly no longer poor—the argument here has some hidden premises that are, based on the points noted earlier, questionable. These premises in turn engage with the ambiguity of what Grubb in his editorial additions alludes to as "development strategies." The causal story justifying the exemption of poor countries from mitigation obligations is precisely that the requirement to pay for emissions reductions would slow down macroeconomic growth, which is necessary to create the possibility to alleviate poverty. At some level, this is a noncontroversial claim, particularly for the poorest countries, where the national average income is on the order of $1,000 to $2,000 per year; equality without growth in these countries would leave everyone equally poor. Yet for this to be the argument to be morally decisive, it seems to depend on the premise that poor countries are already doing as much as they can to reduce poverty—that is to say, that their "development strategies" are in fact intended to maximally benefit their poorest citizens—and thus that a reduction in economic growth would necessarily hurt the poor. Yet the former is empirically an untenable claim; straightforwardly, in all but the poorest countries, poverty could be substantially reduced by a reduction in inequality, which could be financed by a reduction in the consumption of the (relatively) wealthy classes, without reducing the investment that is presumed (in this causal story) to be the driver of economic growth.

A more honest appraisal of the situation, therefore, is that the *practical* consequence of the exemption of developing countries from mitigation obligations is first of all the protection of the consumption of their middle and upper classes and the rate of growth of their consumption. These are people whose per capita incomes and emissions are comparable with, and in some cases much higher than, those of average citizens in rich countries. Any exemption for them from the requirement to pay for mitigation depends for its moral legitimacy on the assumption that the imposition of costs even on the wealthier people in poor countries harms the poor in those countries, which, as we've argued, is questionable.

But the global system is in fact structured so that primary responsibility for poverty alleviation—that is to say, development—lies with national governments. Furthermore, rich countries have by example legitimated (and arguably, through, for example, the structural adjustment programs of the International Monetary Fund, actually mandated) the pursuit of macroeconomic growth over the reduction of inequality. Thus, the claim by poor countries that requiring them to pay for mitigation would harm their necessary prioritization of poverty alleviation and economic development is politically legitimate, even if it's empirically false.

Indeed, a more appropriate reading of the *moral* claims of poor countries would be something like the following: since the rich people in rich countries, who could easily address their countries' mitigation requirements through the reduction of consumption by their own wealthiest citizens (that is to say, by the reduction of inequality), have shown no inclination to do so, it would be unfair to force the rich in poor countries to do so; and, in the absence of a reduction of inequality, expenditure on mitigation *would* hurt the poor in poor countries.

The important point here is simply that, in general, the moral arguments that justify particular forms of burden sharing attach primarily to people rather than countries, and the assumption that the average properties of a country's citizens provide an adequate justification for nationally based rights and responsibilities is very problematic. In the GDRs framework, we have begun to grapple with this by calculating "capacity" and "responsibility" in ways that take account of the distribution of income and emissions within countries. Nonetheless, as critics have pointed out, an international framework such as GDRs can't straightforwardly require sovereign states to allocate the costs of a national obligation in any particular way conformant with an international standard of equity (Moellendorf 2008). This, of course, presents a challenge to all kinds of policy proposals that attempt to ensure fairness based on conceptions of human rights rather than national average indicators, not just GDRs. Shue's (1999) support for national allocations that guarantee "subsistence emissions" faces this problem, and many (e.g., Singer 2002) have worried that the sale of "surplus" emissions permits allocated to poor countries on the basis of equal per capita emissions rights would result in the capture of the revenue by corrupt elites. We offer no magic solutions to this problem but note only that by embedding an analysis of inequality within countries at the core of our approach, we might make it easier for advocates of equitable domestic policies.

From this starting point, we now turn to the most common framings of the "allocation problem" in the work of philosophers writing about climate change.

Ethics and the Allocation Problem

There is at this point a small but significant body of scholarly writing on ethics and climate change, largely but not exclusively focused on the "allocation problem," which is the primary concern of this chapter. As we suggested previously, there is a broad consensus among philosophers that the ethical salience of responsibility, capacity, equality, and need all justify placing the obligation to pay for climate mitigation primarily on the wealthy countries, at least initially.[6] Furthermore, there has been substantial though not universal support for the principle of equal per capita emissions rights as a simple and practical solution to operationalize this consensus.

It's important in this context to recognize that philosophers considering what ethically desirable climate policies might also be politically feasible are necessarily going beyond their disciplinary expertise. As we will show in our somewhat selective review, however, many have done exactly that. In particular, the advocates of equal per capita emissions rights typically argue that it is both "fair enough" and "as fair as possible," both judgments that are fundamentally political as well ethical. Yet, as we will show, there are good reasons to doubt that under very steep emissions reduction targets, equal per capita allocations can adequately protect the right to development.

Henry Shue (1999) argued that controversial ethical theories need not be invoked but rather that responsibility, ability to pay, and the right to an adequate minimum standard of living are all commonsense ethical principles leading to the same conclusion, that the industrialized states have the initial and primary responsibility to pay for mitigation. Shue chooses not to endorse, or even really engage with, the argument for equal per capita rights, suggesting instead that less controversial principles—particularly the right to subsistence emissions—all lead to the same practical conclusion; he does, however, note in passing that pure egalitarianism (and, implicitly, equal per capita allocations) is harder to argue against than often assumed. Important for our purposes is that Shue does not address how one would appropriately determine when a country would stop being counted as "developing."

Dale Jamieson (2001) looked at the question of obligations to reduce emissions from the point of view of the positions presented by actual countries or groups of countries, noting in particular the moral justifications that each gave for why what they preferred should be considered fair. Jamieson comes down in favor of a principle of equality and an equal per capita allocation of tradable emissions permits, identifying the former (equal allocations) with developing-country preferences and the latter (tradable permits) with industrialized-country preferences. He gives some legitimacy to the principle of historical accountability, but he concludes that emissions in the further past should at the very least be counted differently from current emissions (because of ignorance of effects) and in the end dismisses them, arguing that the developing countries shouldn't expect politically to get anything better than equal per capita rights (an example of how political "realism" frames even philosophically based policy evaluations).

Peter Singer (2002), taking the most explicit look at ethical theories, examines four principles, some of which are identifiable with specific ethical traditions: responsibility, equal shares (egalitarianism), aiding the least well off (Rawlsianism), and maximizing happiness (utilitarianism). While he generally concludes, as does Shue, that all would have the same basic consequence—the majority of burdens going to rich countries—he chooses, as does Jamieson, to advocate on practical grounds for an egalitarian principle of equal per capita emissions rights. Like Jamieson, but with a slightly different emphasis, he notes that this may be *more generous* to the industrialized countries than an allocation of emissions rights based on principles of historical responsibility.

Steve Gardiner (2004), in the most comprehensive survey to date, doesn't explicitly advocate any particular allocation principles but seems to support the broad approaches and conclusions of Shue, Singer, and Jamieson. In particular, he argues *against* at least one proposal (Traxler 2002) that, by focusing on the equalization of marginal costs, would offer no particular protection to the rights and interests of poor countries.

Finally, Darrel Moellendorf (2009) reviews many of these same sources and the underlying questions in light of the latest science (the IPCC's Fourth Assessment Report, in any case) and, with Singer and Jamieson, concludes that convergence to equal per capita allocations over time is the most defensible practical proposal.

Most of the other specifically ethical analyses of the "allocation problem" made by nonphilosophers or quasiphilosophers[7] have also concluded that equal per capita rights is the fairest solution to the problem of allocating emissions rights that is at all practical. These claims have been made on a variety of bases, but in general, they turn on an argument something like Singer's: other reasonable principles (responsibility, capacity, need) make an equal per capita allocation more fair than any of the other proposals that (primarily because of the interests and strategic power of the rich and high-emitting countries) have any hope of being implemented; therefore, equal per capita is "fair enough."[8] And indeed, historical endorsements of the equal per capita approach by spokespersons for developing countries, including India and China, have lent some credibility to this argument.

However, the assumption that equal per capita rights—and in particular the "contraction and convergence" (C&C) variant,[9] in which equal rights are phased in over time—adequately protects the interests of poor countries breaks down under stringent mitigation targets. Put simply, the assumption has been that poor countries will either have at least enough permits to avoid incurring significant mitigation costs or even an excess of permits to sell to support "clean development," until they have "developed" so that they are no longer poor. Under even moderately stringent mitigation targets, however, such as an emissions trajectory aimed at a 450 ppm CO_2-equivalent stabilization target, this is arguably not true.

Figure 12.1 shows such a trajectory. Note first that the global per capita pathway is fairly flat. It peaks in 2013, at a level just barely over 4 tCO_2 per capita, before falling to about 1.5 tCO_2 per capita in 2050. There are no surprises in this graph, except perhaps that China's emissions already significantly exceed the global mean in 2010 and that it therefore receives only a tiny increase in emissions before its per capita allocation begins to fall, while its per capita

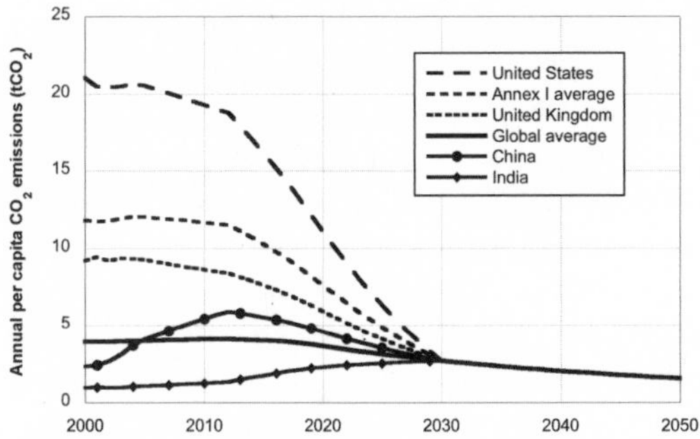

Figure 12.1. Per capita emissions allocated according to "contraction and convergence" (2030 convergence year) under an emissions pathway (based on Den Elzen et al. 2007) designed to stabilize atmospheric GHG concentrations at 450 ppm CO_2-equivalent.

emissions (and per capita income) are still far below the Annex I average.

Prima facie, this alone is enough to suggest that C&C is not fair to China and that it would be unlikely to be accepted by the Chinese and thus to be a workable global solution. And this is an emissions trajectory that has at best a 50 percent chance of keeping global temperature increase below 2°C, and that is significantly less stringent than the one we use elsewhere in this chapter as a model of sufficient precaution.

So far, none of the philosophers that we have cited has addressed these concerns.[10] Some policy analysts have attempted to modify the basic C&C formula to allow per capita emissions in poor countries to grow for a longer time than otherwise allowed (Höhne et al. 2006). But the fundamental problem is that the allocation of emissions permits to countries based on equal emissions *today* (or at some point in the future) ignores the fact that there is a much lower per capita budget available going forward than was used by the rich countries during the course of their development. Unless and until low- or no-carbon energy is cheaper than fossil energy, this fundamentally disadvantages developing countries.

One additional response to this problem has been to assert a principle of equal *cumulative* per capita emissions.[11] This proposal has interesting properties, not least being the possibility of negative emissions allocations, a possibility that appears again in the GDRs framework. However, it raises difficult questions about how to account for emissions at different points in time when (for technological and social reasons) they would have different relative values; and it still fails to address the vast disparity in capacity (wealth, income) among countries, such as Russia and Germany, with similar historical emissions but vastly different levels of wealth today.

We suggest that what is required instead is a more explicit look at how the concepts of responsibility, capacity, and need—already included in the UNFCCC—can be applied to the problem at hand. In particular, this requires further consideration of the complex relationship between *causal responsibility* and *moral responsibility*, an exploration of the *moral significance* of the sacrifices in consumption that would be required of various parties under different burden-sharing arrangements (effectively an inquiry into the definition of *capacity*), and an examination of the ways in which the allocation of emissions rights to *countries* translates into impacts on *individuals*. Not coincidentally, the GDRs framework provides quantified definitions of responsibility and capacity and proposed institutions linking national to individual (or at least class-based) rights and responsibilities.

At the center, we suggest, is the problematization of ethical principles based on individualized conceptions of ethics in a world in which rights and responsibilities are applied to countries and negotiated by persons nominally representing countries but actually significantly representing the interests of particular classes.

These ethical questions will, we hope, provide fruitful grounds for research and discussion in the coming years. Next, however, we turn to the quantified examples of the GDRs framework as it has been elaborated to date.

Quantifying the GDRs Framework

The following discussion is intended primarily to emphasize and explain the basic aspects of the GDRs framework and only secondarily to demonstrate the particular results from following the calculations through to their end. The GDRs framework has been evolving steadily since its origin in 2004 and will likely continue to evolve, and thus this should be considered a snapshot of a work in progress.[12]

As we suggested previously, the GDRs framework has two fundamental elements: the allocation of obligations in proportion to capacity (income) and responsibility (historical pollution) and the calculation of those indicators (capacity and responsibility) in a way that takes into account the distribution of income within countries and is relative to a *development threshold* defined for individuals, not countries. We elaborate these fundamental principles in greater detail, developing specific formulas and data sets that allow us to produce indicative calculations of national "shares" of global climate obligations, shares that could be applied to the obligation to reduce emissions or to pay for adaptation or compensation.

Elsewhere (Baer et al. 2008), we also elaborate two different possible ways that the GDRs framework might be implemented. In the first, global obligations are considered as a financial burden to be divided, and in the second, global emissions reductions (defined relative to a precautionary target) are considered as a "mitigation requirement" to be divided. In this way, we engage with debates about the institutions through which such a principle-based framework for burden sharing might be realized. Nonetheless, in this chapter, we treat these implementation models only very briefly, enough to show how they could be used to model obligations for countries of different kinds.

Defining Capacity, Responsibility, and the RCI

Capacity is a moral term and is not meaningful outside of a relatively concrete context. However, there is little disagreement that monetary income is a close correlate of capacity in the context we are discussing: the ability to pay for climate policies without sacrificing consumption of greater moral priority. In that light, we feel it is reasonable to use monetary income as our indicator of capacity.[13] And since our focus is on this moral priority as it relates to individuals, we define a development threshold as the level below which an individual's income is excluded from consideration as capacity. In this sense, income under the development threshold is conceptually equivalent to the money required to buy "necessities," and capacity itself—income above that threshold—is closely related to what is often called "disposable income."

As should be obvious, there can be no single "correct" place to put such a development threshold, as there can be no unique definition of "necessities," no less their "correct" prices. Nonetheless, our model requires such a line, and there are a variety of ways in which it might be chosen and justified. We set our development threshold at $7,500 per capita annual income, purchasing power parity (PPP) adjusted.[14]

The particular number we use is anchored to a study (Pritchett 2006) that argues, based on the examination of development indicators of health and other factors, for a global poverty line of $6,000, PPP adjusted. We multiply this by 1.25 to make the point that the world can, and should, continue to prioritize human development rather than climate policy for many more than those just barely escaping poverty. We are the first to admit that this will be controversial, but we believe it is defensible and even reasonable.[15]

To use this development threshold in a capacity calculation, we also need a numerical estimate of the income distribution within each country. Robust empirical estimates are not easy to come by, as the collection and interpretation of income statistics are fraught with difficulties. Nevertheless, such surveys are done regularly, and the data are reported as income fractiles—for example, deciles (each 10 percent) or quintiles (each 20 percent)—or as Gini coefficients, which summarize income distributions in a single number.[16] Various research efforts have shown that a lognormal distribution—a distribution that is, relative to the familiar "normal" (Bell curve) distribution, pushed to the left and stretched to the right—is a reasonable approximation of national income distributions (Lopez and Servén 2006). Combining reported Gini coefficients with each country's per capita income in a lognormal distribution thus gives a model of the income distribution as a continuous function, from which the number of people above, below, or between any arbitrary income levels can be calculated.

Figure 12.2 shows charts that use this model of income distribution, together with the most recent reported Gini coefficients and per capita income data projected to 2010, to display the income distribution and associated "capacity" for India, China, and the United States. These charts display the population of each country along the *x* axis from the poorest to the richest and plot their income on the *y* axis (measured in U.S. dollars per capita, PPP adjusted). The charts are scaled so that the length of the *x* axis is proportional to each country's population, so the areas of the sections representing capacity can be graphically compared in absolute terms. This figure shows the estimated number of people with incomes over the development threshold in 2010 to be about 6 percent, 23 percent, and 95 percent for India, China, and the United States, respectively, numbers that show the differences between the countries to be very large relative to the uncertainty of the estimates.[17]

We calculate "responsibility" in a similar way. First, we assume that emissions are linearly proportional to income within a country;[18] then, based on the total CO_2 emissions of a country since 1990, we divide them into an "above the threshold" fraction proportional to capacity, which is counted as responsibility, and a "below the threshold" fraction, which is excluded.[19] The 1990 threshold is controversial, since the historical emissions of the

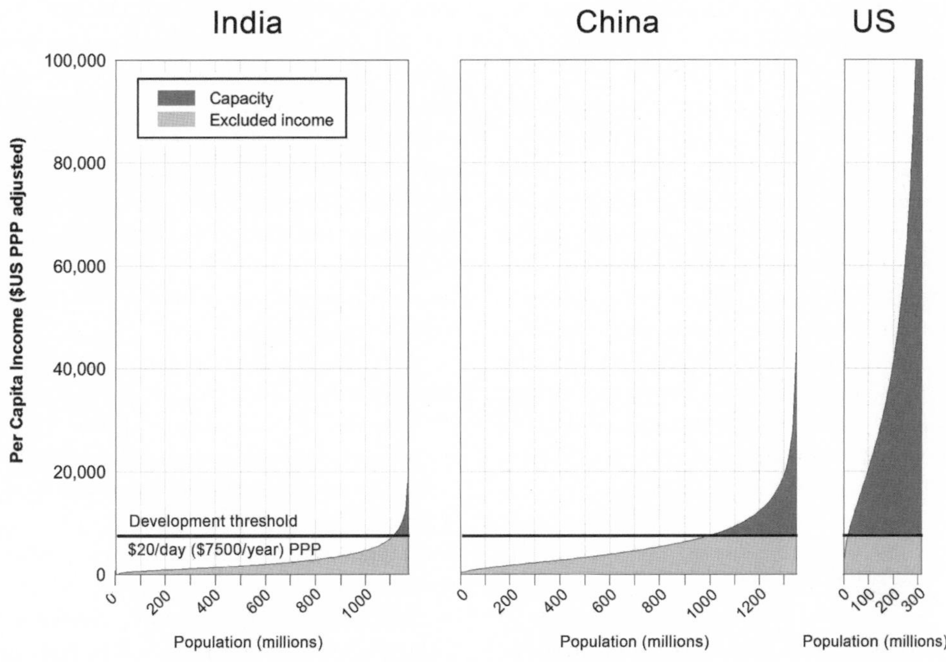

Figure 12.2. Capacity: income above the development threshold. These curves approximate income distributions within India, China, and the United States. Thus, the darker gray areas represent national incomes above the ($20 per person per day, PPP) development threshold, our definition of national capacity. Chart widths are scaled to population, so these capacity areas are correctly sized in relation to each other. Based on projected 2010 data.

rich and poor countries began to diverge dramatically long before (around the time of the industrial revolution), and the 1990 cutoff shifts responsibility substantially toward the poor countries compared with, for example, an 1850 or even 1950 cutoff date. Nonetheless, as many have argued (see, for example, Jamieson 2001), the fact that knowledge of the risks from GHG pollution was not widespread before around 1990, the year of the first report of the IPCC, means that the conditions for moral responsibility do not easily apply before then. There are still many reasonable arguments for applying historical accountability before 1990, but they are beyond our scope here;[20] we do, however, hope to incorporate additional historical data sets in the near future so that the sensitivity of the calculations to earlier accounting dates can be tested numerically.

Finally, we combine capacity and responsibility into a single "responsibility and capacity index" (RCI) using a simple weighted sum:

$$\text{RCI} = \text{a}\,R + \text{b}\,C$$

We specify that a and b sum to 1, so that, as the paired weights go from a = 1 and b = 0 at one extreme to a = 0 and b = 1 at the other, the RCI goes from being exactly equal to responsibility (*R*) to being exactly equal to capacity (*C*). In the standard case, we weight the two equally by setting a = 0.5 and b = 0.5, but this is not the only plausible weighting.[21]

Using this formula, we can now straightforwardly estimate the RCI for any nation (or group of nations) and compare it with the global total to calculate its percentage share. The results of this calculation for selected countries and groups of countries are shown in table 12.1. Note that in the final two columns, we show estimates of the RCI for our list of countries projected out to 2020 and 2030. While the reliability of these estimates becomes less over time, we include them because they show a crucial point: because of the expected rapid growth of GDP, energy use, and emissions in China and other developing countries, they will have a larger (and in some cases much larger) share of global obligations in 2030 than they would in the short run.

How might such obligations be implemented? Consider two complementary examples. First, imagine a single grand international fund to support both mitigation and adaptation.[22] The RCI could serve as the basis for determining each nation's financial contribution to that fund. So, for example, if the 2020 climate-transition funding requirement amounted to $1 trillion (roughly 1 percent of the projected 2020 gross world product), then in 2020, the United States, with about 29 percent of the global RCI, would be obligated to pay $290 billion, the EU's share would be about $230 billion, China's share would be about $100 billion, and India's share would be $12 billion. The RCI thus serves as the basis of a progressive global "climate tax"—but a responsibility and capacity tax rather than a carbon tax. Note that in this example, the funding requirement could apply to any mix of mitigation, adaptation, and compensation.

Another way the GDRs framework could potentially be implemented is as national emissions-reduction obligations, similar to the national targets established under the Kyoto Protocol. In figure 12.3, we show a global "business as usual" trajectory at the top and an "emergency pathway" at the bottom that peaks in 2013 and declines at more than 5 percent annually to 80 percent below 1990 levels in 2050. This "mitigation gap" is then divided into wedges, where the top wedge represents "no regrets" mitigation options (reductions that produce net cost savings or are cost-neutral), and the remaining wedges represent the obligations of countries or groups of countries, changing over time from 2010 to 2030 in proportion to the RCIs shown in table 12.1.

Figure 12.3 shows the very large obligations accruing to the United States and the EU and the substantial obligation accruing to China, especially at later years. However, the significance of these obligations is most visible when these "obligation wedges" are shown compared not to the global mitigation requirement but, rather, to national business-as-usual trajectories. In figure 12.4, we show the emissions reductions for the United States and China. U.S. emissions under BAU are about 6,300 megatons of CO_2-equivalent ($MtCO_2$-e) in 2020, yet in that same

Table 12.1.
GDRs results for representative countries and groups. Percentage shares of total global population, GDP, capacity, responsibility, and RCI for selected countries and groups of countries. Based on projected emissions and income for 2010, 2020, and 2030.

	2010					2020	2030
	Population (percent of global)	GDP per capita ($U.S. PPP)	Capacity (percent of global)	Responsibility (percent of global)	RCI (percent of global)	RCI (percent of global)	RCI (percent of global)
EU 27	7.3	30,472	**28.8**	**22.6**	**25.7**	**22.9**	**19.6**
EU 15	5.8	33,754	**26.1**	**19.8**	**22.9**	**19.9**	**16.7**
EU +12	1.5	17,708	**2.7**	**2.8**	**2.7**	**3.0**	**3.0**
United States	4.5	45,640	**29.7**	**36.4**	**33.1**	**29.1**	**25.5**
Japan	1.9	33,422	**8.3**	**7.3**	**7.8**	**6.6**	**5.5**
Russia	2.0	15,031	**2.7**	**4.9**	**3.8**	**4.3**	**4.6**
China	19.7	5,899	**5.8**	**5.2**	**5.5**	**10.4**	**15.2**
India	17.2	2,818	**0.7**	**0.3**	**0.5**	**1.2**	**2.3**
Brazil	2.9	9,442	**2.3**	**1.1**	**1.7**	**1.7**	**1.7**
South Africa	0.7	10,117	**0.6**	**1.3**	**1.0**	**1.1**	**1.2**
Mexico	1.6	12,408	**1.8**	**1.4**	**1.6**	**1.5**	**1.5**
LDCs	11.7	1,274	**0.1**	**0.04**	**0.1**	**0.1**	**0.1**
Annex I	18.7	30,924	**75.8**	**78.0**	77	69	61
Non–Annex I	81.3	5,096	**24.2**	**22.0**	23	31	39
World	100	9,929	**100%**	**100%**	**100%**	**100%**	**100%**

Projections based on International Energy Agency, *World Energy Outlook 2007.* (Reproduced from Baer et al. 2008.)

year, its overall emissions-reduction obligation would be about 5,100 $MtCO_2$-e. This implies a 50 percent reduction target relative to 1990 levels, and by 2030, estimated reduction obligations exceed projected emissions. Obviously, then, not all of these reductions can be made domestically (the line for "indicative domestic emissions" tracks annual emissions reductions of more than 6 percent, about as fast as one can possibly imagine); the rest must be made in other countries, for example, through emissions trading.

For China, we see the comparable situation of a (relatively) high-emitting developing country. The lowest line illustrates a reduction level of about 5.5 percent annually, roughly equivalent to the global rate shown in figure 12.3. China's obligations, while substantial, are significantly less; thus, it can reasonably expect other countries to provide support for additional mitigation within their borders if the global target is to be met.

This situation reflects both the nature of national obligations and the obvious truth of the greenhouse world: even if the wealthy countries reduce their domestic emissions to zero, they must still enable large emissions reductions elsewhere—in countries that lack the capacity (and responsibility) to reduce emissions fast enough and far enough, at least without significant assistance from others. This is to say that much of the mitigation that takes place within Southern countries must be enabled by the North. Thus, in developing countries, domestic obligations are coupled with the (typically larger) international obligations of other countries to ensure that development can proceed along a decarbonized pathway.

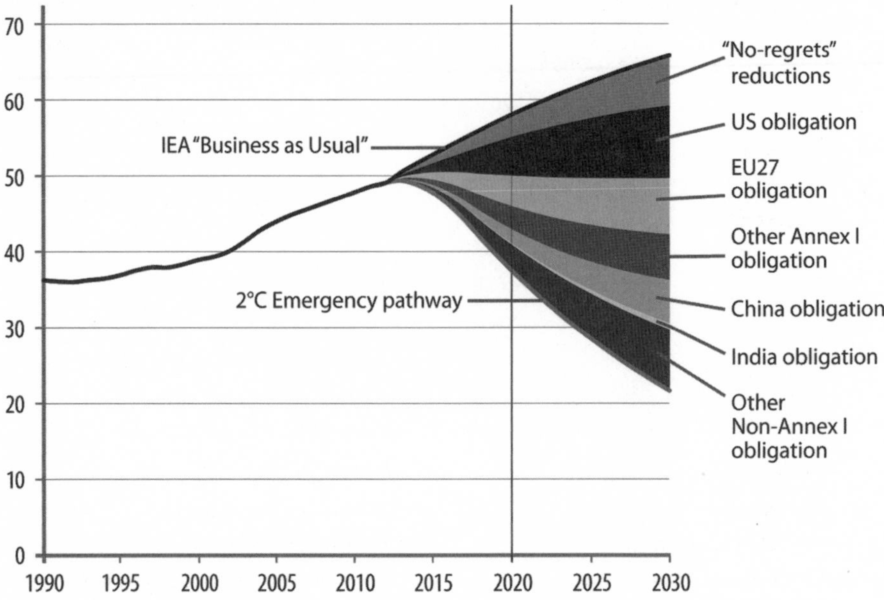

Figure 12.3. Total global mitigation requirement, divided into "national obligation wedges." The widths of the wedges reflect the shares of the global mitigation burden that would be borne by particular nations (or groupings) in proportion to their share of the total global RCI. (Modified after Baer et al. 2008.)

Again, this is only a very high-level view of the quantitative results derived from the GDRs framework so far and the data and assumptions that drive them. Nonetheless, we believe that this is adequate to show that it is possible to define responsibility and capacity in a reasonable and quantitatively rigorous manner and to define obligations for both rich and poor countries that appropriately protect the right to development even under very steep global emissions reductions.

Conclusion

We do not offer this analysis and proposal because we think it will quickly become a prototype for a post-Kyoto agreement. Rather, we believe that it offers a direct look at the fundamental conditions and conflicts that such an agreement must address and resolve, a "reference framework" by which the adequacy and fairness of any proposal can be judged. Perhaps it might also become more than that. Crucially, it moves away from both proposals narrowly defined around the average properties of countries and their interests thus defined and from simple "equal per capita rights" approaches that fail on a variety of grounds adequately to protect the "right to development."

From a philosophical perspective, we believe this can lead to a more nuanced exploration of a variety of questions regarding global justice and climate change. The idea of a "development threshold" based on individuals raises one set of questions, as does the appropriate treatment of historical responsibility. We have not yet even begun to address additional questions raised by the requirement to include emissions from land use and/or deforestation, which have a variety of different social and political implications from the emissions from fossil-energy use. Similarly, the question of who should bear responsibility for the net emissions embodied in goods or

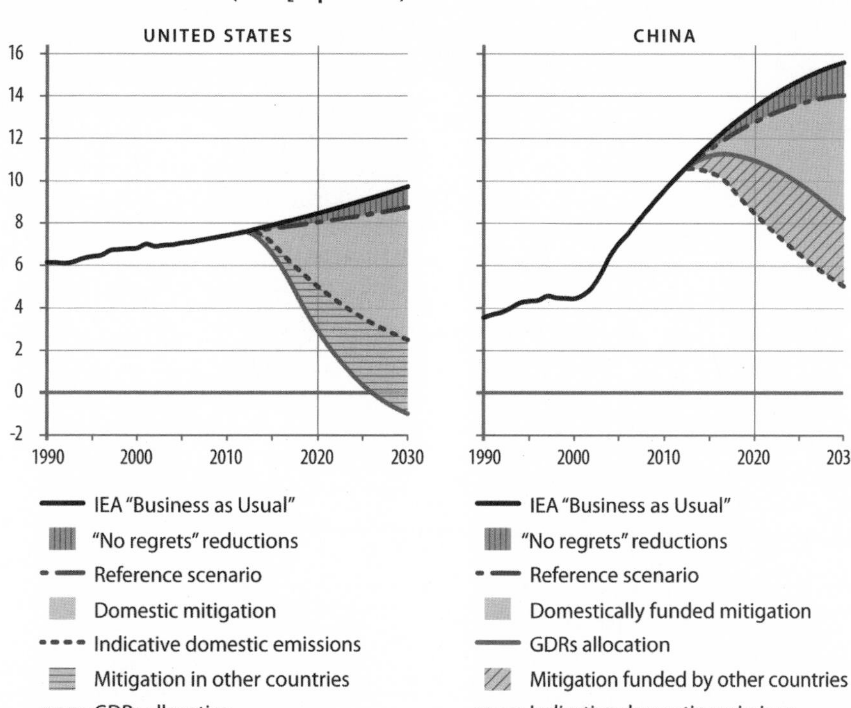

Figure 12.4. U.S. (left) and Chinese (right) obligations. No-regrets reductions (zero or negative cost) are shown with vertical hatching. For the United States, indicative domestic reductions are in gray, with additional, internationally discharged reduction obligation shown with horizontal hatching. For China, its domestic mitigation obligation is in gray, while mitigation in China that is funded by other countries is shown with diagonal hatching. (Modified after Baer et al. 2008.)

services that are traded internationally—which might make up close to 20 percent of emissions for China (Peters and Hertwich 2008)—is philosophically as well as politically important. And of course, questions will need to be asked about whether the obligation to pay for mitigation should in fact be based on the same principles as the obligation to pay for adaptation and/or compensation for climate damages.

Whether and how the GDRs framework can play a significant role in the negotiations, of course, remains to be seen; as noted above, it provides a way to move beyond the division between Annex I and non–Annex I countries, at a time when the negotiations are not yet at that stage. Nonetheless, the general case remains plausible that a global climate treaty must be *fair enough,* and we remain convinced that with the steep rate of global emissions reductions that are necessary to have a high likelihood of avoiding catastrophic climate change, something like GDRs will be necessary.

Acknowledgments

The authors are grateful to the organizers and participants at the Ethics and Climate Change Conference at the University of Washington, where this chapter was originally presented in May 2007, and to Karen Litfin, who offered extensive comments. Tyler Kemp-Benedict produced most of the graphics. All remaining errors of fact and opinion are our own.

Notes

1. Paragraph 4 of Article 3 ("Principles") of the UNFCCC puts it thus: "The Parties have a right to, and should, promote sustainable development." The various operational sections of the UNFCCC, which give effectively all financial burdens to the wealthiest countries, attempt to embody this principle.

2. Criqui and Kouvaritakis (2000) defined an effort-sharing system based on a "CR-Index" (capacity and responsibility index) that combined current per capita emissions and per capita income. Also, the South-North Dialogue proposal (Climate Protection Program 2004) used capacity, responsibility, and a measure of "mitigation potential" to group countries in their multistage framework.

3. The World Bank's estimate for 2005 is 2.6 billion (World Bank Development Indicators 2008, cited in Shah 2009).

4. Notably, the OPEC countries and the "Asian Tigers" are considered non–Annex I "developing country parties," although many are wealthier than some Annnex I countries.

5. This is most extreme in the economics and political science literature based on game-theoretical analyses of the climate problem, but it pervades nondisciplinary policy analysis as well.

6. It's important to note that there is another strand of scholarship on "equity," which eschews drawing any such conclusion, arguing instead in almost relativist terms that "there are many views on equity" and (for instance) there is no non-self-interested reason to prefer equal per capita allocations to purely grandfathered allocations. While this is more common outside philosophy (e.g., Rose et al. 1998), Benito Müller (1999), one of the more visible philosophers writing on climate change, seems to take this view.

7. I count here particularly myself and my EcoEquity collaborators (e.g., Baer 2002, Athanasiou and Baer 2002), Donald Brown (2002), Michael Grubb (1989, 1995), and Matthew Paterson (1996), but there are others.

8. A notable exception is Neumayer (2000), whose paper is aptly titled "In Defense of Historical Accountability."

9. "Contraction and Convergence" is a brand name promoted by the Global Commons Institute in the United Kingdom; see www.gci.org.uk.

10. One exception is Benito Müller, who has noted that the "convergence" aspect of C&C implies that poor countries would not get enough excess permits to sell soon enough for it to really help them make the transition to clean energy (Müller 2000).

11. This allocation principle, modeled by Bode (2004), has recently been endorsed by a variety of Chinese researchers (Chris Buckley, "Top China Think Tank Proposes Greenhouse Gas Plan," Reuters, March 25, 2009).

12. Up-to-date publications and online support, including a "calculator" that allows the user to experiment with alternative parameter choices, are available at http://greenhousedevelopmentrights.org.

13. But see Oxfam (2007) for an analysis in a similar spirit that uses the Human Development Index (HDI) as the indicator of capacity in a joint capacity/responsibility index.

14. The PPP adjustment is intended to convert units in any local currency into a standard unit equivalent to U.S. dollars, with direct reference to the basket of goods and services that this amount of money will buy. Thus, at this threshold, one is living at roughly the poverty-line level of consumption of goods and services in a developed country such as the United States, and someone who has this income level but lives in a developing country is still not very well off.

15. In fact, we settled on this number as more reasonable than the $9,000 figure—1.5 times Pritchett's global poverty line—used in our initial calculations.

16. The technical definition of a Gini coefficient is beyond our scope, but, briefly, a Gini coefficient can theoretically be between 0 (perfectly equal income) and 1 (all of the income held by one person). In practice, they range between 0.20 and 0.30 for egalitarian countries such as many western European states and 0.50 to 0.70 for the least egalitarian states. The latest reported Gini coefficient for the United States is 0.46.

17. It's not possible to give formal uncertainties for these estimates, but they are probably accurate to plus or minus 2 or 3 percent.

18. Again, this is an assumption that is plainly "false" yet defensibly close enough, but see Chakravarty et al. (2009) for an alternative estimation of the relationship.

19. Note that the method we use calculates these "above the threshold" and "below the threshold" values for responsibility for each year since 1990; thus, the proportion of the two will differ from the proportion of capacity to

excluded income, which is calculated for a single year. For this and other details, see Baer et al. 2008 and the online appendices at http://www.greenhousedevelopmentrights.org.

20. See Shue (1999), Neumayer (2000), and Baer (2006).

21. Indeed, in an earlier version, we weighted capacity slightly higher than responsibility, which was criticized by Page (2008) for having no substantial justification for the choice of weights. Further exploration of this question and the relevant principles and precedents remains a substantial future project.

22. Note that this idea is no longer completely far-fetched; Mexico introduced exactly such a proposal into the UNFCCC negotiations in 2008. See http://unfccc.int/files/kyoto_protocol/application/pdf/submission_mexico.pdf.

References

Athanasiou, T., and P. Baer. 2002. *Dead Heat: Global Justice and Global Warming*. New York: Seven Stories Press.

Baer, P. 2002. "Equity, Greenhouse Gas Emissions and Global Common Resources." In *Climate Change Policy: A Survey,* ed. S. H. Schneider, A. Rosencranz and J. O. Niles (Washington, D.C.: Island Press), pp. 393–408.

———. 2006. "Adaptation to Climate Change: Who Pays Whom?" In *Fairness in Adaptation to Climate Change,* ed. W. N. Adger, J. Paavola, S. Huq, and M. J. Mace (Cambridge, Mass.: MIT Press), pp. 131–153. Also chapter 14 in this volume.

Baer, P., T. Athanasiou, S. Kartha, and E. Kemp-Benedict. 2008. "The Greenhouse Development Rights Framework: The Right to Development in a Climate Constrained World," 2d ed. November. Available at http://www.greenhousedevelopmentrights.org.

Bode, S. 2004. "Equal per Capita Emissions over Time: A Proposal to Combine Responsibilty and Equity of Rights for Post 2012 GHG Emission Entitlement Allocation." *European Environment* 14: 300–316.

Brown, D. A. 2002. *American Heat: Ethical Problems with the United States' Response to Global Warming*. Lanham, Md.: Rowman & Littlefield.

Chakravarty, S., A. Chikkatur, H. de Coninck, S. Pacala, R. Socolow, and M. Tavoni. 2009. "Sharing global CO_2 emission reductions among one billion high emitters." *Proceedings of the National Academy of Sciences of the United States of America* 106: 11885–11888.

Climate Protection Programme. 2004. South-North Dialogue on Equity in the Greenhouse. Eschborn: Deutsche Gesellschaft für Technische Zusammenarbeit (GTZ). http://www.wupperinst.org/uploads/tx_wiprojekt/1085_proposal.pdf.

Criqui, P., and N. Kouvaritakis. 2000. "World Energy Projections to 2030." *International Journal of Global Energy Issues* 14.1–4: 116–36.

Den Elzen, M., M. Meinshausen, and D. van Vuuren (2007). Multi-gas emission envelopes to meet greenhouse gas concentration targets: Costs versus certainty of limiting temperature increase. *Global Environmental Change-Human and Policy Dimensions* 17: 260–280.

Frank, A. G. 1972. *Lumpenbourgeoisie: Lumpendevelopment*. New York: Monthly Review Press.

Gardiner, S. M. 2004. "Ethics and Global Climate Change." *Ethics* 114.3: 555–600. Also chapter 1 in this volume.

Grubb, M. 1989. *The Greenhouse Effect: Negotiating Targets*. London: Royal Institute of International Affairs.

———. 1995. "Seeking Fair Weather: Ethics and the International Debate on Climate Change." *International Affairs* 71.3: 463–496.

Höhne, N., M. den Elzen, and M. Weiss. 2006. "Common but Differentiated Convergence (CDC): A New Conceptual Approach to Long-term Climate Policy." *Climate Policy* 6: 181–199.

International Energy Agency (2007). *World Energy Outlook 2007: China and India Insights*. Paris: International Energy Agency.

Jamieson, D. 2001. "Climate Change and Global Environmental Justice." In *Changing the Atmosphere: Expert Knowledge and Environmental Governance,* ed. C. A. Miller and P. N. Edwards (Cambridge, Mass.: MIT Press), pp. 287–308.

Lopez, J. H., and L. Servén. 2006. "A Normal Relationship? Poverty, Growth and Inequality." World Bank Policy Research Paper 3814.

Moellendorf, D. 2008. "Comments on 'Greenhouse Development Rights.'" Princeton University, Princeton, N.J., November 25.

———. 2009. *Global Inequality Matters*. Basingstroke, U.K.: Palgrave Macmillan.

Müller, B. 1999. *Justice in Global Warming Negotiations: How to Achieve a Procedurally*

Fair Compromise, 2d rev. ed. (Oxford: Oxford Institute for Energy Studies).

———. 2000. "Is 'Per Capita Convergence' a Viably Fair Notion for the Developing World?" Personal communication, June 8, 2000.

Neumayer, E. 2000. "In Defence of Historical Accountability for Greenhouse Gas Emissions." *Ecological Economics* 33.2: 185–192.

Oxfam 2007. "Adapting to Climate Change: What's Needed in Poor Countries, and Who Should Pay." Briefing Paper 104, May 29. http://www.oxfam.org/en/policy/briefingpapers/bp104_climate_change_0705.

Page, E. A. 2008. "Distributing the Burdens of Climate Change." *Environmental Politics* 17: 556–575.

Paterson, M. 1996. "International Justice and Global Warming." In *The Ethical Dimensions of Global Change,* ed. B. Holden (New York: St. Martin's Press), pp. 181–201.

Peters, G. P., and E. G. Hertwich. 2008. "CO_2 Embodied in International Trade with Implications for Global Climate Policy." *Environmental Science & Technology* 42: 1401–1407.

Pritchett, L. 2006. "Who Is Not Poor? Dreaming of a World Truly Free of Poverty." *World Bank Research Observer* 21: 1–23.

Rose, A., B. Stevens, J. Edmonds, and M. Wise. 1998. "International equity and differentiation in global warming policy—An application to tradeable emission permits." *Environmental & Resource Economics* 12: 25–51.

Shah, A. 2009. "Poverty Facts and Stats." *Global Issues.* Available at http://www.globalissues.org/article/26/poverty-facts-and-stats. Accessed 01/22/2010.

Shue, H. 1992. "The Unavoidability of Justice." in *The International Politics of the Environment,* ed. A. Hurrell and B. Kingsbury (Oxford, Clarendon Press), pp. 373–397.

———. 1999. "Global Environment and International Inequality." *International Affairs* 75.3: 531–545. Also chapter 5 in this volume.

Singer, P. 2002. *One World: The Ethics of Globalization.* New Haven, Conn.: Yale University Press. Also chapter 10 in this volume.

Traxler, M. 2002. "Fair Chore Division for Climate Change." *Social Theory and Practice* 28: 101–134.

Selling Environmental Indulgences

Robert E. Goodin

According to a common and currently influential diagnosis, the environmental crisis has essentially economic roots. The problem is not just that there are too many people, or even that they are on average enjoying too high a standard of living. All that is true, too, of course. More fundamentally, however, problems of environmental despoliation are said to derive from skewed incentives facing agents as they pursue their various goals.

For some things, people must pay full price. For others, they pay only partially or indirectly or belatedly. To an economist, it goes without saying that the lower the costs, the more people will consume of any particular commodity. Where some of the costs of their activities will be borne by others, agents looking only to their own balance sheets will overengage in those activities. Because some of the costs are "external" (which is to say, are borne by others, rather than themselves), agents will undertake more of those activities than they would have done had they been forced to pay their full costs. They will do more of them than is socially optimal, taking due account of costs and benefits to everyone concerned (Pigou 1932).

Environmental despoliation poses problems of economic externalities of just that sort. Environmental inputs are typically "common property resources." Clean air and water, fisheries, the ozone layer, the climate are everyone's business—and no one's. No one "owns" those things. There is no one with standing to sue you if you take them without paying; nor is there anyone you could pay for permission to impinge on them, even if you wanted to do so. That fact inevitably gives rise to a divergence between the full social costs created by your actions and the portion of those costs sheeted back to you as private costs, to be entered on your own ledger. It is, of course, only the latter sorts of costs to which economically rational agents can be expected to respond (Freeman et al. 1973; Fisher 1981; Pearce et al. 1989, esp. p. 5).

Either of two prescriptions might follow from that economistic diagnosis of the environmental problem. Both would put government in control of—cast it in the role of "owner" of—common property resources. Both vest in government the power to authorize the use of environmental resources, and to punish people for using them without authorization. The two prescriptions differ, principally, over the form that those authorizations and punishments would take.

The standard "legalistic" approach operates by manipulating rights and duties. It is essentially a command-and-control strategy, specifying what people may or must or must not do and attaching penalties to violation of those commands. The newer "economistic" approach works by manipulating incentives. In the limiting case, nothing is required or prohibited: everything just has a higher or lower price; and so long as you are willing to pay the price, you are perfectly welcome to do just as you please. Any actual control system may well combine both modes, of course, but for analytic purposes it behooves us to treat them separately.

The most dramatic form of the economistic strategy is to sell transferable permits to pollute, which permit holders can then resell to others in turn.[1] Imposing "green taxes," conceived essentially as charges for using the environment in certain ways, constitutes a less dramatic and politically more acceptable form of the same basic strategy. In that form, economistic logic attracts the endorsement of a surprisingly wide range of political players: from, on the one side, the OECD (1975; Opschoor and Vos 1988), national treasuries and their advisers (Pearce et al. 1989, chap. 7), and economic think tanks (Epstein and Gupta 1990; Weimer 1990); to, on the other side, various Green Parties across Europe (Die Grunen 1983, sec. IV.1; European Greens 1989, sec. 1; Spretnak and Capra 1986). The "carbon tax" in particular is now the instrument of choice among the widest possible range of policy makers for controlling emissions of greenhouse gases.[2]

In all variations on that economistic strategy, the highest aspiration is to set the price of licenses/permits/fees/taxes at a rate that would force polluters to internalize, in their own cost calculations, the full measure of environmental damage that they do. Of course, calculating that price will never be an easy; many of the complaints with these economistic strategies will amount to little more than the (often perfectly proper) complaint that the price has been set too low (Pearce et al. 1989, chaps. 3, 4, 7). But once we have calculated total social costs correctly, and once we have forced the creators of those costs to internalize them fully, then objections to environmental despoliation should (on the economistic diagnosis of the problem) cease. Once despoilers have been made to repay fully the environmental costs of their activities, there would be no further reason to stop them from proceeding with those activities.

From an economic point of view, that case for "green taxes" seems well nigh indisputable.[3] Environmental economists are therefore frankly dumbfounded when such "unassailable" proposals nonetheless come under attack from fellow environmentalists. The latter, in turn, have proven particularly inept at articulating exactly what they see wrong with green taxes, though. The exchange amounts to a veritable dialogue of the deaf (Kelman 1983, 1985; Frey 1986).

This chapter is thus devoted essentially to bridging a gap within environmentalist discourse. While fully acknowledging all the advantages that environmental economists see in green taxes, I hope to explain in terms congenial to them what other environmentalists have against them. The running analogy that will figure centrally in my discussion is that between green taxes and medieval indulgences. The former amount, in effect, to "selling rights to destroy nature"; the latter amount, in effect, to "selling God's grace."

This analogy, like all analogies, is far from perfect in various respects. Perhaps the most obvious and important point of disanalogy is just this. There is a general ban on sin: a sin is always wrong. Environmental emissions, in contrast, are not always necessarily thought to be wrong. There is, or it is generally thought there should be, no general ban on them. (On the contrary, green taxes and such are offered as *alternatives* to bans.) Religious indulgences are set against the background assumption of a prohibition on sin, the function of indulgences there being to forgive sinners their lapses. Environmental indulgences seem to be set against the background assumption that some pollution will be permitted, the function of environmental indulgences being to allocate those permissions to particular people.

Equating environmental pollution with sin seems to suggest that zero emissions should be our ideal goal. To most practical people, that seems plainly crazy—just the sort of thing that gives philosophy in general a bad name. But it pays to pause to reflect exactly why we think that a zero-emissions standard is so plainly crazy. In part, that is merely because a zero-emissions standard seems unrealistic (so, too, are the Ten Commandments, but they are nonetheless attractive as ideal standards for that). In part, it is because some emissions—ones in certain circumstances, or below certain levels—actually do no harm. The implication there is merely that it is environmental despoliation (an outcome) rather than environmental emissions (an act) that should be counted as the sin. Finally, even some genuinely despoiling emissions seem misdescribed as sins because, though harmful, they cause harm unavoidably and in the service of some greater good. The implication there, however, is not that certain genuinely despoiling emissions are perfectly all right. The implication is, instead, that we are there operating in the realm of "tragedy": even if we have done "the right thing on balance," we will nonetheless have committed a wrong.[4]

These are only preliminary remarks, designed not so much to motivate the analogy as to defuse any strong initial sense of disanalogy, between religious indulgences and environmental ones. Arguments of a more positive sort for treating (certain sorts of) environmental despoliation as akin to a sin will be offered in section II below.[5] There it will be shown that, while less than perfect, the analogy is closer that one might initially suppose. Despite the various points of (often important) disanalogy, it nonetheless remains a telling way into this troubled debate.

The upshot of those arguments is to dash the highest hopes of economistic advocates of green taxes. As I hope will be clear from those discussions, we may not legitimately use green taxes and cognate economistic mechanisms as optimizing devices in directly guiding "policy choice." Such techniques may nonetheless retain a secondary use as tools of "policy enforcement". There, they would be serving merely to provide incentives and disincentives for people to achieve certain "target" levels of maximum permissible environmental damage—levels that have been set elsewhere, by other means, in the political system.[6] That fallback position has much to be said for it. But as shown in section III, that more modest case for green taxes must be sharply distinguished from the other, for it amounts to falling back a very long way indeed from those bolder claims often made on behalf of green taxes.

I. Religious Indulgences: A Potted History

The function of indulgences, in Catholic theology, is to remit time to be served by a sinner in purgatory. Indulgences were granted *by* church officials (originally popes, latterly bishops). They were granted *to* those who had sinned (by definition the only ones in need of them).

The practice of *granting* such indulgences goes back to the early history of the Church. The practice of *selling* them can be traced, fairly precisely, to the need of popes to provide incentives for Crusades—in the first instance for people to participate in them, in the second instance for people to pay for them (Purcell 1975). From the eve of the Third Crusade in 1187 to the Council of Trent which finally abolished the practice in 1563, selling indulgences became an increasingly common phenomenon. Indulgences were increasingly awarded in exchange for assistance, of an increasingly crassly material sort, rendered to the Church and, increasingly, its temporal allies (Boudinhon 1940).

Increasingly, in turn, the practice became the subject of controversy among theologians of all stripes. Notable critics included Jan Hus, who in 1412 crossed King Wenceslas on the matter (Boudinhon 1940). Most famous of all was Martin Luther ([1517] 1963), whose "Ninety-five Theses" nailed to the door of the

Wittenberg Cathedral were largely devoted to an attack on the practice.

This is no place for a detailed examination of either the history or the theology of the matter, though. (On that, see Eliade 1987.) Present purposes will be better served by a more stylized account of generic sorts of possible objections to the sale of indulgences. As is only to be expected, these generic styles of objections track actual Church history only very imperfectly. But this being an exercise in moral philosophy rather than in theology, still less in Church history, that is just as it should be.

II. Grounds for Objecting to the Sale of Indulgences

Surveying the many possible grounds for objecting to the sale of indulgences for sin in religious affairs, surprisingly many of them might apply, *mutatis mutandis*, to the sale of indulgences (in the form of "green taxes" or "pollution permits") for activities degrading the natural environment.

Many environmentalists, of course, would take a vaguely spiritual attitude toward nature (Spretnak 1986). For them, the analogy between the sacrilege of selling nature's benefice and that of selling God's grace might be felt particularly powerfully. It would be wrong, however, to think that this analogy literally works only by implicitly or explicitly giving environmental values a spiritual twist.

1. Selling What Is Not Yours to Sell

One of the recurring themes in opposition to the selling of religious indulgences, even by popes, was that they were selling what was not truly theirs to sell. The item on auction was God's grace, his forgiveness. When it comes to grace and forgiveness, what is at issue is not God's commandments (which popes are indeed empowered to interpret) but rather the exercise of his discretionary powers. Those are for him alone to exercise. It is simply presumptuous—preempting prerogatives properly reserved to him—for others, however high their churchly station, to act on his behalf.[7]

There are important elements of this sort of logic at work within objections to selling environmental indulgences. Those elements figure particularly importantly in the objections of those who take a vaguely mystical view of nature, of course. It is not our place to grant (much less to sell) indulgences for violations of what, on this view, would be regarded as almost literally Mother Nature's physical integrity. It would be simply presumptuous of any human agents to grant indulgences on behalf of Mother Nature. Forgiveness is the prerogative of the party who has been wronged.

There is no need to give environmental ethics a spiritual twist to find an echo of this objection to the selling of indulgences, however. Many, for example, suppose merely that we have "stewardship" responsibilities—either toward nature or perhaps just toward future generations and their interests in the natural environment (Passmore 1980, chaps. 4, 5; Goodin 1985, pp. 169–186; Barry 1989, chaps. 17–19; Sax 1970). It would be objectionable for such stewards to sell environmental indulgences in much the same way, and for much the same reason. They, too, would be selling something that is not theirs to sell. Stewards would then be permitting people to destroy irrevocably that which those stewards are duty-bound to preserve, either for its own sake or for the sake of future generations.

2. Selling That Which Cannot Be Sold

The objection just canvassed deals in terms of a breach of stewardship responsibilities per se. As such, it applies with equal force whatever the reasons for allowing those responsibilities to be breached or whatever form the breach takes. Stewards are bound to protect that which indulgers would allow to be destroyed. Hence granting indulgences, for whatever reason, would seem equally illegitimate.

There is, however, a variation on that objection that applies with peculiar force to the *sale* of indulgences—to the indulging of wrongful behavior for reason of money. The objection there is not (or not just) that the impermissible is permitted. It is instead that the impermissible is permitted for a peculiarly sordid (pecuniary) motive. The objection is to the sale of the unsaleable, more than (and, indeed, often instead of) to the permitting of the impermissible.

The spiritual analogy is again illuminating here. It is not unreasonable to suppose, someone like Luther might say, that God forgives people their sins. It is not even unreasonable for those versed in God's words and his ways to second-guess (in a way that is, of course, utterly nonbinding on him) the circumstances in which he might do so. What *is* unreasonable, however, is to suppose that God's grace can be bought. What counts with him is the purity of the heart, not the size of the purse (Luther [1517], prop. 27).

By the same token in the environmental case, it might be thought that there are indeed circumstances in which it is perfectly proper for the environment to be despoiled. Suppose that were the only way of securing a decent life (or, indeed, life at all) for a great many people who would otherwise lead miserable lives or face even more miserable deaths. Then chopping down large portions of the Amazonian rain forest might well be forgivable, if nonetheless unfortunate. But what makes it forgivable has nothing to do with (or, as in the case here sketched, may even be negatively related to) the size of the purse of those chopping down the forests. Certainly, permission to chop down the forests should not be publicly auctioned to the highest bidder, any more than should remission of time in purgatory for sins committed.

A religious indulgence is granted upon condition of the indulged feeling true contrition for their sins. The environmental indulgence may be granted, by the same token, upon condition of the indulged showing that they have no other choice and that they have made good-faith (albeit unsuccessful) efforts to avoid damage to the environment.[8] The objection here in view is not to conditionality as such, but rather to making the granting of the indulgence conditional upon payment of hard, cold cash. God may grant his favors freely and simply; but God cannot be bought. By the same token, we might forgive people who despoil the environment for certain sorts of reasons—but the pursuit of pure profit (as represented in "willingness to pay" green taxes) is not one of them.

Why that should be so is an open question that admits of various different styles of answer. One might have to do with distributive justice. We might suppose that the present distribution of cash holdings is without justification or that it is positively unjustified. For that reason, we might be reluctant to let one person's environmental quality be determined, in part, by another's unwarranted riches.

Alternatively, the argument might work in terms of "blocked exchanges." We may think that, even if the distribution of cash is morally unexceptionable, there are nonetheless certain things that money ought not be able to buy. Why the category should exist at all is, perhaps, philosophically mysterious; what falls into it certainly is sociologically variable (Simmel [1907] 1978; Tobin 1970; Douglas and Isherwood 1979; Walzer 1983). Still, that the category exists seems both sociologically undeniable and ethically (not just ethnographically) interesting.

One way of justifying the category—and of rationalizing much of its sociological content—is this. It is a clear affront to practical reason to engage in an exchange that secures you cash only at the cost of depriving you of the material and nonmaterial prerequisites for making use of that cash. So selling yourself into slavery is wrong (irrational) because once a slave, you will no longer have the legal capacity to dispose of the money thereby acquired. By the same token, trading all your foodstuffs for money is self-defeating insofar as without sustenance you will not survive to spend the money.

Perhaps the objection to trading environmental quality for money derives from a similar thought. If you trade away (all) the environmental prerequisites for human existence, then the money acquired in exchange for that will do you no good; the trade makes no sense, at least in that limiting case. Perhaps, by extension, it makes little sense in many other much less extreme cases for something of the same sorts of reasons. Private affluence, of certain sorts, anyway, may simply be pragmatically impossible to enjoy, under circumstances of sufficiently severe public squalor (Barry 1989, chap. 20).

Another way of justifying a category of things that ought not be bought and sold is in terms of the corruption of public morals. There is a well-known tendency, firmly established in the literature of empirical social psychology, for extrinsic rewards to drive out otherwise strong intrinsic motivations to perform the same actions. The precise psychological mechanisms at work are many and varied, and the precise nature of the interactions among them is none too clear. What is nonetheless clear is that there are many worthy actions that people would originally have done "for their own sake" but which they will no longer do simply for their own sake once extrinsic material (especially monetary) rewards are also offered for doing them. Putting certain sorts of good deeds on the auction block, so to speak, demeans them and diminishes their intrinsic value in the eyes of those initially most sensitive to such intrinsic moral values (Goodin 1981, 1982, chap. 6; Lane 1991, chaps. 19–21; Frey 1986, pp. 552–556, 1992, 1993).

What people value in that very special way is, as I have said, sociologically variable. Still, insofar as any appreciable part of the population does regard the value of the environment in that special way—and there seem to be reasons to think both that they should and that they do (Sagoff 1988)—then there is likely to be an efficiency cost that will potentially offset any efficiency gain in offering material incentives for environmental protection. Just as in Titmuss's (1971) famous case of blood donation, so, too, with the "free" supply of voluntary environmental protection: it is likely to dry up the more we pay people (those same people or others) for undertaking the same or similar actions.

3. Rendering Wrongs Right

Environmentalists sometimes say that they have no objection to *fining* despoilers of the environment; their objection is merely to charging, licensing, or taxing them. Economists scratch their heads at that. In terms of corporate balance sheets, there is no important difference between fines, charges, and taxes. In strictly economic terms, exactly the same disincentive is provided by a $100,000 fine as a $100,000 charge as a $100,000 tax on any given activity. To careful watchers of the profit-and-loss statements, it is a distinction without a difference.

To others, however, the difference is very real. With a fine, the wrongness remains even after the payment of a fine. It is wrong to have done what you have been fined for doing; you may have "paid your debt to society" and be a member in good standing once again after having done so; but what you did nonetheless remains wrong. Not so with a mere license fee or charge. If you buy a pollution permit, then you are permitted to do what you have paid for permission to do; there is nothing wrong with it. The same is true of a "charge." There is nothing wrong with people dumping wastes in a sanitary landfill, once they have paid the charge for doing so. Similarly, with a "tax," there is nothing wrong with doing most of the things for which we are ordinarily taxed. Quite the contrary, the ordinary activities giving rise to tax liability—like turning a profit or earning a wage—are very much socially approved.[9]

The problem with green taxes or pollution charges or permits, on this model, is that they seem to say, "It is okay to pollute, provided you pay," when the proper message is instead, "It is wrong to pollute, even if you can afford to pay." (The reasons that is the right message are elaborated in section II.6 below.) On the religious analogy, this comes through very strongly. There, an indulgence is forgiveness of sins. The sins clearly remain wrong things to have done. It is the punishment that is being remitted, not the wrongness of the action that is being canceled, by the indulgence.[10]

The bottom line, here, is that putting indulgences up for sale makes them too easy to come by. In the religious case, remission for sin is granted too easily, and consequently sins are taken insufficiently seriously (Luther 1517, prop. 40). Much the same objection applies there as in the case of selling environmental indulgences (Frey 1992, pp. 170–172; see similarly McCarthy 1990). In both cases, the problem with being able to buy your way out of the consequences of a nefarious activity is that anyone with sufficient ready cash is consequently led to take the nefariousness of the activity insufficiently seriously.

4. Making Wrongs All Right

Maybe the point of buying an indulgence is not to make a wrong right but merely to make it all right. Advocates of environmental charges emphasize that we license and tax all sorts of things we vaguely disapprove of, including gambling, smoking, drinking. What payment of the requisite price has done is not to make wrongs right but, rather, to make them "all right"—permissible, if still undesirable in some ideal world.[11]

Classic religious indulgences did something less than that, though. Religious indulgences granted forgiveness for sins past, on condition of penance and a genuine intention not to do it again. Indulgences once granted made it all right to *have* sinned, but that indulgence stopped well short of making it all right to *sin*. The religious formula offered a mechanism for forgiving past wrongs without encouraging future ones. Through religious indulgences, past wrongs were rendered all right, but present or future ones were not.

The whole point (in religious, if not necessarily pragmatic terms) of buying a religious indulgence was backward-looking, to wipe one's slate clean of past sins. The whole point of buying an environmental indulgence is forward-looking, to secure permission to despoil the environment now and in the future. Whereas the religiously indulged are seeking merely forgiveness for things past, the environmentally indulged seek permission for future actions. If buying an environmental indulgence is tantamount to buying a permission to commit a wrong, it is continuing permission (conditional on continuing payment) to commit continuing wrongs.

The reason the wrong remains wrong, even after payment, is simply that the wrong done to the environment and to people using it is not an economic wrong. It is not as if it (or we) are "poorer" for those acts, at least not in any way that can be made good by any transfer of financial resources. Yet while the wrong remains, even after payment of taxes, that wrong is nonetheless permitted on a continuing basis, on continuing payment of the taxes. I return to these themes in section II.6 below.

5. Indulging Some but Not All

In granting indulgences there is a further problem of fairness to confront. Crudely put, it might be thought unfair, somehow, to indulge some but not all sinners. If not all can be (or, anyway, not all will be) indulged, then perhaps it is wrong—unfair—to indulge any at all. And that unfairness might be felt to be especially strong when indulgences are being sold in situations in which some but not all are willing or able to pay the asking price. Less crudely put, it might be thought a matter of elementary fairness that if any sinners are to be indulged, then all with relevantly similar characteristics should be. Of course, not all sinners should be indulged; some are unreconstructed reprobates who really ought to be punished. But all who are in the same boat ought, in fairness, to be treated similarly.[12]

In the religious case, the issue of fairness arguably does not arise. There, indulgences merely reflect God's grace, understood as his purely discretionary whimsy. He can choose to indulge whomsoever he pleases, without a thought for constraints of consistency (although few would be attracted to a vision of so purely capricious a God, perhaps). Insofar as we are making a *social* practice of granting indulgences (environmental or otherwise), however, the practice surely ought to be grounded in principles that are more regular and publicly defensible than that.

The particular problem of fairness arises, in the environmental case, from the fact that we can often afford a few—but only a few—environmental renegades (Kennan 1970). A few countries can continue to hunt whales, for example, without causing the extinction of any species, just so long as not all do. A few countries can continue generating greenhouse gases or emitting CFCs without altering the climate or destroying the ozone layer, just so long as most countries do not. In short, nature can tolerate some but not all misbehaving (Goodin 1995, chap. 18).

In such cases, the question immediately becomes how to choose who gets to play this role of environmental renegade. Advocates of green taxes suggest that these slots should be sold to the highest bidder; others suggest other ways in which this determination might be made (Taylor and Ward 1982). Behind all such schemes, however, is an unspoken assumption that we ought to make sure that all those slots are taken—that we ought to allow just as many renegades as nature itself will tolerate.

Critiques couched in terms of fairness query precisely that proposition. The root idea there is that if we cannot allow everyone to do something, then we ought not to allow anyone to do it. That may not appeal much as a general principle: it seems perfectly reasonable that I should be able to allow some people to share my house without allowing everyone to do so. But that principle seems considerably more apt when it comes to the exploitation of genuinely collective goods: it seems far less reasonable to allow some co-owners of a common property resource to use it in certain ways, without allowing all co-owners to use it similarly.

The impetus to economic efficiency leads us to regard such opportunities to exploit common property resources (by some but not all) as things to be allocated—somehow, to someone. The impetus to fairness leads us to regard such opportunities as things to be eschewed, rather than being allocated at all. Granting environmental indulgences, upon payment of a suitable price, is essentially an allocation device. On the fairness critique, it allocates what ought not to be allocated at all. Those are efficiency gains that, in all fairness, we ought not to pursue.

That rejoinder is not always compelling. Efficiency ought not always be eschewed in the interests of fairness. The fact that we do not have food enough to feed everyone does not mean that we should let such food as we do have go to waste, with the consequence that everyone starves. Rather, we ought to ration scarce necessities in such a way that they do as much good as possible.

Still, in circumstances of rationing, we are characteristically highly sensitive to the precise mechanisms employed for allocating rights to use hyperscarce, necessary resources. We generally want to ensure that the distribution of those rationed commodities is more equal than the distribution of cash holdings or of other commodities in general. That is reflected in the fact that the buying and selling of ration coupons is almost invariably prohibited—and necessarily so, if the ration coupons are to serve their social function as an independent "second" currency, distinct from and restraining on the operation of ordinary economic forces (Tobin 1952, 1970; Neary 1987; Hirshleifer 1987, esp. chap. 1).

6. Grounds for Indulgence

Many of those objections ultimately turn on questions of appropriate (relevant) criteria for granting indulgences. In the religious case, and on one telling of the environmentalist case, it is an inerasable wrong that is being indulged. In the economistic telling of the environmentalist tale, it is merely a previously uncompensated external cost that is being indulged, upon condition of payment of some sum adequate to compensate those who would otherwise have to bear that cost.

In the case of religious indulgences, what is wrong with their being bought and sold is that money payments are the wrong basis for granting them. It is not how much people pay, but rather their regret for what they have done, that is there relevant. Not every penitent can afford to pay, nor are all those who can afford to pay truly penitent. Granting indulgences only to those who pay (or even to all those who pay) would result in a maldistribution of indulgences, by the only standard that is really relevant there.

What ought to be the relevant standards for granting indulgences depends upon the nature of the wrong being indulged, though. The salient feature of sin, in this regard, is that it can never be undone. It is a blot that can never be erased or wiped clean. So the most we can be looking for, in deciding whether to indulge any particular sinner, is a genuinely penitent attitude: sincere regret and a deep commitment not to sin again.

On one account, the wrong done in despoiling the environment is just like that: a presumptuous intrusion into, and destruction of, the creation of another's hand. On that view, it would be wrong for the same reasons to grant environmental indulgences to all, or to any, who were prepared to pay for them. There, as in the religious case, what we should be looking for in granting indulgences is genuine remorse and a firm commitment not to harm nature again. There, as in the religious case, granting indulgences in return for monetary payment would be to grant them for wrong (anyway, irrelevant) reasons. Worse, it would encourage the continued wronging of nature, since knowing you can always buy your way out of trouble tempts you to do it again.

Others, environmental economists conspicuously among them, take a different view of the nature of the wrong done by despoiling the environment. They would say that the wrong is an economic wrong. The wrong is the destruction, or diminution, of a collective good. It is a cost that one person's activities impose upon everyone sharing in those collective goods. Furthermore, environmental economists tend to conceive of that harm as a cost or "welfare loss" which can in principle be recompensed.

That conjunction of attitudes carries important implications for one's view of the power of environmental indulgences to rectify wrongs. If environmental despoilers can and do fully compensate others for the harm that they have done them, then that on this view wipes the slate clean. On that view of the sort of wrong done by environmental despoilers, ability or willingness to pay for indulgences would indeed be a relevant criterion for allocating indulgences. By paying the price, despoilers would—quite literally—have undone the wrong.

Clearly, there are some deep issues at stake in deciding between these interpretations of the wrongs done by environmental despoilers. For those who view the wrong as being done to nature, righting the wrong requires recompensing nature, somehow; and if it is naturalness that is of value, most obvious forms of recompense are not viable—or anyway are not as valuable—options (Goodin 1992, pp. 26–41). Only those who are prepared to view the wrong as one done to other people might conceivably regard paying them a suitable price in exchange for an indulgence as suitable recompense.

It is important to note, however, that not everyone who views the wrong as one done to other people would necessarily regard paying the right price as suitable recompense. That is to say, this difference of opinion does not map easily onto the difference—easily relegated to the "too hard" basket—between deep and shallow ecologists (Devall and Sessions 1985; Sylvan 1985). Even anthropocentric analysts might regard cash transfers, of the sort entailed in buying indulgences through payment of green taxes and such, as inadequate recompense for environmental harms. Anthropocentric analysts might, for example, have a more nuanced notion of human interests, such that people cannot be compensated for losses in one category (e.g., environmental quality) by gains in another (e.g., money, or even any of the things that money can buy) (Goodin 1989, 1995, chap. 11).

My own view of the value of nature is very much like that. The value of natural processes is to provide a context, outside of ourselves (individually, or even collectively), in which to set our lives (see similarly Hill 1983). What is wrong with environmental despoliation is that it deprives us of that context; it makes the external world more and more one of our own (perverse) creation. That is ultimately a wrong to humans, rather than to nature as such, to be sure. It is, nonetheless, a wrong that cannot be recompensed by cash payments. The humans wronged by such practices might be made better off in some sense or another by such payments. But they will be better off, if at all, in dimensions altogether different from those in which their losses have been sustained. The cash offered in payment for environmental indulgences—through green taxes and such—cannot possibly recompense them for the loss of that context that provided meaning, of a sort, to their lives (Goodin 1992, pp. 41–54).

III. Economistic Backtracking

From an economist's point of view, making environmental despoilers pay for indulgences might serve two quite distinct functions. The first and more modest function is one of "policy enforcement." The idea here is to use green taxes simply to provide a disincentive for despoiling. The higher the charges are, for whatever reason, the greater the disincentive effect; that is the end of the story (Baumol and Oates 1971). How much despoliation we want to tolerate, and how much we want to deter, is a matter for determination by other noneconomistic means (by politicians, theologians, or moralists).

I shall return to that more modest version of the tale shortly. First, however, let us consider the more ambitious function that might be served by selling environmental indulgences—one of actual "policy choice." The aim here would be to use green taxes and associated economistic techniques to determine "optimal" levels of despoliation. This second argument subsumes the first; the whole idea is to provide an incentive for (certain) would-be despoilers to desist. But this second argument transcends the first, in acknowledging that certain despoilers ought to be allowed to persist and in providing some mechanism for determining who they should be and how much they should be allowed to despoil.

For purposes of this more ambitious argument, we are required to make the following assumptions:

(1) the price of indulgences fully reflects social costs of the activity;
(2) the activity occurs only upon payment of that price; and
(3) that payment is actually used to compensate or correct for the harm done.

Under those assumptions, environmental despoliation would be a socially optimal activity which actually ought to be engaged in by anyone who can afford to pay for the indulgence out of the proceeds of that despoliation. The sense in which it would be optimal is the weak and unexceptionable Paretian sense of no one being worse off, thanks to the compensation in clause (3) above, and at least one person being better off (that is, despoilers who want to persist even after having had to pay the price of the indulgence).[13]

It is, of course, the latter, more ambitious defense of the sale of environmental indulgences that is more attractive to defenders of the economistic faith. It is the self-same defense that is the greater anathema to their detractors. What is problematic, in particular, is the presumption that money payments can ever correct or compensate for environmental despoliation.

For those who attach great importance to environmental integrity, "correcting" environmental despoliation is simply not a feasible option. No doubt, even for them, restoration and reclamation might be preferable to letting a despoiled bit of nature remain utterly despoiled. But for those who attach great importance to authenticity, the process by which a bit of nature came to be as it is matters greatly; and restored or reclaimed bits of nature, however effective the restoration or reclamation, will necessarily be of less value than they would have been had they never been despoiled in the first place. The reason, quite simply, is that they will have come to be as they are in part through artificial human interventions rather than through more purely natural processes.[14]

Neither, many would say, can cash transfers of the sort received in payment for environmental indulgences compensate for the harms involved in environmental despoliation. If, as just argued, they cannot be used to correct the damage, they could "compensate" if at all only by making people better off in some other respect altogether. Their environment might be worse, but their wine cellar better; and, on balance, they think themselves better off in consequence. Surely it is true that overall well-being is a composite of roughly that sort; and surely some of its components are tradeable, at the margins, for one another in just that way. The question is whether environmental quality is of that character. On at least some of the arguments canvassed above, it is not; it is more fundamental; it is a precondition for valuing, rather than merely a source of values which can be set alongside and traded against other values. If so, neither money nor anything that money can buy can compensate for its loss.[15]

Of course, there are also a great many practical difficulties in calculating (or, rather, in defending any particular calculation of) the cash value of environmental quality. More in deference to those practicalities than in deference to any matters of high principle, environmental economists are sometimes prepared (and governments are often keen) to fall back onto the first "policy-enforcement" defense of green taxes alone.[16]

This fallback position amounts to using a "market-based incentive system to meet preordained environmental quality standards" (Pearce et al. 1989, p. 165, after Baumol and Oates 1971). The basic idea goes something like this. Let there be some independent social determination of the environmental standards that we want to attain. Let those be given by the political process, rather than by any economistic calculation of "social cost" or "optimal" despoliation. Let us merely use the price system to enforce that standard, floating the price up or down until the desired level of environmental quality has been achieved.

There is, on this model, no independent justification of the particular price charged. It is all just a matter of what it takes to get people to cut back on their activities sufficiently to achieve our environmental targets.[17] While there is nothing special about the particular price being charged, however, there is nonetheless a good economistic reason to use price mechanisms to enforce those standards. The rationale is just that, insofar as the standards can be attained in ways that admit of partial noncompliance, pricing mechanisms evoke compliance from those whom economists would regard as the "right" people—those who gain relatively less from environmental despoliation or whom it costs relatively less to desist from it.

This fallback position effectively insulates economists against the criticism that in selling environmental indulgences, they are auctioning off nature's bounty too cheaply. If too much environmental despoliation is occurring, they would say, then that can only be for either of two reasons: either the price has been set too low to achieve the desired standard (and advocates of green taxes and such would be the first to agree that it should be raised as high as necessary to achieve that goal), or else the standard has been set too low (which is the fault of politicians and ought not be taken as criticism of the price mechanism as a way of securing compliance with the standard). Either way, the complaint seems not to touch the practice of selling environmental indulgences, as such.

Of course, efficiency gains from using the pricing mechanism even in this minimal way arise only in very particular circumstances. They presuppose that we can afford to tolerate some people, but not all people, acting as environmental despoilers. Sometimes, though, the situation is such that we cannot—or cannot be sufficiently confident that we can—afford any slackers at all. (Whaling negotiations are often like that: we do not know just how close we are to the limits of a successful breeding population, and given the real risks of destroying the whole species, we do not want to take any chances.) Other times, considerations of fairness of the sort discussed in section II.5 above would lead us to say that, purely as a matter of principle, we should not tolerate any slackers even if practicalities would allow. For reasons either of practicality or principle, we might thus set the desired standard at zero despoliation. And if that is the goal, there is no advantage to pursuing it through the price mechanism.

The more fundamental point to be made here, though, is that in retreating to this fallback position, environmental economists really have given away their strongest claims on behalf of green taxes. Their proudest boast was that the buying of an environmental indulgence made despoliation not merely all right but actually right—socially optimal. But that boast was predicated on the assumption that the price was right, that it was a true reflection of the full social costs of environmental despoliation. If there is no social-cost-based rationale for the particular price being set for environmental indulgences, then their sale cannot perform that role of serving as a solvent turning wrongs into rights.[18]

IV. Conclusions

How attractive we find green taxes and the "polluter pays" principle more generally depends, in large part, upon what we see as their alternative. If, realistically, the alternative is polluters not paying, then the "polluter pays" principle looks to be the relatively more restrictive option. Most of us would probably prefer a regime in which polluters at least are made to pay something—however inadequate that sum (or any sum) might be—if the alternative realistically in view was that otherwise they be allowed to continue polluting with gay abandon.

Suppose, however, the alternative in view was instead that polluters desist from polluting altogether. Then a rule that the "polluter pays" looks to be the relatively more permissive option. If absent the option to pay the alternative is that people not pollute, then giving them permission to pollute upon payment is actually a mechanism for allowing more pollution than would otherwise occur. Seen in that light, many of us may well hesitate to endorse the "polluter pays" principle that, in that other light, looked relatively attractive.

Which is the correct comparison—which, realistically, is the alternative to polluters paying (their not paying or their not polluting)—is essentially a political question. As such, it varies according to time and place, policy arena, and issue area. That in itself is an important lesson. Perhaps it is right that environmentalists should endorse green taxes, in circumstances where stronger prohibitions are not yet in sight. But they ought not turn that into blanket endorsement or an unalterable policy commitment, debarring them from the pursuit of stronger measures should they ever come politically into view.

Therein lies, perhaps, a larger lesson for green politics. True, perhaps environmentalists ought to be realists. They ought not go tilting at windmills; they ought not let the best be the enemy of the good; they ought to get what they can, here and now, rather than holding out in all-or-nothing fashion when doing so only guarantees that nothing will be achieved. Be all that as it may, it is nonetheless equally true that environmentalists ought not be so sensitive to current political realities as to render them insensitive to shifting political realities. Shifting alliances and provisional policy commitments—regarding green taxes—ought to be very much part of the environmentalist's political repertoire.

Acknowledgments

Earlier versions were read at the Universities of Queensland and Melbourne and at various venues around the Australian National University. I am particularly grateful for valuable comments, then and later, from John Dryzek, Patrick Dunleavy, Bruno S. Frey, Rene L. Frey, Daniel Hausman, Max Neutze, Alan Ryan, Rob Sparrow, and Cass Sunstein.

Notes

1. For discussions of such proposals, see Dales 1968, pp. 93–97; Hahn 1982; Hahn and Hester 1987, 1989; Ackerman and Stewart 1988; and Pearce et al. 1989, pp. 165–166). Note that, by setting upper limits on the amount of allowable pollution, these permits are more limiting than charges or taxes, which in principle dictate no such upper limit on the amount of allowable pollution (although in practice, of course, they price it out at some point).

2. Elaborating such proposals, see Epstein and Gupta 1990; Weimer 1990; and Pearce et al. 1989, pp. 165–166. On the political uptake, see Palmer 1992 and Taylor 1992.

3. This is the cumulative conclusion of, e.g., Dales 1968, chap. 6; Kneese and Schultze 1975; Schultze 1977; Schelling 1983; Rhoads 1985, pp. 40–56; Pearce et al. 1989, chap. 7.

4. Poisoning one person so that thousands may live is, by most standards, the obviously right thing to do in the desperately unfortunate circumstances. But there is something obviously wrong with someone who is not even vaguely apologetic to

the bereaved family for the sad necessity of that sacrifice (Nussbaum 1984).

5. See in particular section 11.2 and 6 below. Another tack, unexplored here, is Hill's (1983) observation that certain sorts of character traits hang together, so the environmentally insensitive are likely also to be morally insensitive in ways classically linked to sin.

6. That is precisely the use made of permits under the most familiar instantiation of these techniques, in the U.S. Emissions Trading Program and the Clean Air Act and Amendments of 1990 (U.S. EPA 1986; U.S. Congress 1990). See Hahn and Hester 1989 for discussion of those policies and Ackerman and Steward 1988 for elaboration of the "democratic" roots underlying their rationale.

7. In the Eastern and the older Western Church, "the priest invoked divine forgiveness but could not himself declare the sinner to be absolved" it was only "after the Papal Revolution" in the 11th century that "a new formula was introduced in the West: *Ego te absolvo* ('I absolve you'). This was at first interpreted as the priest's certification of God's action.... In the twelfth century, however, it was interpreted as having a performative, that is, a sacramental as well as a declarative, effect" (Berman 1983, p. 173). Luther's fifth and sixth propositions especially hark back to this older understanding.

8. Some such "good-faith" condition is built into the U.S. offset policy: emissions permits can be sold only by those who have controlled their own emissions more than they are legally required to do; and potential buyers must as a precondition of purchase demonstrate that they have already installed the best available control technology, and they must buy 20 percent more permits than they will actually use (U.S. EPA 1986; Hahn and Hester 1989).

9. Not always; see section II.4 below.

10. Even God might not be able to make wrongs right or bads good. On the so-called Euthyphro argument, his will does not make things good; rather, he wills what he does because it is good independently of his will. In Socrates' formulation. "Is what is holy holy because the gods approve it, or do they approve it because it is holy?" (Plato 1961, sec. 10a, p. 178).

11. This may have to do with limits of criminal sanction. Much that we regard as morally wrong remains legally permissible, because it would be wrong (inappropriate, given the limits of the criminal sanction) literally to outlaw it—so we merely tax it instead.

12. What counts as "the same boat" comes down to a matter of what are the characteristics that would make them "relevantly similar". The question of fairness, posed that way, quickly transforms itself into one of appropriate (relevant) criteria for granting indulgences. That issue is taken up in section 11.6 below.

13. The more standard welfare-economic phrasing of that point would substitute hypothetical compensation for actual in clause (3) above; the test, there, is whether gainers *could* compensate losers (Kaldor 1939; Hicks 1939). Nothing is lost rephrasing the arguments of this chapter in those terms. Nor is anything gained by advocates of optimal despoilation, for the whole point of those opposing such optimization is that certain forms of damage could not even in principle to corrected or compensated by cash payments of any sort.

14. Elliot 1982. Humanity is part of nature, too, of course; but surely we ought not infer from that that any human intervention, however destructive of the rest of creation, is acceptable because it is just part of a natural process. Those intuitions seem firm and clear. How to justify them—and with them, any sharp distinction between human and nonhuman parts of nature—is less straightforward, perhaps. One way is to say that humans derive value from being able to set their lives in some context outside of themselves, either individually or collectively; and for this purpose, it is precisely the nonhuman part of nature that is crucial (Goodin 1992, chap. 2).

15. Even the OECD (1975, p. 28), acknowledges that "direct controls" of a more legalistic, command-and-control sort are preferable to incentives of a "green tax" sort as a "means of preventing *irreversible effects* or *unacceptable pollution* (mercury, cadmium, etc.)."

16. Even the OECD's (1975) *Polluter Pays Principle* proceeds in this way. Its "Guiding Principles" state that "the polluter should bear the expenses of carrying out... measures decided by public authorities to ensure that the environment is in an *acceptable* state" (pp. 12–13; emphasis added). A rather confused "Note" glossing those guidelines elaborates, "The notion of an 'acceptable state' decided by public authorities." The "collective choice" of what is "acceptable" should be made with due regard to comparative social costs of the pollution and of its abatement, but those determinations are to be made politically rather than literally economically; it therefore follows that "the Polluter-Pays Principle is no more than an efficiency principle for allocating costs and does not

(necessarily) involve bringing pollution down to an optimum level of any type" (p. 15). See further Pearce et al. 1989, pp. 157–158; cf. chap. 3.

17. In similar vein, the U.S. Comptroller General (1979) reported to Congress that the then-existing limits on fines that the Nuclear Regulatory Commission could impose on operators of nuclear power plants for safety violations were inadequate deterrents: allowing a maximum penalty of $5,000 for each violation up to a maximum of $25,000 for all violations over a period of 30 consecutive days is a derisory deterrent, when it would cost the operators of the powerplant something on the order of $300,000 to purchase power from the grid every day it is shut down to make repairs.

18. Environmental economists, of course, see themselves retreating to this fallback position purely for reasons of pragmatism—purely because of practical difficulties in calculating costs. The imposition of "standards" which the price mechanism is then used to enforce is nonetheless justified, at root, in terms of social costs, even if they cannot be calculated precisely. Even on this minimal understanding of why the retreat was necessary, however, paying the price still cannot right wrongs. As those environmental economists themselves would be the first to concede (indeed, insist, as a criticism of standard-setting ungrounded in hard economic calculations more generally), standards will only accidentally if at all correspond to what is socially optimal, defined as the level of environmental despoliation that would follow from a proper calculation of social costs. (See Pearce et al. 1989, chap. 7.) Hence paying a price set merely to achieve those standards will only accidentally if at all provide recompense for the damage done. The price is essentially arbitrary, even from the environmental economist's point of view; and paying an *arbitrarily* high price cannot, even from their point of view, serve to right any wrongs.

References

Ackerman, Bruce A., and Richard B. Stewart. 1988. "Reforming Environmental Law: The Democratic Case for Market Incentives." *Columbia Journal of Environmental Law* 13: 171–199.

Barry, Brian. 1989. *Democracy, Power and Justice.* Oxford: Clarendon Press.

Baumol, William J., and Wallace E. Oates. 1971. "The Use of Standards and Prices for Protection of the Environment." *Swedish Journal of Economics* 73: 42–54.

Berman, Harold J. 1983. *Law and Revolution: The Formation of the Western Legal Tradition.* Cambridge, Mass.: Harvard University Press.

Boudinhon, A. 1940. "Indulgences." In *Encyclopedia of Religion and Ethics*, ed. James Hastings (Edinburgh: T. & T. Clark), vol. 7,252–7,255.

Dales, J. H. 1968. *Pollution, Property & Prices.* Toronto: University of Toronto Press.

Devall, Bill, and George Sessions. 1985. *Deep Ecology.* Salt Lake City, Utah: Peregrine Smith Books.

Die Grünen. 1983. *Programme of the German Green Party*, trans. Hans Fernbach. London: Heretic Books.

Douglas, Mary, and Brian Isherwood. 1979. *The World of Goods.* London: Allen Lane.

Eliade, Mirce, ed. 1987. *The Encyclopedia of Religion.* New York: Macmillan.

Elliot, Robert. 1982. "Faking Nature." *Inquiry* 25: 81–94.

Epstein, Joshua M., and Raj Gupta. 1990. *Controlling the Greenhouse Effect: Five Global Regimes Compared.* Washington, D.C.: Brookings Institution.

European Greens. 1989. *Common Statement of the European Greens for the 1989 Elections to the European Parliament.* Brussels: European Greens.

Fisher, Anthony C. 1981. *Resource and Environmental Economics.* Cambridge, U.K.: Cambridge University Press.

Freeman, A. Myrick III, Robert H. Haveman, and Allen V. Kneese. 1973. *The Economics of Environmental Policy.* New York: John Wiley.

Frey, Bruno S. 1986. "Economists Favour the Price System—Who Else Does?" *Kyklos* 39: 537–633.

———. 1992. "Tertium Datur: Pricing, Regulating and Intrinsic Motivation." *Kyklos* 45: 161–184.

———. 1993. "Motivation as a Limit to Pricing." *Journal of Economic Psychology* 14: 635–664.

Goodin, Robert E. 1981. "Making Moral Incentives Pay." *Policy Sciences* 12: 131–145.

———. 1982. *Political Theory and Public Policy.* Chicago: University of Chicago Press.

———. 1985. *Protecting the Vulnerable.* Chicago: University of Chicago Press.

———. 1989. "Theories of Compensation." *Oxford Journal of Legal Studies* 9: 56–75.

———. 1992. *Green Political Theory.* Oxford: Polity Press.

———. 1995. *Utilitarianism as a Public Philosophy*. Cambridge, U.K.: Cambridge University Press.

Hahn, Robert W. 1982. "Marketable Permits: What's All the Fuss About?" *Journal of Public Policy* 2: 395–412.

Hahn, Robert W., and Gordon L. Hester. 1987. "The Market for Bads: EPA's Experience with Emissions Trading." *Regulation* 3/4: 48–53.

———. 1989. "Marketable Permits: Lessons for Theory & Practice." *Ecology Law Quarterly* 16: 361–406.

Hicks, John R. 1939. "The Foundations of Welfare Economics." *Economic Journal* 49: 696–712.

Hill, Thomas E. Jr. 1983. "Ideals of Human Excellence and Preserving Natural Environments." *Environmental Ethics* 5: 211–224.

Hirshleifer, Jack 1987. *Economic Behaviour in Adversity*. Brighton, U.K.: Harvester-Wheatsheaf.

Kaldor, Nicholas. 1939. "Welfare Propositions of Economics and Interpersonal Comparisons of Utility." *Economic Journal* 49: 549–552.

Kelman, Steven. 1981. *What Price Incentives? Economists and the Environment*. Boston: Auburn House.

———. 1983. "Economic Incentives and Environmental Policy: Politics, Ideology and Philosophy." In *Incentives for Environmental Protection*, ed. Thomas C. Schelling (Cambridge, Mass.: MIT Press), pp. 291–332.

Kennan, George F. 1970. "To Prevent a World Wasteland." *Foreign Affairs* 48: 401–413.

Kneese, Allen V., and Charles L. Schultze. 1975. *Pollution, Prices and Public Policy*. Washington, D.C.: Brookings Institution.

Lane, Robert E. 1991. *The Market Experience*. Cambridge, U.K.: Cambridge University Press.

Luther, Martin. 1963. "Ninety-five Theses (1517)." In *Luther's and Zwingli's Propositions for Debate*, trans. Carl S. Meyer (Leiden: E. J. Brill), pp. 3–21.

McCarthy, Eugene J. 1990. "Pollution Absolution." *New Republic* 3: 9.

Neary, J. Peter. 1987. "Rationing." In *The New Palgrave: A Dictionary of Economics* (London: Macmillan), vol. 4, pp. 92–96.

Nussbaum, Martha C. 1984. *The Fragility of Goodness*. Cambridge, U.K.: Cambridge University Press.

Opschoor, J. B., and H. Vos. 1988. *The Application of Economic Instruments for Environmental Protection in OECD Member Countries*. Paris: OECD.

OECD. 1975. *The Polluter Pays Principle*. Paris: OECD.

Palmer, John. 1992. "Community Plans Carbon Fuel Tax." *Guardian Weekly*, vol. 146, no. 21, 11.

Passmore, John. 1980. *Man's Responsibility for Nature*. 2d ed. London: Duckworth.

Pearce, David, Anil Markandya, and Edward B. Barbier. 1989. *A Blueprint for a Green Economy: A Report to the UK Department of the Environment*. London: Earthscan.

Pigou, A. C. 1932. *The Economics of Welfare*, 4th ed. London: Macmillan.

Plato. 1961. *Euthyphro*. In *The Collected Dialogues of Plato*, ed. E. Hamilton and H. Cairns (Princeton, N.J.: Princeton University Press).

Purcell, Maureen. 1975. *Papal Crusading Policy, 1244–1291*. Leiden: E. J. Brill.

Rhoads, Steven E. 1985. *The Economists's View of the World*. Cambridge, U.K.: Cambridge University Press.

Sagoff, Mark. 1988. *The Economy of the Earth*. Cambridge, U.K.: Cambridge University Press.

Sax, Joseph L. 1970. "The Public Trust Doctrine in Natural Resource Law" *Michigan Law Review* 68: 471–566.

Schelling, Thomas C. ed. 1983. *Incentives for Environmental Protection*. Cambridge, Mass.: MIT Press.

Schultze, Charles. 1977. *The Public Use of Private Interest*. Washington, D.C.: Brookings Institution.

Simmel, Georg. 1978. *The Philosophy of Money* trans. T.B. Bottomore and D. Frisby. London: Routledge & Kegan Paul (originally published 1907).

Spretnak, Charlene. 1986. *The Spiritual Dimension of Green Politics*. Santa Fe, N.M.: Bear & Co.

Spretnak, Charlene, and Fritjof Capra. 1986. *Green Politics: The Global Promise*. Santa Fe, N.M.: Bear & Co.

Sylvan, Richard. 1985. "A Critique of Deep Ecology" *Radical Philosophy* 40:2–12 and 41:10–22.

Taylor, Jeffrey. 1992. "Global Market in Pollution Rights Proposed by U.N.," *New York Times*, January 31, C1 and C12.

Taylor, Michael, and Hugh Ward. 1982. "Chickens, Whales and Lumpy Public Goods: Alternative Models of Public-goods Provision." *Political Studies* 30: 350–370.

Titmuss, R. M. 1971. *The Gift Relationship*. London: Allen & Unwin.

Tobin, James 1952. "A Survey of the Theory of Rationing" *Econometrica* 20: 521–553.

———. 1970. "On Limiting the Domain of Inequality." *Journal of Law and Economics* 13: 363–378.

U.S. Comptroller General. 1979. *Higher Penalities Could Deter Violations of Nuclear Regulations.* Washington, D.C.: General Accounting Office.

U.S. Congress. 1990. Clean Air Act, Amendments. Public Law 101–549 (S. 1630), November 15 1990. *Statutes*, Vol. 104, 2399–2712.

U.S. Environmental Protection Agency. 1986. Emissions Trading Policy Statement. *Federal Register*, Vol. 51, 43–814.

Walzer, Michael. 1983. *Spheres of Justice.* Oxford: Martin Robertson.

Weimer, David L. 1990. "An Earmarked Fossil Fuels Tax to Save the Rain Forests." *Journal of Policy Analysis and Management* 9: 254–259.

Adaptation to Climate Change

Who Pays Whom?

Paul Baer

The problem of adaptation to climate change is complex and multifaceted. At its core, however, are two simple questions: what actions should be taken to prevent or reduce harm that will be caused by anthropogenic climate change, and who should pay for those actions that have costs? In this chapter I focus on the latter question, concerning *liability* for the funding of adaptation. I argue that obligations for funding adaptation are based on ethical principles governing just relationships between individuals in a "life-support commons," which are essentially the same as the norms of justice governing other forms of harm. Simply, it is wrong to harm others by abusing a commons, and if one does, one owes compensation.

In this view, ethics and justice address the rights and responsibilities of individuals; obligations between countries are derivative, based on the aggregate characteristics of their populations, and pragmatic, given the existing state system. Furthermore, liability can be disaggregated in other ways; as I argue, it is equally important that the distribution of liability can be differentiated between classes within nations. A simple quantitative exercise applying these principles of justice to the adaptation problem suggests net liability from the North to the South but also net liability for adaptation from wealthy classes in the South.

Adaptation in the Negotiations

The United Nations Framework Convention on Climate Change (UNFCCC) devotes a small but significant amount of attention to adaptation to climate change.[1] Only in the last few years, however, with the creation of the Least Developed Countries (LDC) Fund and the Special Climate Change Fund (SCCF) under the UNFCCC, the creation of an Adaptation Fund under the Kyoto Protocol, as well as the support for the development of National Adaptation Plans of Action (NAPAs), have delegates and advocates begun to focus seriously on the problems of adaptation and adaptation funding.

Given the disproportionate share of current and past emissions from the industrialized countries of the North and the evidence that the developing countries of the South are more vulnerable to climate damages,[2] almost any plausible interpretation of "common but differentiated responsibilities"[3] implies that the North should shoulder the major part of the costs of adaptation. Funding by industrialized nations for the SCCF and the LDC Fund, whose mandates include adaptation-related activities, is a *de facto* recognition of the validity of this argument. However, contributions to those funds are voluntary, small, and not directly tied to any metrics of responsibility.

To the extent that these arrangements represent precedents for adaptation funding, they partially address the problem of common but differentiated responsibility; all the donors to the SCCF and LDC fund are wealthy countries with significant responsibility. However, the funds' voluntary nature allows other countries with equal or greater wealth or responsibility to avoid paying for adaptation, and the Adaptation Fund itself applies only to Kyoto parties, allowing UNFCCC parties which have not ratified Kyoto off the hook completely.

It seems likely that Northern governments are resistant to explicit claims for "polluter pays" liability for adaptation investments because there is a clear link between current responsibility for adaptation and eventual liability for compensation for actual climate damages. Northern governments might reasonably fear that acknowledging such claims would obligate Northern countries to the largest share of a potentially enormous financial liability.

Direct "polluter pays" liability has been avoided so far by pragmatically emphasizing ability to pay rather than responsibility for climate change when it comes to adaptation funding decisions, while continuing to give rhetorical support to the importance of responsibility. The strong correlation between responsibility and capacity (that is, between historical emissions and wealth) has allowed this compromise to justify an initial round of adaptation-related funding. However, reliance on capacity as the ultimate basis for assigning economic burdens implies that the magnitude of funding is primarily determined by limits to capacity, which in practical terms is the same as the limit on willingness to pay. When legitimate claims for adaptation funding—to say nothing of actual damages attributable to climate change—exceed the magnitude of the roughly $1 billion (U.S.) per year that is currently on the table,[4] the questions of ethical and legal liability will come to the forefront. My goal in this chapter is to contribute to this emerging discussion by framing the problem as one of liability for harm in a commons, and showing that there are established principles in ethics and law which should guide the development of the relevant rules and institutions.

The Climate System As a Life-Support Commons

References to "the global commons" and "the atmospheric commons" are fairly frequent these days.[5] Those who use the terms tend to have fairly strong normative perspectives. In much of the social sciences, however, efforts to analyze "commons problems" have developed a specialized analytic language which downplays normative judgment. In this literature on social interdependence, heavily influenced by economics, a unifying focus has been the concept of *externalities,* which broadly refers to processes by which purposive human activities bring benefits or harms to "outsiders" not involved in the decision process.

Scholars have created a variety of typologies of systems in which externalities are important, which can be brought together under the heading of "common goods", "public goods" or "collective action problems" (Cornes and Sandler, 1996; Kaul et al., 1999; Oakerson, 1992). The classic "commons," such as the village-controlled sheep meadow made famous by Garret Hardin (1968), represents only one type, albeit the best known, of such common goods. Analysis of such physical "common-pool resources" has led to a variety of important distinctions between "open access" (unregulated) and "common property" (regulated, whether formally or informally) systems, and to extensive empirical and theoretical work concerning their governance (e.g., Burger et al., 2001, Ostrom, 1990).

However, the problem of adaptation to (and compensation for) climate harm focuses not on sharing the sink capacity of the "atmospheric commons,"[6] but rather on the climate as a system whose modification brings harm through various pathways. Economists and others have generalized this type of common-goods problem as "public bads," of which air and water pollution are prototypical examples. This in turn is a subset of a broader category of "public goods," which are defined as goods (in the broad sense) which are *nonrival*—their consumption by one party does not decrease the value to another party—and *non-excludable*—parties cannot be prevented from making use of them.[7] From this perspective, it is the service of "climate stability" which is identified as a public good.[8]

Prototypical public goods, however, such as street lights and national defense, are generally seen as those that must be actively produced at some cost; thus the "problem" of public goods is funding the "optimal" provision of the goods in a situation where there is a strong incentive to free-ride. Climate stability, and similar "pollutable" goods such as clean air and clean water, are significantly different. They exist prior to human intervention but are degraded by human actions, causing actual, physical harm to other people who depend on them in various ways. To highlight this aspect, I suggest that the climate system, like air and water in the context of toxic pollution, be considered a *life-support commons.*[9]

From an abstract perspective on individual (or corporate) incentives, the traits of a life-support commons are the same as those of public goods more generally—an actor who pollutes a life-support commons can avoid costs while capturing private benefits unless a regulatory system of some kind exists to sanction "free riders". Thus in economic terms, the problems all appear to be the same: balancing the sum of private benefits with the sum of social costs.[10] But the moral structure of the problem is quite different. In particular, in a life-support commons, the question of *the right not to be harmed* comes to the forefront. And it is thus that we move from economics and political science to ethics and law.

Ethics and Law in a Life-Support Commons

The ethical structure of the climate problem is, to a first order, quite simple: deliberate acts that create greenhouse pollution for one party's benefit will inevitably cause some amount of harm to others.[11] Limiting this harm by reducing greenhouse pollution—the primary objective of the UNFCCC—is one response. Another—my focus in this chapter—is establishing responsibility, and thus legal liability, for the harm that will be caused by climate change.

That it is wrong to harm others (or risk harming them) for one's own gain is as close to a universal ethical principle as I am aware of. It is a principle that can be justified in all kinds of ethical or moral frameworks, from divine revelation to deontological ethics to social contract theory and probably many more.[12] It is in fact a prime example of what some philosophers call common (or commonsense) morality (Beauchamp, 2003; Portmore, 2000). And there is a strong corollary principle, which is that if one does such harm to another, one owes compensation (Shue, 1999).

Of course, these principles are not absolute—in practice, there are many other relevant considerations, such as the relative size or importance of the benefit and harm, the cost or difficulty of preventing the harm, or the social relations between the parties. Laws and customs that govern particular people and activities represent the working out of these principles in the real world. One aspect of this dynamic is the creation of *rights*, which can include both the right to carry out an activity and the right to protection from harm from the actions of others. These rights are embodied in criminal law, which establishes penalties for violations of rights to protection from harm to person or property, and in common (civil) law, which allows injured parties to obtain compensation from the responsible parties.

The legal perspective is important because law embodies ethical norms, and because there are critical legal precedents which bear directly on the climate problem. In particular, Western law has developed strategies for dealing with several aspects of the climate problem that are common to harms caused by many kinds of pollution: the separation in time and space of cause and effect, the scientific uncertainty of the causal chain, the collective nature of both the causal agents and the harmed parties, and perhaps most critically, the question of intent and the predictability of subsequent harm at the time that the actions were taken (Brennan, 1993; Penalver, 1998).

One of the most important ways in which modern law has dealt with harm caused by pollution of a life support commons is through the creation of standards of acceptable risk. Environmental laws and regulations which have health-based standards (e.g., the requirement to protect the public health with an "adequate (or ample) margin of safety") are in fact referred to as "rights-based" regulations (Powell, 1999). Those that create a numerical standard for risk[13] create what I call a "statistical right" to protection from environmental harm, and violation of the standards can be subject to criminal penalties similar to other violations of rights to protection from harm to person or property.

Even where criminal sanctions are not in place, harm caused by pollution is subject to *tort law*—civil law allowing harmed parties to obtain compensation from the party causing the harm. Such compensation can be *fault-based* (based on the intent or the negligence of the party causing the harm), but it can also be based on *strict liability* that does not require the finding of fault, requiring instead simply that the product or process of a given enterprise be shown to have caused the harm in question. Case law and legal reasoning in the US concerning environmental issues (so-called "toxic torts") have evolved to address the issues of probabilistic and fractional causation that are directly relevant to harms from climate change (Brennan, 1993).

Fault-based liability has at least two justifications: a utilitarian basis in the social-welfare gains from the deterrent effect and a fairness basis in the "corrective justice" achieved by having the party causing the harm directly reimburse the victim (Perry, 1992). While these two conceptions have some different consequences (for example, punitive damages make sense only under a deterrence-based justification), they are not our main concern here; the important point is simply that there are several ethical principles that coexist in our society that support financial liability for fault-based injuries. Such fault-based liability would clearly apply to damage caused by greenhouse pollution emitted since the time when the risks of anthropogenic climate change were widely recognized.

The moral logic of strict liability is less obvious, and in fact the counter-intuition is fairly reasonable—why should I be responsible for harms I couldn't know I was causing, and thus could not have prevented? This is of course precisely the basis of the argument against liability for climate harm based on historical emissions. However, the positive argument for strict liability is also quite reasonable: if there are unexpected harms from some activities, shouldn't the party that benefited from the actions bear the costs of the harm, rather than the victims (Keating, 1997)? Strict liability has been upheld in a wide variety of tort cases, including environmental cases (Brennan, 1993), and has been codified legislatively in the U.S. in Superfund legislation, which holds polluters liable for the cleanup costs of hazardous waste sites even if the dumped materials were not known at the time to be harmful.[14]

The preceding discussion is meant to establish three main points: that the underlying structure of the climate-change problem, that of regulating polluting behavior in a life-support commons, is well understood; that the ethical norms for behavior in such a commons are fairly clear; and that these ethical norms are well established in national contexts and codified in national law, in part through regulation that may involve creation of "statistical rights" and in part through tort law which includes both fault-based and strict liability. The question that arises is whether these ethical and legal principles should apply across borders in a global life-support commons. My claim, which I won't defend in detail here, is that they should, and indeed that to some extent they do.[15]

International law in fact reflects these ethical norms for governing a life-support commons in its rhetoric (so-called "soft law"—see Dimento, 2003). The Stockholm Declaration of 1972 declares in the famous Principle 21 (reaffirmed in Principle 2 of the Rio Declaration) that states have "the responsibility to ensure that activities within their jurisdiction or control do not cause damage to the environment of other States or of areas beyond the limits of national jurisdiction." The importance of liability is further recognized in Principle 22, which declares that "States shall cooperate to develop further the international law regarding liability and compensation for the victims of pollution and other environmental damage caused by activities within the jurisdiction or control of such States to areas beyond their jurisdiction" (UN Conference on the Human Environment, 1972). And indeed, the UNFCCC itself is an example of the recognition of the ethical principles governing a global life-support commons.

However, there has been essentially no implementation of international law regarding compensation for pollution, and the UNFCCC specifically avoids establishing any legal liability for compensation from anthropogenic climate harm.[16] Given the ethical consensus reflected in soft law, the reasonable conclusion is simply that the economic and political interests of polluters have so far outweighed ethical norms when it comes to the practical establishment of rights in a global commons. This paper is meant to contribute to the arguments for making those norms and rights legally enforceable.

Operationalizing Liability and Responsibility

Creating a framework for liability for climate damages and adaptation funding requires moving from the general discussion of ethical and legal principles to specific definitions and indices. In this section I first address the question of *legitimate claims* (the costs that must be paid) and then the question of *responsibility*.

Defining Legitimate Claims

What would constitute legitimate claims for payments related to current or future climate harms? This question defies simple answers. Costs associated with climate damages can be divided into at least two types: adaptation costs and residual damages (Tol and Verheyen, 2004). Conceptually, residual damages are easier to identify, while in practice the difficulties of attributing particular climate damages (e.g., storm or drought damage) to anthropogenic climate change pose very difficult problems. In comparison, it's difficult to identify conceptually what the appropriate level of investment in adaptation should be, but it's relatively straightforward to identify the investments that are actually made.[17]

There are few precedents in ethical or legal frameworks for proactive and defensive investments that can provide protection from the effects of pollution. There have been some legal settlements which have relocated entire communities when it has been impractical to adequately clean up a toxic hazard.[18] But in part because most pollution operates through biological mechanisms for which there is no obvious means for "adaptation," most legal settlements rely on various forms of cleanup which are analogous to the mitigation side of climate protection.[19]

However, because climate change creates harm primarily through macroscale physical impacts rather than through biological toxicity, there are many obvious actions that can be taken to reduce the risks of harm associated with any given level of anthropogenic climate change.[20] Thus where such risk-reducing actions have costs, it makes sense to make such investments in advance, rather than waiting until the harm has occurred to provide compensation (when some of those requiring compensation may be dead). This is of course the whole point of proactive climate adaptation.

It's important to note that there are two different motivations for such proactive adaptation. One is the premise that there is a right to be protected from harm by the actions of others, which would imply that all adaptation expenses required to reduce risk to an "acceptable" level (to provide a given "statistical right" to protection) are justified.[21] The other is a cost-effectiveness argument: it is cheaper to invest proactively than to allow harm to occur and compensate the victims after the fact (a logic parallel to that embodied in insurance policies that require, for example, the installation of sprinkler systems).

Negotiating an acceptable definition of legitimate adaptation claims will be a difficult exercise, but it will no doubt draw upon both of these justifications for the level of funding. Similarly, it will be difficult to establish a mechanism that balances "polluter pays" responsibility for residual damages with the difficulty of attributing such damages to anthropogenic GHG emissions. For the purposes of this paper, however, what matters is simply that there are ethical bases and legal precedents for negotiating some level of financial liability from polluters to potential and actual victims, which together I call legitimate claims.

Defining Responsibility

Above I argued that causal responsibility and moral responsibility are both recognized as justifications for "polluter pays" liability in a life-support commons. It is not a legally or philosophically complete argument, but I believe its conclusions are robust, and provide the justification for further development of climate law that takes account of historical emissions.

There are many different ways historical emissions could be accounted for in a responsibility index. These considerations have been extensively discussed in debates over using a responsibility index to allocate reduction targets,[22] rather than liability for adaptation or compensation.[23] First, how should we estimate *causal* responsibility? That is, what is the appropriate mapping between emissions and impacts? The causal chain is quite complex and suffused with uncertainties. Greenhouse gases cause increases in radiative forcing, the effects of which depend on the extent and duration of the gases' accumulation in the atmosphere. The remaining atmospheric stock from a country's emissions, which can be estimated using a carbon cycle model, is thus one plausible proxy for causal responsibility. Another possible proxy is the increase in global temperature from the time-integrated increase in radiative forcing.[24]

A second question is how to treat *moral* responsibility. Although I argued that ignorance of harmful effects does not eliminate ethical and legal liability, there is little doubt that it is a relevant factor in considering exactly how liability should be limited. In discussing responsibility for climate change, this has often meant choosing a cutoff date, prior to which emissions are not counted in a responsibility index. The year 1990 is often mentioned as a date beyond which knowledge of the harm from greenhouse-gas (GHG) emissions was no longer plausibly deniable (e.g., Tol and Verheyen, 2004); selecting an earlier cutoff date can be thought of as making a compromise between strict and fault-based liability.

A third question is which gases and sources should be counted. This has practical aspects, such as the availability and reliability of data on the emissions of various greenhouse gases, as well as other more subjective aspects such as choices among methods for comparing various GHGs.[25] How to count emissions from land-use change also raises difficult questions – some countries deforested centuries ago and now have regrowing forests, while others are now extensively clearing land for agriculture, energy and timber. Simple arguments that such emissions should or should not be counted are unlikely to be persuasive.

In addition to these questions, other issues such as the attribution of emissions to importers or exporters, changes in national borders over time, and the distinction between luxury and subsistence emissions, all bear on the question of an appropriate index of responsibility. But perhaps the single greatest issue is the imperfect correlation between responsibility and capacity (ability to pay)—that is to say, between historical emissions and current wealth. Some countries with high cumulative per capita emissions have remained (or become) relatively poor; some countries have become wealthy with relatively low emissions.

There are many ways in which an index of responsibility could be modified to account for capacity; for example, a "pure" index of responsibility (in which the emissions from all countries or regions are treated in the same fashion) could be scaled by a factor such as the percentage over or under the global mean per capita income (Hayes, 1993; Smith et al., 1993). In practice, developing an acceptable index of greenhouse liability from metrics of responsibility and capacity will not be easy; as with all such negotiations, countries will put forward formulas that favor their self interest and find ways to justify them as fair. However, if there is to be anything approximating the necessary funding for adaptation and fair compensation for harm, such a formula will be necessary.

A Quantitative Model Using Real Data

In the final section of this chapter, drawing on the preceding arguments, I develop a framework for calculating *net greenhouse liability* based on indicators for responsibility and legitimate claims. I present the framework at a general level, with an indication of its flexible application across different definitions of the indicators and for different groups. I then develop a specific example looking at net North-South liability (using the Annex I/non-Annex I definitions from the UNFCCC), at the net liability of different income groups both North and South, and at selected regions and countries.

Estimation of the legitimate claims (adaptation and residual damage costs) of a given community, class or country is, as I suggested earlier, going to both very difficult and very controversial. For the sake of my indicative calculation, I'm going to make the assumption that *claims are to a first order equal per capita*. This flies in the face of conventional wisdom that poor people and poor countries are more vulnerable to climate change than rich people or rich countries.[26] However, one could also argue that to the extent that property is also eligible for protection through adaptation (and for compensation if damaged), the greater amount of property at risk in rich countries would counterbalance the greater bodily risk of death and injury in poor countries.[27]

The point however is that even if one makes such a "conservative" assumption of equal per capita legitimate claims, the vast disparity in responsibility results—as the calculations below will demonstrate—in large net liability from wealthy to poor, whether at the country, regional or class level. As the climate negotiators begin discussing exactly how much might be reasonably spent on adaptation globally, this calculation can provide a method for estimating a reasonable lower bound for justified adaptation funding.

Definitions

As I argued in the previous section, responsibility for climate change is complex and controversial, and a wide variety of indicators of responsibility are defensible. For the reasons I outlined, however, I believe that cumulative emissions should be the dominant indicator of responsibility, and I will use it as the basis for estimating liability in the following calculations, allowing various modifications in a sensitivity analysis.

Responsibility

The general framework allows the various indices to be calculated for arbitrary groups, whether they be countries, regions, classes, or something else. For each group I define the responsibility r_i as the fraction of total responsibility, based on appropriately modified cumulative GHG emissions. For my reference case I will use cumulative emissions of CO_2 since 1950 from fossil fuel combustion and cement manufacturing, with various modifications as a sensitivity analysis.

Legitimate Claims

For each of the selected groups, define the total legitimate claims C as

$$C = \Sigma c_i.$$

where the claims c_i of each group would be identified by a common process or formula. As I suggested above, for this exercise I will assume claims are equal per capita.

Net Liability

In general terms, where c_i is the monetary value of a group's legitimate claims, then the group's net liability L_i is simply equal to its responsibility-weighted share of the total claims minus the value of its own legitimate claims:

$$L_i = r_i \bullet C - c_i$$

If we accept the premise of equal per capita claims, then a group's claim c_i is simply equal to its population share p_i times the total claims C. Thus the net liability L_i can be straightforwardly calculated as

$$L_i = r_i \bullet C - p_i \bullet C$$

or

$$L_i = (r_i - p_i) \bullet C.$$

Straightforwardly, this implies that a group will have positive liability if its share of responsibility is greater than its share of population, or alternatively, if its per capita responsibility is greater than the global mean.

Results

With a given definition of responsibility, the fractions r_i and p_i are empirically determined, and a group's net liability as a share of total claims will be determined independently of the magnitude of total claims. To make a concrete, if necessarily highly indicative, calculation, I took data for my example from the World Resource Institute's Climate Analysis Indicators Tool (CAIT).[28] Results are shown in table 14.1. Cumulative CO_2 emissions since 1950 from fossil energy and cement manufacture are shown in the first row. The second row translates this into the fraction of responsibility. The third row gives the population in 2000, and the fourth row the fraction of total population. The fifth row calculates net liability as defined above by subtracting the fraction of claims (by assumption the same as the population fraction) from the responsibility fraction.

This aggregated net liability, together with actual population, gives the per capita share of net liability. However, since this would be a very small number and very hard to interpret, I have used an arbitrary estimate of total claims of $50 billion per year for an interpretation of both aggregate and per capita net liability.[29] With the assumptions made, the table shows that Annex I countries have a net liability equal to 53% of total claims, which would translate into $26 billion dollars per year; while the non-Annex I countries would be owed those amounts. In per capita terms it would come to about $22 owed per northerner, and about $6 owed to each southerner.

The difference between the $26 billion in net flows and the $50 billion in total claims is money owed from northerners to northerners and from southerners to southerners, which is to say, "internal" funding of adaptation. I take up this issue further in the section on liability by income class. However, first I want to show the results of a sensitivity analysis based on alternative definitions of responsibility.

Sensitivity to Alternative Responsibility Indices

In table 14.2, I show for both Annex I and non-Annex I as a whole, and for the United states, the European Union,[30] China, and India, net liability based on five additional possible responsibility indices. In addition to cumulative CO_2 from fossil energy[31] (the reference case in table 14.1), I calculate (1) cumulative CO_2 emissions since 1850 for fossil CO_2; (2) cumulative fossil CO_2 emissions since 1990; (3) cumulative CO_2 emissions from all sources (including land use) since 1950; (4) contribution to atmospheric CO_2 stocks from fossil CO_2 emissions since 1950 (which takes account of absorption by sinks), and temperature change from fossil CO_2 emissions since 1950. Calculations

Table 14.1.
Net liability calculation for Annex I and non-Annex I regions.

	Annex I	Non-Annex I
Cumulative emissions (GtC from energy and industry, 1950–2000)	154	56
Fraction of responsibility r_i	0.72	0.28
Population 2000 (billions)	1.17	4.73
Fraction of population p_i	0.20	0.80
Net liability as a fraction of vulnerability $(r_i - p_i)$	0.53	−0.53
Net liability based on total adaptation claims of \$50 billion/year (\$billion/year)	26	−26
Net per capita liability (\$/year)	22.5	−5.6

See text for discussion of row entries.

for remaining carbon stocks and temperature change come from WRI's CAIT spreadsheets and use the algorithms included described in the CAIT documentation.

The figures in table 14.2 are net liability as defined above, i.e., net of equal per capita claims; adding back the population fraction (shown in the first row of table 14.2) to each of the values in the subsequent rows gives the value of the responsibility index itself (share of global emissions on the specified basis). Although I show two significant figures and no range of uncertainty, the combination of data uncertainty and (in the last two cases) model uncertainty should make it clear that these numbers are really educated guesses.

The resulting pattern is unsurprising given the global distribution of emissions over time. Extending responsibility back in time shifts the burden to the North, whereas shifting it closer to the present shifts the burden to the South; including additional sources of CO_2 (i.e., land use change) shifts the burden to the South; accounting for absorption by sinks shifts the burden to the South. Values based on cumulative temperature change are indistinguishable from values based on cumulative emissions for the period from 1950.

This is by no means an exhaustive list of the permutations of even the selected factors, to say nothing of accounting for non-CO_2 gases, or for more subjective definitions of responsibility like subsistence vs. luxury emissions.[32] The maximum and minimum values shown at the bottom vary by about a factor of two for all selected regions/countries except India, but importantly the sign is the same for all formulae. That is to say, the net direction of financial flows from countries and regions are not sensitive to the definition of responsibility.

Disaggregation by Class

I suggested above that the underlying principles of responsibility that are relevant for greenhouse liability are not based on nations except as a matter of pragmatism. The same distributional principles that apply between nations should apply within nations, with increased liability for those that are more responsible.

Information on the distribution of emissions within countries is fairly scarce. Furthermore, when accounting for historical responsibility, problems of aggregation will become even more difficult at the subnational scale. However, as there is a strong correlation between income and emissions, and between present income and past income, I will for the sake of illustration use current income distribution as a proxy for historical emissions.[33]

Table 14.2.
Sensitivity of net liability to alternative definitions of responsibility index.

	Net liability as a percentage of total global claims					
Definition of Responsibility	Annex I	Non-Annex I	U.S.	EU-15	China	India
Population Fraction 2000 (percent)	19%	78%	5%	6%	21%	17%
Cumulative CO_2 from Energy, 1950–2000 (%)	52%	−52%	22%	12%	−12%	−14%
Cumulative CO_2 from Energy, 1850–2000 (%)	57%	−56%	25%	17%	−14%	−15%
Cumulative CO_2 from Energy, 1990–2000 (%)	42%	−41%	19%	8%	−7%	−13%
CO_2 from all sources incl. land use, 1950–2000 (%)	32%	−31%	12%	7%	−11%	−15%
Change in Concentrations, CO_2 from Energy 1950–2000 (%)	50%	−50%	21%	11%	−11%	−14%
Change in Temperature, CO_2 from Energy 1950–2000 (%)	53%	−53%	22%	12%	−12%	−15%
MAX	57%	−31%	25%	17%	−7%	−13%
MIN	32%	−56%	12%	7%	−14%	−15%

Information on income inequality is reported on a national basis. Income distribution by quintile was available for 116 countries,[34] including all but the smallest Annex I countries and most of the large non-Annex I countries. To estimate the income by quintile for the Annex I and non-Annex I regions, I took the median value for each grouping (which vary only slightly from the non-population-weighted mean) of the countries for which data were available. In addition, to estimate the relative responsibility of the highest income groups in each region or nation, I also estimated the income fraction of the highest decile, with a simple assumption that the highest decile receives two thirds of the income of the highest quintile. The results, based on my reference case of cumulative fossil/industrial CO_2 emissions since 1950, are shown in tables 14.3a–14.3c.

The first table (14.3a) shows the income distribution data itself, including the estimated aggregated values for Annex I and non-Annex I groups and for all the top deciles. The second table (14.3b) shows the net liability for each income group in each country or region, assuming equal per capita claims, and the reference-case responsibility index. The third table (14.3c) shows for each income group the dollar liability per capita associated with $50 billion in total claims, as in my earlier example. And again, as with the first example, the two-significant figure percentages (and the to-the-dime calculation of monetary liability) hide extremely large uncertainties from many sources.

The example shows that with the given definition of responsibility, the highest decile of the global South has a small but positive net liability, while in the North, the poorest quintile has very low liability. In the United States, liability increases more rapidly with income than in Annex I overall. Among Southern countries, the top quintile in China has (very small) positive liability while in India, even the top decile has negative liability.

Table 14.3a.
Percentage of total regional or national income by quintile or decile.

Region or country	Quintile 1	Quintile 2	Quintile 3	Quintile 4	Quintile 5	Top Decile
Annex I	7.9%	13.1%	17.3%	22.6%	38.9%	25.9%
Non-Annex I	5.6%	10.1%	14.8%	21.3%	48.0%	32.0%
US	5.2%	10.5%	15.6%	22.4%	46.4%	30.9%
China	5.9%	10.2%	15.1%	22.2%	46.6%	31.1%
India	8.1%	11.6%	15.0%	19.3%	46.1%	30.7%

Table 14.3b.
Net liability as a percentage of total global claims, by income quintile or decile.

Region or country	Quintile 1	Quintile 2	Quintile 3	Quintile 4	Quintile 5	Top Decile
Annex I	2%	6%	9%	12%	24%	17%
Non-Annex I	−14%	−13%	−12%	−10%	−3%	1%
US	0%	2%	3%	5%	11%	8%
China	−4%	−3%	−3%	−2%	0%	1%
India	−3%	−3%	−3%	−3%	−2%	−1%

Data are based on cumulative fossil CO_2 emissions since 1950 and the assumption that emissions responsibility is proportional to current income.

Table 14.3c.
Dollar value of net liability per capita, based on $50 billion in total claims, by income quintile or decile.

Region or country	Quintile 1	Quintile 2	Quintile 3	Quintile 4	Quintile 5	Top Decile
Annex I	3.8	11.8	18.1	26.1	50.8	70.5
Non-Annex I	−7.4	−6.8	−6.2	−5.3	−1.6	0.6
US	3.9	16.2	28.1	43.9	99.6	135.6
China	−7.1	−6.4	−5.5	−4.2	0.1	2.9
India	−7.7	−7.5	−7.4	−7.1	−5.6	−4.7

Interpretation and Caveats

As I noted above, these calculations are sensitive to a large number of subjective assumptions and should be taken as strictly indicative. It is beyond the scope of this chapter to run a more comprehensive sensitivity analysis, though this provides a framework within which such an analysis could easily be done. However, the general patterns I expect to remain fairly robust.

It should be clear from the two main examples I've given that, while one can specify one's assumptions to allow arbitrary precision in calculation, it would be a mistake to seek a single "correct" formula and attempt to precisely parameterize it. Rather, the framework allows for comparison of liability "scenarios." It is the purpose of the background concerning ethics and law to provide arguments for why some scenarios may be normatively preferable to others, arguments which are only meaningful in the context of the real world of international politics.

Conclusion

The arguments and calculations in this paper no doubt raise more questions than they answer. Yet in the face of the clearly inadequate measures so far, they strongly support the development of a regime of international legal liability for adaptation to climate change and compensation for climate harm. The basis for legitimate claims for adaptation funding, and for the establishment of liability tied to responsibility for emissions, is shown to lie in ethical and legal norms for behavior in a life-support commons. These principles underlie domestic pollution regulation and environmental tort law, and have been endorsed rhetorically in international law; it is now time to begin to implement them globally.

On the basis of an assumption that the legitimate claims for adaptation funding are distributed globally on an equal per capita basis—a conservative assumption compared to usual conclusions that developing countries are more vulnerable and will require greater investment to adapt—net North-to-South liability is shown to lie in the range of 30–60% of total claims. A reference case using cumulative emissions from fossil and industrial CO_2 limited to the period 1950-present shows the net liability of the North to be equal to 53% of legitimate adaptation needs. Using a reasonable value of $50 billion in annual claims, this is a debt of about $22 per year for each northerner and a credit of about $6 per year for each southerner.

Critically, however, this liability (both positive and negative) is shown to be unequally divided between classes in the North and South. In the reference case, using the first-order approximation that emissions are proportional to income, the wealthiest decile in the North has 17 percent of net liability, versus 2 percent for the lowest quintile in the North, while in the South, the top decile has a small positive liability (1% of total claims) in spite of the South's net negative liability, while the poorest quintile has a credit equal to 17 percent of total claims.

As I have noted, these numbers are based on simple and plausible but nonetheless controversial assumptions, and given the uncertainty in the data, should not be considered to be "precise" even for those assumptions. I offer them as a contribution to efforts to conceptualize, and eventually negotiate, an adequate and equitable adaptation and compensation regime, in which "polluter pays" becomes more than an empty promise.

Acknowledgements

The author wishes to acknowledge the contributions of Neil Adger, Tom Athanasiou, Steve Gardiner, Barbara Haya, Richard Howarth, Saleemul Huq, M. J. Mace, Richard Norgaard, Jouni Paavola, Richard Plevin, Steve Schneider, Roda Verheyen, two anonymous reviewers, and all the participants in the Justice in Adaptation to Climate Change conference and the Stanford Environmental Forum where I presented earlier versions of this paper. All remaining errors of fact and value are my own.

Notes

1. Article 4, Parargraphs 1(b), 1(e), 1(f), 3, 4, and 8 all specifically address adaptation and funding for adaptation. See Mace (2004) for details.

2. The Annex I countries-a close proxy for "the North"–have about a fifth of global population and account for half to three-quarters of anthropogenic greenhouse forcings, depending on how you account for "responsibility" (see

Section IV). The disproportionate vulnerability of the South due to (for example) increased dependence on agriculture and lack of resources for adaptation is a general consensus asserted by the IPCC in its Third Assessment Report (McCarthy et al., 2001).

3. The phrase "common but differentiated responsibilities" is most frequently cited because it appears in Article 3 (Principles) of the UNFCCC, but it also appears in Article 4 (Commitments) in the context of measures including support for adaptation.

4. At COP6-bis in Bonn in 2001, the EU, Canada, Iceland, New Zealand, Norway and Switzerland pledged to contribute €450 million annually by 2005 to the voluntary funds; with estimates of total CDM revenue running well under $1 billion (e.g., den Elzen and de Moor, 2002), the 2% surchange for the Adaption Fund will not add much to this.

5. Citing examples is hardly necessary, but there are not less than four books since 1990 titled "The Global Commons."

6. In the context of greenhouse pollution, the "atmospheric commons" is shorthand for the whole system of sinks (including oceanic and terrestrial) which remove or store GHGs. I provide an ethical analysis of the problem of equitable sharing of global GHG sinks ("pollution rights") in Baer (2002), and a more political analysis in Baer et al. (2000) and Athanasiou and Baer (2002).

7. There is also an extensive sub-typology of "public goods" in which they can be considered "pure", "impure, "club goods," "joint products" etc. See for example Cornes and Sandler (1996).

8. Of course here "climate stability" is a relative term, meaning stability of the means and variabilility within historical experience.

9. I believe that this is an original formulation, although this concept is implicit in the literature, and at least one author has referred specifically to "common pool life support functions" (Rees, 2002, p. 39).

10. For example, Cornes and Sandler (1996) use the same basic framework to analyze problems ranging from military security to agricultural research to transportation to climate change.

11. At this point I emphasize only that the polluting activities were deliberate, not that pollution itself was deliberate or that the harm they cause was understood at the time. I return to this important issue below. Furthermore, one is not immune to the impacts of one's own pollution.

12. One ethical system in which this principle does not hold is utilitarianism, which would permit (even encourage) harming others for my own benefit as long as it produces more benefit for me than the harm it causes. This aspect of the use of utilitarianism to justify cost-benefit analysis of problems like climate change is one reason many people reject it on ethical grounds.

13. For example, Section 112 of 1990's Clean Air Act amendments, which regulates hazardous air pollutants (HAPs), authorizes the EPA to delete a category of stationary sources of HAPs if the risk of premature mortality to the most exposed individual is less than one in one million (Powell, 1999).

14. Superfund legislation is technically known as CERCLA, The Comprehensive Environmental Response, Compensation and Liability Act of 1980. See for example Revesz and Stewart (1995) or Switzer and Bulan (2002).

15. I believe a strong *prima facie* case exists that the rightness or wrongness of imposing harm on another person shouldn't depend on their geographical location or political community. More robust philosohpical arguments for the extension of ethical obligations to people in other countries have a long history; see for example Shue (1996) or Singer (2002).

16. In an outstanding recent article, Tol and Verheyen (2004) discuss the general basis for the responsibility of states for compensation for pollution under international law. They note that in the UNFCCC negotiations the industrialized nations specifically resisted state-based responsibility for compensation for climate damages, but also that at the time the small island states issued a declaration that signing the treaty did not renounce their rights under international law concerning state responsibility.

17. I defer for now the problem of "additionality" or "incremental costs"–separating the costs of an investment into the parts which are necessary specifically to reduce the harm from anthropogenic climate change, and the parts which are primarily motivated by other factors (including protection from natural climate variability).

18. Love Canal is the most famous case; see for example Fletcher (2002).

19. There are of course *behavioral* adaptations to toxic pollution–like not eating contaminated fish—that parallel many of the possible behavioral adaptations to climate change. To my knowledge there has been little discussion of the costs of such actions or related liability.

20. As noted in the recent Third Assessment Report of the IPCC, "There are many arbitrary lists of possible adaptation measures, initiatives or strategies that have a potential to moderate impacts..." (Smit and Pilifosova, 2001). See Fankhauser (1996) and other chapters in the same volume for a start.

21. Note that it is the combination of global mitigation investment and local adaptation investment that would serve to reduce risk to the acceptable level; that is to say, legitimate adaptation funding is dependent on the level and effectiveness of mitigation.

22. An early contribution to this discussion can be found in Hayes and Smith (1993). Following Brazil's pre-Kyoto proposal that burden sharing among Annex I countries be based on historical contribution to temperature (UNFCCC, 1997), there have been a variety of analyses assessing both ethical justification (e.g., Neumayer, 2000) and technical difficulties (e.g., den Elzen and Schaeffer, 2002). A good typology of various responsibility indicators has recently been produced by the World Resources Institute (Baumert and Markoff, 2003).

23. Though Tol and Verheyen (2004) is an important exception.

24. This is the basis of the Brazilian Proposal (see note 22 above).

25. The IPCC has established a method for comparing GHGs on the basis of "Global Warming Potentials" (Houghton et al., 1995), which has spawned a lengthy literature of critiques, defenses and alternatives (see for example Shackley and Wynne (1997), or Fuglestvedt et al. (2003) and the associated commentaries in the June 2003 issue of *Climatic Change*.)

26. The IPCC's Third Assessment Report notes that "The effects of climate change are expected to be greatest in developing countries in terms of loss of life and relative effects on investment and the economy" (McCarthy et al., 2001, p.8). Some recent estimates (e.g., Tol, 2002) even show net positive GDP impacts in developed countries compared to net negative impacts for most developing countries (at 1 °C of additional warming).

27. As more than one commentator has pointed out, putting the property of the wealthy in apposition to the lives of the poor is, to be polite, deeply problematic. However, while I believe strongly that protecting lives and health should in fact be the priority of climate adaptation, I believe that property deserves protection from climate damage as well. I ask my readers then to bear with me, and allow me to assume that in poor countries, this equal per capita share of the money to be used for adaptation will go foremost to reducing bodily risks, with sufficient remaining to protect a greater proportion of property in poor countries than will be protected in rich countries.

28. WRI's Climate Analysis Indicators Tool (WRI 2003) is available on their website at http://cait.wri.org, and is based on data from the Carbon Dioxide Information and Analysis Center, the Energy Information Administration of the US Department of Energy, the World Bank, and other sources.

29. The $50 billion per year figure is arbitrary but not unreasonable. Consider the following: Approximately 200 GtC of carbon were emitted globally from fossil fuel combustion between 1950 and 2000. Suppose we were to retroactively tax these emissions for the marginal damages they will cause - the classic Pigouvian justification for pollution taxes. Estimates for this "social cost of carbon" vary widely, but in a recent paper, respected British environmental economist David Pearce (2003) argued that the "base-case" estimate was in the range of $4–9 per ton of carbon. This would value the damages from the historical emissions at $800–1800 billion dollars. If this were to be paid off over 20 years (during which time another 200 GtC may have been emitted), that would be $40–90 billion dollars per year. And it should be noted that Pearce's estimates were a response to a paper prepared for the British government (Clarkson and Deyes, 2002) whose best-guess estimate was much higher: £70/ton—about $130/ton by today's exchange rates—which would value the cumulative emissions since 1950 at over 26 trillion dollars.

30. Based on the 15 nations of the "Old" EU.

31. Here and in the rest of this paragraph I include CO_2 from cement manufacturing in "fossil energy".

32. See Agarwal and Narain (1991) and Agarwal et al. (1999) for equity issues such as luxury vs. subsistence emissions, accounting for fair use of global oceanic sinks, etc., which are framed in the context of the allocation of emissions rights but are highly relevant to this discussion.

33. I realize this is a problematic assumption, and it will be important in the future to perform a sensitivity analysis. Furthermore, it would be desirable to acquire additional data concerning the distribution of current and historical emissions within countries.

34. My income distribution data comes from the World Bank, but was accessed through the World Resource Institute's online interface, http://earthtrends.wri.org.

References

Agarwal, Anil, and Sunita Narain. 1991. *Global Warming in an Unequal World: A Case of Environmental Colonialism*. New Delhi: Centre for Science and Environment.

Agarwal, Anil, Sunita Narain, and Anju Sharma, eds. 1999. *Green Politics: Global Environmental Negotiations 1*. New Delhi: Centre for Science and the Environment.

Athanasiou, Tom, and Paul Baer. 2002. *Dead Heat: Global Justice and Global Warming*. New York: Seven Stories Press.

Baer, Paul. 2002. "Equity, Greenhouse Gas Emissions and Global Common Resources." In *Climate Change Policy: A Survey*, ed. Steven H. Schneider, Armin Rosencranz, and John Niles, 393–408. Washington, DC: Island Press.

Baer, Paul, John Harte, Barbara Haya, Antonia V. Herzog, John Holdren, Nathan E. Hultman, Daniel M. Kammen, Richard B. Norgaard, and Leigh Raymond. 2000. "Equity and Greenhouse Gas Responsibility in Climate Policy." *Science* 289: 2287.

Baumert, Kevin A., and Matthew Markoff. 2003. *Indicator Framework Paper*. Washington D.C.: World Resources Institute. Available at http://cait.wri.org.

Beauchamp, T. L. 2003. "A defense of the common morality." *Kennedy Institute of Ethics Journal* 13: 259–274.

Brennan, T. A. 1993. "Environmental Torts." *Vanderbilt Law Review* 46: 1–73.

Burger, Joanna, Elinor Ostrom, Richard B. Norgaard, David Policansky, and Bernard D. Goldstein, eds. 2001. *Protecting the Commons: A Framework for Resource Management in the Americas*. Washington, D.C.: Island Press.

Clarkson, R., and K. Deyes. 2002. *Estimating the Social Cost of Carbon Emissions*. GES Working Paper 140. London: HM Treasury. Available at http://www.hm-treasury.gov.uk/Documents/Taxation_Work_and_Welfare/Taxation_and_the_Environment/tax_env_GESWP140.cfm.

Cornes, Richard, and Todd Sandler. 1996. *The Theory of externalities, pubblic goods, and club goods*. Cambridge, UK: Cambridge University Press.

den Elzen, Michel G. J., and André P. G. de Moor. 2002. "Evaluating the Bonn-Marrakesh agreement." *Climate Policy* 2: 111–117.

den Elzen, Michel, and Michiel Schaeffer. 2002. "Responsibility for past and future global warming: Uncertainties in attributing anthropogenic climate change." *Climatic Change* 54: 29–73.

Dimento, Joseph F. C. 2003. *The Global Environment and International Law*. Austin, TX: University of Texas Press.

Fankhauser, Samuel. 1996. "The Potential Costs of Climate Change Adaptation." In *Adaptating to Climate Change: Assessments and Issues*, ed. Joel B. Smith, Neeloo Bhatti, Gennady Menzhulin, Ron Benioff, Mikhail I. Budyko, Max Campos, Bubu Jallow, and Frank Rijsberman, 80–96. New York: Springer-Verlag.

Fletcher, T. 2002. "Neighborhood change at Love Canal: contamination, evacuation and resettlement." *Land Use Policy* 19: 311–323.

Fuglestvedt, J. S., T. K. Berntsen, O. Godal, R. Sausen, K. P. Shine, and T. Skodvin. 2003. "Metrics of climate change: Assessing radiative forcing and emission indices." *Climatic Change* 58: 267–331.

Hardin, Garrett. 1968. "The Tragedy of the Commons." *Science* 162: 1243–1248.

Hayes, Peter. 1993. "North-South." In *The Global Greenhouse Regime: Who Pays?*, ed. Peter Hayes and Kirk Smith, 144–168. London: Earthscan, for the United Nations University Press.

Hayes, Peter, and Kirk Smith, eds. 1993. *The Global Greenhouse Regime: Who Pays?* London: Earthscan, for the United Nations University Press.

Houghton, J. T., L. G. Meira Filho, J. Bruce, Hoseung Lee, B. A. Callander, E. Haites, N. Harris, and K. Maskell, eds. 1995. *Climate Change 1994*. Cambridge: Cambridge University Press.

Kaul, Inge, Isabel Grunberg, and Marc A. Stern. 1999. "Defining Global Public Goods." In *Global Public Goods: International Cooperation in the 21st Century*, ed. Inge Kaul, Isabel Grunberg, and Marc A. Stern, 2–19. New York: Oxford University Press for the United Nations Development Program.

Keating, G. C. 1997. "The idea of fairness in the law of enterprise liability." *Michigan Law Review* 95: 1266–1380.

Mace, M. J. 2004. "Just Adaptation under the UNFCCC: The Legal Framework." In *Justice in Adaptation to Climate Change*, ed. W. N. Adger and J. Paavola, 53–76. Cambridge, MA: MIT Press.

McCarthy, James J., Osvaldo F. Canziani, Neil A. Leary, David J. Dokken, and Kasey S. White, eds. 2001. *Climate Change 2001: Impacts,*

Adaptation and Vulnerability. Cambridge, UK: Cambridge University Press.

Neumayer, Eric. 2000. "In Defence of Historical Accountability for Greenhouse Gas Emissions." *Ecological Economics* 33: 185–192.

Oakerson, R. J. 1992. "Analyzing the Commons: A Framework." In *Making the Commons Work: Theory, Practice and Policy*, ed. D. W. Bromley, D. Feeny, M. A. McKean, P. Peters, J. L. Gilles, R. J. Oakerson, C. F. Runge, and J. T. Thomson, 41–59. San Francisco: Institute for Contemporary Studies.

Ostrom, Elinor. 1990. *Governing the Commons: The Evolution of Institutions for Collective Action*. Cambridge, UK: Cambridge University Press.

Pearce, David 2003. "The social cost of carbon and its policy implications." *Oxford Review of Economic Policy* 19: 362–384.

Penalver, E. M. 1998. "Acts of god or toxic torts? Applying tort principles to the problem of climate change." *Natural Resources Journal* 38: 563–601.

Perry, S. R. 1992. "The Moral Foundations of Tort Law." *Iowa Law Review* 77: 449–514.

Portmore, D. W. 2000. "Commonsense morality and not being required to maximize the overall good." *Philosophical Studies* 100: 193–213.

Powell, Mark R. 1999. *Science at EPA: Information in the Regulatory Process*. Washington DC: Resources for the Future.

Rees, W. E. 2002. "An ecological economics perspective on sustainability and prospects for ending poverty." *Population and Environment* 24: 15–46.

Revesz, Richard L., and Richard B. Stewart, eds. 1995. *Analyzing Superfund: Economics, Science and Law*. Washington, D.C.: Resources for the Future.

Shackley, Simon, and Brian Wynne. 1997. "Global Warming Potentials: ambiguity or precision as an aid to policy?" *Climate Research* 8: 89–106.

Shue, H. 1996. *Basic Rights: Subsistence, Affluence and U.S. Foreign Policy*. Princeton, NJ: Princeton University Press.

———. 1999. "Global environment and international inequality." *International Affairs* 75: 531–545. Also chapter 5 in this volume.

Singer, Peter. 2002. *One World: The Ethics of Globalization*. New Haven, CT: Yale University Press.

Smit, Barry, and Olga Pilifosova. 2001. "Adaptation to Climate Change in the Context of Development and Equity." In *Climate Change 2001: Impacts, Adaptation and Vulnerability*, ed. James J. McCarthy, Osvaldo F. Canziani, Neil A. Leary, David J. Dokken, and Kasey S. White, 877–912. Cambridge, UK: Cambridge University Press.

Smith, Kirk R., Joel Swisher, and Dilip R. Ahuja. 1993. "Who pays (to solve the problem and how much)?" In *The Global Greenhouse Regime: Who Pays?*, ed. Peter Hayes and Kirk Smith, 70–98. London: Earthscan, for the United Nations University Press.

Switzer, Carole Stern, and Lynn A. Bulan. 2002. *CERCLA: Comprehensive Response, Compensation and Liability Act*. Chicago: American Bar Association.

Tol, R. S. J. 2002. "Estimates of the damage costs of climate change. Part 1: Benchmark estimates." *Environmental & Resource Economics* 21: 47–73.

Tol, R. S. J., and R. Verheyen. 2004. "State responsibility and compensation for climate change damages—a legal and economic assessment." *Energy Policy* 32: 1109–1130.

UN Conference on the Human Environment. 1972. Stockholm Declaration. *United Nations Conference on the Human Environment*. United Nations, Stockholm, Sweden.

UNFCCC. 1997. *Proposed Elements of a Protocol to the United Nations Framework Convention on Climate Change, Presented by Brazil in Response to the Berlin Mandate*. UNFCCC/AGBM/1997/MISC.1/Add.3.

Adaptation, Mitigation, and Justice

Dale Jamieson

In this chapter I claim that climate change poses important questions of global justice, both about mitigating the change that is now under way and about adapting to its consequences.[1] I argue for a mixed policy of mitigation and adaptation, and defend one particular approach to mitigation. I also claim that those of us who are rich by global standards and benefit from excess emissions have strenuous duties in our roles as citizens, consumers, producers, and so on, to reduce our emissions and to finance adaptation.

The Unavoidability of Adaptation

When I began my research on global climate change in the mid-1980s, it was commonly said that there were three possible responses: prevention, mitigation, and adaptation. Even then we were committed to a substantial climate change, although this was not widely known. This realization began to dawn on many people on June 23, 1988, a sweltering day in Washington, D.C., in the middle of a severe national drought, when climate modeler James Hansen testified before a U.S. Senate committee that it was 99 percent probable that global warming had begun. Hansen's testimony was front-page news in the *New York Times*, and was extensively covered in other media as well. Whether or not Hanson was right, his testimony made clear that we were entering a new world, what Schneider (1989) called "the greenhouse century."

Once it became clear that prevention was no longer possible, mitigation quickly moved to center stage. One week after Hansen's testimony, an international conference in Toronto, convened by the World Meteorological Organization (WMO), called for a 20 percent reduction in greenhouse-gas (GHG) emissions by 2005. In November, the World Congress on Climate and Development, meeting in Hamburg, called for a 30 percent reduction by 2000. Later that same year, acting on a proposal by the United States, the WMO and the United Nations Environment Programme (UNEP) established the

15

Intergovernmental Panel on Climate Change (IPCC) in order to assess the relevant scientific information and to formulate response strategies.[2] In December 1989, the United Nations General Assembly adopted a resolution, proposed by Malta, that essentially authorized the negotiation of a climate-change convention. The following year the IPCC published its first report, and the International Negotiating Committee (INC) was established. In 1992 the Framework Convention on Climate Change (FCCC) was officially opened for signature at the Rio Earth Summit. It came into force on March 21, 1994, and by May 24, 2004, had been ratified by 189 countries.

The main objective of the FCCC is to stabilize "greenhouse gas concentrations in the atmosphere at a level that would prevent dangerous anthropogenic interference with the climate system." This goal is consistent with accepting some degree of climate change so long as it is not "dangerous." In the negotiations leading up to the adoption of the FCCC, all the developed countries except the United States and the Soviet Union favored binding targets and timetables for emissions reductions as a way of reaching this goal. However, in the end the FCCC embodied voluntary commitments on the part of developed countries to return to 1990 levels of GHG emissions by 2000.

It soon became clear that while some European countries might succeed in keeping this commitment, the United States, Australia, New Zealand, Japan, Canada, and Norway would not. In 1995, at the first Conference of the Parties (COP 1), the "Berlin Mandate" was adopted. The parties pledged that by the end of 1997 an agreement would be reached establishing binding, "quantified, emission limitation reduction objectives" for the industrialized countries, and that no new obligations would be imposed on other countries during the compliance period. In December 1997, the parties agreed to the Kyoto Protocol, which in its broad outlines satisfied the Berlin Mandate. However, many of the most important details regarding the rules of implementation were left for future meetings.

Almost immediately the Kyoto Protocol came under fire from several different directions. It was simultaneously attacked as too weak, too strong, unworkable, and, at least in the United States, politically unacceptable, Meeting in the Hague in November 2000, a lame-duck American administration and its allies, Japan, Russia, Canada, Australia, and New Zealand (collectively known as JUSCAN), argued that countries should be able to satisfy up to 80 percent of their reductions by emissions trading and by establishing carbon sinks.[3] The Europeans rejected this, and the meeting seemed headed for disaster. However, rather than admitting defeat, the conference was suspended until July 2001. In the interim, in March 2001, the new Bush administration caught the world by surprise by renouncing the Protocol. Ironically, this improved the negotiating position of America's JUSCAN partners. In order to come into force the Protocol had to be ratified by at least 55 countries, including Annex 1 countries responsible for 55 percent of Annex 1 country emissions in 1990.[4] Since the U.S. share of such emissions is about 36 percent, it became imperative to keep the rest of JUSCAN in the Protocol. In addition, some hoped that by offering concessions, the United States could be persuaded to climb down from its extreme position and rejoin the negotiation. The result was that in July 2001, in Bonn, the European Union (EU) acceded to most of the demands that the Americans had made earlier in the Hague. The Protocol was further weakened in Marrakech in November 2001, when negotiators gave in to Russia's demand that its transferable credits for sinks be doubled. After two more years of study and negotiation. Russia finally ratified the Kyoto Protocol on November 18, 2004. On February 16, 2005, the Kyoto Protocol came into force, binding virtually every country in the world except the United States and Australia.

It is not completely clear what will be the effect of the Kyoto Protocol. While once it was envisioned that it would reduce developed-country emissions by about 14 percent between 2000 and 2010, it now appears that in the wake of the Bonn and Marrakech agreements it could countenance as much as a 9 percent increase in emissions from these countries.[5] Were that to occur, there would be little difference between the Kyoto path and a "business as usual" scenario, at least with respect to GHG emissions over the next decade.

Essentially what has occurred is that the vague loopholes that were embedded in the text of the Kyoto Protocol, rather than being eliminated, have been quantified and transformed into central features of an emissions control regime. In order to convey the flavor of these loopholes I will mention only the example of Russian "hot air." As a result of the post-communist economic collapse. Russian GHG emissions have sharply declined since 1990. What has happened, in effect, is that Russia is being allowed to sell the rights to emissions that would not have occurred, to countries that will in fact use them. Thus, more GHGs will be emitted than would have been the case under a regime that simply established mandatory emissions limits without such flexible mechanisms as emissions trading and credits for carbon sinks. Russia benefits economically, countries with high levels of GHG emissions are allowed to carry on business more or less as usual, and politicians can take credit for having addressed the problem. Meanwhile, global climate change continues largely unabated.

At the eighth Conference of the Parties (COP 8) meeting in Delhi in October 2002, the United States, once the foremost advocate of bringing developing countries into an emissions-control regime, joined with the Organization of Petroleum Exporting Countries (OPEC), India, and China in blocking the attempts of the EU to establish a more inclusive regime after the Kyoto commitments expire in 2012.[6] At COP 10, meeting in Buenos Aires in December 2004, the United States did everything it could to block even informal discussion of a post-2012 emissions regime. In retrospect, COP 8 may be seen as our entrance into an era in which the world has given up on significantly mitigating climate change, instead embracing a de facto policy of "adaptation only." Indeed, the most public pronouncement of COP 8, the Delhi Ministerial Declaration on Climate Change and Sustainable Development, emphasized adaptation almost to the exclusion of mitigation.

As should be clear already, the climate-change discussion has its own vocabulary, and it is important to understand exactly what is meant by such terms as *adaptation*. One influential characterization is this: "adaptation refers to adjustments in ecological-social-economic systems in response to actual or expected climate stimuli, their effects or impacts."[7] Various typologies of adaptation have been developed,[8] but for the present purposes it is sufficient to mark distinctions on two dimensions.

Some adaptations are conscious responses to climate change while others are not. For example, plans that are currently under way to evacuate low-lying Pacific islands are conscious adaptations, while adaptations by plants, animals, and ecosystems, and also those by farmers who incrementally respond to what they see as climate variability and changes in growing season, are nonconscious adaptations. Intuitively, this distinction is between climate-change policy adaptations and those responses that are autonomous or automatic. On another dimension, some adaptations are anticipatory while others are reactive. An example of an anticipatory adaptation is constructing sea walls in order to minimize the impact of an expected sea-level rise. An example of a reactive adaptation is the efforts of a coastal community, damaged by a hurricane, to rebuild to a more secure standard. This dimension marks the intuitive distinction between adaptations based on foresight and those that are responses to immediate events. Taking these dimensions together, we can say that climate-change adaptations can be driven by policy or by autonomous responses, and they can be based on predictions or stimulated by events.

There are, of course, other dimensions on which one might distinguish adaptations, and the categories that I have characterized admit of degrees of membership. These complications need not concern us for the present purposes, however.[9]

From the beginning of the climate-change controversy, some in the research community have been concerned about the place of adaptation on the policy agenda.[10] There were several sources of this concern.

First, the community that studies climate and weather impacts is greatly influenced by the natural-hazards community, which has long been committed to the idea that human societies are to a great extent maladapted to their environments. Researchers point to ongoing failures to adapt to such predictable features of a stable climate regime as droughts, storms, and hurricanes. For people who suffer from these events it matters little if they are part of normal variability, associated with various long-term natural cycles, or consequences of anthropogenic climate change. What people experience is weather, not the statistical abstractions constructed by climatologists. An increasing focus on adaptation would help vulnerable people whether or not climate change is occurring.

A second source of concern, often expressed by anthropologists and those influenced by the social movements of the 1960s, is rooted in opposition to scientistic, top-down, managerial approaches to human problems. Here the concern is that focusing primarily on mitigation (i.e., reducing GHG emissions) transforms problems of human survival and livelihood into technical problems of "carbon management," best approached by scientists with their formal methods of prediction and their economistic approaches to evaluating policy options. With this view, subsistence farmers in the developing world would do better by adjusting and adapting to changing environmental conditions based on their indigenous knowledge than waiting for the right sort of policy to emerge in New York, Geneva, or Washington and then filtering down through a panoply of national institutions, subject to who knows what kinds of distortions and revisions.

In the discussion surrounding the Kyoto Protocol some researchers seemed to suggest that adaptation was a neglected option as a response to climate change.[11] Yet concern for adaptation is both implicit and explicit in the FCCC.[12] The sentence that follows the statement of the objective quoted earlier states that "such a level should be achieved within a time-frame sufficient to allow ecosystems to adapt naturally to climate change, to assure that food production is not threatened, and to enable economic development to proceed in a sustainable manner." Article 4, which specifies the commitments undertaken by the parties to the Convention, mentions adaptation on several occasions. The parties agree to implement national or regional adaptation measures, to cooperate in preparing for adaptation to the impacts of climate change, and to take adapting to climate change into account in their relevant social, economic, and environmental policies and actions. In 1994, the IPCC published technical guidelines to assist nations in performing "vulnerability and adaptation assessments," and in 1995 at COP 1 in Berlin, explicit guidance was provided on adaptation planning and measures. The second IPCC report published in 1996 observed that many societies are poorly adapted to climate, and emphasized the importance of adopting "no regrets" policies to better adapt to both the prevailing climate regime and what may come next.

More recently, in July 2003, the strategic plan of the United States government's Climate Change Science Program listed, as one of its goals, understanding "the sensitivity and adaptability of different natural and managed ecosystems and human systems to climate and related global changes."[13] No comparable goal regarding mitigation figured in the plan.

Once it became clear that prevention was not possible, adaptation had to be part of the portfolio of responses. The logic of the U.S. government's *Climate Action Report* 2002 is unassailable: "because of the momentum in the climate system and natural climate variability, adapting to a changing climate is inevitable."[14] The adaptations may be clumsy, inefficient, inequitable, or inadequate, but it has been clear for some time that human beings and the rest of the biosphere will have to adapt to climate change or they will perish. What is in question is not whether a strategy of adaptation should and will be followed, but whether in addition there will be any serious attempt to mitigate climate change.[15]

The Importance of Mitigation

My claim is that a policy of adaptation without mitigation, the one we may be slouching toward, runs serious practical and moral risks. The practical risk, which itself has moral dimensions, is that a GHG forcing may quite suddenly drive the climate system into some unanticipated, radically different state to which it is virtually impossible to adapt. Such a catastrophic climate surprise could occur through climate change setting off a series of positive feedbacks, for example warmer temperatures leading to lower albedo (surface reflectancy), leading to warmer temperatures, leading to lower albedo, and so on—or through the flipping of a climate "switch." The current climate regime depends on regular circulation systems in the oceans and atmosphere that at various times have turned on, shut down, or been radically different. At the end of the Younger Dryas, about 11,500 years ago, global temperatures rose up to 8°C in a decade and precipitation doubled in about three years.[16] The GHG forcing that is now occurring increases the probability of such an abrupt change. As a recent report from the National Academy of Sciences (2002, p. 107) states.

> In a chaotic system, such as the earth's climate, an abrupt change could always occur. However, existence of a forcing greatly increases the number of possible mechanisms. Furthermore, the more rapid the forcing, the more likely it is that the resulting change will be abrupt on the time scale of human economies or global ecosystems.

Indeed, there is some evidence that abrupt changes may already be occurring. The Arctic circulation appears to be slowing,[17] and since the 1980s the Arctic oscillation has been stuck in its positive phase, causing lower pressures to persist over the Arctic. This has led to warmer summers and stormier springs, resulting in the greatest contraction of Arctic sea ice since modern measurements began, and perhaps much longer if anecdotal and anthropological reports are to be believed.[18] The recent Arctic Climate Impact Assessment sponsored by the Arctic Council, a high-level intergovernmental forum that includes the United States, found that the warming in the Arctic is much more extreme than that in the mid-latitudes, with some Arctic regions having warmed 10 times as much as the mid-latitude average.[19] Perhaps most telling, in the summer of 2000 a Canadian ship succeeded in transiting the legendary, once impassable Northwest Passage, the elusive goal of mariners since the 16th century.

Even without abrupt climate change, an "adaptation only" policy runs serious moral risks. For such a policy is likely to be an application of the "polluted pay" principle, rather than the "polluter pays" principle. Some of the victims of climate change will be driven to extinction (e.g., some small island states and endangered species), and others will bear the costs of their own victimization (e.g., those who suffer from more frequent and extreme climate-related disasters).

Consider what happens when a climate-related disaster strikes a developing country. Often large amounts of aid are pledged and commitments are made to provide both humanitarian assistance and support for transforming the society in order to reduce its vulnerability to future disasters, but little meaningful change actually occurs. Consider an example.[20]

In 1998 Hurricane Mitch struck Honduras, killing at least 6,500 people and causing $2 billion to $4 billion in damage, an amount equivalent to 15 percent to 30 percent of gross domestic product. At the height of the emergency, donors pledged $72 million to the World Food Program for immediate humanitarian assistance. More than a year later, less than one-third of the promised funds had been delivered. At a donors' conference in Stockholm in 1999, $9 billion was pledged for the reconstruction and transformation of Central America. The conference report stated that "the tragedy of Hurricane Mitch provided a unique opportunity to rebuild not the same, but a better Central America."[21] Many of the resources that were provided were reprogrammed funds or "in kind" contributions. Much of the promised aid was not delivered in any form. Still, a significant amount of aid did find its way into the country, especially compared to pre-Mitch levels of assistance.

The three-year reconstruction period is now over, and we can ask what has been accomplished. There are success stories trumpeted by various governments and nongovernmental organizations, and it would be incorrect to say that no improvements have been made. Still, Honduras remains extremely poor and vulnerable to climate-related disasters. One observer writes that even

> after Mitch, we see many environmentally bad habits on replay. People are moving back into high-risk zones, farming practices degrade upper watersheds, illegal logging damages forests, trash dumping and sediment stop up storm drains (50 percent are out of order...), new buildings weaken river channels; lack of educational campaigns, poor emergency readiness, forest burning.[22]

Tragically, we have lived through this story before. In 1974, Hurricane Fifi swept through Honduras, killing about 8,000 people and causing about $1 billion in damages. Shortly after this event, studies showed that the destruction was exacerbated by various social, economic, and political conditions. These included deforestation, as well as the displacement of campesinos into isolated valleys and onto steep hillsides by foreign-owned banana plantations and large-scale beef ranches. After Hurricane Mitch, studies again implicated these same factors. The report of the 1999 donors' conference states that the tragedy "was magnified by man-made decisions due to poverty that led to chaotic urbanization and soil degradation."[23] This cycle of vulnerability is made vivid by the following description:

> On the North Coast, the Aguan River flooded big after Fifi. It is a closed basin and dumps huge amounts of water straight into the ocean. Not only did the same flooding occur with Mitch, but it carried the village of Santa Rosa de Aguan out to sea, drowning dozens. There was no effort in the headwaters to do something to avoid this repeat catastrophe.[24]

What I am suggesting is that the moral risk of a policy of "adaptation only" is that it will hit the poor the hardest, yet it is they who have done the least to bring about climate change. They will suffer the worst impacts, and they have the least resources for adaptation.

Some people would deny that the poor are most vulnerable, pointing to the long history of mutual accommodation between indigenous peoples and their environments. However, underdevelopment is not the same as lack of development. In some regions of the world people are less able to feed themselves and to manage their environments than they were in the distant past.[25] In some cases contact with the Northern-dominated global economy has brought the risks of capitalism without the benefits. Traditional ways of coping have been lost or driven out, while modern approaches are not available. From this perspective underdevelopment should be thought of as something that has been produced by the global economy rather than as some point of origination from which development proceeds. This, however, is not to endorse any "myth of merry Africa" in which all was paradisiacal before European contact. No doubt, in many regions "capitalist scarcity [has simply] replaced precapitalist famine."[26]

Whatever is true about the details of these speculations, it is clear that poor countries will suffer most from climate change just as poor countries suffer most today from climate variability and extreme events. Honduras suffers more from hurricanes than Costa Rica, Ethiopia suffers more from drought than the United States, and probably no country is more affected by floods than Bangladesh. In 1998, 68 percent of Bangladesh's landmass was flooded, affecting about 30 million people, and this was only one of seven major floods that occurred over a 25-year period. Generally, 96 percent of disaster-related deaths in recent years have occurred in developing countries.[27]

The vulnerability of poor countries to climate change has been widely recognized in international reports and declarations, including the most recent IPCC report.[28] The Johannesburg Declaration, issued on the 10th anniversary of the 1992 Rio Earth Summit, declared that "the adverse effects of climate change are already evident, natural disasters are more frequent and more devastating and developing countries more vulnerable."[29] The Delhi Declaration, cited earlier, expressed concern at the vulnerability of developing countries, especially the least developed countries (LDCs) and small island developing states (SIDSs), and identified Africa as the region suffering most from the synergistic effects of climate change and poverty.

One response to the fact that it is the poor countries that will suffer most from climate change would be to internationalize the costs of adaptation. This is favored by many of those in the research community who have championed adaptation and was also envisioned in Article 4.4 of the FCCC, which commits developed countries to "assist the developing country Parties that are particularly vulnerable to the adverse effects of climate change in meeting costs of adaptation to those adverse effects."

Discussions about providing such assistance did not begin until COP 1 in Berlin in 1995, and only recently have begun to move to the center stage. The 2001 Marrakech Accords established three new funds to assist developing countries with adaptation. The Least Developed Countries Fund supports the development of adaptation action plans. The Special Climate Change Fund assists all developing countries (not only the LDCs) with adaptation projects and technology transfer. The Kyoto Protocol Adaptation Fund finances concrete adaptation projects and programs. The latter fund is resourced by an adaptation levy placed on transactions under the clean development mechanism, the program under which greenhouse-gas reductions are traded between companies in the developed and developing world. The other two funds are supported by voluntary contributions. Canada and Ireland have committed $10 million to the Less Developed Country Fund, and various nations have pledged to contribute a total of $450 million per year to the Special Climate Change Fund. These funds were supposed to begin operation in 2005, but they were stalled at the COP 10 meeting in December 2004, in part due to demands by Saudi Arabia that it receive compensation if the world turns away from the use of fossil fuels.

While I am in favor of establishing these funds, many practical problems must be overcome before significant resources are invested, and even on the most optimistic scenarios there are clear limitations on what these funds can accomplish.[30] Parry et al. (2001) have shown that on "business as usual" emissions scenarios, hundreds of millions of additional people will be at risk from hunger, malaria, flooding, and water shortages. Economists standardly estimate the damages of climate change on such scenarios at 1.5 to 2 percent of GDP.[31] This implies damages of between $705 billion and $940 billion per year in current dollars once the full impacts of climate change are felt. The damages from sea-level rise alone have been estimated at $2 trillion over the next 50 years.[32] Although more than half of global GDP is in the developed countries, the damages of climate change are likely to be significantly higher than 2 percent of GDP in the LDCs.

These numbers have an air of unreality about them, and the cost of adaptation would presumably be less than the damages that climate change would entail. Still, even if the Marrakech mechanisms were fully funded, it seems quite unlikely that they would begin to approach the level of resources required to fully finance adaptation to climate change in the poor countries. Moreover, even if these mechanisms would significantly defray the costs of adaptation for the poor, another injustice would be entailed. The United States is the largest emitter of GHGs; yet it is outside the Kyoto framework, thus not a contributor to the funds established by that agreement. It is difficult to see any system as just in which the world's largest emitter of GHGs does nothing to pay for the damages it causes.

Even more troubling than the fact that poor countries suffer more from climate-related impacts than rich countries is the fact that poor people suffer more from such impacts than rich people, wherever they live. The disproportionate impact on the poor was specifically cited in the donors' report on Hurricane Mitch, but this pattern of the poor suffering most from extreme climatic events has been documented as far back as the "little Ice Age" that occurred in Europe from 1300 to 1850.[33]

A recent example is the Chicago heat wave of July 14–20, 1995. In a fascinating book, Klineberg (2002) documents in detail the victims of this event; they were disproportionately low-income, elderly, African-American males living in violence-prone parts of the city. A total of 739 people died in the heat wave, more than four times as many as in the Oklahoma City bombing that occurred three months earlier although it received much less media attention. This pattern of the poor suffering disproportionately from climate-related impacts, even in rich countries, occurred once again in the wake of Hurricane Katrina, which struck the Gulf Coast of the United States in September 2005. As I write these words the damages have not yet been assessed, but it is clear that they are quite catastrophic.

Poor people suffer more than rich people from climate-related impacts, wherever they live, but poor people in poor countries suffer most of all. A recent report from a consortium of international organizations concluded that

> climate change will compound existing poverty. Its adverse impacts will be most striking in the developing nations because of their geographical and climatic conditions, their high dependence on natural resources, and their limited capacity to adapt to a changing climate. Within these countries, the poorest, who have the least resources and the least capacity to adapt, are the most vulnerable.[34]

This conclusion should not be surprising since the poor suffer more from "normal" conditions, and often only need a good shove to plunge into catastrophe.

Climate change and variability have enormous and increasing impacts on developing countries, yet very little has been done to integrate these considerations with overall development objectives. At the United Nations Millennium Summit in September 2000, the world's governments committed themselves to eight Millennium Development Goals (MDGs), the achievement of which is supposed to result in a 50 percent reduction in global poverty by 2015. Despite the fact that one of these goals is "ensuring environmental sustainability," the MDGs make no mention of climate change or climate-related disasters as threats to environmental sustainability or to the overall goal of poverty reduction. Yet the report from the African Development Bank et al. (2003) quoted earlier states that "climate change is a serious threat to poverty reduction and threatens to undo decades of development effort." A similar conclusion was reached in a recent review of the United Nations International Decade for Natural Disaster Reduction, which stated that "millennium development targets cannot be reached unless the heavy human and economic toll of disasters is reduced."[35] It is clear that climate change and variability should be thought of not only as environmental problems but also as major influences on the development process itself.[36]

These claims are borne out by a brief look at some examples. Climate change is expected to increase the incidence of malaria in some regions. While malaria is a human health problem, it is also an obstacle to development. Gallup and Sachs (2000) found that between 1965 and 1990, a high incidence of malaria was associated with low economic growth rates and that a 10 percent reduction in malaria was associated with a 0.3 percent increase in economic growth. Freeman, Martin, Mechler, Warner, and Hausmann (2002) showed that in Central America over the next decade, exposure to natural disasters could shrink a growth rate of 5 to 6 percent per year to one that is virtually flat. This would have the effect of consigning millions to poverty which they might otherwise escape.

It is the poor who suffer most from climate-related disasters, and in the end they are largely on their own. International assistance is typically inadequate, and many of the changes required to reduce vulnerability can be made only by affected communities themselves in conjunction with their governments. In turn local, regional, and national decision makers are often constrained by the economic and political realities of the global order. There is little reason to expect this pattern to shift as a changing climate increasingly makes itself felt in climate-related disasters.

Grand proposals have been made for addressing these problems. For example, Al Gore (1992) proposed a "Global Marshall Plan" aimed at "heal[ing] the global environment." Even if there were popular support for such proposals, there would not be much reason to be optimistic. Rich countries, perhaps especially the United States, have the political equivalent of attention deficit disorder. A "Global Marshall Plan," or even a conscientious effort to finance adaptation to climate change on a global scale, would require a level of sustained commitment that most Western societies seem incapable of maintaining, especially now when the war on terrorism presents similar challenges and is perceived as much more urgent. Indeed, if we had the moral and political resources to internationalize adaptation and distribute the costs fairly, it seems likely that the attempt to control emissions would succeed and we could effectively mitigate the effects of climate change. A just approach to adaptation is not really an alternative to a just approach to mitigation, since it would mobilize the same resources of respect and reciprocity. Just as we must acknowledge the necessity of adaptation, so a just approach to climate change cannot escape the challenge of mitigation.[37]

Mitigating climate change by reducing GHG emissions is important for a number of reasons. First, slowing down the rate of change allows humans and the rest of the biosphere time to adapt, and reduces the threat of catastrophic surprises.[38] Second, mitigation, if carried out properly, holds those who have done the most to produce climate change responsible, at least to some extent, for their actions. It is a form of moral education. As President Bush has said in other contexts, it is important for actions to have consequences. As I have said, mitigation as envisioned by the FCCC embodies aspects of the "polluter pays" principle. By bearing some costs to reduce GHG emissions, those who have been most instrumental in causing climate change bear some of the burdens. An exclusive focus on adaptation is an instance of the "polluted pay" principle. Those who suffer from climate change bear the costs of coping with it.[39]

Mitigation: A Modest Proposal

There are various mitigation schemes that could plausibly be seen as both just and economically efficient, including what I have elsewhere called a "modest proposal."[40] The proposal is modest in that it conjoins two ideas that are very much alive in the policy world, each of which has influential supporters. However, the conjunction of these ideas has not been forcefully advocated because those who support one conjunct typically oppose the other. Still, the elements of the proposal have been discussed by a number of authors in varying degrees of detail.[41]

The United States government, especially during the Clinton administration, made a very strong case for the idea that a GHG mitigation regime should be efficient, and that emissions trading is a powerful instrument for realizing efficiency.[42] Developing countries, led by India, have convincingly argued that a GHG mitigation regime must be fair, and that fairness recognizes that the citizens of the world have equal rights to the atmosphere.[43] In my view both the United States and the developing countries have a point. The emphasis on efficiency promoted by the United States is potentially good for the world as a whole. The emphasis on equality promoted by the developing countries seems to me to be morally unassailable. The challenge is to construct a fair system of emissions trading.[44]

The main problem with emissions trading as it is developing is that not enough thought is being given to what might be called the end game and the start game: the total global emissions that we should permit and how permissions to emit should be allocated. I propose that we give the Americans what they want, an unrestricted market in permits to emit GHGs, but that we distribute these permits according to some plausible principle of justice.

What would be such a principle? I can think of the following general possibilities.

1. Distribute permissions on a per capita basis.
2. Distribute permissions on the basis of productivity.
3. Distribute permissions on the basis of existing emissions.
4. Distribute permissions on the basis of some other principle.
5. Distribute permissions on the basis of some combination of these principles.

Principles 4 and 5 are principles of last resort,[45] and principle 3 is implausible. The existing pattern of emissions primarily reflects temporal priority in the development process, rather than any moral entitlement. In general, it is hard to see why temporal priority in exploiting a commons should generate any presumptive claim to continue the exploitation. Suppose that I started grazing a large herd of cows on some land that we own together before you were able to afford any cows of your own. Now that you have a few cows you want to graze them on our land. But if you do, some of my cows will have to be taken off the land and as a result I will be slightly less rich. Therefore, I demand compensation. Surely you would be right in saying that since we own the land in common you have a right to your fair share. The fact that you haven't been able to exercise that right does not mean that you forfeited it.

Principle 2 has a point. Surely we would not want to allocate emissions permissions toward unproductive uses. If the world can only stand so many GHG emissions, then we have an interest in seeing that they are allocated toward efficient uses.[46] But what this point bears on is how emissions should be allocated, not on how emissions permissions should initially be distributed. Markets will allocate permissions toward beneficial uses. But it is hard to see why those who are in a position to make the most productive use of GHGs should therefore have the right to emit them for free. This is certainly not a principle that we would accept in any domestic economy. Perhaps, if you owned my land, you would use it more productively than I do. For this reason you have an incentive to buy my land, but this does not warrant your getting it for free.

In my opinion the most plausible distributive principle is one that simply asserts that every person has a right to the same level of GHG emissions as every other person. It is hard to see why being American or Australian gives someone a right to more emissions, or why being Brazilian or Chinese gives someone less of a right. The problem with this proposal is that it provides an incentive for pro-natalist policies. A nation can generate more permissions to emit simply by generating more people. But this problem is easily addressed. For other purposes the FCCC has recognized the importance of establishing baseline years. There is no magic in 1990 as the reference year for emissions reductions. But if 1990 is a good year for that purpose, let us just say that every nation should be granted equal per capita emissions permissions, indexed to its 1990 population. If you do not like 1990, however, then index to another year. It is important to my proposal that per capita emissions be indexed to some year, but exactly which year is open to negotiation.[47]

Three problems (at least) remain. First, in indexing emissions to 1990 populations I am in effect giving the developed countries their historical emissions for free. But don't the same considerations that suggest that everyone who was alive in 1990 should have equal permissions apply to everyone who has ever lived? There is some force to this objection. But knowledge of the consequences of GHG emissions does to some extent seem morally relevant. Suppose that when my mother grazed her cows on our common property, the world was very different. Neither of us thought of what we were doing as eroding common property. Indeed, neither of us thought of the area on which the cows were grazing as property at all. I benefited from the activities of my mother, but neither your mother nor mine was aware of any harm being produced. If my mother had been cleverer perhaps she would have asked your mother for the exclusive right to graze cows on this piece of land. Perhaps your mother would have acceded because she had no cows and didn't think of land—much less this land—as property (much less as her property). Suppose that I say that since we now have different understandings, I'm going to set matters right, and that from this point on you have an equal right to graze cows on our land. I acknowledge that if I am to graze more cows than you I will have to buy the right.

I think many people would say that I have done enough by changing my behavior in the light of present knowledge. Perhaps others would say that there is still some sort of unacknowledged debt that I owe you because of the benefits I reaped from my mother's behavior.[48] But what I think is not plausible to say is that what my mother did in her ignorance is morally equivalent to my denying your right to use our land to the same extent that I do. For this reason I don't think that historical emissions should be treated in the same way as present and future emissions. The results of historical emissions are also so much a part of the fabric of the world that we now presuppose that it is difficult to turn the clock back. At a practical level, countries such as Canada, Australia, and the United States have had a difficult time determining what compensation they owe their indigenous peoples. Determining the effects of unequal appropriation of the atmosphere through history would be even more difficult.

The second problem is that some would insist that it matters where GHG emissions occur, not because of their impact on climate, but because of their effects on quality of life. A high quality of life, it is argued, is associated with high levels of GHG emissions. What this objection brings out is that a bad market in emissions permissions would be worse than no market at all. In a properly functioning market, nations would only sell their emissions permissions if the value of the offer was worth more to them than the permission to emit. But while no international market in emissions permissions could be expected to run perfectly, there is no reason to think that such a market cannot run well enough to improve the welfare of both buyers and sellers.

This leads to the problems of monitoring, enforcement, and compliance. These are difficult problems for any climate regime. Perhaps they are more difficult for the regime that I suggest than for others, but I think that it is clear that any meaningful emissions control will require a vast improvement in these areas.[49]

The scheme that I suggest has many advantages. It would stabilize emissions in a way that would be both efficient and fair. It would also entail a net transfer of resources from developed to developing countries, thus reducing global inequality.

Agents and Beneficiaries

Thus far I have argued that it is important to mitigate climate change both in order to reduce the risks of a climate surprise and because a policy that involves mitigation is more likely to distribute the costs fairly than a policy of "adaptation only." I have also briefly sketched and defended one approach to mitigation that is both fair and efficient. However, it is one thing to say how the world ought to be and it is another to give an account of whose responsibility it is to bring that world about. When it comes to the specification of moral agents and beneficiaries at the global scale, there are three important models in play.[50]

The first model is the familiar one of state sovereignty that goes back at least to the Treaty of Westphalia in 1648. This model sees states as morally decisive over their own people, and the international order as constructed from agreements or conquests among these sovereigns. In this view states are both the agents and beneficiaries of any duties that might exist to address climate change. While this view continues to have strong advocates, in a world in which people and states are tied together by a single environment, a globalized economy, and common threats, this model seems less plausible than it once did.[51] Indeed, it is rejected both by those who seek to establish a global order based on human rights and environmental protection, and by those who want to establish the hegemony of a single power based on its unique commitment to some set of preferred values.[52]

A second model, the sovereignty of peoples, has been developed by Rawls (1999), arguably the leading political theorist of the 20th century. Rawls characterizes a people as having the following three features: a reasonably just government that serves its interests in various ways, including protecting its territory; a common culture, usually in virtue of speaking the same language and sharing historical memories; and finally, having a moral conception of right and justice that is not unreasonable. A society of peoples is established when decent peoples agree to adopt the law of peoples, codified in eight principles that express a commitment to keep agreements and to honor human rights, and to go to war only in self-defense and then to abide by the laws of war.

While Rawls is a liberal and his account of the law of peoples is sometimes called "a theory of liberal sovereignty," he specifically rejects the idea that a theory of distributive justice applies globally. The main reason for this is that the purpose of the negotiation that leads to establishing of the law of peoples is to arrive at "fair terms of political cooperation with other peoples."[53] Representatives of peoples would accept duties to contribute to the welfare of other peoples, but they would only be instrumental to the larger purpose of assisting other peoples to play their proper role in the society of peoples. Either as peoples or individuals we do not, according to Rawls, have direct duties to the individuals who constitute other peoples.

Rawls's distinction between peoples and states is central to his view; yet it is difficult to maintain. "Peoples," insofar as this concept is well defined, seem suspiciously statelike. One way that peoples are supposed to be importantly different from states is that, unlike states as traditionally conceived, peoples can only wage defensive wars and must honor human rights. However, these features do not clearly distinguish states from peoples, since they can be seen as moral restrictions on the sovereignty of states rather than as indicating a change of subject from states to peoples. If peoples are not states, then it is unclear what they are or whether they behave coherently enough to star in a theory of international justice.

Rawls speaks as if peoples are well defined, self-contained, and as if they map onto territories and the Law of the Excluded Middle applies to membership in them. None of this is true. We need only to contemplate the claims of Palestinians, Kurds, or Orthodox Jews, or consider various national laws that attempt to legislate a people's identity in order to see that the very attempt to define a people is a problematical and highly political act. The fact that peoples are not self-contained and do not map onto specific territories is evidenced by several recent wars, notably in the Balkans. That the Law of the Excluded Middle does not apply to membership in a people can be seen by Mexican-Americans, Irish-Americans, or any number of other claimed, hyphenated identities. Indeed, individuals may shift their identities, depending on their purposes.[54] These considerations suggest that either Rawls's law of peoples is at heart a "morality of states," which he denies, or it is founded on a vague and unstable concept.

One particularly objectionable feature of Rawls's views is that because he thinks of peoples as normally occupying territories, he invests national boundaries with a moral significance that they do not have.[55] It is unjust, if anything is, that a person's life prospects should turn on which side of a river she is born, or where exactly an imaginary line was drawn decades ago by a colonial power. But for Rawls, there is nothing morally objectionable about the arbitrariness of borders or the differential life prospects that they may engender. When a pregnant woman in Baja California (Mexico) illegally crosses the border to San Diego, California (United States), so that her child will be born an American citizen with all the advantages that brings, there is for Rawls nothing troubling about the circumstances that motivate her action. Peoples have the right to control the borders of their own territories, but how can we fault a woman for doing what she thinks is best for her child?[56]

Problems such as these lead people to embrace a third view, cosmopolitanism, which holds that it is individual people who are the primary agents and beneficiaries of duties.[57] In this view duties, including duties of distributive justice, project across national boundaries, connecting individuals with each other, regardless of citizenship and residency.

While there are real differences between Rawls and his cosmopolitan critics, I believe that they can be brought closer together than one might think. Perhaps we can begin to see this when we realize that Rawls and his critics are to some extent motivated by different concerns. Cosmopolitans are concerned with what we might call moral or social "ontology." They insist that it is individual people who are the fundamental grounds of moral concern, not collectives or abstractions such as peoples or nations. Rawls is concerned with the question of how peoples with different views of the good can cooperate fairly with each other, and move together toward a peaceful future in which human rights prevail.[58] From the perspective of a person in a developing country who is being provided with a micro-loan (for example), it makes little difference whether she is being aided because she is the direct beneficiary of a moral obligation, or because the people of which she is a part is being aided so that it can become part of the society of peoples.[59]

Rather than adjudicating between these views, I want to offer another perspective. We do not have to choose between being individuals who have duties to other individuals, or being members of a people that owes duties to other peoples. Both are true, and more besides. We are parents, students, members of NGOs, Irish-Americans, Muslims, citizens of towns and states, stockholders, consumers, patrons of the arts, sports fans, home owners, commuters, and so on. We occupy multiple roles that have different responsibilities and causal powers attached to them. It is from these roles and powers that duties flow.

For example, I may have duties to reduce my consumption of energy, encourage my acquaintances to do the same, join organizations and support candidates that support climate-stabilization policies, disinvest in Exxon, support NGOs and projects in developing countries that assist people in adapting to climate change, and contribute to organizations that protect nonhuman nature. Exactly what duties I have depends on many factors including my ability to make a difference, how these duties compete with other moral demands, and so on. In the picture that I am urging, our duties form a dense web that crosses both institutional and political boundaries. We do not have to choose between accounts that privilege particular levels of analysis.[60]

A full account would have to explain exactly how the clear, urgent duties relating to adaptation and mitigation that I have described map onto us as individuals in the various roles that we occupy. Indeed, it is here where much of the slippage occurs between the abstract recognition of what ought to be done and what I am motivated to do. In fact, a kind of "shadow" collective-action problem can break out within each of us. I may agree that as a consumer I am responsible for intolerable amounts of GHGs, yet it may be very difficult to disaggregate this responsibility to me in my various roles as father, teacher, little league baseball coach, and so on. Many questions remain, but my central claim is clear: we have strenuous duties to address the problem of climate change, and they attach to us in our various roles and relationships.

Objections

The simplest objection to what I have said would involve denying that there are any such things as duties that transcend national boundaries.[61] Whatever plausibility such a claim might have would rest on supposing that it is neutral in applying to all countries and their citizens equally. For example, this claim would imply both that Americans have no duties to Sierra Leoneans and that Sierra Leoneans have no duties to Americans. However, while this claim may be formally neutral it certainly is not substantively neutral.[62] Americans, acting both as individuals and through their institutions, can greatly influence the welfare of the citizens of Sierra Leone, but Sierra Leoneans are virtually powerless to influence the welfare of Americans. Thus, the apparently reciprocal nature of the duties involved can easily be seen as a mere charade.[63]

However, it is easy to see why in the past some may have thought that duties do not transcend national boundaries. Famines and other disasters have occurred throughout history, but in many cases it was not known outside the affected regions that people were dying. Even when it was known and people were willing to provide assistance, little could be done to help those in need. When people are not culpably ignorant and they are not in a position to be efficacious, there is little point in ascribing duties to them. But today things are very different with respect to information and causal efficacy. We live in an age in which national boundaries are porous with respect to almost everything of importance: people, power, money, and information, to mention a few. These help to make obligations possible. If people, power, money, and information are so transnational in their movements, it is hard to believe that duties and obligations are confined by borders.[64] The view that duties do not transcend national boundaries (unlike lawyers, guns, and money—not to mention drugs and immigrants) is really equivalent to denying people in the developing world a place at the table. It is the global equivalent of the domestic denial of rights to women and minority populations.

While most philosophers and theorists these days would not challenge the very existence of transnational duties, some would hold that there are very few such duties and that they are comparatively weak. Such a view is sometimes expressed by granting the existence of transnational duties but denying that they are duties of justice. There are two distinct grounds for such a view.

The first ground, which is broadly based in the tradition of the 17th-century philosopher Thomas Hobbes, is based on denying that there is any such thing as "natural justice." On this view justice is entirely a matter of convention: justice consists in conforming to enforceable agreements: injustice consists in violating them. Since there is little by way of enforceable, international agreements, there are few transnational duties.

The second ground for such a view is based on a communitarian account of justice. While this view may grant that enforceable agreements across communities can generate duties of justice, it holds that such duties typically arise within, rather than among, communities, and do not require explicit agreements. Since the world is characterized by a plurality of communities rather than by a single global community, the necessary condition for a dense network of transnational duties of justice is not satisfied. Thus, communitarians come to the same conclusion as Hobbesians: there is little ground for supposing that there is a panoply of transnational duties of justice.[65]

I will not mount a systematic refutation of these views here but instead restrict myself to a single observation about the view that while transnational duties may exist, they are not duties of justice. As I have indicated. there are different grounds for such a denial. Such a denial may rest on the view that some transnational duties are distinct from duties of justice because they do not originate in agreement, are not owed to specific beneficiaries, or are less urgent than duties of justice. What I want to insist on is that that there are urgent duties to respond to climate change, that those of us who are part of the global middle class contribute significantly to causing the problem, and that we can identify generally those who will suffer from our actions.[66] If this much is granted, then I am not sure that anything of significance turns on either asserting or denying that the duties in question are duties of justice.[67]

The second objection has been raised most consistently and forcefully by Schelling (1992, 1997, 2000), who argues in the following way. Suppose that it is true that we have duties to improve the welfare of those who are worse (or worst) off. There are other, more efficient and efficacious, ways of doing this than by reducing our GHG emissions. For example, we could invest in clean water systems, vaccinations, literacy programs, and so on. Or we could simply give money to those who are worse off. Schelling concludes that

> it would be hard to make the case that the countries we now perceive as vulnerable would be better off 50 or 75 years from now if 10 or 20 trillions of dollars had been invested in carbon abatement rather than economic development.[68]

While this objection has some force, plausible responses can be given.

First, for any actual transfer from the rich to the poor, there is likely to be another possible transfer that is more beneficial. However, this does not imply that every such transfer we make is wrong, irrational, or ill advised. This is because the alternative policies we choose between are not all those that are logically or physically possible, but those that have some reasonable chance of actually being implemented. Some of our duties with respect to climate change have a reasonable chance of being implemented because they involve controlling our own behavior or taking action in a democratic society. Even if the results of our discharging these duties were not optimal relative to the set of logically or physically possible actions that we might perform, their consequences would be very good indeed and this is sufficient for making it at least morally permissible to carry them out.[69]

Furthermore, the duty to mitigate climate change does not depend on some general duty to benefit the worse (or worst) off. Such a principle might generate this duty, but so would more modest principles that require us to refrain from imposing serious risks on others. Indeed, the modesty of the principles required to ground such duties is part of what makes action on climate change both possible and urgent, despite the obstacles hindering such action.[70]

Finally, transferring resources to the worse (worst) off rather than mitigating our carbon emissions would do nothing to reduce the risk of catastrophic climate change. Nor would it provide comfort to those morally considerable aspects of nature that are vulnerable to climate change. There is no guarantee that transforming the poor into the rich would in itself protect environmental values, such as respect for what is wild and natural, that are at the heart of many people's concern about climate change.

For these reasons, despite the power of Schelling's objection, the idea that we have a duty to mitigate climate change is not defeated.

The Problem of Motivation

Even if what I have said is correct, a problem may linger. Morality is fundamentally directed toward action. Many would say that it seems clear that we are not motivated to address this problem. What is the point of seeing climate change as posing moral questions if we are not motivated to act? To this I have four related responses.

First, outside the United States, especially in Europe and the developing world, the problem of climate change is widely seen as a moral issue. Much of the anger at the American withdrawal from the Kyoto Protocol can only be understood by appreciating this fact. Seeing climate change as posing moral questions is part of appreciating others' points of view. Of course, having appreciated how climate change can be viewed in this way, we are free to reject this perspective. However, I believe that once we appreciate climate change as a moral problem, this view is virtually irresistible.[71]

Moreover, rejecting the moral framing of the climate-change problem and instead approaching it from the perspective of self-interest does not lead to solutions. Although I think we could get farther on this ground than we have gotten thus far, ultimately acting on the basis of narrow self-interest locks us into collective-action problems that lead to worse outcomes overall. This is borne out by the current state of climate-change negotiations and also helps explain why we as individuals often feel so powerless in the face of this problem.[72]

Third, a moral response to climate change is difficult to escape. For the challenge of climate change is not only global and abstract, but also local and intimate. Once obligations are seen in the way described in the previous section—as forming a dense web of connections that link us in our myriad roles and identities to people all over the world—then it becomes clear that virtually everything we do is morally valenced. When we bike instead of drive or donate money to Oxfam, we issue moral responses to the problem of climate change. Denying responsibility, dissembling, and ignoring the problem are themselves moral responses.

Finally, I think that it is a plain fact that climate change poses moral questions. While I do not want to argue in detail here about the concept of morality or defend the idea that there is a simple and direct relation between grasping the way the world is and being motivated to act, surely there is some connection between seeing an act as morally right and performing it. That something is the morally right thing to do is a powerful consideration in its favor. It may not always carry the day, but it cannot easily be ignored.

Taken together, these considerations go some way toward demonstrating the utility of viewing climate change as a moral problem.

Concluding Remarks

There are some reasons to be hopeful that the global community is beginning to wake up to the problem of climate change. The Kyoto Protocol came into effect in 2005, and the European Union is eager to take more aggressive action after 2012, when the first Kyoto commitment period expires. American corporations that do business outside the United States will be governed by the Kyoto system, and many are increasingly receptive to the idea of a single global system for managing GHG emissions. Even the northeastern states and California, largely ruled by Republican governors, are moving toward adopting their own GHG emissions policies. Meanwhile, the Inuit peoples are preparing a case to present to the Inter-American Commission on Human Rights, charging that the United States is threatening their existence through its contributions to global warming.

Despite these signs of hope, climate change is a scientifically complex issue that is difficult to address effectively and, in the United States at least, politicians can safely ignore this issue without fear of punishment. It is in part another victim of the war on terrorism. While climate change may be far from the public mind, GHGs continue to build up in the atmosphere, and the risks of climate change continue to magnify. When it comes to responding to fundamental changes in the systems that control life on earth, denial, distortion, and spin are not viable long-term strategies.[73] Eventually, concern about climate change will emerge as an important public issue, and a movement toward creating a law of the atmosphere will gain momentum.

In the meantime it is important to recognize that those who suffer from extreme climatic events are often the victims of greed, indifference, and mendacity. It is human beings and their societies that are largely responsible for the climate change now under way, not nature or fortune. People and nations who willfully evade taking responsibility for the consequences of their actions may one day be called to account.

Acknowledgments

I have lectured on this material at many universities and conferences around the world and I regret that I cannot acknowledge all those from whom I have learned. However, I would like to thank the participants in the workshop at Dartmouth College for which this chapter was originally prepared, and the following for their written comments on earlier versions of the manuscript: Kier Olsen DeVries, Steve Gardiner, Roger Pielke Jr., Dan Sarewitz, Peter Singer, Walter Sinnott-Armstrong, Christine Thomas, and Leif Wenar.

Notes

1. In discussions of climate change "mitigation" refers to policies or actions directed toward reducing greenhouse-gas emissions: "adaptation" refers to how plants, animals, and humans respond to climate change (excluding, of course, their mitigation responses). The meaning of these terms is further elaborated later.

2. For an account of the formation of the IPCC, see Agrawala 1998.

3. Emissions trading is a scheme in which an entity (such as a nation) whose emissions of some substance are limited by a binding agreement can purchase the right to emit more of the substance in question from an entity that will limit its emissions by the same amount in exchange for the payment (emissions trading is discussed in detail below). Carbon sinks are biological or geological reservoirs (such as forests) in which carbon is sequestered, the idea being that nations can "offset" their emissions by sequestering carbon that would otherwise be in the atmosphere.

4. Annex 1 countries are the industrialized countries of North America and Europe, Japan, Australia, and New Zealand (a full list can be found at http://unfccc.int/resource/docs/convkp/conveng.pdf); together they were responsible for more than two-thirds of global GHG emissions in 1990.

5. Babiker, Jacoby, Reilly, and Reiner 2002.

6. For a list of OPEC member states see www.opec.org.

7. Smit, Burton, Klein, and Wandel 2000, p. 225. It should be noted that the term *adaptation* is typically used positively in opposition to the negative term *maladaptation*.

8. See, for example, Abramovitz et al. 2002. Smithers and Smit 1997; Kates 2001; Kelly and Adger 2000; Reilly and Schimmelpfennig 2000; and Smit, Burton, Klein, and Wandel 2000.

9. Still, it is worth observing that adaptations can stand in feedback relations to the climate change to which they are a response. For example, one possible adaptation to a warmer world is more extensive use of air conditioning, which itself contributes to greater warming. Thus, we must be careful that in trying to live with climate change, we do not make it worse. I owe this point to Steve Gardiner.

10. For example, see Jamieson 1990, 1991.

11. For example, Rayner and Malone 1997; Pielke Jr. 1998; Parry, Arnell, Hulme, Nicholls, and Livermore 1998; and Pielke Jr. and Sarewitz 2000.

12. Because he has a definition of the term different from the one employed in the FCCC, Pielke Jr. (2005) claims that adaptation is a neglected option, despite the occurrence of the word in the treaty and in many subsequent official documents. This way of putting the point seems to transform an important substantive critique into what appears to be a linguistic dispute. The core of Pielke Jr.'s challenge is that focusing on adaptation to climate variability and extreme events, whatever their causes, would be much more effective than focusing on climate change, with the emphasis on scientific knowledge and mitigation strategies that this approach brings along, and the attendant policy gridlock that follows. While I am sympathetic to this view, it raises important questions about how to determine relevant alternatives when faced with policy questions. Why not, for example, abandon questions of weather and climate altogether and focus instead on global poverty? I have more to say about this in my response to Schelling below.

13. From http://www.climatescience.gov/Library/stratplan2003/vision/default.htm.

14. From http://www.epa.gov/oppeoeel/globalwarming/publications/car/ch6.pdf.

15. The idea that climate change poses a dichotomous choice between adaptation and mitigation may stem from Matthews (1987), who drew a sharp distinction between those she called "adaptationists" and "preventionists"; but already by 1991 Crosson and Rosenberg (1991) were treating this as a mistaken dichotomy that had been bypassed by the policy discussion.

16. National Academy of Sciences 2002, p. 27.

17. Häkkinen and Rhines 2004.

18. Thompson and Wallace 2001.

19. Available at http://amap.no/workdocs/index.cfm?dirsub=%2FACIA%2Foverview.

20. The following discussion is based on Glantz and Jamieson 2000.

21. Summary Report of Proceedings: Inter-American Development Bank Consultative Group Meeting for the Reconstruction and Transformation of Central America (May 1999), Stockholm. Available at http://www.iadb.org/regions/re2/consultative_group/summary.htm.

22. *Honduras This Week*, May 29, 2000. Available at http://www.marrder.com/htw/special/environment/70.htm.

23. Summary Report, Inter-American Development Bank Consultative Group.

24. *Honduras This Week*, May 29, 2000.

25. Davis 2001.

26. Iliffe 1987, p. 3.

27. See African Development Bank et al. 2003 and the sources cited therein for documentation of the claims made in this paragraph.

28. IPCC 2001.

29. Available at http://www.johannesburgsummit.org/html/documents/summit_docs/1009wssd_pol_declaration.doc.

30. One problem is that these funds are intended to finance adaptation to climate change, not adaptation to natural climate variability. This requires a successful applicant to identify the incremental risk posed by climate change and show that the benefit that the proposed project would provide would address only this increment. This burden is not only almost impossible to discharge in many cases, but it is an absurd requirement for reasons explained below.

31. IPCC 2001.

32. Ayres and Walters 1991, as cited in Spash 2002, p. 164.

33. Fagan 2001.

34. African Development Bank et al. 2003, p. 1.

35. From http://www.id21.org/society/S10aisdr1g1.html.

36. See also Jamieson 2005a.

37. For reasons discussed in the next section and suggested in note 30, it is also easier to specify and quantify duties related to mitigation than those related to adaptation. Carbon dioxide emissions are directly measurable; success in adapting to climate change is not.

38. However, we should bear in mind that, though they are importantly related, reducing

emissions is not exactly the same as slowing down the rate of climate change (Pielke Jr., Klein, and Sarewitz 2000).

39. For more on justice in adaptation see Adger, Huq, Mace, and Paavola 2005.

40. Jamieson 2001.

41. For example, Athanasiou and Baer 2002; Brown 2002; Cazorla and Toman 2001; Clausen and McNeilly 1998; Grubb 1995; Meyer 2000; Sachs et al. 2002; Shue 1995; Singer 2002; and the papers collected in Toth 1999. Of course, these ideas also have their detractors. For a critique of emissions trading see various papers by Larry Lohmann at www.thecornerhouse.org.uk. For an excellent survey of the issues see Gardiner 2004.

42. For a thorough defense of emissions trading in a GHG control regime see Stewart and Wiener 2003; for a contrary view, see Schelling 2002.

43. For a defense of this view see Agarwal and Narain 1991.

44. The following nine paragraphs are revised from Jamieson 2001.

45. Principle 4 is a principle of last resort because my list includes all the principles that I can think of that are attractive, and principle 5 because it does not have the theoretical economy of the other principles on the list.

46. While this principle is one that is often associated with the American position and there are different ways of understanding the data, it is clear that the United States is an inefficient producer of GDP relative to most European countries and Japan. Thus, this principle might imply that some American emissions permissions should be transferred to France (for example).

47. For a defense of 2050 as the index year, see Singer 2002; generally, for a discussion, see Gardiner 2004.

48. For example, Gardiner 2004 and Shue 1992.

49. See Stewart and Wiener 2003 for further discussion of these issues.

50. See Held 2002.

51. For an argument that some transnational corporations are more powerful than many states, and hence *de facto* more sovereign, see Korten 1995 and Hutton 2002.

52. For the first view see Singer 2002; for the second see Boot 2002.

53. Rawls 1999, p. 69.

54. For more on these points see O'Neill 1994.

55. Pogge (1994) vigorously argues this point; I have learned much from his critical discussion of Rawls.

56. For further objections along these lines see Beitz 2000, Buchanan 2000, and Kuper 2000.

57. There are more expansive ways of characterizing cosmopolitanism (e.g., Jones 1999, p. 15), and less expansive ways (e.g., dropping the requirement that individual people are the primary agents); this will do for the present purposes.

58. Here I have benefited from discussions with Leif Wenar and from reading Wenar 2002.

59. For further discussion, see Crisp and Jamieson (2000).

60. Related views have been put forward by Kuper (2000) and Sen (2002). In Jamieson 2005c, I have discussed this view in some detail from a utilitarian perspective.

61. Dobson (1998) chides me for largely ignoring this view in Jamieson 1994. I have been helped by his discussion.

62. Anatole France derided the claim that laws against sleeping under bridges apply equally to the rich and poor.

63. I have selected Sierra Leone for my example since it ranks dead last in the United Nations Development Program's Human Development Index (UNDP 2000).

64. While philosophers often draw technical distinctions between duties and obligations, for present purposes I use these terms interchangeably.

65. Of course a Hobbesian or communitarian could consistently hold that there are extensive and rigorous transnational duties but that they are not duties of justice. This sort of Hobbesian or communitarian could agree with much that I say.

66. See Sachs (1993, p. 5) on the idea of the global middle class.

67. A clarification (at the behest of Walter Sinnott-Armstrong): my claim is that (everything else being equal) X's contributing significantly to causing a problem that harms a generally identifiable moral patient is a sufficient (not a necessary) condition for supposing that X has a duty with respect to the contribution.

68. Schelling 1992, p. 7.

69. Indeed, it may be obligatory to carry some of them out. There are a number of ways of defending such a claim in detail; one such way is by recourse to a moral theory that I call "progressive consequentialism" in unpublished work.

70. Because climate change involves actions in which some identifiable people and corporations are involved in inflicting harms on other people, there is beginning to be interest in viewing these

actions as candidates for legal remedies. There has been discussion of such litigation in the pages of The *New York Times*, The *Economist*, and the *Financial Times*, as well as in the offices of various reinsurance companies and multinational corporations (or so it is said). However, the most severe consequences of climate change will be suffered by those in the further future, and there are serious philosophical problems about how duties to such beneficiaries should be understood. See Parfit 1984 and Howarth's 2005 essay in this volume.

71. Indeed, I believe that there is generally a movement toward environmental justice becoming the key organizing concept of environmentalism (see Jamieson 2005b).

72. See Jamieson 2005c and Gardiner 2003.

73. Melissa Carey of Environmental Defense remarks: "The earth is round, Elvis is dead, and yes, climate change is happening."

References

Abramovitz, J., T. Banuri, P. Girot, B. Orlando, N. Schneider, E. Spanger-Siegfried, J. Switzer, and A. Hammill. 2002. *Adapting to Climate Change: Natural Resource Management and Vulnerability Reduction*. Gland, Switzerland: IUCN—The World Conservation Union.

Adger, N., S. Huq, M. Mace, and J. Paavola, eds. 2005. *Fairness in Adapting to Climate Change*. Cambridge, Mass. MIT Press.

African Development Bank; Asian Development Bank: Department for International Development. United Kingdom; Directorate-General for International Cooperation, the Netherlands; Directorate General for Development, European Commission; Federal Ministry for Economic Cooperation and Development, Germany; Organisations for Economic Development. United Nations Development Programe. United Nations Environment Program, and the World Bank. 2003. *Poverty and Climate Change: Reducing the Vulnerability of the Poor through Adaptation*. Washington, D.C.: The World Bank.

Agarwal, A., and S. Narain. 1991. *Global Warming in an Unequal World*. New Delhi: Centre for Science and Development.

Agrawala, S. 1998. "Context and Early Origins of the Intergovernmental Panel on Climate Change." *Climatic Change* 39: 605–620.

Athanasiou, T., and P. Baer. 2002. *Dead Heat: Global Justice and Global Warming*. New York: Seven Stories Press.

Ayres, R. U., and J. Walters. 1991. "The Greenhouse Effect: Damages, Costs and Abatement." *Environmental and Resource Economics* 13: 237–270.

Babiker, M. H., H. D. Jacoby, J. M. Reilly, and D. M. Reiner. 2002. "The Evolution of a Climate Regime: Kyoto to Marrakech." *Environmental Science and Policy* 2/3: 195–206.

Beitz, C. 2000. "Rawls's Law of Peoples." *Ethics* 110.4: 669–696.

Boot, M. 2002. *The Savage Wars of Peace: Small Wars and the Rise of American Power*. New York: Basic Books.

Brown, D. 2002. *American Heat: Ethical Problems with the United States' Response to Global Warming*. Lanham. Md: Rowman & Littlefield.

Buchanan, A. 2000. "Rawls's Law of Peoples: Rules for a Vanished Westphalian World. *Ethics* 110.4: 697–721.

Cazorla, M., and M. Toman. 2001. "International Equity and Climate Change Policy. In *Climate Change Economics and Policy*, ed. M. Toman, (Washington, D.C.: Resources for the Future), pp. 235–247.

Clausen, E., and L. McNeilly. 1998. *Equity and Global Climate Change*. Washington, D.C.: Pew Center on Global Climate Change.

Crisp, R., and D. Jamieson. 2000. "A Global Resources Tax: On Pogge on Rawls." In *The Idea of Political Liberalism: Essays on Rawls*, ed. V. Davion and C. Wolf (Lanham, Md.: Rowman and Littlefield), pp. 90–101.

Crosson, P., and N. Rosenberg, 1991. "Adapting to Climate Change." *Resources* 103: 17–21.

Davis, M. 2001. *Late Victorian Holocausts: El Nino Famines and the Making of the Third World*. London: Verso.

Dobson, A. 1998. *Justice and the Environment*. New York: Oxford University Press.

Fagan, B. 2001. *The Little Ice Age: How Climate Made History, 1300–1850*. New York: Basic Books.

Freeman, P., L., Martin, R., Mechler, K., Warner, and P. Hausmann. 2002. *Catastrophes and Development: Integrating Natural Catastrophes into Development Planning*. Disaster Management Facility, World Bank, Working Paper Series No. 4, Washington, D.C.

Gallup, J. L., and J. D. Sachs. 2000. *The Economic Burden of Malaria*. CID Working Paper 52, Center for International Development, Harvard University, Cambridge, Mass.

Gardiner, S. 2003. "The Pure Intergenerational Problem." *Monist: Special Issue on Moral Distance* 86.3: 481–500.

———. 2004. "Ethics and Global Climate Change." *Ethics* 114: 555–600. Also chapter 1 in this volume.

Glantz, M., and D. Jamieson. 2000. "Societal Response to Hurricane Mitch and Intra Versus Intergenerational Equity Issues: Whose Norms Should Apply?" *Risk Analysis* 20.6: 869–882.

Gore, A. 1992. *Earth in the Balance*. Boston: Houghton Mifflin.

Grubb, M. 1995. "Seeking Fair Weather." *International Affairs* 71.3: 463–496.

Häkkinen, S., and P. Rhines. 2004. "Decline of Subpolar North Atlantic Circulation during the 1990s." *Science* 304: 555–559.

Held, D. 2002. "Law of States, Law of Peoples: Three Models of Sovereignty." *Legal Theory* 8: 1–44.

Howarth, R. B. 2005. "Against High Discount Rates." In *Perspectives on Climate Change: Science, Economics, and Politics*, ed. W. Sinnott-Armstrong and R. B. Howarth (Amsterdam: Elsevier).

Hutton, W. 2002. *The World We're In*. London: Little, Brown.

Iliffe, J. 1987. *The African Poor: A History*. Cambridge, U.K.: Cambridge University Press.

Intergovernmental Panel on Climate Change (IPCC). 2001. *Climate Change 2001: Impacts, Adaptation and Vulnerability: A Contribution of Working Group II to the Third Assessment Report of the Intergovernmental Panel on Climate Change*. Cambridge, U.K.: Cambridge University Press.

Jamieson, D. 1990. "Managing the Future: Public Policy, Scientific Uncertainty, and Global Warming." In *Upstream/Downstream: Essays in Environmental Ethics*, ed. D. Scherer. (Philadelphia: Temple University Press), pp. 67–89.

———. 1991. "The Epistemology of Climate Change: Some Morals for Managers. *Society and Natural Resources* 4: 319–329.

———. 1994. "Global Environmental Justice." In R. Attfield & A. Belsey (Eds), *Philosophy and the Natural Environment*, ed. R. Attfield and A. Belsey (Cambridge, U.K.: Cambridge University Press), pp. 199–210. Reprinted in Jamieson 2002.

———. 2001. "Climate Change and Global Environmental Justice. In C. Miller & P. Edwards (Eds), *Changing the Atmosphere: Expert Knowledge and Environmental Governance*, ed. C. Miller and P. Edwards (Cambridge, Mass.: MIT Press), pp. 287–307.

———. 2005a. "Duties to the Distant: Aid, Assistance, and Intervention in the Developing World." *Journal of Ethics* 9: 151–170.

———. 2005b. "The Heart of Environmentalism." In *Environmental Justice and Environmentalism: Contrary or Complementary?* ed. P. Pezzullo and R. Sandler (Cambridge, Mass.: MIT Press).

———. 2007. "When Utilitarians Should Be Virtue Theorists." *Utilitas* 19.2: 160–183. Cf. Gardiner, chap. 1, in this volume.

Jones, C. 1999. *Global Justice: Defending Cosmopolitanism*. Oxford: Oxford University Press.

Kates, R. 2001. "Cautionary Tales: Adaptation and the Global Poor." *Climate Change* 45: 5–17.

Kelly, P. M., and W. N. Adger. 2000. "Theory and Practice in Assessing Vulnerability to Climate Change and Facilitating Adaptation." *Climatic Change* 47: 325–352.

Klineberg, E. 2002. *Heat Wave: A Social Autopsy of Disaster in Chicago*. Chicago: University of Chicago Press.

Korten, D. C. 1995. *When Corporations Rule the World*. West Hartford, Ct. Kumarian Press.

Kuper, A. 2000. "Rawlsian Global Justice: Beyond *The Law of Peoples* to a Cosmopolitan Law of Persons." *Political Theory* 28.5: 640–674.

Matthews, J. 1987. "Global Climate Change: Toward a Greenhouse Policy." *Issues in Science and Technology* 3: 57.

Meyer, A. 2000. *Contraction and Convergence*. Dartington. U.K.: Green Books.

National Academy of Sciences. 2002. *Abrupt Climate Change: Inevitable Surprises*. Washington, D.C.: National Academy Press.

O'Neill, O. 1994. "Justice and Boundaries." In *Political Restructuring in Europe: Ethical Perspectives*, ed. C. Brown (London: Routledge), pp. 69–88.

Parfit, D. 1984. *Reasons and Persons*. Oxford: Oxford University Press.

Parry, M., N. Arnell, M. Hulme, R. Nicholls, and M. Livermore. 1998. "Adapting to the Inevitable." *Nature* 395 (6704): 741.

Parry, M., N. Arnell, T. McMichael, R. Nicholls, P. Martens, S. Kovats, M. Livermore, C. Rosenzweig, A. Iglesias, and G. Fischer. 2001. "Millions at Risk: Defining Critical Climate Change Threats and Targets." *Global Environmental Change* 11.3: 181–183.

Pielke, R. Jr. 1998. "Rethinking the Role of Adaptation in Climate Change Policy." *Global Environmental Change* 8.2: 159–170.

———. 2005. "Misdefining Climate Change: Consequences for Science and Action." *Environmental Science and Policy* 8.6 (December): 548–561.

Pielke, R. Jr., and D. Sarewitz. 2000. "Breaking the Global-Warming Gridlock." *Atlantic Monthly* 286.1: 55–64.

Pielke, R. Jr., R. Klein, and D. Sarewitz. 2000. "Turning the Big Knob: An Evaluation of the Use of Energy Policy to Modulate Future Climate Impacts." *Energy and Environment* 11: 255–276.

Pogge, T. 1994. "An Egalitarian Law of Peoples." *Philosophy and Public Affairs* 23.3: 195–224.

Rawls, J. 1999. *The Law of Peoples*. Cambridge, Mass.: Harvard University Press.

Rayner, S., and E. Malone. 1997. "Zen and the Art of Climate Maintenance." *Nature* 390.27 (November): 332–334.

Reilly, J., and D. Schimmelpfennig. 2000. "Irreversibility. Uncertainty, and Learning: Portraits of Adaptation to Long-term Climate Change." *Climatic Change* 45: 253–278.

Sachs, W. 1993. "Global Ecology and the Shadow of development." In *Global Ecology: A New Arena of Political Conflict*, ed. W. Sachs (London: Zed Books), pp. 3–21.

Sachs, W., H. Acselrad, F. Akhter, A. Amon, T. B. G. Egziabher, H. French, P. Haavisto, P. Hawken, H. Henderson, A. Khosla, S. Larrain, R. Loske, A. Roddick, V. Taylor, C. Von Weizsäcker, and S. Zabelin. 2002. *The Jo'burg Memo: Fairness in a Fragile World, Memorandum for the World Summit on Sustainable Development*. Berlin: Heinrich Boll Foundation.

Schelling, T. 1992. "Some Economics of Global Warming." *American Economic Review* 82.1: 1–14.

———. 1997. "The Cost of Combating Global Warming: Facing the Tradeoffs." *Foreign Affairs* 76.6: 8–14.

———. 2002. "What Makes Greenhouse Sense?" *Foreign Affairs* 81.3: 2–9.

Schneider, S. H. 1989. *Global Warming: Are We Entering the Greenhouse Century?* San Francisco: Sierra Club Books.

Sen, A. 2002. "Justice across Borders." In *Global Politics and Transnational Justice*, ed. P. de Greiff and C. Cranin (Cambridge, Mass.: MIT Press), pp. 37–51.

Shue, H. 1992. "The Unavoidability of Justice. In A. Hurrell & B. Kingsbury (Eds), *The International Politics of the Environment*, ed. A. Hurrell and B. Kingsbury (Oxford: Oxford University Press), pp. 373–397.

———. 1995. "Avoidable Necessity: Global Warming, International Fairness and Alternative Energy." In I. Shapiro & J. W. Decew (Eds), *Theory and Practice: NOMOS XXXVII*, ed. I. Shapiro and J. W. Decew (New York: NYU Press), pp. 239–264.

Singer, P. 2002. *One World: The Ethics of Globalization*. New Haven, Ct.: Yale University Press.

Smit, B., I. Burton, R. Klein, and J. Wandel. 2000. "An Anatomy of Adaptation to Climate Change and Variability." *Climatic Change* 45: 223–251.

Smithers, J., and B. Smit. 1997. "Human Adaptation to Climatic Variability and Change." *Global Environmental Change* 7: 251–264.

Spash, C. 2002. *Greenhouse Economics: Value and Ethics*. New York: Routledge.

Stewart, R. B., and J. B. Wiener. 2003. *Reconstructing Climate Policy*. Washington, D.C.: AEI Press.

Thompson, D. W. J., and J. M. Wallace. 2001. "Regional Climate Impacts of the Northern Hemisphere Annular Mode." *Science* 293.5527: 85–89.

Toth, F. 1999. *Fair Weather? Equity Concerns in Climate Change*. London: Earthscan.

United Nations Development Program (UNDP). 2000. *Human Development Report 2000*. New York: Oxford University Press.

Wenar, L. 2002. "The Legitimacy of Peoples." In P. de Greiff & C. Cronin (Eds), *Global Politics and Transnational Justice*, ed. P. de Greiff and C. Cronin (Cambridge, Mass.: MIT Press), pp. 53–76.

Is "Arming the Future" with Geoengineering Really the Lesser Evil?

Some Doubts about the Ethics of Intentionally Manipulating the Climate System

Stephen M. Gardiner

"So convenient it is to be a reasonable Creature, since it enables one to find or make a Reason for everything one has a mind to do."

—BENJAMIN FRANKLIN

16

I. An Idea That Is Changing the World

The term *geoengineering* lacks a precise definition but is widely held to imply the intentional manipulation of the environment on a global scale.[1] For most of the last 30 years, there has been a wide consensus that such manipulation would be a bad idea. However, in August 2006, Paul Crutzen, the climate scientist and Nobel laureate, published an article that reignited debate about whether we should explore geoengineering "solutions" as a response to the escalating climate-change problem.[2] This was soon followed by other contributions and proposals,[3] and now interest in geoengineering has become widespread, in both academia and the world of policy. As a result, *Time* magazine recently listed geoengineering as one of its "Ten Ideas That Are Changing the World."[4]

Geoengineering is a relatively new and underexplored topic. This is true both of the science and the ethics. Just as we are not close to fully understanding exactly how to geoengineer if we were to choose to do so, or what the impacts of any geoengineering scheme would be, so we are also not sure how to understand the normative dimensions of undertaking geoengineering. Indeed, at this point almost no moral and political philosophy has even been attempted.[5] In such a setting, it is useful to get some sense of the moral terrain: of what the major issues might be, of how they might be investigated, and so of how understanding might move forward. This is the main aim of this chapter. To pursue it, I shall focus on one prominent argument for geoengineering, raising a number of serious challenges that have wider application. In my view, these challenges are sufficient to seriously threaten the argument, at least in its most prominent and limited form, and so shift the burden of proof back onto proponents of geoengineering. Still, I want to make clear from the outset that my purpose is not to determine whether the pursuit of geoengineering can, in the end, be morally justified.[6] Instead, my concern is with the moral implications of such a pursuit, and the deceptiveness of some arguments being offered for it. Given that (I suspect) the scientific and political momentum is such that serious research is almost certain, and ultimate deployment also probable (at least on moderately pessimistic assumptions about what the future holds), this is a serious issue. The argument I consider concedes that geoenginnering is some kind of "evil." But if we are to set an evil course, the moral costs should be exposed. This is so even if, as I shall argue, some of these costs are debts that few may have any intention of paying.

Ethical discussion of geoengineering is made more difficult by the complexity of the terrain. First, a number of interventions are already being proposed for combating climate change, and it is not clear that all of them should be classified together. For example, some suggest deflecting a small percentage of incoming radiation from the sun by placing huge mirrors at the Legrange point between it and the earth, some advocate fertilizing the oceans with plant life to soak up more carbon dioxide, some suggest a massive program of reforestation, and some propose capturing vast quantities of emissions from power plants and burying them in sedimentary rock deep underground. But do these interventions raise the same issues? Should we count all of them as "geoengineering"?[7]

Second, different arguments can be (and often are) offered in favor of the same specific intervention. For example, some advocate a given geoengineering "solution" because they think it much more cost-effective than mitigation, others say that it will "buy time" while mitigation measures are implemented, and still others claim that geoengineering should only be implemented as a last resort, to stave off a catastrophe. Such differences in rationale are important because they often make for differences in research and policy implications. For example, they can affect what kinds of geoengineering should be pursued, to what extent, and with what safeguards.

In this chapter, I focus on one specific intervention and one rationale being offered for it. The intervention is that of injecting sulfate aerosols into the stratosphere in order to block incoming solar radiation by modifying the earth's albedo.[8] The rationale is a certain kind of "lesser evil" argument. It begins by conceding both that mitigation—direct and substantial reductions in anthropogenic emissions—is "by far" the best approach to climate policy and that there is something morally problematic about geoengineering proposals. However, it goes on to claim that so far, progress on the preferred policy has been minimal,[9] so that there is reason to revisit geoengineering options. In particular, it is argued, if the failure to act aggressively on mitigation continues, then at some point (probably 40 years or more into the future), we might end up facing a choice between allowing catastrophic impacts to occur or engaging in geoengineering. Both, it is conceded, are bad options. But engaging in geoengineering is less bad than allowing catastrophic climate change. Therefore, the argument continues, if we end up facing the choice, we should choose geoengineering. However, if we do not start doing serious research on geoengineering now, then we will not be in a position to choose that option should the nightmare scenario arise. Therefore, we should start doing that research now. (I call this the "Arm the Future" argument, or AFA.)

I focus on this combination of intervention and rationale for three reasons. First, it is currently the most popular proposal under consideration and the one that most strongly motivates Crutzen.[10] (For this reason, I shall label it the "Core Proposal."[11]) Second, the focus on sulfate injection helps us to sidestep the definitional worries about what constitutes geoengineering: such direct intervention into the chemistry of the stratosphere appears to be a clear case.[12] Third, appeals to the lesser evil are attractive to a wide audience, including those who are otherwise strongly against technological intervention. Indeed, in the current context, they are often seen as almost irresistible, constituting a straightforward and decisive move that no sane person could reject. Hence, such arguments seem among those most likely to justify geoengineering.

As I have indicated, the main aim of the chapter is to explore the moral context of the decision to pursue geoengineering. Still, as a secondary matter, I argue here for three more specific conclusions. First, the arm the future argument is far from straightforward or decisive. Instead, it assumes much that is contentious and is overly narrow in its conclusions. Second, the argument obscures much of what is at stake in the ethics of geoengineering, including what it means to call something an evil and whether doing evil has further moral implications. Third, the argument arises in a troubling context, and this implies that it should be viewed with suspicion. Climate change constitutes an especially serious challenge to ethical behavior because it involves the intersection of global, intergenerational, and theoretical obstacles to action. Because of this, *we*—the current generation, and especially those in the affluent countries—are particularly vulnerable to moral corruption, that is, to the subversion of our moral discourse to our own ends. In such a setting, we should be especially cautious about arguments that appear to diminish our moral responsibilities. As Benjamin Franklin suggests, we must beware the "conveniences" of being "reasonable creatures."

The discussion proceeds as follows. Sections II and III set out the context in which the core proposal emerges. Section IV presents some internal challenges to the "arm the future" argument. Sections V and VI consider some more general challenges that face "lesser evil" arguments considered as such and discuss why they may arise in the case of geoengineering. Section VII summarizes the main conclusions of the chapter and asks what lessons should be drawn for future discussions of geoengineering.

II. The Problem of Political Inertia

"The rise in global carbon dioxide emissions last year outpaced international researchers' most dire projections."

—Juliet Eilperin (2008)

Before we turn to the core proposal itself, it is worth examining its context. Crutzen's position is largely motivated by what I have called the "Problem of Political Inertia."[13] He asserts that, despite the fact that mitigation is "by far the preferred way"[14] to address climate change, so far, efforts to lower carbon dioxide emissions have been "grossly unsuccessful."[15] The grounds for Crutzen's skepticism are easy to see. Since 1990, when the threat of global climate change was firmly established by the first report of the Intergovernmental Panel on Climate Change, humanity's overall response to climate change has been pretty disappointing. One sign of this is that both global emissions and the emissions of most major countries, such as the United States, have been increasing steadily during this period. For example, from 1990 to 2005, global emissions rose by almost 30 percent (from 6.164 to 7.985 billion metric tons of carbon), and U.S. emissions rose by just over 20 percent.[16] Another sign is that global emissions have been growing even more rapidly in the recent past (from an average of 1.5 percent to 2 percent per annum to around 3 percent in 2007). Indeed, this growth is so rapid that the numbers are currently at the very high end of projected emissions given back in 1990.[17] Given such inertia, Crutzen infers that "there is little reason to be optimistic" about future reductions;[18] indeed, he asserts that the hope that the world will now act decisively is "a pious wish."[19] This is his ultimate reason for proposing geoengineering.

If political inertia is the key problem, what causes it? Crutzen does not say. However, in my own view, a good part of the explanation is that global climate change constitutes "a perfect moral storm":[20] the convergence of three nasty challenges (or "storms") that threaten our ability to behave ethically. These three storms arise in the global, intergenerational, and theoretical dimensions.

The global challenge is familiar. Both the sources and the effects of anthropogenic emissions are spread throughout the world, across local, national, and regional boundaries. According to many writers, this creates a tragedy of the commons, because the global system is not currently set up to govern this kind of situation. Worse, there are skewed vulnerabilities: those who are most vulnerable and least responsible will probably bear the brunt, at least in the short to medium term. This is because whereas the developed nations are, by and large, responsible for the bulk of emissions to this point, they appear much less vulnerable to the more immediate impacts than the less developed countries, where most of the world's poor reside. This mismatch of vulnerability and responsibility is exacerbated by the fact that the developed countries are more powerful politically and therefore more capable of bringing about a solution, but the less developed are poorly placed to call them to account.

The intergenerational challenge is less familiar. The impacts of climate change are subject to major time lags, implying that a large part of the problem is passed on to the future. One reason for this is that emissions of the main anthropogenic greenhouse gas, carbon dioxide, persist in the atmosphere for very long periods of time; even the typical carbon dioxide molecule remains for several hundred years, but 10 percent to 15 percent remains for 10,000 years and 7 percent for 100,000 years. Given this, the full cost of any given generation's emissions will not be realized during that generation's lifetime. This suggests that each generation faces the temptation of intergenerational buck passing: it can benefit from passing on the costs and/or harms of its behavior to future people, even when this is morally unjustified. Moreover, if the behavior of a given generation is primarily driven by its concerns about what happens during its own lifetime, then such overconsumption is likely.[21]

The third challenge is theoretical. We do not yet have a good understanding of many of the ethical issues at stake in global-warming policy. For example, we lack compelling approaches to issues such as scientific uncertainty, international justice, intergenerational justice, and the appropriate form of human relationships to animals and the rest of nature. This causes special difficulties given the presence of the other storms. In particular, given the intergenerational storm and the problem of skewed vulnerabilities, each generation of the affluent is susceptible to arguments for inaction (or inappropriate action) that shroud themselves in moral language but are actually weak and self-deceptive. In other words, each generation of the affluent is vulnerable to moral corruption: if members of a generation give undue priority to what happens within their own lifetimes, they will welcome ways to justify overconsumption and give less scrutiny that they ought to arguments that license it. Such corruption is easily facilitated by the theoretical storm and obscured by other features of the global storm.[22]

Since the perfect moral storm makes us vulnerable to moral corruption, we should be on our guard. Naturally, then, the general question we should ask about any geoengineering proposal is whether it provides a way out of the perfect moral storm or whether, instead, it amounts only to a serious manifestation of moral corruption. Hence, in the present case, the issue becomes: Is the core proposal (and the growing clamor in its favor) a solution or part of the problem?

III. Two Preliminary Arguments

"The economics of geoengineering are—there is no better word for it—incredible."

—Scott Barrett (2008, p. 49)

The core proposal acknowledges that geoengineering is a bad thing. But why concede this? Why consider geoengineering an evil at all? To motivate this idea, it is useful to consider briefly two other arguments for geoengineering that lurk in the background.

1. The Cost-Effectiveness Argument

The first argument claims that geoengineering ought to be pursued simply because it is the most cost-effective solution to the climate crisis. Hence, some enthusiasts claim that albedo modification is relatively cheap and administratively simple to deploy. It is said to be relatively cheap because (it is claimed) the basic mechanism for inserting sulfur into the stratosphere, though expensive in absolute terms, is orders of magnitude cheaper than switching whole economies to alternative energy. It is said to be administratively simple because action need not require international agreement; in theory, the actual deployment could be done by one country or corporation acting alone.[23]

The cost-effectiveness argument has not (yet) proven persuasive to many people. This is presumably because a number of important considerations seem to count against it. First, since the albedo-modification plan does not remove emissions from the atmosphere but rather allows their accumulation to continue accelerating, some important effects of carbon dioxide emissions—such as ocean acidification and its implications for marine organisms and systems—remain untouched. Thus, at best, this intervention only deals with one part of the problem; and at worst, it implicitly assumes the deployment of further technological fixes, so that sulfate injection turns out to be only the tip of a geoengineering iceberg.[24]

Second, the claim that albedo modification is cheap appears to focus only on the costs of actually delivering sulfur into the stratosphere, using cannons mounted on ships or specially modified airliners. But this seems curiously myopic. (One doesn't decide whether to embark on brain surgery by focusing on the price of the knife.) In particular, it appears simply to *assume* that this kind of geoengineering will have no expensive side effects. But worries about side effects are, of course, many people's central reason for rejecting all geoengineering proposals.[25]

Third, the claim that geoengineering is administratively simple appears morally and politically naïve.[26] Can we really imagine that major countries will happily stand aside while a single power or corporation modifies the climate without their input and oversight? At the very least—given that the effects of geoengineering are likely to vary across different countries and regions—won't there be debate about which kind of geoengineering should be pursued and to what extent? Aren't there major issues of liability to be resolved? In short, isn't this the kind of issue on which international agreement will be absolutely necessary if serious social, economic, political, and military conflict is to be avoided?

Finally, the basic cost-effectiveness argument ignores important issues about the human relationship to nature. Given the wider context of escalating species extinction, rampant deforestation, dramatic population increases, and so on, is it not cavalier to assume that the *only* issue that arises with climate change is whether to employ a "quick" and "cheap" technological fix? Moreover, some have even gone so far as to suggest that, even if successful, adopting a geoengineering "solution" might turn out to be worse for humanity in the long run than the problem it is supposed to solve; perhaps it would be better, all things considered, to endure a climate catastrophe than to encourage yet more risky interventions in, and further domination of, nature.[27]

For these and other reasons, most people have concluded not only that the cost-effectiveness argument does not justify deliberate albedo modification but also, on the contrary, that such intervention is something we have serious reason to avoid—an "evil" in the most modest sense.[28] This is an important claim, since it imposes a burden of proof on other arguments for this kind of geoengineering. They must show that its merits are, all things considered, serious enough to override the "evils" involved.

2. The "Research First" Argument

The second lurking argument comes from Ralph Cicerone, president of the National Academy of Sciences. Cicerone believes that we should separate out questions about research on geoengineering from those concerning actual deployment. On the one hand, he supports allowing research and peer-review publication, since this will help us to "weed out bad proposals" and "encourage good proposals" and because knowledge is worthwhile for its own sake, a consideration that (he says) backs the normal presumption in favor of freedom of inquiry. On the other hand, Cicerone concedes that deployment raises special issues. Hence, he proposes that scientists get together and agree on a moratorium on testing or deploying geoengineering. Once some good concrete proposals have emerged from research, he believes that the process should be opened up to public participation.

There is something attractive about Cicerone's proposal and about the model it implies of science and its role in society. However, there are serious concerns about how good that model really is, and in particular how it holds up in the real social and political world in which we live. To begin with, although almost everyone will like the idea of "weeding out" bad geoengineering proposals, Cicerone's aim of "encouraging" the good ones is contentious. So, much depends on his third rationale, that we should promote the acquisition of knowledge for its own sake. But there are some significant issues here.

The first is that it is not obvious that any particular research project should be supported just because it enhances knowledge. To begin with, in the real world, there are limited resources for research. Since we cannot fully fund everything, projects compete with one another for finance and expertise. Given this, the claim that geoengineering research increases knowledge is insufficient to justify our pursuit of it. If we prioritize geoengineering, other knowledge-enhancing projects will be displaced. Some rationale is needed for this displacement.

Second, some kinds of knowledge enhancement seem trivial. Suppose, for example, that someone proposes a project to count (not estimate) the number of blades of grass in each individual backyard in Washington State. Do we really have a reason to support this research? Presumably not. Similarly, some experts claim that geoengineering research may turn out to be in some sense trivial. For example, they suggest that it is highly unlikely to yield the kind of results needed to justify action on the time scale envisioned[29] and that the rate of technological progress is so fast that it may make little sense even to try.[30]

Third, there are such things as morally bad projects. Consider, for example, research whose aim is to find the maximally painful way in which to kill someone or the cheapest way to commit genocide against a specific minority population. Arguably, if such projects succeed, they increase our knowledge. But it is not clear that this alone gives us reason to support them. Similarly, if, as we have suggested above, geoengineering really is some kind of evil, why encourage the pursuit of "good" ways to do it? Why not promote research with better aims (e.g., green technology)?[31]

The second issue about the knowledge-enhancement argument concerns Cicerone's conclusion: there is a crucial ambiguity in the notion of "supporting research." Specifically, "support" is not an all-or-nothing affair. There are major differences between, for example, individual scientists and journals being willing to review and publish papers, major funding agencies encouraging geoengineering proposals, and governments providing massive resources for a geoengineering "Manhattan Project." The kind of support Cicerone emphasizes is that of the participation of reviewers and journals in publishing work on geoengineering. This is a very limited kind of support. But others want something much more substantial, amounting to a significant shift in the existing research effort. Surely, giving this kind of preeminence to the cause of geoengineering research cannot be justified merely by appealing to the value of knowledge for its own sake. Instead, a much more robust argument is needed.

The final issue with Cicerone's argument is that it is not clear that geoengineering activities can really be limited to scientific research in the way that he suggests. First, there is such a thing as institutional momentum. In our culture, big projects that are started tend to get done.[32] This is partly because people like to justify their sunk costs; but it is also because starting usually creates a set of institutions whose mission it is to promote such projects.[33] For such reasons, sometimes the best time to prevent a project from proceeding is before the costs are sunk and the institutions created. Second, there are real concerns about the idea of a moratorium. After all, if the results of research are to be published in mainstream journals that are freely available online or in libraries across the world, what is to stop some rogue scientist, engineer, or government from deciding to use that research? Third, there are worries about who gets to make such decisions and why, and about how they are enforced. If the future of the planet is at stake, why is it that the rest of humanity should cede the floor to a "gentleman's agreement" among a specific set of scientists? Fourth, there are issues about conducting geoengineering research in isolation from public input, and in particular divorced from discussions about the ethics of deployment. The background assumption that is being made seems to be that such input and discussion has *nothing to tell us* about the goals of geoengineering research or how it should be conducted. But it is not clear why we should accept this assumption.[34] After all, many people do not accept it in the case of other important scientific issues, such as research on stem cells, genetic enhancement, and biological warfare.

In summary, stronger arguments are needed for considering substantial investment in geoengineering research, and a more robust account of the conditions under which deployment would be considered is also necessary. This is where "lesser evil" arguments enter the discussion.

IV. Arming the Future

"Life's toughest choices are not between good and bad but between bad and worse. We call these choices between lesser evils. We know that whatever we choose, something important will be sacrificed. Whatever we do, someone will get hurt. Worst of all, we have to choose. We cannot wait for better information or advice or some new set of circumstances. We have to decide now, and we can be sure that there will be a price to pay. If we do not pay it ourselves, someone else will."

—Michael Ignatieff (2004, p. vii)

If there is a presumption against geoengineering, how might this be met?[35] One promising approach is based on the general idea that "we may reach the point at which [geoengineering] is the lesser of two evils."[36] This idea has been influential in discussions about geoengineering for climate change since the earliest days and has appealed to both its enthusiasts and its detractors.[37]

1. The Basic Argument

The core proposal offers one kind of lesser evil argument, and so appears to fit neatly into this framework.[38] As we have seen, the basic structure of this argument seems to be as follows:

(AFA1) Reducing global emissions is by far the best way to address climate change.

(AFA2) In the last 15 years or so, there has been little progress on reducing emissions.

(AFA3) There is little reason to think that this will change in the near future.

(AFA4) If very substantial progress on emissions reduction is not made soon, then at some point (probably 40 years or more into the future), we may end up facing a choice between allowing catastrophic impacts to occur or engaging in geoengineering.

(AFA5) These are both bad options.

(AFA6) But geoengineering is less bad.

(AFA7) Therefore, if we are forced to choose, we should choose geoengineering.

(AFA8) But if we do not start to do serious scientific research on geoengineering options soon, then we will not be in a position to choose it should the above scenario arise.

(AFA9) Therefore, we need to start doing such research now.

The arm the future argument is complex. But, on the surface at least, it does seem to be the right *kind* of argument. For one thing, it acknowledges that geoengineering is problematic and that there is a burden of proof against it. For another, it offers a weighty moral reason to endorse geoengineering—that of preventing a catastrophe—and it is easy to see why this reason addresses the deficiencies of the cost-effectiveness and "research first" arguments. The threat of catastrophe appears both to meet the burden of proof against geoengineering and to justify prioritizing research on it over other kinds of research. Finally, the AFA appears to address one significant part of the perfect moral storm. Under the scenario it sketches, geoengineering research emerges as one way of assisting future generations. If the world really isn't going to do very much about reducing emissions, then substantial investment in geoengineering research emerges as an alternative way to meet our intergenerational obligations.

At first glance, then, the AFA appears to make a very strong, even overwhelming, case for geoengineering research and also (under the stated circumstances) ultimate deployment. However, I will now argue that matters are not as straightforward as they initially seem. To begin with, we would do well to proceed with caution. In general, arguments from moral emergency are perennially popular in both private and public life, and for an obvious reason. Clearly, part of the point of claiming that one is in morally exceptional circumstances is in order to secure an exemption from the usual norms and constraints of morality. But this fact should give us pause. After all, there will always be those who would prefer that morality not apply to them or their projects, and all of us are vulnerable to such thoughts at some time or other. Morality sometimes seems inconvenient to us (like truth, as Al Gore reminds us), and in such cases, we'd often like to have an exemption. Hence, we should be wary of arguments from emergency; clearly, they are open to manipulation.[39] Moreover, as we have seen, in the case of climate change, we have additional reason for caution: if climate change is a perfect moral storm, the incentives for moral corruption will be high.

2. Five Challenges

Given all of this, the core proposal should be subjected to special scrutiny. In the remainder of this section, I will focus on five challenges that face the AFA. In the following two sections, I will raise some wider worries that apply to "lesser evil" arguments considered more generally.

(i) Which Nightmare?

The first challenge concerns whether the nightmare scenario is the relevant emergency. In general, we should not simply accept *as a stipulation* that some policy that is said to be an evil (such as geoengineering) should be endorsed because under some circumstances it would be a lesser evil than some other policy (such as allowing a catastrophic climate change). Instead, we should ask important questions such as: How likely is this emergency situation (where one has actually to decide between these two options) to arise? Is it the most relevant emergency situation? Is it true that the two evils are the only alternatives? Is the lesser evil really lesser, all things considered?

As it happens, the answers to these questions seem very much in doubt in the present case. In particular, there are serious concerns about the salience of the "nightmare scenario" where a decision must be made between embarking on geoengineering or allowing catastrophic climate change to occur. Consider the following.

First, for a group of decision makers actually to face this emergency situation, they would need to know at least the following: that the planet was on the verge of very serious climate impacts, that geoengineering was very likely to—and the only thing likely to—prevent them, and that the side effects of deployment (including not just the physical and ecological effects but also the human and political effects) would be minor in relation to the harm prevented. But this, I submit, would be a pretty unusual scenario. Moreover, it is questionable whether it makes sense to organize policy around it. For one thing, the scenario might be so unusual that it makes sense to ask whether it is even *worth* preparing for. (After all, it does not seem to make sense to prepare for every possibility.[40]) For another, there may be other emergency situations that are more salient, and if so, it may be better to prepare for these emergencies. For example, perhaps the more salient emergency situation is one where choices have to be made about how to cope with, or reverse, a catastrophic change that has already occurred. (This scenario might be more salient for a number of reasons. For example, perhaps it is simply more likely, or perhaps preparing for it would help us to deal with the nightmare scenario as well, at least to some extent.) In general, the claim that the nightmare scenario described by the AFA is *the* nightmare that we should be concerned to address requires further support.[41]

Second, in one respect the core proposal may not be neutral here. The AFA proceeds as if the decision to do research will have no influence on the likelihood of the nightmare situation's arising. But it is not clear what justifies this assumption. Many people worry that substantial research on geoengineering will itself encourage political inertia on mitigation and bring on the nightmare scenario and deployment.[42] If this is so, we might have strong reason to limit or resist such research at this stage. We do not want to create a self-fulfilling prophecy.

These points illustrate a weakness in hypothetical "lesser evil" arguments such as the AFA. Even if one accepts in principle that one should make a "lesser evil" choice in some highly stylized case, such as the nightmare scenario, this fails to justify a policy of preparing to make that choice.[43] The salience of the scenario to current policy still needs to be demonstrated. By itself, this kind of hypothetical "lesser evil" argument is not enough.

(ii) Other Options?

The second challenge to the core proposal concerns its account of the current options. The AFA does not involve a straightforward appeal to moral emergency, since it explicitly concedes that the nightmare scenario is not *yet* upon us. According to the argument, we are not *now* in the relevant "lesser evil" situation, having to choose between the evils of allowing catastrophe and pursuing geoengineering;[44] instead, the decision currently to be made is about whether and how to prepare for such a situation.[45]

This shift is important because it puts questions about how the emergency is supposed to arise back into play. One of the usual effects of actually being in an emergency is to make many of the background conditions much less salient. For example, if I see a small child drowning in a pond whom I could easily save just by reaching down to pick him out, we do not normally think that I should to stop to mull over questions such as how he came to be there, and who is officially responsible for saving him. The relevant question is what to do here and now. But none of this is the case if one is *anticipating* an emergency. Then it is perfectly appropriate to consider how the emergency might arise.

First, sometimes the best way to plan for an emergency is to *prevent* its arising. In the case of the pond, for example, one might erect a small wall to prevent toddlers from falling in. Similarly, suppose—as the AFA suggests—that we are interested in preventing a catastrophic climate change brought on by the failure to reduce emissions directly, through regulation and political leadership. Even given this failure, we still have other options. For example, perhaps we can prevent the emergency by indirect means, such as by investing in a massive "Manhattan Project" that produces very cheap alternative energy by 2030.[46] The general point here is that if a good option is available that will prevent the emergency situation from arising, the fact that we would choose a (lesser) evil if it did arise might be irrelevant to what to do now.[47] Again, the nightmare scenario loses its salience.

Second, considering how the emergency might arise can also help us to put other options on the table for dealing with it even if it does ultimately come about. In the present case, the AFA implicitly suggests that the *very best* we can do now to help future people faced with the threat of an imminent climate catastrophe is to research geoengineering. But this claim is unsupported and open to challenge. Most conspicuously, there are other ways in which we might aid future people on the brink of such a calamity. For example, perhaps we could prepare them for a massive emergency deployment of existing alternative-energy technology (e.g., we could establish a Strategic Solar Panel Reserve), or perhaps we could establish a robust international climate assistance and refugee program, or perhaps we could do both of these things together with any number of other alternatives. In addition, there may be a presumption in favor of alternatives that are (by contrast with geoengineering) not "evil" in any sense. In any case, their relative merits should at least be discussed.

(iii) Additional Liabilities?

The third challenge facing the core proposal concerns additional liabilities. The AFA concedes that it is probably not *us*—our generation—who will actually make the decision to deploy the lesser evil. Most writers appear to assume that the nightmare scenario will not unfold until the second half of this century at the earliest, if at all.[48] There are probably two basic reasons for this. First, mainstream scientists suspect that the kind of threshold effects most likely to produce the nightmare scenario are some way off, if they are plausible at all.[49] Second, many believe that the basic research needed on the possible methods and impacts of geoengineering will take a similar time period to emerge. Discussion of the problem is very much in its infancy, much of the relevant work is at a highly speculative stage, and robust scientific guidance will take a long time to appear.[50] Hence, Stephen Schneider, for example, emphasizes in a recent review that "strong caveats, which suggest that it is premature to contemplate implementing any geoengineering schemes in the near future, are stated by all responsible people who have addressed the geoengineering question."

Given these things, it seems highly likely that if the nightmare scenario arises, it will confront future generations, not the current generation.[51] The AFA tends to obscure this point by referring to what "we" will be forced to choose, where this refers to some temporally extended sense of "we," such as humanity as such or a given country considered across time. But once the point is made clear, the role of the argument becomes to imply that the responsibility of the current generation is (merely) to aid future generations in choosing the best kind of geoengineering possible. Unfortunately, this conclusion tends to obscure a vital moral feature of the situation: the potential crisis is to be brought about by our (the current generation's) failure to pursue better climate policies.[52] Acknowledging this matters because there seems to be an important moral difference between (on the one hand) preparing for an emergency and (on the other hand) preparing for an emergency that is *to be brought about by one's own moral failure*.

Many things might be said about this, but I will make just a couple of remarks here. First, if someone puts others in a very bad situation through a moral failure, we usually do not think it enough for her to respond merely by offering the victims an evil way out. Instead, we believe that the perpetrator has substantial obligations to help the victims find better alternatives and also, if the alternatives are costly or harmful, to compensate them for making this necessary. If this is right, then even if the AFA were correct in other respects, we should not conclude from it that current people owe future generations *only* research on geoengineering; much more seems required. For example, we might owe them a very substantial compensation fund, or we might be obliged to run graver risks ourselves on their behalf. These are potentially very serious implications. For example, if we force a risky geoengineering project onto future people, we might have to compensate them with a massive climate assistance and refugee program, potentially amounting to a global safety net.[53] Similarly, if the threat of catastrophe is extreme, we may be required to forestall it by attempting risky geoengineering on ourselves.[54]

Second, concerns about additional liabilities are heightened in circumstances where we fail to do what we should to prevent a catastrophic evil *partly because* we know in advance that a solution of lesser evil will still be available to others. For example, suppose that we knowingly allow a crisis to unfold, which we could prevent by taking a nonevil option open to us. Suppose also that we do this partly because we know that others will eventually be forced to step in to prevent the coming catastrophe, even though they will have to accept significant evils in order to do so. Finally, add to this that we act in this way simply because we want to secure some modest benefits for ourselves.[55] Surely, such *calculated moral failure* would make us liable for even greater burdens, both compensatory and punitive.

(iv) Fatal Silence?

The fourth challenge to the core proposal aims to broaden the remit of geoengineering policy still further. The key idea is that the political issues raised by any decision for geoengineering would be profound, so that the proposal's silence on this topic is fatal to its moral acceptability. To motivate this idea, I offer the following simple argument, which I shall call the "stalking horse" argument."[56]

The argument comes in three parts. The first concerns legitimacy:

(SH1) The climate system is a basic background condition of human life and social organization on this planet.

(SH2) To engage in geoengineering would alter the human relationship to this basic background condition and the relationship between humans subject to that condition.

(SH3) Hence, geoengineering raises new and profound issues of global governance.

(SH4) Institutions of global governance must be politically legitimate.

(SH5) Hence, any argument for the permissibility of geoengineering has to explain the politically legitimacy of those institutions charged with making the decision to geoengineer.

At first glance, the AFA appears to run afoul of the concern for political legitimacy. Because it is silent on the topic, it fails to establish that geoengineering would be permissible.[57]

This brings us to the second part of the argument, which concerns norms:

(SH6) A basic principle of modern political thought is that institutions of governance are legitimate only if they can be justified to those who are subject to them.

(SH7) Hence, geoengineering institutions must be justified to those who are subject to them.

(SH8) If a set of institutions is to be justified to those subject to them, it must explicitly or implicitly invoke appropriate norms of justice and community. (For example, it must not be seriously unfair or parochial in its concerns.)[58]

(SH9) Therefore, any successful argument for the permissibility of geoengineering must invoke appropriate norms of justice and community.

Again the AFA is silent on these matters. But it also faces a specific threat:

(SH10) A good part of the political inertia on climate change is caused by resistance to such norms.

(SH11) Hence, there is good reason to suspect that the attempt to establish legitimate geoengineering institutions will face similar resistance.

(SH12) Hence, unless the roots of political inertia can be addressed, any decision to geoengineer is likely to be illegitimate, because it violates norms of justice and community.

The import of this last part of the argument is as follows. Since the AFA not only fails to address the problem of political inertia but also tries to operate within its constraints, it is likely to license illegitimate geoengineering and therefore violations of norms of justice and community.[59] This is reason to reject that argument.[60]

(v) Lingering Inertia?

The fifth and final challenge to the core proposal concerns moral corruption and political inertia. The AFA suggests that geoengineering research is a kind of insurance policy. But presumably, there are many such policies—that is, many ways in which we might try to aid the future if we think that serious reductions in emissions will not occur. Now, as we have already seen, one issue is that only some of these policies involve geoengineering; there are other options (e.g., the "Manhattan Project" for alternative energy or the climate refugee project). But it is also true that there are many policies that might include geoengineering research as a component. These run the gamut from various "geoengineering research only" proposals (e.g., ranging from merely tolerating very limited research to launching a truly massive geoengineering "Manhattan Project") to more general approaches, where geoengineering research is included within a much more robust package (e.g., ranging from including substantial compensation to future people and the world's poor to proposing the creation of a new global order for a geoengineered world). Given this plethora of geoengineering policies, there is a real question about which one to choose.

This generates a serious worry. As stated, the AFA advocates only for geoengineering research; it does not even mention wider considerations. Moreover, as put forward by Crutzen, the proposal seems to imagine only a moderate redirection of scientific resources. In short, in context, the core proposal tends to suggest that the relevant policy is "modest geoengineering research only." But why think that this is the salient backstop policy? The worry is that "modest geoengineering research only" gains prominence only because it is the approach most compatible with continued intergenerational buck passing. In essence, we'd be happy to spend a few million dollars on research that our generation will probably not have to bear the risks of implementing, and we'd be even happier to think that in doing so, we were making a morally serious choice in favor of protecting future generations. But thinking so hardly makes it the case. What makes us think that our preference for "modest geoengineering research only" is not just another manifestation of moral corruption? Specifically, doesn't it seem likely that the same forces that oppose substantial mitigation measures will also oppose any other policies that involve serious costs or commitments for the current generation of the world's richer countries, including (but not limited to) substantial compensation proposals, the running of extra risks by the current generation on behalf of the future, the setting of punitive damages, and (even) huge investment in geoengineering research and deployment, if that were required?[61] More generally, doesn't the focus of the AFA on scientific research conveniently obscure this problem?

3. Refining the AFA?

The last three challenges all rely on the idea that the conclusion of the AFA is too narrow. But perhaps this charge is uncharitable. Specifically, although it is true that the AFA does not explicitly mention things that need to be done other than geoengineering research, as a matter of logic, it does not exclude them either. So perhaps all that is being asserted in the AFA is that we owe the future *at least* geoengineering research.

(i) The Neutrality Interpretation

Are the narrow interpretations uncharitable? It is difficult to say; the answer depends partly on how far one is willing to press the principle of charity. To begin with, while it is true that the AFA does not explicitly exclude more robust geoengineering policies, it also does not mandate them. At best, the overall lesson of the argument is underdetermined. This is worrying in itself. Given the threat of moral corruption, we should be wary of allowing such room for maneuver. More importantly, there are some indications that narrow interpretations are not uncharitable. Consider first the most natural reading of the objection, which we might call the "neutrality interpretation." According to this reading, the AFA establishes only that we owe the future research on geoengineering, and it *simply takes no position* on whether we owe them anything else as well. Is this interpretation plausible?

Two considerations suggest not. First, in the public and political discussion of geoengineering, there is virtually no mention of compensation, global justice, and the like. *Time* magazine, for example, does not list either "geoengineering with compensation" or "reforming the global order to facilitate geoengineering" as one of its "Ten Ideas That Are Changing the World." Instead, the implicit proposal is geoengineering alone, on the assumption that nothing much else changes. Moreover, to the extent that wider considerations are mentioned in the public debate, they are usually seen as obstacles only to mitigation, not to a robust geoengineering policy. For example, the *Time* article concludes: "unless the geopolitics of global warming change soon, the Hail Mary pass of geoengineering might become our best shot";[62] there is no thought that the geopolitics might have to change before geoengineering should be seriously considered.

Second, the AFA is under internal pressure. It already requires that political inertia precludes some better options, including at least substantial mitigation, but also probably any radical alternative-energy revolution. So severe political obstacles must be assumed if geoengineering is to seem like a serious option at all. But then there is a real worry that these obstacles will be so severe that "modest research only" really is the only (politically) viable geoengineering policy. If this is not their view, proponents of the AFA need to explain why it is not.

(iii) Core Component Interpretation

Perhaps, then, friends of the AFA should embrace the dark view that more robust geoengineering policy is unlikely but still maintain that the evil of climate catastrophe is so severe that research should be done on geoengineering regardless. The guiding idea here would be that even if more robust geoengineering policies would be better than the modest approach, the urgency of the nightmare scenario means that geoengineering research itself has absolute priority. In essence, the claim is that the moral imperative in favor of at least modest research is quite central and decisive: any such research under any conditions is better than no research at all, because in the nightmare scenario, we should deploy geoengineering whatever else we do, whatever the wider circumstances, even acknowledging the other moral costs. Call this the "core component interpretation."

There is some evidence for the core component interpretation in the leading scientific work. The key advocates of research recognize that geoengineering raises broader concerns. Crutzen, for example, acknowledges that "ethical, and societal issues, regarding the climate modification scheme are many."[63] Similarly, Cicerone states: "While a strong scientific basis is necessary for geoengineering, it is far from sufficient. Many ethical and legal issues must be confronted and questions arise as to governance and monitoring."[64] Still, since these are practically the only remarks that these authors address to ethical constraints, both seem to operate under the assumption that whatever the broader concerns are, they are either insufficient to blunt the case for geoengineering research or can be dealt with later, once research is under way.[65]

How plausible are such views? Can we really isolate the ethical and political considerations in this way? Are they really some kind of "afterthought" that can be safely deferred? I am not so sure. Let me begin with three preliminary thoughts. First, it would be misleading to suggest that the AFA defers *all* ethical considerations; on the contrary, the "lesser evil" claim is itself a moral one, and it is central to the argument. So the real question is not whether ethics can be left until later but whether, given the central ethical argument, some related moral considerations can be safely ignored or deferred.

Second, the suggestion that the problem of political inertia is so bad that we should organize our policy around geoengineering research alone (deferring or ignoring other ethical considerations) embodies a profound skepticism that should not be conceded without argument. After all, the thought is that neither mitigation, nor adaptation, nor alternative energy, nor compensation, nor geopolitical reform, nor even more extensive geoengineering research has a realistic chance of political success. But why accept this? And if things are really so bad, why think that "moderate geoengineering research only" has better prospects?

Third, not only is such profound skepticism questionable, but its truth would have further important moral implications. If a large number of alternative policies would be preferable, but none is available because of *our own* political inertia, the scale of our moral failure in choosing modest research at this point would be immense. But this suggests that the sense in which we are now *morally required* to pursue such a policy is sharply attenuated.[66] How are we to understand the force of the obligation to facilitate the lesser evil when we are so conspicuously refusing all prior (and many nonevil) moral demands? Is there not a worrying moral schizophrenia underlying this proposal?[67]

More substantively, we should be careful about the further presuppositions of the core component interpretation. First, the claim that scientific research should be the *sole* central component of any geoengineering policy requires further support. It is not clear that the ethical and geopolitical concerns with geoengineering are any less central than the scientific ones or that there are good pragmatic reasons to defer them. For example, arguably, we have at least as strong a reason to make sure that any given geoengineering policy does not set off a major geopolitical conflict as to start preparing such a policy in the first place. Severe climate change is not the only catastrophe to be avoided, after all. Global nuclear war would also count; so, presumably, would any geoengineering intervention designed to systematically destroy the less developed countries in order to spare the developed.

Second, it may be that the moral and political concerns turn out to be *more central* than the scientific ones. On the one hand, there are reasons not to prioritize geoengineering science now. As we have seen, some claim that we simply cannot do the research necessary in the time envisioned, and others believe that the rate of technological progress is so fast that it makes little sense to try.[68] Such worries may be more pronounced if we plan to do only modest research.[69] Moreover, it may be that the best science to be doing now involves continuing to work on the details of how the climate system works. If future generations do need to consider geoengineering, this research may be more useful to them than anything else we can deliver.

On the other hand, it may be that failure to deal with the moral and political considerations is more likely to thwart the effort to aid the future than failure to do the science. For example, countries will (rightly) be concerned that geoengineering science and technology might be misused. In particular, they will worry about the possibility of *predatory geoengineering*: intervention to further political goals beyond those of stabilizing the climate, particularly those contrary to the interests of some of the nations affected.[70] Hence, if we supply the future only with improved possibilities for geoengineering and no account of how to implement them in an ethical way, then such concerns may paralyze deployment. This may be so even if the world's people are otherwise persuaded of the importance of geoengineering to climate stability. In short, geoengineering research may only facilitate a different "lesser evil" scenario, one where decision makers must choose between climate catastrophe and geopolitical catastrophe. This is a nastier nightmare scenario than that envisioned by the AFA, but it is not clear that it is any less likely or relevant to policy.

At this point, it might be good to review. In this section, I identified five specific challenges facing the core proposal: first, it is not clear that the nightmare scenario it envisages is salient; second, there are other ways in which we could prepare; third, if the scenario did arise, we would owe the future more than geoengineering; fourth, the argument ignores concerns about political legitimacy; and fifth, its narrow focus is suggestive of lingering inertia. In addition, I considered the objection that some of these challenges uncharitably assume that the AFA is too narrow. Specifically, in one interpretation, the argument is simply silent on wider considerations, and in another, it holds that research should be the focus of our policy even if it is the only thing that can be done to aid the future. Against the former interpretation, I argued that it is implausible in context and fails to appreciate the internal pressure placed on the AFA by its own claims about political inertia. Against the second interpretation, I claimed that it assumes a profound skepticism that ought not to go unchallenged, that it threatens a serious form of moral schizophrenia, that it falsely prioritizes scientific research over other forms of preparation for climate emergency, and that it fails to appreciate the salience of other nightmare scenarios, such as those where the choice is between climate and geopolitical catastrophe. For such reasons, I conclude that, as it stands, the AFA is seriously underdetermined and that efforts to rectify this face substantial obstacles. Because of this, the case for both research on and ultimate deployment of geoengineering is far from being straightforward or irresistible. I will now consider some more general worries about "lesser evil" arguments, which strengthen this conclusion.

V. Underestimating Evil?

"One might have the idea that the unthinkable was itself a moral category . . . in the sense that [a man] would not entertain the idea of doing [such actions]. . . . Entertaining certain alternatives, regarding them indeed as alternatives, is itself something that he regards as dishonourable or morally absurd."

—Bernard Williams (Smart and Williams 1973)

What does it mean to choose the lesser evil? What is at stake? We can begin by acknowledging that there is something morally appealing about the notion of choosing the lesser evil in a situation of grave crisis. Such a choice can seem heroic, even to display a deep moral seriousness. One reason for this is that most people seem to believe that there are circumstances when the consequences are so severe that normal rules must be overridden. Another is that a strong rigorism about moral rules often seems morally unattractive, perhaps even an irrational fetish.

To illustrate the attractiveness of these thoughts, consider the case of the inquiring murderer famously discussed by Kant. In one version of this case, you are confronted with a Nazi stormtrooper asking whether you are hiding Jews in your house. As it happens, you are. Since lying is normally immoral, are you morally bound to tell the stormtrooper the truth? Most people think not. Sticking to the normal rules in such cases, they believe, would be deeply bizarre; a morally serious person could not do such a thing. Similarly, the "lesser evil" argument can seem overwhelmingly appealing in the case of geoengineering. Faced with a possible catastrophe, why wouldn't one try geoengineering? Wouldn't failure to do so constitute an irrational fetish?

Clearly, such concerns are important. But matters are not as simple here as they initially seem. To see this, consider the following three obstacles that a "lesser evil" argument must seek to overcome.

1. Opacity

The first is the problem of opacity. In the abstract form in which they are usually presented, "lesser evil" arguments are often inscrutable.[71] For one thing, we are asked simply to compare two bad options and rank one as lesser; but we are not usually asked for the reasons for our rankings. For another, the options themselves are frequently underdescribed. Such opacity creates concerns. Perhaps people's conceptions of the options differ: they implicitly fill in the details of the lesser and greater evils in ways that pick out what features would be most salient to them, and these are not the same. Moreover, perhaps their underlying concerns are at odds: even where they agree on the salient features, they take them to be salient for very different reasons, and this has different implications.

Such things matter for two reasons. First, any apparent consensus in favor of a "lesser evil" argument may turn out to be dangerously shallow. Although there is outward agreement that some generic form of action (such as geoengineering) would be permissible under some circumstances, there is deep, but implicit, disagreement about what those circumstances would be. Suppose, for example, that some scientists believe that geoengineering would be permissible in order to prevent the "greater evil" of a mass extinction, but some economists believe that it would be permissible to prevent the "greater evil" of a short-term drop in economic growth. In that case, their apparent consensus on the need to pursue geoengineering research might turn out to be shallow: assent to the "lesser evil" argument would mask deep disagreements about the appropriate goals of geoengineering policy.[72]

The second reason that opacity matters is that it is likely to obscure the real moral arguments. The true justificatory work is done by the underlying reasons together with whatever features of the underdescribed options the person is regarding as salient; it is these that underlie that person's assent to the basic form of "lesser evil." Hence, appealing to the lesser evil functions not as an independent argument in favor of some policy but rather as a convenient umbrella term that covers a number of different considerations. But if this is the case, then such an appeal might fail to do real normative work; indeed, it might hinder that work by drawing attention away from the real justifications for policy.[73]

2. Denial

The second obstacle facing "lesser evil" arguments is the problem of denial. Some may simply refuse to accept that the lesser evil should be chosen under any circumstances: a lesser evil is still an evil, they will say, and therefore not to be chosen. This, of course, is Kant's attitude to the inquiring-murderer case. One ought not to lie *simpliciter* is his position, and let the chips fall where they may.

Most people do not find Kant's position compelling in this case. But we should be wary of simply rejecting it out of hand. For one thing, even in the case of the inquiring murderer, it is difficult to *show* how or why an uncompromising attitude is irrational or otherwise in error. More importantly, even if most of us do not agree with Kant in that case, there are situations in which the same kind of attitude seems more plausible. For example, suppose some great evil could be prevented if you would just kill your own grandmother in cold blood. (If necessary, embellish the case. For example, imagine that your grandmother is morally innocent and that the killing would be against her wishes.) Is it so obvious that you should do this? Surely, one can understand why a person might resist, and for reasons that seem at least possibly morally appropriate.

The possibility of resistance has important implications. First, it suggests that "lesser evil" arguments might turn out to be logically invalid: one cannot infer from the fact that an evil is "lesser" in some sense that it ought to be chosen. Second, it implies that rival attitudes to the relevant evil will be at the heart of many disputes about "lesser evil" cases. Those who resist "lesser evil" arguments are likely to protest that such arguments typically assume an *impoverished* account of evil—such as the earlier "something one has serious reason to avoid"—and that it is only because of this that they begin to look plausible at all.[74] In short, "lesser evil" arguments underestimate what it is to call something an evil.

3. The Unthinkable

At first glance, it may seem that these points stand or fall with the assertion of the strong and uncompromising view that evil ought never to be done. But in fact, one need not go this far. First, finer-grained distinctions are possible. Consider, for example, what Bernard Williams says about the category he calls the "unthinkable": "entertaining certain alternatives, regarding them indeed as alternatives, is itself something that [someone] regards as dishonourable or morally absurd."[75] Perhaps not all evils are also unthinkable, and those that are not might sometimes be chosen. Still, if some evils are unthinkable, then one cannot be confident that "lesser evil" arguments will always go through. Perhaps some evils are lesser than others in some respects but still nonetheless unthinkable.[76] In that case, merely showing that an evil is lesser will not be enough to justify action.

Second, Williams's focus is not on what should be done but rather on what options should be entertained. His central claim is that it is dishonorable to regard certain options as legitimate alternatives. (Note that even if one thinks that this claim is too uncompromising, it might be weakened to say that *under some circumstances* it is morally shameful to regard an evil option as a legitimate alternative, *even if perhaps in other situations it is not*. The basic point would remain.) This thought seems pertinent in the current case. One can certainly see someone arguing that *advance planning* for a nightmare scenario is itself morally inappropriate when that nightmare is to be brought on by *one's own future moral failure*. Hence, some will say that it is morally inappropriate to start planning for geoengineering when mitigation and adaptation are still on the table; instead, all of our energies and efforts should go into preventing the nightmare scenario—where geoengineering starts to look acceptable—from arising.

To illustrate the appeal of this attitude, consider a related "lesser evil" argument. Call this the "Survival Argument":

> If very substantial progress on emissions reduction is not made soon, then the world may plunge into chaos because

> of catastrophic climate change. If this happens, my family may face a choice between starvation and fighting for its own survival. Both starvation and fighting for survival are bad options. But fighting for survival is less bad. Therefore, if we are forced to choose, we should choose fighting for survival. But if we do not begin serious preparations for fighting for survival now, then we will not be in a position to choose that option should the circumstance arise. Therefore, my family needs to commence serious preparations for fighting for survival now.

What do we think of this argument? Should we arm ourselves, build fortified camps in the boonies, withdraw our children from school and train them instead in wilderness survival and combat, and so on? Wouldn't this be a lesser evil than entering the world of climate chaos unprepared? Perhaps. Still, it seems plausible to say that devoting ourselves to such a strategy at this point in time is not merely unwarranted but also an unacceptable evasion of moral responsibility. The survival argument—with its focus on the lesser evil—ignores this, and so is to be criticized. This suggests a general flaw in hypothetical "lesser evil" arguments and one that the AFA may share.

Finally, Williams goes on to suggest a further worry about the limits of moral reasoning. Perhaps there are some situations so extreme that it would be insane to plan for them, because morality somehow gives out:

> [Someone might] find it unacceptable to consider what to do in certain conceivable situations. Logically, or indeed empirically conceivable they may be, but they are not to him *morally conceivable*, meaning by that that their occurrence as situations presenting him with a choice would represent not a special problem in his moral world, but something that lay beyond its limits. For him, there are certain situations so monstrous that the idea that the processes of moral rationality could yield an answer in them is insane: they are situations which so transcend in enormity the human business of moral deliberation that from a moral point of view it cannot matter any more what happens. Equally, *for him to spend time thinking what one would decide if one were in such a situation is also insane, if not merely frivolous*.[77]

It seems at least possible that some "lesser evil" situations are of this sort. (Consider, for example, the one where you must choose to kill your own grandmother in cold blood.) Still, whether the decision for geoengineering is one of these is a more difficult question. Presumably, the answer depends in part on how risky one thinks a particular method of geoengineering is likely to be and what kinds of obligations to the future and to other species one thinks we have. In my own view, the nightmare scenario envisioned by the core proposal is not nearly so extreme that "from a moral point of view it cannot matter any more what happens." Still, something related to Williams's concern is relevant. What are we to say about "monstrous" situations that strain normal moral deliberation?

3. Marring Evils

This thought leads to the third obstacle facing "lesser evil" arguments. As it happens, many people—Williams included—believe that even actions that are normally "unthinkable" must sometimes be done.[78] Yet even when there is agreement that certain evils are of this sort, people have different attitudes to the relevant moral emergencies. One might be aptly described in terms of the well-known bumper-sticker slogan "Shit Happens." On this view, the occurrence of a lesser evil situation is an unfortunate fact about the world, more serious than, but otherwise akin to, other shifts in empirical circumstances. But another attitude is quite different. Consider the following classic case. In his novel *Sophie's Choice*, William Styron tells the tragic tale of a mother who is put in a situation where she must choose between saving one of her children or submitting both to be killed by the Nazis. Sophie chooses to save her son but relinquish her daughter. The novel explores her subsequent life as she deals with the fact of her choice and its consequences. Ultimately, Sophie kills herself, unable to come to terms with the decision she made.

Sophie's Choice is a modern literary classic. But it is also of philosophical interest. Most people agree that Sophie's suicide is tragic. For many, this is because they believe that she wrongly blames herself for the death of her daughter. The situation in which she found herself was, it is said, monstrously difficult. Nevertheless, she did the right thing in choosing and ought not to be wracked by guilt. Others are to blame, not Sophie. She should recognize that and feel better about herself. Perhaps she should even praise herself for being able to make the decision to save at least one person's life (her son's) under such emotionally difficult circumstances. (After all, "shit happens.") For others, however, Sophie's suicide is tragic in a more traditional sense. Sophie does not make a moral mistake. Even though she makes a defensible (perhaps even "the best"[79]) decision in that terrible situation, and even though she bears no responsibility for being in it, still she is right to think that her choice carries negative moral baggage. Although she is not to be blamed for the decision in the usual way, it is nevertheless true that her life is irredeemably *marred* by it.[80] We might admire Sophie in certain respects, but no one would say that she lives the kind of life that is desirable for a human being to live. No one would want to be Sophie.[81] Interestingly, this second attitude seems to be Sophie's own, and the one that ultimately leads her to suicide. She says: "In some way I know I should feel no badness over something I done like that. I see that it was—oh, you know—beyond my control, but it is still so terrible to wake up these many mornings with a memory of that, having to live with it. When you add it to all the other bad things I done, it makes everything unbearable. Just unbearable."[82]

The idea that a life can be marred from a moral point of view, and possibly irredeemably, is a controversial one in moral theory. So, let me make three quick points about how I'm understanding the claim. First, I propose using the phrase "marring evil" in a special, technical sense, to refer to a negative moral evaluation of an agent's action (or actions), which is licensed when the agent (justifiably) chooses the lesser evil in a morally tragic situation and which results in a serious negative moral assessment of that agent's life considered as a whole.[83] Second, I propose this because I assume that the evil that torments Sophie is a special instance of a more general category of ills. People's lives are subject to serious negative evaluation even when their choices are not "forced" by circumstance in the way that a "lesser evil" decision is said to be. For one thing, normally evil actions—such as those of premeditated murder or genocide—stain or tarnish lives, too. For another, some believe that a person's life can be compromised by circumstances beyond his control even if he himself makes no evil choices. (Aristotle, for example, claims that even though Priam of Troy was virtuous, his life was not a flourishing one. The tragedy that befell his family and city in the Trojan War was sufficient to undermine that claim.[84]) In short, marring is just one way in which a life may be morally compromised, or "tarnished."[85] Third, we need not assume that all tarnishings (or marrings) are irredeemable. Perhaps some can be outweighed or expunged (e.g., by other good actions), with the result that a positive (or neutral) overall evaluation of the agent's life is restored. Still, some tarnishings may be irredeemable. I will call these "blighting evils."

With these points in mind, let us return to Sophie. The dispute over how to understand her choice is sometimes described as turning on the question of whether or not there are genuine moral dilemmas, situations in which an agent cannot help but act in a way that is morally reprehensible in at least some sense. Those in the first camp say that there are no genuine moral dilemmas—and so no marring evils in my sense—and those in the second say that there are. Now I suspect that many of you are already thinking that a discussion of *Sophie's Choice* seems oddly (perhaps even shockingly) out of place in a paper on climate change and geoengineering. I admit that Sophie's choice is an extreme case. Nevertheless, I mention it because attention to the dispute about genuine moral dilemmas helps us to see some important issues within the ethics of geoengineering.

First, the dispute helps to make sense of some of the angst present within the debate. Consider the contrast between those who see geoengineering as merely one among a set of possible policy options—to be chosen simply on the basis of a set of normal policy criteria, such as technical feasibility, likely side-effects and cost-effectiveness—and those who are reluctant to consider geoengineering even as a last resort and even then are unhappy about having to do so. My suggestion is that there may be a connection between, on the one hand, the first group and those who deny the existence of genuine moral dilemmas and, on the other hand, the second group and those who accept that such dilemmas exist. This connection might explain why, when the first group goes on about technical feasibility and the like, these arguments do not really seem to address the core concerns of the second. Even when it is said that geoengineering is a necessary evil, the second group is not happy—they don't seem to process the term "necessary evil" in the same way.[86]

Second, introducing the categories of tarnishing, marring, and blighting evils enriches the debate. These senses of evil are distinct from, and much more morally loaded than, the modest sense of evil as "something one has serious reason to avoid." But they are also less uncompromising than that implied by the claims that evils ought never to be done, that they should never even be considered, or that they exceed the bounds of moral deliberation. Tarnishing, marring, and blighting evils can often be thought about, and perhaps sometimes ought to be chosen. Still, they come with considerable moral baggage, which a morally serious person cannot ignore. If we suppose for a moment that there are, or might be, such evils, then precisely how to categorize the "evils" at the center of a lesser evil argument becomes an important issue.

Third, if such evils exist, this raises a question about whether putting someone in a marring situation—one where they might be required (or have strong reason) to incur such an evil—constitutes a special kind of moral wrong, or at least one that greatly increases the moral gravity of the action. Surely, the thought goes, there is a significant moral difference between putting others in a situation where they must choose between (normally) bad options and putting them in a situation where their choice will tarnish or even blight their lives. Other things being equal, we have much stronger reason to avoid the latter situation and so are liable to greater censure if we fail to do so. Indeed, this reason may be so strong that ignoring it—and unnecessarily inflicting a marring choice on others—itself counts as an evil that blights *our* lives.

This third issue might turn out to be especially serious in the case of climate change. Consider just two kinds of cases. First, perhaps the inaction of some countries (e.g., the high-emitting developed countries) will inflict marring choices on the people of other countries (e.g., the lower-emitting developing countries). For example, if current and past emissions cause Bangladesh to flood and force its people to migrate, it is not beyond the realm of possibility that some parents (or the Bangladeshi government or other agencies) might be placed in situations similar to the one confronting Sophie. Second, the current generation, by exploiting its temporal position, may put some future generation in a position where it must make a marring choice. For example, perhaps our actions will cause that future generation to confront an abrupt climate change so severe that they must choose to burn a large amount of fossil fuel in order to prevent an immediate humanitarian disaster, even knowing that this will then impose further catastrophes on some later generation.[87] In this case, we are responsible for putting the first future generation in a position where it must inflict a great harm on the second, and so mar itself. This seems to be a serious moral wrong on our part. It might also be a blighting evil.

VI. A Climate of Evil

"The arrogance of human beings is just astounding."
—Oceanographer Sallie Chisholm
(Monastersky 1995)

Of course, none of the above explains why geoengineering specifically might bring on a marring evil. Presumably, successfully answering this question would require a much larger project. Moreover, since here we are only trying to survey the moral terrain, an answer is not strictly necessary. Still, since the very idea of marring is controversial in itself, and perhaps especially so when applied to geoengineering, it might be worth at least gesturing at the shape the relevant reasons might take.

Some possibilities emerge from considering how many climate scientists (some specifically responding to the core proposal) argue against geoengineering. First, it is common to imply that pursuing geoengineering manifests arrogance and recklessness. For example, Jeff Kiehl writes: "On the issue of ethics, I feel we would be taking on the *ultimate state of hubris* to believe we can control Earth. We (the industrially developed world) would essentially be telling the (rest of the) world *not to worry* about our insatiable use of energy."[88] Similarly, Stephen Schneider argues: "Rather than pin our hopes on the *gamble* that geoengineering will prove to be inexpensive, benign and administratively sustainable over centuries—*none of which can remotely be assured now*—in my value system I would prefer to start to lower the human impact on the Earth through more conventional means."[89]

Second, climate scientists frequently claim that pursuing geoengineering represents a kind of blindness, a failure on the part of humanity to address the underlying problem. For example, Kiehl says, "In essence we are treating the symptom, not the cause. Our species needs to begin to address the cause(s) behind the problem."[90] Moreover, it is often suggested that this reluctance to address the underlying problem is somehow shortsighted, obstinate, or even bizarre. For example, Schneider likens the climate-change problem to heroin addiction and compares the decision to pursue geoengineering to choosing "a massive substitution of [planetary] methadone" over "slowly and surely" weaning the addict.[91] Similarly, Gavin Schmidt offers the analogy of a small boat being deliberately and dangerously rocked by one of its passengers. Another traveler offers to use his knowledge of chaotic dynamics to try to counterbalance the first but admits that he needs huge informational resources to do so, cannot guarantee success, and might make things worse. Schmidt concludes: "So is the answer to a known and increasing human influence on climate an ever more elaborate system to control the climate? *Or should the person rocking the boat just sit down?*"[92]

Two features of these criticisms of geoengineering strike me as especially intriguing. The first is the hint that at least one core wrong associated with geoengineering is best captured by gesturing at certain character traits, such as hubris, recklessness, and an obstinate resistance to look at the central problem. The second feature is the tendency to see the moral issue as one that faces us as members of collectives—whole societies, the industrialized world, or even humanity considered as such. In short, one worry that these scientists have about the decision to pursue geoengineering concerns what it might show about us—our lives, our communities, our generation, our countries, and ultimately our species. What kind of people would make the choice to geoengineer? Would they be reckless, hubristic, and obstinate people? Would this be a generation or country consumed by its own (perhaps shallow) conception of its own interests and utterly indifferent to the suffering and risks imposed on others? Would it be a species that was failing to respond to a basic evolutionary challenge?

Such concerns are relevant to political inertia over climate change in general. On one natural way of looking at things, groups with which many of us identify are predominantly responsible for creating the problem, are currently largely ignoring the problem, and are also refusing to address the problem in the best way possible because of a strong attachment to lesser values. These are serious moral concerns and give rise to substantial moral criticism. Who would want to be associated with such groups and implicated in such behavior? Are we not saddened, even ashamed? Is this not a tarnishing evil?

Perhaps. But what about geoengineering specifically? Why might choosing it tarnish a life? Again, let me emphasize that this is not the place for a full account and that such an account is not necessary for current purposes. Still, we can point to three worries that give us some sense of what such an account might look like.

Consider first those who cause the nightmare scenario to arise. One way in which our lives might be tarnished would be if the commitment to geoengineering becomes a vehicle through which we (e.g., our nation and/or our generation) try to disguise our exploitation of other nations, generations, and species. Specifically, our willingness to facilitate (or engage in) geoengineering might show that we have failed to take on the challenge facing us and instead have succumbed to moral corruption. Indeed, the decision to geoengineer might reveal just how far we are prepared to go to avoid confronting climate change directly, and this might constitute a tarnishing, even blighting, evil. Think about what people mean when, in tragic circumstances, they say, "Has it really come to this?"[93]

Consider now those who choose geoengineering as the lesser evil in some nightmare scenario. Why might this be marring? One reason is that through their choice, they inflict grave harms on innocents that otherwise might not have occurred. Suppose, for example, that geoengineering really does cause less harm than climate catastrophe but that this harm accrues to different individuals.[94] In that case, when we choose geoengineering, innocents are harmed through our agency, and this may be a marring evil (even if it is a lesser evil overall). One can certainly imagine it being something that people find, as the expression goes, "hard to live with." Indeed, this is a prominent feature of other marring cases.[95]

Finally, and more controversially, consider the position of humanity more generally. Pursuing geoengineering may be taken as a sign that we, as a species, have failed to meet a basic challenge and should be saddened or ashamed for that reason. One thought is this. Humanity is, in geological and evolutionary terms, a recent arrival on the planet and is currently undergoing an amazingly rapid expansion, in terms of sheer population numbers, technological capabilities, and environmental impact. A basic question that faces us as humans, then, is whether, amid all of this, we can meet the challenge of adapting to the planet on which we live. In this context, the decision to geoengineer might be taken to show that we have, to a significant extent, failed; and such a failure might be blighting.[96]

More specifically, suppose the basic idea is that as a species, we already had a perfectly serviceable planet to live on, but now we are undermining that; we have, in elementary terms, "fouled the nest." We could clean it up—that would be the most direct approach, the one most likely to work—but so intent are we on continuing our messy habits that we will pursue any means to avoid that, even those that impose huge risks on others and involve further alienation from nature.[97] In this case, so the thought goes, the decision to geoengineer constitutes the crossing of a new threshold on the spectrum of environmental recklessness and therefore embodies a recognition of our continued and deepening failure. On this view, it is natural to think that it will be a sad and shameful day in the life of humanity when such a decision is made, that (if the choice is forced as a lesser evil) such a decision mars the lives of those who make it, blights those who bring about the nightmare situation, and perhaps even tarnishes humanity as such.

In summary, I have pointed to three reasons for thinking that the decision to engage in geoengineering might involve tarnishing. Some of these reasons are, of course, highly controversial and embody distinctive perspectives on global environmental issues. Still (I emphasize again), the purpose of this discussion is not to defend such views but merely to survey the moral terrain. The general point here is only that simple "lesser evil" arguments fail even to consider the possibility that there might be such things as tarnishing evils. Thus, such arguments are too quick and obscure important ethical issues. More specifically, if we focus on the core proposal in its most abstract and simplistic form, we might miss much of what is at stake in the decision to geoengineer.

VII. But... Should We Do It?

In conclusion, the purpose of this chapter has been to survey some of the moral landscape relevant to geoengineering. This has been done through an exploration of one popular proposal, at the heart of which is the arm the future argument. This argument is often presented as offering a straightforward and decisive case for research on geoengineering. I have argued that this is not so. First, the argument assumes much that is contentious, including that geoengineering is our only (or most central) option, that it is less risky than other options, and that any consensus on the need to geoengineer will not be shallow but will be deep enough to guide policy. Second, it is overly narrow in its focus. For example, it does not even consider the serious issues of compensation, political legitimacy, and the role of political inertia in framing geoengineering policy. Third, the argument obscures deeper moral considerations, such as what is at stake in calling something evil, whether evils ought sometimes to be chosen, whether there are marring choices, and whether putting others in situations where they are forced to choose a marring evil is an extra and special kind of moral wrong. Finally, the presence of such an underdeveloped argument is especially worrying in the context of the perfect moral storm, where moral corruption is likely. In such a setting, the arm the future argument runs the risk of being glib, cavalier, and even perhaps morally irresponsible.

What lessons should we draw from this discussion? One we cannot draw is that no "lesser evil" argument for research on, or deployment of, geoengineering can ever succeed. Our survey of the terrain raises serious difficulties for such arguments but does not show that these cannot be overcome. Still, progress has been made. First, the survey dispels the illusion of irresistibility surrounding the arm the future argument and hence shifts the burden of proof back onto proponents of geoengineering. Second, it strongly implies that if we pursue geoengineering at all, then a broad range of obligations—far beyond mere scientific research—must be considered. These flow from our responsibility for the climate problem, our failure to choose nonevil solutions, our creation of nightmare scenarios that are potentially marring for others, and our infliction on the future of the special liabilities and political realities associated with geoengineering. Third, the discussion suggests that these extended obligations are likely to be demanding, involving not just technological assistance but also substantial compensation and wider commitments to norms of global justice and community.

Some, I suspect, will want to craft such points into a new case for geoengineering, suggesting that the survey reveals some shapes that such a defense might take. One obvious thought is that perhaps geoengineering can be justified as part of some broad climate policy portfolio that includes many of the alternative policies I mention, suitably embedded in wider ethical and political concerns.[98] Perhaps this is correct; still, we must be cautious. First, the urge to find *some* kind of argument for geoengineering should give us pause. Is this a policy in search of a rationale? Are we simply looking for an argument that will justify geoengineering, rather than seeing where the arguments lead? In the perfect moral storm, this is a worrying thought. Second, the idea that a fully-moralized lesser evil argument might justify the pursuit of geoengineering may be more interesting in theory than in practice. Politically, such an approach seems likely to curb the current enthusiasm for the core proposal in many quarters, restoring many of the same motivational obstacles that face conventional climate policies and introducing further moral and political objections. Indeed, for many, the mere mention of wider and more demanding obligations will be enough to undermine (for them) geoengineering's status as any kind of *lesser* evil, all things considered. Given this, we must take seriously the possibilty that moralized "lesser evil" solutions will be even less available politically than the nonevil options. This, of course, tends to shift the focus back to "modest geoengineering research only." But now it has been revealed that such an approach counts as the lesser evil only in a severely attenuated sense. "Modest geoengineering research only" is likely to be *far down the list even of evil options.* Talk of the lesser evil covers this up. In a perfect moral storm, this is an important conclusion. As Franklin might put it, "how convenient."

Acknowledgments

Earlier versions of this chapter were presented to the "Ethics and Climate Change" conference at the University of Washington, the "Global Justice and Climate Change" conference at Oxford University, the conference "Human Flourishing, Restoration and Climate Change" at Clemson University, a European Science Foundation exploratory seminar at the University of Oslo, the Department of Philosophy at the University of California, San Diego, and the annual meetings of the American Association for the Advancement of Science in San Francisco, the Association for Legal and Social Philosophy in the United Kingdom, and the American Philosophical Association (Pacific Division) in Vancouver, BC. I am grateful to those audiences, and especially to Richard Arneson, Cecilia Bitz, David Brink, Simon Caney, Ralph Cicerone, James Fleming, Espen Gamlund, Michael Gillespie, Mathew Humphrey, Dale Jamieson, Monte Johnson, Richard Miller, Alan Robock, William Rodgers, Henry Shue, Richard Somerville, Behnam Taebi, and Mike Wallace. I'd also like to thank Jeremy Bendik-Keymer, Don Maier, Christopher Preston, and Allen Thompson for their careful written comments.

Notes

1. Schelling 1996; Keith 2000.

2. Crutzen 2006. Crutzen's piece appeared in *Climatic Change*, accompanied by a set of responses from other distinguished scientists, including Bengtsson 2006, Cicerone 2006, and Kiehl 2006.

3. Such as Wigley 2006.

4. Walsh 2008.

5. In ethics, the exception is the groundbreaking Jamieson 1996. Other early articles with something to say about ethics include Bodansky 1996, Keith 2000, Schelling 1996, Schneider 1996, and Kellog and Schneider 1974.

6. After all, I focus on only one argument for geoengineering when many others might be offered, I consider only a fairly limited version of that argument, and I admit in advance that the challenges I raise may not be decisive.

7. For an overview, see Keith 2000.

8. This approach is appealing in large part because it has a natural precedent whose implications are generally understood: the aim is to simulate the known cooling effects of a large volcanic eruption. However, it should be noted that whereas volcanic eruptions are usually isolated events whose effects on temperature last only a year or two, the geoengineering proposal involves continuous injections of aerosols for a period of at least decades and possibly centuries. Not only is this a different proposition—amounting to continuous sustained eruption rather than an isolated event—but there are worries that it soon becomes effectively irreversible. First, because the sulfate particles only mask the effects of increasing greenhouse-gas concentrations in the atmosphere and because they dissipate quickly, any attempt to halt the experiment would probably commit the earth to a swift rebound effect. Second, if the masking effect is large, then the rebound effect will likely also be large, and this is so in the current case. Under the circumstances imagined by the current proposal, the masking effect is of comparable magnitude (2°C to 6°C) to the kind of catastrophic climate change that the intervention is trying to prevent. Third, since the speed of the change is itself a factor, this would probably make unmasking worse than allowing the original climate change. Hence, many scientists believe that once we have been doing sulfate injection for a while, we will, in effect, be committed to continuing indefinitely (e.g., Matthews and Caldeira 2007).

9. We could add to this that there has been a similar lack of progress on the other necessary policy: adaptation.

10. The attribution to Crutzen requires some interpretation, since his claims are not fully explicit. Still, even if it is disputed, this should not undermine the interest of the present chapter. The AFA is clearly one major argument for geoengineering, and anecdotal evidence suggests that it is widespread. Moreover, the points I make about it have wider relevance.

11. Crutzen is far from the first to advocate the core proposal. As Stephen Schneider put it: "In this case, the messenger is the message" (Morton 2007, 133). More recently, similar claims appear in Victor et al. 2009.

12. The role of the sulfate example is simply to focus our attention on clear cases of geoengineering. Presumably, other clear cases would serve just as well; hence, the chapter has wider application. Still, in the remainder of the paper the reader should assume that when I use the term "geoengineering" this is the intervention I have in mind.

13. The centrality of the problem of political inertia can be obscured by the fact that Crutzen initially gives most prominence to a different aerosol problem: as policy makers try to tackle normal air-pollution problems by reducing sulfur dioxide emissions, they will thereby increase global warming, and the increase may be dramatic. However, despite the prominence Crutzen gives this "Catch-22" situation, the problem of inertia appears more fundamental for him. He explicitly claims that the aerosol problem could be solved through mitigation and indeed that this would be the best solution: "By far the preferred way to resolve the policy makers' dilemma is to lower the emissions of the greenhouse gases" (Crutzen 2006, pp. 211–212; see also p. 217). Hence, his view is *not* that the aerosol problem as such makes geoengineering necessary (e.g., because it puts us into new territory, where mitigation alone will not be enough, so that geoengineering must be considered as well).

14. Crutzen 2006, p. 211.

15. Ibid., p. 212.

16. Marland et al. 2008.

17. Moore 2008.

18. Crutzen 2006, p. 217.

19. Ibid.

20. Gardiner 2006; Gardiner forthcoming.

21. In my view, such buck passing is already manifest in recent climate-change policy. See Gardiner 2004 and Gardiner forthcoming.

22. I say more about moral corruption in Gardiner forthcoming.

23. Schelling 1996; Barrett 2008.

24. See Lovelock 2008, p. 3887.

25. For example, some are concerned that sulfate injection may lead to further destruction of stratospheric ozone. Crutzen (himself a pioneer in examining the ozone problem) is optimistic that this problem is small, given the quantity of sulfate to be injected, and also suggests that alternatives to sulfates might be tried. But, as he acknowledges, this requires more research. (See Crutzen 2006, pp. 215–216).

26. Bodansky 1996.

27. Jamieson 1996.

28. I consider less modest senses in section IV below.

29. Bengtsson 2006, p. 233.

30. Thomas Schelling warns that if we are preparing for intervention that is 50 years or more off, this may be pointless preparation; technological change over such a period may be so profound as to make the preparation worthless. The precise import of this claim is unclear. (Perhaps we should prepare less than we might otherwise do? Perhaps we should do comparatively more basic climate research for geoengineering and less technical research?) But it does cast doubt on the claim that the best we can do for future generations is geoengineering research. See Schelling 1996.

31. Some scientists sympathetic to Cicerone's argument are confident that in the end, there are no good geoengineering proposals to be had. Hence, they support research on the grounds that it will reveal this "fact" more clearly and prevent geoengineering strategies from being implemented merely for political reasons. But this is a different rationale. Note that it assumes not only that good proposals will not emerge but also that further science will be enough to circumvent the political forces in favor of geoengineering (even when existing science has not) and that it is worth "wasting" scarce scientific resources in this effort.

32. Jamieson 1996.

33. Don Maier also suggests to me that (i) often such institutions compete, so that we should expect geoengineering institutions to discourage those that promote mitigation, and (ii) such institutions create psychological momentum—individuals do not like to abandon projects in which they have invested time, energy, money, and emotion.

34. Jamieson argues for just the opposite conclusion: that geoengineering research could only be justified if accompanied by research into the ethics of geoengineering (Jamieson 1996).

35. Some parts of this section draw on Gardiner 2007.

36. Jamieson 1996, pp. 332–333.

37. For example, Stephen Schneider, himself generally an opponent of geoengineering, reports that back in 1992, the concerns of a National Academy of Science panel were "effectively countered" by the following argument: "Let us assume . . . that . . . the next generation of scientific assessments . . . converged on confidently forecasting that the earth had become committed to climate change . . . serious enough to either require a dramatic retrenchment from our fossil fuel based economy . . . or to endure catastrophic climatic

changes. Under such a scenario, *we would simply have to practice geoengineering as the 'least evil'* (Schneider 1996, pp. 295–296; emphasis added). Schneider attributes the argument to Robert Frosh.

38. Crutzen specifically cites the Schneider passage, with approval.

39. Indeed, this helps to explain why arguments from emergency—and declarations of states of emergency when normal political processes and rights are suspended—are often employed by political despots.

40. Perhaps it is possible that you will win $10 million in the lottery this year. But that doesn't mean that you should *now* hire an investment banker to develop a plan for how to use it.

41. One response to this argument would be simply to concede it and argue merely that the nightmare scenario is at least one among a number of emergencies that we should prepare for. For more on this kind of argument, see below.

42. It might also facilitate inertia on adaptation and increase the severity of any given climate catastrophe by undermining people's ability to cope.

43. The basic ideas here are familiar from another context. Proponents of torture try to force their opponents to admit that in the case of a ticking bomb—where you, the authorities, know that your prisoner has hidden a nuclear device under the streets of a major city but don't know where it is—torture is permissible and then to infer from this that torture is justified. In a classic move against this kind of "lesser evil" argument, Henry Shue concedes that torture may be permissible in the ticking-bomb case but argues that this does not imply anything about what the policies of those not confronting such a case should be (Shue 1978, 2005). For one thing, the case may be theoretically possible but in practice so very improbable as to make planning for it irrational; for another, actually planning for the case—for example, creating a bureau for torture and training torturers—may have such profound and predictable negative consequences that this is a decisive reason to reject it.

44. Crutzen is explicit about this: the idea is that we must prepare for the possibility of an emergency, not that we are actually in one right now. Hence, his core position is that we should develop geoengineering to serve as a backstop technology, to deploy if the situation eventually deteriorates. The AFA is explicitly a "backstop argument."

45. One could embellish the AFA to claim that we are already in a different "lesser evil" situation. Suppose the argument is: (1) current research on geoengineering is an evil, because it really does increase the probability of deployment; but (2) given the possibility of the nightmare scenario, we must take the risk and choose this evil; and (3) we must do so *now* or else risk being too late. This embellishment probably makes the original argument more promising. Nevertheless, it seems to be threatened by many of the other considerations raised here. For one thing, we (now) have alternatives to geoengineering, including mitigation and investment in alternative energy, to name just two; for another, there are serious questions about whether geoengineering research will succeed and about whether this is even a good time to begin.

46. Some have argued that such an approach is not only more feasible than geoengineering, but also secures a better outcome. Consider Bengtsson's response to Crutzen, where he concludes: "I do consider it more feasible to succeed in solving the world's energy problem, which is the main cause to the present concern about climate change, than to successfully manage a geo-engineering experiment on this scale and magnitude, which even if it works is unable to solve all problems with the very high concentration of greenhouse gases in the atmosphere" (Bengtsson 2006, 233).

47. Schelling 1996.

48. Schelling, an economist, explicitly assumes that the decision was at least fifty years off in the mid-1990s (Schelling 1996). Moreover, Crutzen assumes that geoengineering will only be necessary if mitigation efforts fail. But such efforts will have almost no impact on temperature rise in the next thirty years, and a limited impact in the next forty to fifty years.

49. See, for example, Lenton et al. 2008. Victor et al. 2009 explictily invoke such "tipping points." However, Alan Robock (2008) has suggested to me the future generation assumption may not be widely shared by scientists, since there will be substantial climate impacts during the next fifty years. The latter claim is highly plausible if one is referring to gradual effects rather than abrupt changes. Still, I suspect that such concerns push towards an "arm the present" argument, and perhaps one that does not focus on catastrophe and suggests fairly quick deployment. My impression is that such arguments are not yet mainsteam. In any case, they require independent treatment.

50. Schneider 2008, 3856. In the current context, although the basic albedo modification idea at the center of the core proposal has been around for a while, little work has been done on

the impacts of sustained modification of this kind, the extent to which it is irreversible, and its regional and ecological impacts. Bengtsson 2006 lists some of the general worries. Several early articles on such issues have been released since Crutzen's paper which call into question some of his main claims. See, for example, Matthews and Caldeira 2007; Rasch et al. 2008; Robock 2008.

51. If this assumption is dropped, then we shift to an "arm the present" argument. This, in my view, is importantly different. For example, it reduces (though it does not eliminate) intergenerational concerns, since now the current generation has to take some of the risks of implementation. In addition, if the nightmare is to arrive sooner than forty years, it is much less likely that mitigation is a plausible alternative: we are probably already committed. For such reasons, "arm the present" arguments require independent treatment. See Gardiner forthcoming.

52. This is something that Crutzen himself is very clear about. He argues that we ought to pursue mitigation, but we probably won't; therefore, he concludes, we should research geoengineering.

53. Kellog and Schneider 1974 make a similar point about unilateral geoengineering.

54. See Wigley 2006. This possibility reveals that not all geoengineering proposals need manifest intergenerational moral corruption. For example, the attempt to "buy time" by geoengineering may pose more threats to current people than to future people. If so, they hardly manifest buck passing. (Of course, there may be other reasons to resist them, especially if they are very risky or if they pose disproportionate threats to the world's poor.)

55. Suppose, for example, that it is my child in the pond, that I let her climb in, that I then just watch the drowning, knowing that you will jump in, and that I do all of this even though you are old, much farther away, and risk a heart attack from the exertion, while I am young and merely concerned about getting my shoes muddy.

56. The name reflects the fact that the argument is intended as a place holder, to stand in for a set of more sophisticated accounts. I assume that such accounts might emerge with a wide variety of views in global political philosophy (including, for example, cosmopolitanism, Rawlsian nationalism, communitarianism, and libertarianism) and that the "stalking horse" argument merely offers a general framework within which they might operate.

57. It is true that the AFA ultimately concludes only that research on geoengineering is justified. But in doing so, it relies on the claim that geoengineering should be chosen in the nightmare situation (AFA7), and no argument about political legitimacy is made there. This suggests that the idea is that *any* decision to geoengineer would be morally appropriate in the nightmare scenario. On this claim, see below.

58. Notice that I make no assumption about how robust those norms must be. This is because the argument is supposed to appeal to a wide spectrum of views in global political philosophy. Moreover, though perhaps in some views, the norms would be weak enough not to conflict with political inertia, this would need to be shown. Remember that we are trying to survey the moral terrain in this chapter. I do not claim to have proven that the "stalking horse" argument ultimately succeeds, only to have suggested that the ethics of geoengineering must address it.

59. Victor et al. 2009, which emerged just as this chapter was going to press, does raise the issue of legitimacy, and claims that work on establishing norms needs to be done. Still, it seems concerned only with the narrow issue of implementation, assuming that existing political arrangements remain more or less as they are. Moreover, in another recent work Victor envisions the norms arising "through an intensive process ... best organized by the academies of sciences in the few countries with the potential to geoengineer" (see Victor 2008). For obvious reasons, such a process raises concerns if the intent is to generate appropriate norms of global justice and community.

60. The obvious retort to this would be to say, "But in the nightmare scenario, any decision to geoengineer would be legitimate." On this, see the core component interpretation below.

61. This point does not require the success of the previous arguments. Perhaps we owe the future some of these things because limited geoengineering research will do no good or because of other past injustices or because minimal humanitarian duties require it.

62. Walsh 2008.

63. Crutzen 2006, p. 217.

64. Cicerone 2006, p. 224.

65. Cicerone, of course, is explicit about this: he thinks that we should do the research first and bring in the broader issues at a later stage.

66. See also the wilderness-survival example offered below.

67. The idea that approaches to important moral questions may exhibit some kind of schizophrenia is pioneered by Michael Stocker (1976). I cannot pursue it here.

68. Schelling 1996.

69. Bengttson 2006.

70. How would the United States feel about geoengineering if it thought that China, Russia, or Iran was going to do it?

71. This worry is especially relevant in the case of geoengineering, where the "lesser evil" claim is typically not so much argued for as simply asserted as decisive in a sentence or two before the discussion moves on. Of course, "lesser evil" arguments need not be opaque; but—for the reasons mentioned below—we should pay special attention to opacity when the threat of moral corruption is high.

72. Even among scientists, there are variations in the description of the catastrophic evil to be averted. Is it runaway temperature change, caused by a convergence of positive feedbacks that make mitigation no longer possible? Is it a major abrupt change, such as a shutdown of the thermohaline circulation or a sudden collapse of the Greenland ice sheet? Is it accelerating extinctions caused by linear climate change? These are distinct scenarios and may call for quite different emergency measures.

73. Of course, the "lesser evil" argument might play an appropriate role in summarizing something like an overlapping consensus on a given policy. But this argument would need to be made independently and would face a substantial burden of proof. The worry is that the "lesser evil" argument as usually stated is an attempt to avoid meeting that burden.

74. This concern resonates with familiar complaints about the way economists tend to reduce moral wrongs to mere "costs," as, for example, when they insist on seeing fines as mere fees. See Goodin 1994.

75. See Smart and Williams 1973.

76. For example, suppose that coldblooded murder is a lesser evil than genocide but still unthinkable.

77. Smart and Willliams 1973.

78. Williams concedes that Jim should shoot the Indian in his classic case. His worry is that utilitarianism comes to this conclusion far too quickly, without realizing what is at stake for the agent in the decision.

79. In my own view, Sophie's choice is probably not the best; still, I do not think that we should blame her for it, nor do I believe that it is the existence of some alternative that produces the marring effect.

80. Hursthouse 1999, pp. 73–75.

81. This is so even if we agree that Sophie did the right thing and perhaps even if we think that there is a sense that she made a heroic choice. Even if we think that our everyday behavior falls morally far short of Sophie's, there is still a clear sense in which we don't want to be Sophie. We'd rather fall short under normal circumstances than make a heroic choice in this one. It is not clear whether this attitude is best characterized as a moral one or one that seeks to restrict the relevance of morality. But it is clearly an evaluative one.

82. Styron 1979, p. 538.

83. I assume that this definition requires refinement. But this is not the place for such work.

84. Aristotle, *Nicomachean Ethics*, Book I.9–10.

85. The distinctions to be made between ways in which lives may be tarnished raises interesting questions in ethical theory, but these cannot be pursued here. For present purposes, the mere signaling of the category, and the fact that it does not automatically disappear in a nightmare scenario, is all that is needed.

86. For whatever it is worth, I have found that when I present the contents of this chapter in public, the audience is divided about whether there are such things as marring evils and whether geoengineering might constitute one. Typically, one-third will have no truck with marring evils at all. Of the rest (who believe that marring is possible), somewhere between two-thirds and one-third think that geoengineering may be a marring evil. This wide range may reflect moderate disciplinary biases. On average, mainstream economists and political scientists seem less friendly to marring arguments for geoengineering, whereas scientists, environmentalists, and the public at large are more so. Within moral philosophy, consequentialists are traditionally opposed to marring arguments, while virtue ethicists and some deontologists are more sympathetic.

87. Gardiner 2009.

88. Kiehl 2006, p. 228; emphasis added.

89. Schneider 1996, p. 300; emphasis added.

90. Kiehl 2006, p. 228.

91. Schneider 1996, pp. 299–300.

92. Schmidt 2006, responding to Crutzen; emphasis added.

93. Another root of tarnishing would be if geoengineering led to the infliction of marring choices on others. See above.

94. Robock 2008.

95. Consider again Williams's example of Jim and the Indians. Williams famously concludes

that Jim should shoot the Indian. But he chastises utilitarianism for reaching the same conclusion too easily, without realizing what is at stake for Jim in such a decision. Perhaps what is at stake is a marring evil.

96. Of course, the charge of failure is controversial. In this context, *adaptation* is a complex and value-laden term. First, the basic survival of the species might be one necessary component. Surely (pace Lenman 2002), it would be grounds for shame if our inaction led to extinction. Fortunately, most scientists do not think that this is likely, even under extreme scenarios. Still, the more credible extremes are not very comforting. For example, James Lovelock (2006) believes that the worst-case scenario is a few hundred thousand humans hunkered down at the poles. Who would want to be implicated in bringing that about? Second, alternatively, some might say that it would count as adaptation if a few million humans survived, living in huge artificial domes atop a desolate planet. But this also seems to miss something. Many believe that part of the human challenge is to develop an appropriate relationship to nature, including to other species inhabiting the earth. Surviving in domes does not satisfy that demand. If this is the best that humanity can manage, it might still be a source of sadness and shame. Although the core proposal is far from a commitment to domes, it begins to enter similar territory.

97. Indeed, perhaps we are pushed in this direction by the very factors (e.g., ways of life, institutions, values) that caused the mess in the first place.

98. Another thought is that perhaps the nightmare scenario is coming much more quickly than the "arm the future" argument suggests, and the catastrophe is to be much more severe than mainstream projections currently imply; hence, geoengineering is not merely our only option, but there is no time to consider the nuances (see notes 49 and 51). But notice that this shifts the ground (Gardiner forthcoming). For example, this rationale is much more pessimistic than the AFA. In particular, it denies that catastrophe is some way off, and that it could still be averted through conventional methods, and hence rejects the claim that geoengineering, and the associated research, would be unnecessary if we would just do the right thing. Such pessimism requires justification. Moreover, the details have implications for policy. Preparing to geoengineer in ten years is a very different proposition for preparing to geoengineer in forty or more.

References

Barrett, Scott. 2008. "The Incredible Economics of Geoengineering." *Environmental Resource Economics* 39: 45–54.

Bengtsson, Lenart. 2006. "Geoengineering to Confine Climate Change: Is it At All Feasible?" *Climatic Change* 77: 229–234.

Bodansky, Daniel. 1996. "May We Engineer the Climate?" *Climatic Change* 33: 309–21.

Cicerone, Ralph. 2006. "Geoengineering: Encouraging Research and Overseeing Implementation." *Climatic Change* 77: 221–226.

Crutzen, Paul. 2006. "Albedo Enhancement by Stratospheric Sulphur Injections: A Contribution to Resolve a Policy Dilemma?" *Climatic Change* 77: 211–219.

Eilperin, Juliet. 2008. "Carbon Is Building Up in Atmosphere Faster Than Predicted." *Washington Post,* September 26, 2008.

Gardiner, Stephen M. 2004. "The Global Warming Tragedy and the Dangerous Illusion of the Kyoto Protocol." *Ethics and International Affairs* 18.1: 23–39.

———. 2006. "A Perfect Moral Storm." *Environmental Values* 15: 397–413. Also chapter 4 in this volume.

———. 2007. "Is Geoengineering the Lesser Evil?" *Environmental Research Web.* Available at http://environmentalresearchweb.org/cws/article/opinion/27600. Accessed April 18.

———. 2009. "Saved by Disaster? Abrupt Climate Change, Political Inertia, and the Possibility of an Intergenerational Arms Race." *Journal of Social Philosophy* 20.4: 140–162.

———. Forthcoming. *A Perfect Moral Storm.* Oxford: Oxford University Press.

Goodin, Robert E. 2004. "Selling Environmental Indulgences." *Kyklos* 47: 573–596. Also chapter 13 in this volume.

Hursthouse, Rosalind. 1999. *On Virtue Ethics.* Oxford: Oxford University Press.

Ignatieff, Michael. 2004. *The Lesser Evil: Political Ethics in an Age of Terror.* Princeton, N.J.: Princeton University Press.

Jamieson, Dale. 1996. "Intentional Climate Change." *Climatic Change* 33: 323–336.

Keith, David. 2000. "Geoengineering: History and Prospect." *Annual Review of Energy and the Environment*: 245–284.

Kellogg, W. W., and Schneider, S. H. 1974. "Climate Stabilization: For Better or Worse?" *Science* 186: 1163–1172.

Kiehl, J. 2006. "Geoengineering Climate Change: Treating the Symptom over the Cause?" *Climatic Change* 77: 227–228.

Lenman, James. 2002. "On Becoming Extinct." *Pacific Philosophical Quarterly* 83 (September): 253–269.

Lenton, Timothy, Hermann Held, Elmar Kriegler, Jim Hall, Wolfgang Lucht, Stefan Rahmsdorf, and Hans Joachim Schnellnhuber. 2008. "Tipping Points in the Earth's Climate System." *Proceedings of the National Academies of Sciences* 105.6: 1786–1793.

Lovelock, James. 2006. *The Revenge of Gaia: Why the Earth Is Fighting Back—and How We Can Still Save Humanity*. New York: Basic Books.

———. 2008. "A Geophysicist's Thoughts on Geoengineering." *Philosophical Transactions of the Royal Society* 366: 3883–3890.

Marland, G., T. Boden, and R. J. Andreas. 2008. "Global CO_2 Emissions from Fossil-Fuel Burning, Cement Manufacture, and Gas Flaring: 1751–2005." Carbon Dioxide Information Analysis Center, U.S. Department of Energy. Available at http://cdiac.ornl.gov/trends/emis/glo.htm.

Matthews, Damon, and Kenneth Caldeira. 2007. "Transient Climate: Carbon Simulations of Planetary Geoengineering." *Proceedings of the National Academy of Sciences* 104.24: 9949–9954.

Monastersky, Richard. 1995. "Iron versus the Greenhouse: Oceanographers Cautiously Explore a Global Warming Therapy." *Science News* 148 (September 30): 220.

Moore, Frances. 2008. "Carbon Dioxide Emissions Accelerating Rapidly." Earth Policy Institute. Available at http://www.earth-policy.org/Indicators/CO2/2008.htm.

Morton, Oliver. 2007. "Is This What It Takes to Save the World?" *Nature* 447 (May 10): 132–136.

Rasch, Philip, Simone Tilmes, Richard Turco, Alan Robock, Luke Oman, Chih-Chieh Chen, Georgiy Stenchikov, and Rolando Garcia. 2008. "An Overview of Geoengineering of Climate Using Stratospheric Sulphate Aerosols." *Philosophical Transactions of the Royal Society* 366: 4007–4037.

Robock, A. 2008. "20 Reasons Why Geoengineering May Be a Bad Idea." *Bulletin of Atomic Scientists* 64.2: 14–18.

Schelling, Thomas. 1996. "The Economic Diplomacy of Geoengineering." *Climatic Change* 33: 303–307.

Schmidt, Gavin. 2006. "Geoengineering in Vogue." Real Climate, June 28. Available at: www.realclimate.org.

Schneider, Stephen. 1996. "Geoengineering: Could—or Should—We Do It?" *Climatic Change* 31: 291–302.

———. 2008, "Geoengineering: Could We or Should We Make It Work?" *Philosophical Transactions of the Royal Society* 366: 3843–3862.

Shue, Henry. 1978. "Torture." *Philosophy and Public Affairs* 7.2 (Winter): 124–143.

———. 2005. "Torture in Dreamland: Disposing of the Ticking Bomb." *Case Western Reserve Journal of International Law* 37: 231–239.

Smart, J. J. C., and Bernard Williams. 1973. *Utilitarianism: For and Against*. Cambridge, U.K.: Cambridge University Press.

Stocker, Michael. 1976. "The Schizophrenia of Modern Ethical Theories." *Journal of Philosophy* 73: 453–466.

Styron, William. 1979. *Sophie's Choice*. New York: Random House.

Victor, David, M. Granger Morgan, Jay Apt, John Steinbruner, and Katherine Ricke. 2009. "The Geoengineering Option: A Last Resort Against Global Warming?" *Foreign Affairs* March/April.

Walsh, Bryan. 2008. "Geoengineering." *Time*, March 12. Available at http://www.time.com/time/specials/2007/article/0,28804,1720049_1720050_1721653,00.html.

Wigley, T. M. L. 2006. "A Combined Mitigation/Geoengineering Approach to Climate Stabilization." *Science* 314: 452–454.

Part V

Individual Responsibility

17

When Utilitarians Should Be Virtue Theorists

Dale Jamieson

1. I begin with an assumption that few would deny, but about which many are in denial: human beings are transforming earth in ways that are devastating for other forms of life, future human beings, and many of our human contemporaries. The epidemic of extinction now under way is an expression of this. So is the changing climate. Ozone depletion, which continues at a very high rate, is potentially the most lethal expression of these transformations, for without an ozone layer, no life on earth could exist. Call anthropogenic mass extinctions, climate change, and ozone depletion "the problem of global environmental change" (or "the problem" for short).[1]

2. Philosophers in their professional roles have by and large remained silent about the problem. There are many reasons for this. I believe that one reason is that it is hard to know what to say from the perspective of the reigning moral theories: Kantianism, contractarianism, and commonsense pluralism.[2] While I cannot fully justify this claim here, some background remarks may help to motivate my interest in exploring utilitarian approaches to the problem.

3. Consider first Kantianism. Christine Korsgaard writes that it is "nonaccidental" that utilitarians are "obsessed" with "population control" and "the preservation of the environment."[3] For "a basic feature of the consequentialist outlook still pervades and distorts our thinking: the view that the business of morality is to *bring something about.*"[4] Korsgaard leaves the impression that a properly conceived moral theory would have little to say about the environment, for such a theory would reject this false picture of the "business of morality." This impression is reinforced by the fact that her remark about the environmental obsessions of utilitarians is the only mention of the environment in a book of more than 400 pages.[5]

It is not surprising that a view that renounces as "the business of morality" the question of what we should bring about would be disabled when it comes to thinking about how to respond to global environmental change. The silence of Kantianism on this issue is related to two deep features of the theory: its individualism and its emphasis on the interior. Some

Kantian philosophers have tried to overcome the theory's individualism, but this is difficult since these two features are closely related.[6] Kant was not so much interested in actions *simpliciter* as the sources from which they spring. But if our primary concern is how we should act in the face of global environmental change, then we need a theory that is seriously concerned with what people bring about, rather than a theory that is (as we might say) "obsessed" with the purity of the will.[7]

4. Contractarianism has difficulties in addressing environmental problems in general and global environmental change in particular for at least three reasons. First, it generally has a hard time coping with large-scale cooperation problems and the difficulties with assurance to which they give rise. Second, contractarianism has a difficult time with negative "externalities"—the consequences for me (for example) when you and another consenting adult agree to produce and consume some substance that pollutes the air. It may be possible to overcome these problems, at least in principle, through various revisions of the core theory. But the deeper problem with contractarianism is that it excludes from primary moral consideration all those who are not parties to the relevant agreements.[8] Yet much of our environmental concern is centered on those who are so excluded—future generations, distant peoples, infants, animals, and so on.

5. Commonsense pluralism is hampered by its intrinsic conservatism.[9] Although commonsense pluralists morally condemn obvious forms of bad behavior, they are ultimately committed to the view that most of what we do is perfectly acceptable. The role of moral philosophy is primarily to explain and justify our everyday moral beliefs and attitudes rather than seriously to challenge them. From this stance they criticize utilitarianism for being too revisionist and utilitarians for being no fun.[10] But what produces global environmental change is everyday behavior that is innocent from the perspective of common sense: building a nice new house in the country, driving to school to pick up the kids, and, indeed, having kids in the first place, to mention just a few examples.[11] By the standards of common sense, a moral theory that would prescribe behavior that would prevent or seriously mitigate global environmental change would be shockingly revisionist.

6. Some may say that the reigning moral theories have little to say about our problem because it is not a moral problem. No doubt climate change (for example) presents all sorts of interesting and important scientific and practical challenges, but this does not make it a moral problem.[12]

The question of what is (and is not) in the scope of morality is itself an interesting and important question worthy of extensive treatment, but here I will confine myself to only a few remarks. Deontologists might not consider global environmental change a moral problem because, on their view, moral problems center on what we intend to bring about, and no one intends to bring about global environmental change.[13] Similarly, Kantians who reject the idea that "the business of morality is to *bring something about*" might also have reason to exclude our problem from the domain of morality. But whatever one's official view about the scope of morality, the question of how we should regulate our behavior in the face of climate change, ozone depletion, and mass extinctions is important for anyone who cares about nature or human welfare—and these concerns have traditionally been thought to be near the center of moral reflection.

7. For present purposes I assume that our problem is a moral problem. I investigate utilitarian approaches to our problem because utilitarianism, with its unapologetic focus on what we bring about, is relatively well positioned to have something interesting to say about our problem. Moreover, since utilitarianism is committed to the idea that morality requires us to bring about the best possible world, and global environmental change confronts us with extreme, deleterious consequences, there is no escaping the fact that, for utilitarians, global environmental change presents us with a moral problem of great scope, urgency, and complexity.

However, I would hope that some of those who are not card-carrying utilitarians would also have interest in this project. Consequences matter, according to any plausible moral theory. Utilitarianism takes the concern for consequences to the limit, and it is generally of interest to see where pure versions of various doctrines wind up leading us. Moreover, I believe that the great traditions in moral philosophy should be viewed as more like research programs than as finished theories that underwrite or imply particular catechisms. For this reason it is interesting to see how successfully a moral tradition can cope with problems that were not envisioned by its progenitors.[14]

8. While Korsgaard castigates utilitarianism for its environmental obsessions, many environmental philosophers see utilitarianism as a doctrine that celebrates consumption rather than preservation. Specifically, it has been accused of preferring redwood decks to redwood trees and boxes of toothpicks to old-growth forests. Other environmental philosophers argue that utilitarianism cannot account for the value of biodiversity, ecosystems, or endangered species, and go on to condemn the theory for "sentientism" and "moral extensionism." According to these critics, rather than presenting us with a new environmental ethic, utilitarianism is the theory that has brought us to the edge of destruction.[15]

But utilitarianism has an important strength that is often ignored by its critics: it requires us to do what is best. This is why any objection that reduces to the claim that utilitarianism requires us to do what is not best, or even good, cannot be successful. Any act or policy that produces less than optimal consequences fails to satisfy the principle of utility. Any theory that commands us to perform such acts cannot be utilitarian.[16]

As I understand utilitarianism, it is the theory that we are morally required to act in such a way as to produce the best outcomes. It is not wedded to any particular account of what makes outcomes good, of what makes something an outcome, or even of what makes something an action.[17] Moreover, having good theoretical answers to these questions does not mean that we will always know what is right when it comes to practical decision making. And even when we think we know what is right we may change our minds in the light of reflection, analysis, or experience. If utilitarianism is true, embracing the theory may be the first step toward doing what is right, but it is certainly not the last.[18]

9. Utilitarianism is a highly context-sensitive moral theory. Since my concern here is with how a utilitarian should respond to an actual moral problem, I need to make some simplifying assumptions in order to produce responses that are more definitive than "It depends." So in what follows, I will assume that the utilitarian in question holds fairly generic and reasonably traditional views about the matters mentioned in the previous paragraph (e.g., that well-being is at least one of the things that are good, that my causing something to occur or obtain is part of what makes something an outcome of my action, etc.). I will also assume that taken together these views imply that, all things considered, global environmental change is bad (or at least not best). Furthermore, I will assume that the utilitarian in question is a person whose psychology is more or less like mine, and that we have roughly the same beliefs about how the world is put together. I do not mean anything fancy by this—only that, for example, our decision making is not decisively affected by our belief that this world is just a training ground for the next, that most of the world's leaders are agents of an alien conspiracy, or that I am as likely to be a brain in a vat as a guy with a job. Given this background, in the face of global environmental change, a utilitarian agent faces the following question: How should I live so as to produce the best outcomes?

10. Part of what should be taken into account in answering this question is that global environmental change presents us with the world's biggest collective-action problem. Together we produce bad outcomes that no individual acting alone has the power to produce or prevent. Moreover, global environmental change often manifests itself in ways that are quite indirect. The effects of climate change (for example) include sea-level rises and increased frequencies of droughts, storms, and extreme temperatures. These effects in turn may lead to food shortages, water crises, disease outbreaks, and transformations of economic, political, and social structures.[19] Ultimately, millions may die

as a result, but climate change will never be listed as the cause of death on a death certificate. Because our individual actions are not decisive with respect to outcomes, and we are buffered both geographically and temporally from their effects, many people do not believe that their behavior has any effect in producing these consequences.[20] Even when people do see themselves as implicated in producing these outcomes, they are often confused about how to respond, and uncertain about how much can reasonably be demanded of them.

For a utilitarian, this much seems clear: agents should minimize their own contributions to global environmental change and act in such a way as to cause others to minimize their contributions as well. However, in principle, these injunctions could come apart. It is possible that the best strategy for a utilitarian agent would be hypocrisy: increasing my own contributions to the problem could be necessary to maximally reducing contributions overall (perhaps because my flying all over the world advocating the green cause is essential to its success). Or asceticism could be the best strategy: paying no attention to anyone's contributions but my own might be the most effective way for me to reduce overall contributions to the problem.[21] There may be particular utilitarian agents for whom one of these strategies is superior to a "mixed" strategy. However, it is plausible to suppose that for most utilitarian agents under most conditions, the most effective strategy for addressing the problem would involve both actions primarily directed toward minimizing their own contributions, and actions primarily directed toward causing others to minimize their contributions.[22] This would seem to follow naturally (but not logically) from the fact that we are social animals who strongly influence others and are strongly influenced by them.

11. In light of these considerations, how should a utilitarian agent live in order to address the problem? I believe that one feature of a successful response would be noncontingency. Noncontingency requires agents to act in ways that minimize their contributions to global environmental change, and specifies that acting in this way should generally not be contingent on an agent's beliefs about the behavior of others.

The case for noncontingency flows from the failure of contingency with respect to this problem. Contingency, if it is to be successful from a utilitarian point of view, is likely to require sophisticated calculation. But when it comes to large-scale collective-action problems, calculation invites madness or cynicism—madness because the sums are impossible to do, or cynicism because it appears that both morality and self-interest demand that "I get mine," since whatever others do, it appears that both I and the world are better off if I fail to cooperate. Indeed, it is even possible that in some circumstances the best outcome would be one in which I cause you to cooperate and me to defect.[23] Joy-riding in my '57 Chevy will not in itself change the climate, nor will my refraining from driving stabilize the climate, though it might make me late for Sierra Club meetings. These are the sorts of considerations that lead people to drive their '57 Chevys to Sierra Club meetings, feeling good about the quality of their own lives, but bad about the prospects for the world. Nations reason in similar ways. No single nation has the power either to cause or to prevent climate change. Thus nations talk about how important it is to act while waiting for others to take the bait. Since everyone, both individuals and nations, can reason in this way, it appears that calculation leads to a downward spiral of noncooperation.[24]

This should lead us to give up on calculation, and giving up on calculation should lead us to give up on contingency. Instead of looking to moral mathematics for practical solutions to large-scale collective-action problems, we should focus instead on noncalculative generators of behavior: character traits, dispositions, emotions, and what I shall call virtues. When faced with global environmental change, our general policy should be to try to reduce our contribution regardless of the behavior of others, and we are more likely to succeed in doing this by developing and inculcating the right virtues than by improving our calculative abilities.[25]

12. This may sound like a familiar argument against act-utilitarianism. Act-utilitarianism is the theory that directs agents to perform that act which brings about the best outcome,

relative to other acts that the agent could perform. Some philosophers have argued on conceptual grounds that agents who are guided by act-utilitarianism would not produce the best outcomes. This is because certain goods (e.g., cooperation, valuable motives, loving relationships) are inaccessible to, or unrealized by, agents who always perform the best act.[26] Thus, rather than being "direct utilitarians" who focus only on acts, we should be "indirect utilitarians" who focus on motives, maxims, policies, rules, or traits.

The first point to notice is that it does not follow that act-utilitarians do not bring about the best world from the fact (if it is one) that certain goods are inaccessible to, or unrealized by, act-utilitarians. The world may be constructed in such a way that the best state of affairs is not one in which these values obtain, however important they may be taken individually. For example, the pleasure of drinking fine wine is inaccessible to, or unrealized by, a teetotaler, but it does not follow from this that the teetotaler's life is not the best life for him to lead, all things considered (i.e., the one that produces the most utility). By declining the pleasures of wine, the teetotaler may mobilize resources (both financial and energetic) that allow him to realize more utility than he otherwise would if he did not abstain from alcohol.[27]

However, what I have said thus far is consistent with the rejection of act-utilitarianism, but my main concern here is not with the architecture of various versions of utilitarianism. My focus is on the moral psychology of a utilitarian agent faced with the problem, rather than on the conceptual structure of value. I agree that such a utilitarian agent should not adopt act-utilitarianism as a decision procedure and try to transform herself into a moment-by-moment, act-utilitarian calculating device. One reason is that it is not possible for the attempt to succeed. We are cognitively and motivationally weak creatures, with a shortage of time, facts, and benevolence. Our very nature as biological and psychological creatures is at war with the injunction "Transform yourself into a moment-to-moment, act-utilitarian calculating device and act on this basis." There is no reason to think that attempting to live an impossible dream will produce more good than any other course of action.

This seems so obvious that I sometimes (darkly) wonder who invented act-utilitarianism, when, where, and for what purpose. As a theoretical construct it has its uses, but the idea that a utilitarian moralist must embrace a psychologically impossible doctrine on pain of inconsistency is to misunderstand the very project of moral theorizing.[28]

Clearly, Bentham and Mill were strangers to this doctrine.[29] They were promiscuous in their application of the principle of utility to acts, motives, rules, principles, policies, laws, and more besides.[30] Rather than beginning with the principle of utility and then demanding that people become gods or angels in order to conform to it, they start from a picture of human psychology which they then bring to the principle. While conforming to the principle of utility is supposed to make us and the world better, embedding the principle in human psychology is what makes the principle practical. Bentham and Mill were aware of the fact that the world comes to people in chunks of different sizes: sometimes we must decide between acts, at other times between rules or policies. Indeed, acts can express rules and policies, and rules and policies are instantiated in acts. One of the most difficult problems we face as moral agents is trying to figure out exactly what we are choosing between in particular cases.[31] Yes, textbook act-utilitarianism is a nonstarter as an answer to our question, but who would have thought otherwise?[32]

Ultimately, the most important problem with act-utilitarianism is also a problem with indirect views that focus on motives, rules, or whatever. All of these accounts are "local," in that they privilege some particular "level" at which we should evaluate the consequences of actions that are open to us. Rather than adopting any such local view, we should be "global" utilitarians and focus on whatever level of evaluation in a particular situation is conducive to bringing about the best state of affairs.[33] Derek Parfit saw this point clearly when he wrote: "Consequentialism covers, not just acts and outcomes, but also desires, dispositions, beliefs, emotions, the color of our eyes, the climate and everything else. More exactly, C covers anything that could make outcomes better or worse."[34]

13. Some may sympathize with my rejection of utilitarian calculation, but think that in appealing to the virtues I have thrown myself into the arms of something worse. There are other, safer, havens for refugees from utilitarian calculation, it might be thought.

Some may say that what is needed to address our problem is coercive state power, not virtuous citizens. I do not see these as mutually exclusive alternatives. Legitimate states can only arise and be sustained among people who act, reason, and respond in particular ways. The mere existence of a collective-action problem does not immediately give rise to an institution for managing it, independent of the values and motivations of actors. Indeed, if it were otherwise, we would not be confronted by our problem. While it is true that our problem cannot fully be addressed without the use of state power, this observation does not answer or make moot the questions that I am asking.

Others may say that the solution to our problem consists in developing collective or shared intentions of the right sort. One version of this view holds that individual agents need to form intentions "to play one's part in a joint act" or to "see themselves as *working together* to promote human well-being".[35] It may be that such intentions would have an important role to play in successfully addressing our problem, but questions remain about what exactly such intentions consist in, how they arise, what sort of people would have them, and exactly why and in what circumstances they would be adopted.[36] My investigation is meant to address these further questions. In this respect my account can be seen as complementary to, or even perhaps as part of, the project of investigating shared or collective intentions as solutions to collective-action problems.

14. It is now time for me to say something more constructive about my conception of a virtue. Julia Driver's account is helpful as a first approximation: a moral virtue is "a character trait that systematically produces or gives rise to the good."[37] Clearly, this account should be supplemented to reflect the fact that the emotions are closely associated with the virtues.[38] Emotions play an important role in sustaining patterns of behavior that express such putative virtues as loyalty, courage, persistence, and so on. Without emotions to sustain them, it is difficult to imagine how parenting, friendship, and domestic partnership could exist among creatures like us.[39]

Even if Driver's account were supplemented in this way, it would still remain quite generic, since there are different understandings of such expressions as "character trait," "systematically," "produces," and "gives rise to." Moreover, this account would leave many important questions unanswered, including those about the relations between the virtues and human flourishing, and about the relations between the virtues themselves. However, answering these questions is not required for my purposes. What matters to me is the contrast between calculative and noncalculative generators of action, and I use "the virtues" as the name for a large class of the latter.[40]

Some virtue theorists will not be very welcoming of this project. They would deny that an account of the sort I want to give constitutes a version of "virtue ethics." For they hold that "What is definitive of virtue ethics . . . is that it makes virtues not just important to, but also in some sense basic in, the moral structure."[41] Perhaps in deference to this view, what I should be understood as exploring is when an account of utility maximizing requires a theory of virtue.[42]

15. Here is a reminder of what I am claiming. Given our nature and the nature of our problem, noncontingency is more likely to be utility-maximizing than contingency. This is because contingency is likely to require calculation, and calculation is not likely to generate utility-maximizing behavior. Thus, in the face of our problem, utilitarians should take virtues seriously. Focusing on the virtues helps to regulate and coordinate behavior, express and contribute to the constitution of community through space and time, and helps to create empathy, sympathy, and solidarity among moral agents.

16. The most serious problem with the idea that noncontingency should be an important part of a utilitarian theory of how to respond to our problem is that it is in tension with an

underappreciated, but extremely important, general feature of utilitarianism: noncomplacency. Noncomplacency refers to the fact that ways of life and patterns of action should be dynamically responsive to changing circumstances, taking advantage of unique opportunities to produce goodness, and always striving to do better.

Consider first how noncomplacency counts against some versions of indirect utilitarianism, especially those motivated by the desire to produce moral judgments that are more closely aligned with commonsense morality than the judgments that act-utilitarianism would seem to deliver.[43] Views motivated by this desire can lead to a kind of moral complacency that is at odds with any theory that is directed toward producing the best outcomes. Consider two examples.

Suppose that I am a motive-utilitarian who acts on the set of motives that produces more utility overall than any other set of motives that I could have. Imagine that in a one-off situation it is clear that I could produce the most good by acting in a way that is horrific from the point of view of commonsense morality, and that this action is not consistent with my set of standing motivations. A conscientious utilitarian should struggle to perform this one-off act. If she fails in her struggle, she should regret her failure—not because a utilitarian should value regret for its own sake, but because feelings of regret are a characteristic response to the failure to do one's duty. Such feelings of regret may also have a role to play in steeling the agent so that in the future she can perform such one-off acts, however repugnant they may seem to her. Someone who complacently comforted herself with the knowledge that her motives are the best ones to have overall ought to be suspect from a utilitarian point of view, for she acts in a way that she knows is wrong and does not even try to do better.

A similar story can be told about someone who knows he ought to save a stranger rather than his brother in some moment of stress. Such a person, insofar as he is a utilitarian, cannot really be satisfied by telling himself that on the whole he does better acting on the intuitive level rather than ascending to the critical level. He would be like a pilot who on the whole does better flying at 30,000 feet rather than ascending to 40,000 feet, comforting himself about the importance of acting on the basis of good rules of thumb while he is headed directly toward a fully loaded 747. He may not be able to bring himself to do the right thing, but more than shoulder shrugging is called for.

Noncomplacency should lead a utilitarian to moral improvement in two ways. First, she should be sensitive to the fact that circumstances change. What is the best motivational set in an analog world may not be best in a digital one. Moving from Minnesota to California may bring with it not only a change of wardrobe, but also a different optimal motivational set. Second, a utilitarian should constantly strive to shape his motivational set in such a way that his behavior is ever more responsive to particular situations. Broad motives and rules of thumb are starting points for a utilitarian agent, but not where he should aspire to end his struggle for moral improvement.

The problem is that noncomplacency, which seems to me to be important and underappreciated by indirect utilitarians, appears to be in tension with noncontingency, which is required in order to address large-scale collective-action problems. Virtues give utilitarians a way of making human behavior inflexible enough to deal with collective-action problems, but outside the context of collective-action problems it is flexible patterns of behavior that generally are needed for utility maximizing.

17. One approach would be to relax the demand of noncomplacency by giving up utilitarianism in favor of progressive consequentialism. Progressive consequentialism requires us (only?) to produce a progressively better world rather than the best world. Abandoning the maximizing requirement of utilitarianism in favor of a diachronic duty to improve the world would help relieve, but not entirely resolve, the tension between noncontingency and noncomplacency. For as long as noncontingency is in the picture there are going to be conflicts between the character traits that it evokes, and the demand of noncomplacency that on at least some occasions we act in ways that are contrary to what these traits would manifest. Relaxing

the demands of duty will make these conflicts rarer but will not eliminate them entirely.[44]

18. Another, complementary, approach is to develop a highly domain-specific account of the virtues. When it comes to global environmental change, utilitarians should generally be inflexible, virtuous greens, but in most other domains they should be flexible calculators.

The problem with this is that life is not very good at keeping its domains distinct. Suppose that my friend Peter asks me to give him a lift to an Oxfam meeting and that this is the only way that he will be able to attend.[45] However, I am an inflexible, virtuous green when it comes to global environmental change. My green dispositions cause my hand to tremble at the very thought of driving, and I cannot bring myself to give Peter a lift to the meeting. If I were a globally flexible calculator instead, I would not care in what domain utilities are located. If driving Peter to the meeting would produce better consequences than my refusing, then I would give Peter a lift. Thus it would seem that noncontingency in the domain of global environmental change may not contribute to realizing what is best overall.

One response would be to say that in this case I should calculate about whether to calculate. In one way this response is correct and in another way it is wrong. As theorists we should try to identify those cases in which calculation is likely to lead to optimal outcomes and those in which it will not, and this requires calculating the utility of calculating in various domains (as indeed we did informally in the previous paragraph). But as utilitarian agents we should not calculate about whether to calculate, for this would defeat the very possibility of inculcating the character traits that make us virtuous greens. And anyway, such higher-order calculation threatens an infinite regress of calculations as well as generally straining psychological credulity.[46]

So what should I say to Peter? First, the problems of global environmental change are so severe and the green virtues so generally benign that the domain over which they should dominate is very large. Second, the green virtues would never take hold if their particular expressions were systematically exposed to the test of utility; so if we think that having green virtues is utility-maximizing overall, then we ought not to so expose their expressions (except in extreme cases, of which, I have been assuming, this is not one). So too bad for Peter and his Oxfam meeting.

But the problem of calculation reappears with the words "except in extreme cases." For a utilitarian, the commitment to noncontingency must include such an "escape clause." If this were an extreme case (suppose that the lives and well-being of the entire population of a medium-sized African country turned on Peter attending the Oxfam meeting) and I could not bring myself to give Peter a lift, then I would be no better than one of those compulsive rule worshipers whom utilitarians love to bash. But without calculation, how can I know whether or not this is an extreme case?

Part of the answer is that we are simply able to recognize some extreme cases as such; we just do it. When the house is on fire, a child is screaming, atrocities are being committed and civilizations threatened, moral mathematics are not needed in order to see that the patterns of behavior that are generally best may not be up to it in the present case. Of course, there may also be cases in which calculation would be needed in order to see that it would be best to break patterns of behavior given to us by the green virtues. But on these occasions the virtuous green will just have to forgo the best, trusting in the overall utility-maximizing power of the green virtues.

19. There is a further challenge to which I have already briefly alluded. If others are having a good time changing climate, destroying ozone, and driving species to extinction, and the green cause is hopeless, then it appears that I am morally obliged to join in the fun. A utilitarian should not, at great cost to herself, plow through the snow on her bike while everyone else is blowing past her in their gas-guzzling "suburban utility vehicles" (SUVs). If the world is to be lost anyway, then the morally responsible utilitarian will try to have a good time going down with the planet. If the best outcome (preventing global environmental change) is beyond my control and the worst outcome would be for me to live a life

of misery and self-denial in a futile attempt to bring about the inaccessible best outcome, then the best outcome that I can produce may involve my living a high-consumption lifestyle. But everyone can reason in this way, and so we may arrive at the conclusion not just that it is permissible to live like a normal American but that utilitarians are morally obliged to do so. This seems truly shocking.

There are really two arguments here. The first argument concerns the decision process of a single agent; the second claims that the first argument generalizes to all similarly situated agents.

Consider the second argument first. This argument trades on equivocating about whether or not the best outcome is in fact accessible to an agent. Imagine a world of only two agents, Kelly and Sean. From Kelly's point of view, if it is clear that Sean will fail to behave in an environmentally friendly way, then it may be best for Kelly to fail to do so as well. But if Sean is in the same position with respect to her decision as Kelly, then it cannot be taken as given that Sean will not engage in the environmentally friendly behavior, for that is just what she is reasoning about. If there is any point to her reasoning about this, then the environmentally friendly behavior must be accessible to her, contrary to what we assumed when we considered Kelly's decision process. The apparent generalization of the first argument introduces an equivocation that is not implicit in the first argument itself.[47]

The first argument should not be confused with what might be called the Nero objection. This objection states that, just as Nero fiddled while Rome burned, so a utilitarian agent should fiddle (or its functional equivalent) while global environmental change ravages the planet. Since Nero's fiddling was morally horrendous, the functionally equivalent utilitarian fiddling must be morally horrendous as well. However, Nero's fiddling and that of the utilitarian are not equivalent in relevant respects. What is horrendous about the image of Nero fiddling while Rome burns is that he probably set the fires, or could have had them put out. Rather than making the best of a bad situation, he was making a bad situation.[48] This is clearly forbidden by utilitarianism.

Here is a better account of the first argument. In the domain of global environmental change-relevant behavior, what we want is inflexible green behavior, but even here it should not be too inflexible. Suppose that there is some threshold of cooperation that must be surpassed if global environmental change is to be mitigated. If this threshold will not be surpassed regardless of what I do, then it might be best for me to act in some other way than to exemplify green virtues. But calculating about whether the threshold has been met seems to defeat the advantage of inflexibility that green virtues are supposed to deliver. Moreover, if the calculation delivers the result that I ought to behave in a way that is environmentally destructive, then this seems to contradict the result that we know morality must deliver. It is for reasons such as these that some people think that moving from a focus on actions to a focus on character does not solve collective-action problems.

Whether or not the shift of focus from actions to character succeeds in solving the problem depends on exactly what the problem is. If utilitarianism really implied that I should throw tequila bottles out of the window while commuting to work in my SUV, this result would not on the face of it be any more shocking than some other possibilities that utilitarianism can countenance in various hypothetical situations:—for example, that in some cases I might be morally obliged to hang innocent people, torture prisoners, or carpet-bomb cities. The reason that these objections do not sway anyone with utilitarian sympathies is that, by hypothesis, all of these cases presuppose that my acting in these horrific ways would produce the best possible world.[49] If the world is in such a deplorable state that hanging innocent people would actually constitute an improvement, that is surely not the fault of utilitarian theory. On the other hand, if the assumption that the contemplated act is optimal is not in play, then the critic is making the ubiquitous error (discussed earlier) of purporting to show that utilitarianism directs agents to act in ways that make the world worse or less good than it could be. As we have seen, utilitarianism can have no such implication.

If the best outcome is truly inaccessible to me, then it is not obviously implausible to suppose that I have a duty to make the best of a bad situation.[50] When I was a kid, growing up in a neighborhood that would certainly have been a first-strike target had there been a nuclear war between the Americans and the Russians, we often seriously discussed the following question. Suppose that you know that They have launched their missiles and that We have retaliated (or vice versa), and that in twenty minutes the planet will be incinerated. What should you do?[51] The idea that we should enjoy the life that remains to us may not be the only plausible response to this question, but it is surely not an implausible one.

What many people find grating about this answer, I think, is the idea that we have a duty to enjoy life in such a situation. Some might agree that it would be prudentially good to do so, but find it outrageous that morality would be so intrusive, right up to the end of the world. When it comes to the case in which the green cause is hopeless, it might be thought that matters are even worse. It is one thing to say that it is permissible or excusable to abandon our green commitments in such circumstances; it is another thing entirely to say that we have an affirmative duty to join the ranks of the enemy, and to enjoy the very activities that destroy the features of nature that we cherish.[52]

This objection has proceeded under the assumption that we might find ourselves in circumstances in which we know that living according to our green values would be entirely ineffectual, and that we would enjoy helping ourselves to the pleasures of consumerism. On these implausible assumptions, the objector is correct in claiming that utilitarianism would require us to join the side of the environmental despoilers. However, there is nothing really new in principle about this kind of case. It is another example of either the demandingness of utilitarianism, or of how utilitarianism holds our "ground projects" and therefore our integrity) hostage to circumstances beyond our control.[53]

It is not my task here to defend utilitarianism as anything more than a plausible research program. However, it is surely old news that utilitarianism can require us to break familiar patterns of behavior that are dear to our hearts when doing so would realize what is best. Of course this would be difficult to do, and most of us, most of the time, would not succeed in doing what is right. (No one said that it was easy to be a utilitarian.) But our failures to do what is right would not count against doing what is best as a moral ideal, any more than the human proclivity for violence should lead us to give up on peace as a cherished moral value. Or so it seems at first glance.

However, the most important point is this. My present concern is not with alternative realities or possible worlds; it is facts about this world that are relevant for present purposes. I am concerned with how a utilitarian agent should respond to the problem of global environmental change that we actually face here and now. Global environmental change is not like the case of an impending interplanetary collision that is entirely beyond our control. Nor is it an "all or nothing" phenomenon. Collectively, we can prevent or mitigate various aspects of global environmental change, and an individual agent can affect collective behavior in several ways. One's behavior in producing and consuming is important for its immediate environmental impacts, and also for the example-setting and role-modeling dimensions of the behavior.[54] It is a fact of life that one may never know how one's long-term projects will fare, or even how successful one has been in motivating and enlisting other people to pursue them, but this is as much grounds for optimism as pessimism. Nor does an environmentally friendly lifestyle have to be a miserable one.[55] Even if in the end one's values do not prevail, there is comfort and satisfaction in living in accordance with one's ideals.[56] All of this taken together suggests that real utilitarian agents here and now should try to prevent or mitigate global environmental change rather than celebrate its arrival.

However, presently there is no algorithm for designing the optimal utilitarian agent.[57] Nor is there an algorithm for constructing the perfect constitution, which constrains majority rule when it should, but does not prevent its expression when it should not.[58] Nevertheless, we have better and worse people and constitutions, and sometimes we know them when we see them. It might be nice to have a calculus that we could apply to constitutions and character, but absent this, we can still go forward living our lives and organizing our societies. These responses may not satisfy those who are concerned with the logic of collective action or who believe that every question must admit of a precise answer. But they should go some way toward satisfying those who, like me, are concerned with the moral psychology of collective action, and are willing to accept Aristotle's view that deliberation can never be completely divorced from practical wisdom.

20. What I have argued thus far is that despite various conundrums and complexities, in the face of global environmental change, utilitarians should be virtue theorists. While it is not my task here to provide a full account of what virtues utilitarians should try to develop and inculcate, I will conclude with a brief, tentative sketch of what might be called the "green virtues."[59] My goal is not to construct a complete account of the ideal utilitarian moral agent, but only to provide a sample of how we might think about the green virtues that such an agent might exemplify.[60] There is a modest literature on this subject, and a fair amount of experience with, and reflection on, green lifestyles, on which we can build.[61]

Abstractly we can say that the green virtues are those that utilitarians should try to exemplify in themselves and elicit in others, given the reality of global environmental change. Practically, it seems clear that green virtues should moralize such behavior as reproduction and consumption. As Alan Durning writes:

> When most people see a large automobile and think first of the air pollution it causes rather than the social status it conveys, environmental ethics will have arrived. Likewise, when most people see excess packaging, throwaway products, or a new shopping mall and grow angry because they consider them to be crimes against their grandchildren, consumerism will be on the retreat.[62]

21. Green virtues fall into three categories: those that reflect existing values; those that draw on existing values but have additional or somewhat different content; and those that reflect new values. I call these three strategies of virtue-identification preservation, rehabilitation, and creation. I will discuss each in turn, offering tentative examples of green virtues that might fall into these various categories.

Thomas Hill Jr. offers an example of preservation.[63] He argues that the widely shared ideal of humility should lead people to a love of nature. Indifference to nature "is likely to reflect either ignorance, self-importance, or a lack of self-acceptance which we must overcome to have proper humility."[64] A person who has proper humility would not destroy redwood forests (for example) even if it appears that utility supports this behavior. If what Hill says is correct, humility is a virtue that ought to be preserved by greens.

Temperance may be a good target for the strategy of rehabilitation. Long regarded as one of the four cardinal virtues, temperance is typically associated with the problem of *akrasia* and the incontinent agent. But temperance also relates more generally to self-restraint and moderation. Temperance could be rehabilitated as a green virtue that emphasizes the importance of reducing consumption.

A candidate for the strategy of creation is a virtue we might call mindfulness. Much of our environmentally destructive behavior is unthinking, even mechanical. In order to improve our behavior, we need to appreciate the consequences of our actions that are remote in time and space. A virtuous green would see herself as taking on the moral weight of production and disposal when she purchases an article of clothing (for example). She makes herself responsible for the cultivation of the cotton, the impacts of the dyeing process, the energy costs of the transport, and so on. Making decisions in this way would be encouraged by the recognition of a morally admirable trait that is rarely exemplified and hardly ever noticed in our society.[65]

Although I have been speaking of individual agents and their virtues, it is easy to see that institutions play important roles in enabling virtue. Many of these roles (e.g., inculcation, encouragement) have been widely discussed in the literature on virtue theory. However, it is also important to recognize that how societies and economies are organized can disable as well as enable the development of various virtues. For example, in a globalized economy without informational transparency, it is extremely difficult for an agent to determine the remote effects of her actions, much less take responsibility for them.[66] Thus, in such a society, it is difficult to develop the virtue of mindfulness.

22. If what I have said is correct, the contrast typically drawn between utilitarianism and virtue theory is overdrawn. Utilitarianism is a universal emulator: it implies that we should lie, cheat, steal, even appropriate Aristotle, when that is what brings about the best outcomes. In some cases and in some worlds it is best for us to focus as precisely as possible on individual acts. In other cases and worlds it is best for us to be concerned with character traits. Global environmental change leads to concerns about character because the best results will be produced by generally uncoupling my behavior from that of others. Thus, in this case and in this world, utilitarians should be virtue theorists.[67]

The central morals of this chapter are these. Philosophically, we should ask when, not whether, utilitarians should be virtue theorists. Practically, we need to develop a catalog of the green virtues and identify methods for how best to inculcate them. Some may consider this an "obsession" produced by allegiance to a particular moral theory, but to my mind this is not too much to ask of those who are philosophizing while human beings are bringing about the most profound transformation of earth to occur in 50 million years.

Acknowledgments

Earlier versions of this chapter were presented at the Utilitarianism Reconsidered conference in New Orleans; the Department of Philosophy at Edinburgh University; the Subfaculty of Philosophy at the University of Oxford; the Center for Values and Social Policy at the University of Colorado; the Australasian Association of Philosophy meeting in Sydney; the International Conference on Applied Ethics at the Chinese University of Hong Kong; the Department of Philosophy at the University of Wisconsin, Madison; the Minnesota Monthly Moral Philosophy Meeting; the Philosophy Program at the Graduate Center of the City University of New York; and the Department of Philosophy at Yale University. I am deeply grateful for all of the interesting discussion provided by these audiences. I thank especially David Copp, Roger Crisp, and James Griffin for helpful comments. The origin of this chapter goes back many years to a conversation with Barbara Herman about the scope and domain of morality; while nothing I say here will settle the differences between us that were expressed that afternoon, I want to thank her for causing me to think so long and hard about this problem.

Notes

1. While "global environmental change" may seem a clumsy or misleading expression, it has come to be the standard way of referring to this cluster of problems in the scientific and policy literatures; see e.g. the Web site for *The Encyclopedia of Global Environmental Change* (http://www.wiley.co.uk/wileychi/egec/). For an overview of these problems see the World Resources Institute, the United Nations Environment Program, and the World Bank, *World Resources 2000–2001* (New York, 2000), also available at http://wristore.com/worres20.html.

2. Some would modify this list of the reigning moral theories by adding or substituting contractualism or virtue ethics.

3. *Creating the Kingdom of Ends* (New York, 1996), p. 300.

4. Korsgaard, *Creating*, p. 275. Annette Baier thinks that contemporary moral philosophers have not yet escaped the clutches of Kant. *Postures of the Mind* (Minneapolis, 1985), p. 235.

5. However Korsgaard does briefly discuss the moral status of plants and animals in *The Sources of Normativity* (New York, 1996), chap. 4, and she extensively discusses Kantian views of animals

in her University of Michigan Tanner Lecture, "Fellow Creatures: Kantian Ethics and Our Duties to Animals," *Tanner Lectures on Human Values*, vol. 25, ed. Grethe B. Peterson (Salt Lake City: University of Utah Press, 2005), pp. 77–110.

6. See for example the work of Onora O'Neill collected in her *Constructions of Reason: Explorations of Kant's Practical Philosophy* (New York, 1989). Korsgaard tries to overcome the interiority of the theory by focusing on "how we should relate to one another" as the subject matter of morality (*Creating*, p. 275).

7. There are interpretations of Kant, perhaps most notably that of R. M. Hare (see e.g. *Freedom and Reason* [Oxford, 1965]), which emphasize the idea of universalizability and deemphasize the notion of the goodwill. This is not the reading of Kant with which I am concerned here, in part because it has become less influential in recent years, but also because (at least in this respect) it blurs the distinction between Kantianism and utilitarianism.

8. This is quite clear in the work of David Gauthier and Jan Narveson, for example. For an early discussion of these problems see my "Rational Egoism and Animal Rights," *Environmental Ethics* 3 (1981).

9. Although there are many differences and disagreements among them, and some would reject the charge of conservatism, I associate this view with British philosophers such as Jonathan Dancy and Stuart Hampshire and American philosophers such as Susan Wolf.

10. Antirevisionists come in different stripes, but for one version see the introduction to Judith Jarvis Thomson, *The Realm of Rights* (Cambridge, 1990); on the second point, see Susan Wolf, "Moral Saints," *Journal of Philosophy* 79 (1982), esp. p. 422. For a utilitarian response to such claims, see Peter Singer, *How Are We to Live? Ethics in an Age of Self-Interest* (Buffalo, N.Y., 1995).

11. On the environmental consequences of American reproductive behavior, see Charles A. S. Hall, R. Gil Pontius Jr., Lisa Coleman, and Jae-Young Ko, "The Environmental Consequences of Having a Baby in the United States," *Wild Earth* 5 (1995).

12. There is room for drawing various subtle distinctions here. Jürgen Habermas claims that "[h]uman responsibility for plants and for preservation of whole species cannot be derived from duties of interaction, and thus cannot be *morally* justified," but goes on to say that "there are good *ethical reasons* that speak in favor of the protection of plants and species." See his *Justification and Application: Remarks on Discourse Ethics*, trans. Ciaran Cronin (Cambridge, Mass., 1993), p. 111.

13. For further discussion of deontology and the role of intentions in shaping moral constraints, see Nancy (Ann) Davis, "Contemporary Deontology," *Companion to Ethics*, ed. Peter Singer (Oxford, 1991), and the references cited therein.

14. I hope it is clear that my intention thus far has been only to show that, on a first approximation, in comparison with its rivals, utilitarianism appears well positioned to address the problem, and in this regard is worthy of detailed investigation. I do not mean to suggest that alternative approaches, however resourceful, are totally incapable of providing interesting responses to our problem.

15. For such criticisms see J. Baird Callicott, "Animal Liberation: A Triangular Affair," *Environmental Ethics* 2 (1980); Holmes Rolston III, "Respect for Life: Counting What Singer Finds of No Account," *Singer and His Critics*, ed. Dale Jamieson (Oxford, U.K., 1999); Eric Katz, *Nature as Subject* (Lanham, Md. 1997); John Rodman, "The Liberation of Nature," *Inquiry* 20 (1977); and Mark Sagoff, "Animal Liberation and Environmental Ethics: Bad Marriage, Quick Divorce," *Osgood Hall Law Journal* 22 (1984).

16. Korsgaard insightfully writes that "[u]sually the 'standard objections' that one school of thought raises against another are question-begging in deep and disguised ways" (*Creating*, p. xiii).

17. In characterizing utilitarianism in this way, I chime with Liam Murphy (*Moral Demands in Nonideal Theory* [New York, 2000], p. 6) rather than with Shelly Kagan who uses the term "consequentialism" for what I call utilitarianism; see his discussion in *Normative Ethics* (Boulder, 1998). For further discussion of these terms, see my "Consequentialism," in "Ethics and Values," *Encyclopedia of Life Support Systems* (EOLSS), ed. R. Elliot, developed under the auspices of UNESCO (Oxford, U.K., 2002), available at http://www.eolss.net. See also my *Ethics and the Environment: An Introduction* (Cambridge, U.K., in press), chap. 4.

18. Indeed it may not even be the first step. Utilitarianism may imply that utilitarianism should be an "esoteric morality." Whether or not it has this implication depends on facts about particular people and societies. For discussion of esoteric morality see Henry Sidgwick, *The Methods of Ethics*, 7th ed. (London, 1907), p. 490; and Derek Parfit, *Reasons and Persons* (New York, 1984), part 1 (esp. chap. 1).

19. For the most recent, authoritative and systematic account of the consequences of climate change, see *Climate Change 2001: Impacts, Adaptation and Vulnerability*, ed. James J. McCarthy, Osvaldo F. Canziani, Neil A. Leary, David J. Dokken, and Kasey S. White (New York, 2001), and the updates found at http://www.ipcc.ch/. See also my "The Epistemology of Climate Change: Some Morals for Managers," *Society and Natural Resources* 4 (1991).

20. On this general issue see Jonathan Glover, "It Makes No Difference Whether or Not I Do It," *Applied Ethics*, ed. Peter Singer (New York, 1986); and Parfit, *Reasons*, chap. 3.

21. It should be obvious that I am using *hypocrisy* and *asceticism* as technical terms; a full-blooded analysis of these concepts would reveal richer and more subtle conditions for application than what is suggested by the text.

22. Since such a strategy may well involve the construction and inculcation of norms, I believe that nothing I say here is inconsistent with Philip Pettit's discussion of norms as responses to collective-action problems in part III of his *Rules, Reasons, and Norms* (Oxford, 2002). One way of relating our accounts would be to say that the account that I develop is a (relatively) thick description of what utilitarian agents would have to be like in order for relevant norms to emerge and to reduce their own contributions to the problem. Although my focus is primarily on individual agents, the argument generalizes to all similarly situated utilitarian agents. Moreover, I believe that the importance of individual agents in addressing collective action problems is not fully appreciated by many theorists (see below for further discussion).

23. I discuss this objection further below.

24. For further argument to this conclusion see Donald Regan, *Utilitarianism and Cooperation* (New York, 1980).

25. While the virtues, as I understand them here, are noncalculative generators of behavior, their exercise does not exclude deliberation. I am indebted to Steve Gardiner and Jerrold Katz for helpful discussion of these points.

26. For some important discussions of these points see Regan, *Utilitarianism*; Allan F. Gibbard, "Rule Utilitarianism: Merely an Illusory Alternative?" *Australasian Journal of Philosophy* 43 (1965); Robert M. Adams, "Motive Utilitarianism," *Journal of Philosophy* 73 (1976); and Peter Railton, "Alienation, Consequentialism, and the Demands of Morality," *Philosophy and Public Affairs* 13 (1984).

27. Some may feel the pull of this example, but find it out of the question that a life without friends could be utility-maximizing. But if we assume that utility-maximizing behavior is frequently associated with acting on agent-neutral reasons, then it is not difficult to see why strong personal relationships might lead us to act in less than optimific ways. Of course, even if this is true there is no question that many of us here and now would do worse by abandoning our friends and setting ourselves up as rootless cosmopolitan utility maximizers. For a recent discussion of some of these issues, see Elizabeth Ashford, "Utilitarianism, Integrity, and Partiality," *Journal of Philosophy* 97 (2000).

28. My quarrel here is not with those who have distinguished act- from rule-utilitarianism as part of an investigation of the varieties of utilitarianism, but rather with the way in which this distinction has subsequently been canonized and then read back into the tradition. For an excellent study in the former spirit see David Lyons, *Forms and Limits of Utilitarianism* (Oxford, 1965).

29. For a contrary view see Henry R. West, *An Introduction to Mill's Utilitarian Ethics* (New York, 2004). But see also Fred Berger, *Happiness, Justice, and Freedom: The Moral and Political Philosophy of John Stuart Mill* (Berkeley, 1984).

30. See Michael Slote's discussion of Bentham in "Utilitarian Virtue," *Midwest Studies in Philosophy Volume XIII Ethical Theory: Character and Virtue*, ed. P. French, T. Uehling Jr., and H. Wettstein (Notre Dame, 1988).

31. Onora O'Neill has written insightfully about this in the context of Kantian ethics (*Constructions*, chap. 9). See also Stanley Cavell, *The Claim of Reason* (New York, 1979), pp. 263–267.

32. In unpublished work I have tried to develop a perspective on the purposes of moral theorizing that I believe are implicit in the tradition of consequentialist moral philosophy. I discuss these ideas under the rubric "naturalized moral theory." For the beginnings of such an account see my "Method and Moral Theory," *Companion*, ed. P. Singer.

33. This distinction between global and local utilitarianism derives from the felicitous distinction between global and local consequentialism drawn by Philip Pettit and Michael Smith, who argue persuasively for the superiority of the global view in their "Global Consequentialism," in *Morality, Rules, and Consequences: A Critical Reader*, ed. B. Hooker, E. Mason, and D. Miller (Edinburgh, 2000). See also Shelly Kagan's "Evaluative Focal Points" in the same volume.

34. Parfit, *Reasons*, p. 25.

35. For the first view see Christopher Kutz, *Complicity: Ethics and Law for a Collective Age* (New York, 2000), p. 11; for the second, Murphy, *Moral Demands*, p. 96 (note, however, that Murphy's remark is in the context of a larger investigation of an individual's moral duty of beneficence under conditions of partial compliance). Other approaches to collective or shared intentions advocate revising our conceptions of agents or of intending, rather than focusing on the content of intentions. For example, John Searle holds that jointly intentional action can only be explained by postulating an irreducible form of intending that he calls "we-intending" in his *Intentionality* (Cambridge, U.K., 1983), chap. 3); for discussion see Kutz, *Complicity*, chap. 3.

36. Christopher McMahon (in his *Collective Rationality and Collective Reasoning* ([New York, 2001]) tells us that the solution to prisoners' dilemmas (a class of problems closely related to our problem) is to treat them as pure coordination problems. However, in prisoners' dilemmas each agent is better off defecting whatever other agents do, while this is not the case in pure coordination problems. Since prisoners' dilemmas have a different structure from pure coordination problems, clear, convincing motivation is needed for why we should view them in the way that McMahon suggests, and some account needs to be provided of what agents would have to be like in order to act in the preferred way. In the absence of such accounts, this gambit seems merely to change the subject. For further discussion, see Gerald Gaus, "Once More unto the Breach, My Dear Friends, Once More," *Philosophical Studies* 116 (2003); and Michael Weber, "The Reason to Contribute to Cooperative Schemes," in the same issue. My brief remarks in this paragraph are not meant to minimize the contributions of McMahon, Kutz, and others, but only to suggest that more detailed work needs to be done.

37. Julia Driver, *Uneasy Virtue* (New York, 2001), p. 108.

38. Here I agree with Rosalind Hursthouse, *On Virtue Ethics* (Oxford, 1999), part 2. Driver also discusses the relations between the virtues and the emotions, but I am not clear on what her considered view is on this matter.

39. On Robert Frank has argued that emotions promote self-interest by solving commitment problems, in his *Passions within Reason* (New York, 1988).

40. However, not all noncalculative generators of action count as virtues. Some are too trivial, others are vices, and still others would be too far from the traditional notion of a virtue even for me to call virtues.

41. James Griffin, *Value Judgment: Improving Our Ethical Beliefs* (New York, 1996), p. 113; see also Michael Slote, *From Morality to Virtue* (New York, 1992). For a more relaxed view about what counts as virtue ethics see Julia Annas, *The Morality of Happiness* (New York, 1993).

42. An objection to virtue theory that is beginning to gain currency draws on results from social psychology that show that contextual factors are stronger predictors of behavior than facts about individual character. For such objections, see Gilbert Harman, "Moral Philosophy Meets Social Psychology: Virtue Ethics and the Fundamental Attribution Error," reprinted in his *Explaining Value and Other Essays in Moral Philosophy* (Oxford, 2000); and John Doris, "Persons, Situations, and Virtue Ethics," *Nous* 32 (1998), and his *Lack of Character: Personality and Moral Behavior* (New York, 2002). Because I am not committed to any particular account of the virtues, much less to one that makes them radically internal to agents rather than relative to contexts, I do not believe that this objection threatens the claims that I advance here.

43. Bernard Williams fastens onto a somewhat similar point in his critique of Hare's "two-level" theory (see his "The Structure of Hare's Theory," *Hare and Critics*, ed. D. Seanor and N. Fotion ([Oxford, 1988]). But while Williams emphasizes the psychological untenability of living simultaneously at both the "intuitive" and "critical" levels, my criticism is specifically aimed at someone who rests content with rules of thumb when she is committed to the view that morality requires her to do what is best.

44. There is quite a lot more to be said about progressive consequentialism. I say a little more in "Consequentialism," and Robert Elliot discusses this view under the rubric "improving Consequentialism" in his *Faking Nature* (New York, 1997).

45. Let us assume that in this case the benefits and harms do not cross domains: the benefits of Peter attending the meeting attach only to famine relief, and the harms of my driving are confined to their contribution to global environmental change.

46. Such problems are much discussed in the economics literature under the rubric of "optimal stopping rules." See, for example, G. J. Stigler's classic "The Economics of Information," *Journal of Political Economy* 69 (1961).

47. There are ways of trying to revive the second argument by casting it in probabilistic terms, but I cannot consider that possibility here.

My understanding of a range of such cases has benefited greatly from discussions with Scott (Drew) Schroeder.

48. The *locus classicus* for this image of Nero is Gibbon, but recent scholarship suggests that Nero has been maligned, that he neither set the fires nor was indifferent to the destruction they caused. See Miriam T. Griffin, *Nero: The End of a Dynasty* (London, 1984).

49. R. M. Hare makes a similar argument with respect to slavery; see his "What Is Wrong with Slavery," *Philosophy and Public Affairs* 8 (1979).

50. Here I break with Christopher Kutz (*Complicity*, pp. 124–132), who rejects what he calls "consequentialism" for failing to explain why it is wrong to participate in a bad practice whose occurrence is overdetermined. For an alternative view to Kutz's, see Frank Jackson, "Group Morality," *Metaphysics and Morality: Essays in Honour of J. J. C. Smart*, ed. Philip Pettit, Richard Sylvan, and Jean Norman (Oxford, 1987). Intuitions about overdetermination cases seem to run in different ways, depending on particular cases and how they are described; a full treatment of this problem is beyond the aspirations of this chapter. I have benefited here from reading unpublished work by Dan Moller.

51. This question is similar to one many of us may face in our future (or, arguably, face now): What should you do knowing that, in some specified amount of time, you will surely die? And, of course, we should not be too confident that the question from my youth may not yet again become relevant.

52. This objection echoes a remark of C. S. Lewis to the effect that if one is about to be swept over a waterfall, one does not have to sing praises to the river gods.

53. This latter objection to utilitarianism was a constant theme in the work of Bernard Williams and has stimulated an enormous literature. To begin at the beginning with the famous case of Jim and the Indians, see his "A Critique of Utilitarianism," in J. J. C. Smart and Bernard Williams, *Utilitarianism: For and Against* (Cambridge, 1973). For an unusually insightful discussion of the "demandingness" objection see Murphy, *Moral Demands*, chaps. 2–3.

54. See Ziva Kunda, *Social Cognition: Making Sense of People* (Cambridge, U.K., 1999), pp. 501–506.

55. Contrary to what one might think reading the newspapers, relationships between subjective reports of well-being and economic measures (such as per capita GDP) are equivocal and complex. An easy way into these issues is through the home page of Ed Diener, one of the leading researchers in the study of subjective well-being (http://www.psych.uiuc.edu/Nediener/).

56. One way of developing this thought in a decision-theoretic context would be to follow Alexander Schuessler (in his *A Logic of Expressive Choice* [Princeton, 2000]) in distinguishing the "expressive" from the "outcome" value of a choice. This distinction may also help explain our intuitions in cases of overdetermined harms (mentioned in note 50). The deepest general philosophical discussion of these issues that I know is Thomas Hill Jr., "Symbolic Protest and Calculated Silence," *Philosophy and Public Affairs* 9 (1979). However, Hill focuses mostly on obviously malevolent acts and practices rather than the apparently innocent ones implicated in global environmental change.

57. David Lyons discusses a similar point when he talks about the "moral opacity" and "moral ambiguity" of utilitarianism (in "The Moral Opacity of Utilitarianism," *Morality*, ed. Hooker et al.), though I'm not certain exactly what conclusion he wants to draw from his discussion.

58. Jon Elster has extensively discussed the analogy between individual and collective precommitment and restraint, most recently in his *Ulysses Unbound* (New York, 2000).

59. James Griffin points out (*Value Judgement*, p. 106) that the problem of calculation returns here to haunt us, since in order to identify virtues it appears that we need to be able to determine exactly which character traits are utility-promoting. To some extent this is a problem that will have to be faced by any theory that takes both character and consequences seriously.

60. A full account of the ideal utilitarian agent facing our problem would have to find a place for vices as well, as I was reminded by Corliss Swain. Indeed, it is plausible to suppose that vices such as greed would be as important in explaining and motivating behavior as the virtues that I mention here.

61. For a start on the literature of green virtue theory, see Ronald L. Sandler, *Character and the Environment: A Virtue-Oriented Approach to Environmental Ethics* (New York, 2007); Ronald Sandler and Philip Cafaro (eds.), *Environmental Virtue Ethics* (New York, 2005); and Louke van Wensveen, *Dirty Virtues: The Emergence of Ecological Virtue Ethics* (Amherst, 1999).

62. Alan Durning, *How Much Is Enough? The Consumer Society and the Future of the Earth* (New York, 1992), p. 138.

63. In his “Ideals of Human Excellence and Preserving the Natural Environment,” *Reflecting on Nature: Readings in Environmental Philosophy*, ed. Lori Gruen and Dale Jamieson (New York, 1994).

64. Hill, “Ideals,” p. 108.

65. Cooperativeness would be another important characteristic of agents who could successfully address our problem (as well as collective-action problems generally). Surprisingly, this characteristic appears to be neglected by both ancient and modern writers on the virtues (Hume may be an exception). Perhaps a virtue of cooperativeness is a candidate for creation, or perhaps, though not itself a virtue, cooperativeness would be expressed by those who have a particular constellation of virtues. For discussion of the importance of cooperativeness to morality, see Robert A. Hinde, *Why Good Is Good: The Sources of Morality* (London, 2002).

66. There is a growing literature on this topic. See, for example, David C. Korten, *When Corporations Rule the World* (West Hartford, Ct., 1995).

67. Roger Crisp reaches a similar conclusion in “Utilitarianism and the Life of Virtue,” *Philosophical Quarterly* 42 (1992).

It's Not *My* Fault

Global Warming and Individual Moral Obligations

Walter Sinnott-Armstrong

18

1. Assumptions

To make the issue stark, let us begin with a few assumptions. I believe that these assumptions are probably roughly accurate, but none is certain, and I will not try to justify them here. Instead, I will simply take them for granted for the sake of argument.[1]

First, global warming has begun and is likely to increase over the next century. We cannot be sure exactly how much or how fast, but hot times are coming.[2]

Second, a significant amount of global warming is due to human activities. The main culprit is fossil fuels.

Third, global warming will create serious problems for many people over the long term by causing climate changes, including violent storms, floods from sea-level rises, droughts, heat waves, and so on. Millions of people will probably be displaced or die.

Fourth, the poor will be hurt most of all. The rich countries are causing most of the global warming, but they will be able to adapt to climate changes more easily.[3] Poor countries that are close to sea level might be devastated.

Fifth, governments, especially the biggest and richest ones, are able to mitigate global warming[4] They can impose limits on emissions. They can require or give incentives for increased energy efficiency. They can stop deforestation and fund reforestation. They can develop ways to sequester carbon dioxide in oceans or underground. These steps will help, but the only long-run solution lies in alternatives to fossil fuels. These alternatives can be found soon if governments start massive research projects now.[5]

Sixth, it is too late to stop global warming. Because there is so much carbon dioxide in the atmosphere already, because carbon dioxide remains in the atmosphere for so long, and because we will remain dependent on fossil fuels in the near future, governments can slow down global warming or reduce its severity, but they cannot prevent it. Hence, governments need to adapt. They need to build sea walls. They need to reinforce houses that cannot withstand storms. They need to move populations from low-lying areas.[6]

Seventh, these steps will be costly. Increased energy efficiency can reduce expenses, adaptation will create some jobs, and money will be made in the research and production of alternatives to fossil fuels. Still, any steps that mitigate or adapt to global warming will slow down our economies, at least in the short run.[7] That will hurt many people, especially many poor people.

Eighth, despite these costs, the major governments throughout the world still morally ought to take some of these steps. The clearest moral obligation falls on the United States. The United States caused and continues to cause more of the problem than any other country. The United States can spend more resources on a solution without sacrificing basic necessities. This country has the scientific expertise to solve technical problems. Other countries follow its lead (sometimes!). So the United States has a special moral obligation to help mitigate and adapt to global warming.[8]

2. The Problem

Even assuming all of this, it is still not clear what I as an individual morally ought to do about global warming. That issue is not as simple as many people assume. I want to bring out some of its complications.

It should be clear from the start that individual moral obligations do not always follow directly from collective moral obligations. The fact that your government morally ought to do something does not prove that you ought to do it, even if your government fails. Suppose that a bridge is dangerous because so much traffic has gone over it and continues to go over it. The government has a moral obligation to make the bridge safe. If the government fails to do its duty, it does not follow that I personally have a moral obligation to fix the bridge. It does not even follow that I have a moral obligation to fill in one crack in the bridge, even if the bridge would be fixed if everyone filled in one crack, even if I drove over the bridge many times, and even if I still drive over it every day. Fixing the bridge is the government's job, not mine. While I ought to encourage the government to fulfill its obligations,[9] I do not have to take on those obligations myself.

All that this shows is that government obligations do not always imply parallel individual obligations. Still, maybe sometimes they do. My government has a moral obligation to teach arithmetic to the children in my town, including my own children. If the government fails in this obligation, then I do take on a moral obligation to teach arithmetic to my children.[10] Thus, when the government fails in its obligations, sometimes I have to fill in, and sometimes I do not.

What about global warming? If the government fails to do anything about global warming, what am I supposed to do about it? There are lots of ways for me as an individual to fight global warming. I can protest bad government policies and vote for candidates who will make the government fulfill its moral obligations. I can support private organizations that fight global warming, such as the Pew Foundation,[11] or boycott companies that contribute too much to global warming, such as most oil companies. Each of these cases is interesting, but they all differ. To simplify our discussion, we need to pick one act as our focus.

My example will be wasteful driving. Some people drive to their jobs or to the store because they have no other reasonable way to work and eat. I want to avoid issues about whether these goals justify driving, so I will focus on a case where nothing so important is gained. I will consider driving for fun on a beautiful Sunday afternoon. My drive is not necessary to cure depression or calm aggressive impulses. All that is gained is pleasure. Ah, the feel of wind in your hair! The views! How spectacular! Of course, you could drive a fuel-efficient hybrid car. But fuel-efficient cars have less "get up and go." So let us consider a gas-guzzling sport-utility vehicle. Ah, the feeling of power! The excitement! Maybe you do not like to go for drives in sport-utility vehicles on sunny Sunday afternoons, but many people do.

Do we have a moral obligation not to drive in such circumstances? This question concerns driving, not buying cars. To make this clear, let us assume that I borrow the gas guzzler from a friend. This question is also not about legal obligations. So let us assume that it is perfectly legal to go for such drives. Perhaps it ought to be illegal, but it is not. Note also that my question is not about what would be best. Maybe it would be better, even morally better, for me not to drive a gas guzzler just for fun. But that is not the issue I want to address here. My question is whether I have a moral obligation not to drive a gas guzzler just for fun on this particular sunny Sunday afternoon.

One final complication must be removed. I am interested in global warming, but there might be other moral reasons not to drive unnecessarily. I risk causing an accident, since I am not a perfect driver. I also will likely spew exhaust into the breathing space of pedestrians, bicyclists, or animals on the side of the road as I drive by. Perhaps these harms and risks give me a moral obligation not to go for my joy ride. That is not clear. After all, these reasons also apply if I drive the most efficient car available, and even if I am driving to work with no other way to keep my job. Indeed, I might scare or injure bystanders even if my car gave off no greenhouse gases or pollution. In any case, I want to focus on global warming. So my real question is whether the facts about global warming give me any moral obligation not to drive a gas guzzler just for fun on this sunny Sunday afternoon.

I admit that I am inclined to answer, "Yes." To me, global warming does seem to make such wasteful driving morally wrong.

Still, I do not feel confident in this judgment. I know that other people disagree (even though they are also concerned about the environment). I would probably have different moral intuitions about this case if I had been raised differently or if I now lived in a different culture. My moral intuition might be distorted by overgeneralization from the other cases where I think that other entities (large governments) do have moral obligations to fight global warming. I also worry that my moral intuition might be distorted by my desire to avoid conflicts with my environmentalist friends.[12] The issue of global warming generates strong emotions because of its political implications and because of how scary its effects are. It is also a peculiarly modern case, especially because it operates on a much grander scale than my moral intuitions evolved to handle long ago when acts did not have such long-term effects on future generations (or at least people were not aware of such effects). In such circumstances, I doubt that we are justified in trusting our moral intuitions alone. We need some kind of confirmation.[13]

One way to confirm the truth of my moral intuitions would be to derive them from a general moral principle. A principle could tell us why wasteful driving is morally wrong, so we would not have to depend on bare assertion. And a principle might be supported by more trustworthy moral beliefs. The problem is, which principle?

3. Actual Act Principles

One plausible principle refers to causing harm. If one person had to inhale all of the exhaust from my car, this would harm him and give me a moral obligation not to drive my car just for fun. Such cases suggest:

> *The harm principle*: We have a moral obligation not to perform an act that causes harm to others.

This principle implies that I have a moral obligation not to drive my gas guzzler just for fun *if* such driving causes harm.

The problem is that such driving does not cause harm in normal cases. If one person were in a position to inhale all of my exhaust, then he would get sick if I did drive, and he would not get sick if I did not drive (under normal circumstances). In contrast, global warming will still occur even if I do not drive just for fun. Moreover, even if I do drive a gas guzzler just for fun for a long time, global warming will not occur unless lots of other people also expel greenhouse gases. So my individual act is neither necessary nor sufficient for global warming.

There are, admittedly, special circumstances in which an act causes harm without being either necessary or sufficient for that harm. Imagine that it takes three people to push a car off a cliff with a passenger locked inside, and five people are already pushing. If I join and help them push, then my act of pushing is neither necessary nor sufficient to make the car go off the cliff. Nonetheless, my act of pushing is a cause (or part of the cause) of the harm to the passenger. Why? Because I intend to cause harm to the passenger, and because my act is unusual. When I intend a harm to occur, my intention provides a reason to pick my act out of all the other background circumstances and identify it as a cause. Similarly, when my act is unusual in the sense that most people would not act that way, that also provides a reason to pick out my act and call it a cause.

Why does it matter what is usual? Compare matches. For a match to light up, we need to strike it so as to create friction. There also has to be oxygen. We do not call the oxygen the cause of the fire, since oxygen is usually present. Instead, we say that the friction causes the match to light, since it is unusual for that friction to occur. It happens only once in the life of each match. Thus, what is usual affects ascriptions of causation even in purely physical cases.

In moral cases, there are additional reasons not to call something a cause when it is usual. Labeling an act a cause of harm and, on this basis, holding its agent responsible for that harm by blaming the agent or condemning his act is normally counterproductive when that agent is acting no worse than most other people. If people who are doing no worse than average are condemned, then people who are doing much worse than average will suspect that they will still be subject to condemnation even if they start doing better, and even if they improve enough to bring themselves up to the average. We should distribute blame (and praise) so as to give incentives for the worst offenders to get better. The most efficient and effective way to do this is to reserve our condemnation for those who are well below average. This means that we should not hold people responsible for harms by calling their acts causes of harms when their acts are not at all unusual, assuming that they did not intend the harm.

The application to global warming should be clear. It is not unusual to go for joy rides. Such drivers do not intend any harm. Hence, we should not see my act of driving on a sunny Sunday afternoon as a cause of global warming or its harms.

Another argument leads to the same conclusion: the harms of global warming result from the massive quantities of greenhouse gases in the atmosphere. Greenhouse gases (such as carbon dioxide and water vapor) are perfectly fine in small quantities. They help plants grow. The problem emerges only when there is too much of them. But my joy ride by itself does not cause the massive quantities that are harmful.

Contrast someone who pours cyanide poison into a river. Later someone drinking from the river downstream ingests some molecules of the poison. Those molecules cause the person to get ill and die. This is very different from the causal chain in global warming, because no particular molecules from my car cause global warming in the direct way that particular molecules of the poison do cause the drinker's death. Global warming is more like a river that is going to flood downstream because of torrential rains. I pour a quart of water into the river upstream (maybe just because I do not want to carry it). My act of pouring the quart into the river is not a cause of the flood. Analogously, my act of driving for fun is not a cause of global warming.

Contrast also another large-scale moral problem: famine relief. Some people say that I have no moral obligation to contribute to famine relief because the famine will continue and people will die whether or not I donate my money to a relief agency. However, I could help a certain individual if I gave my donation directly to that individual. In contrast, if I refrain from driving for fun on this one Sunday, there is no individual who will be helped in the least.[14] I cannot help anyone by depriving myself of this joy ride.

The point becomes clearer if we distinguish global warming from climate change.

You might think that my driving on Sunday raises the temperature of the globe by an infinitesimal amount. I doubt that, but even if it does, my exhaust on that Sunday does not cause any climate change at all. No storms or floods or droughts or heat waves can be traced to my individual act of driving. It is these climate changes that cause harms to people. Global warming by itself causes no harm without climate change. Hence, since my individual act of driving on that one Sunday does not cause any climate change, it causes no harm to anyone.

The point is not that harms do not occur from global warming. I have already admitted that they do. The point is also not that my exhaust is overkill, like poisoning someone who is already dying from poison. My exhaust is not sufficient for the harms of global warming, and I do not intend those harms. Nor is it the point that the harms from global warming occur much later in time. If I place a time bomb in a building, I can cause harm many years later. And the point is not that the harm I cause is imperceptible. I admit that some harms can be imperceptible because they are too small or for other reasons.[15] Instead, the point is simply that my individual joy ride does not cause global warming, climate change, or any of their resulting harms, at least directly.

Admittedly, my acts can lead to other acts by me or by other people. Maybe one case of wasteful driving creates a bad habit that will lead me to do it again and again. Or maybe a lot of other people look up to me and would follow my example of wasteful driving. Or maybe my wasteful driving will undermine my commitment to environmentalism and lead me to stop supporting important green causes or to harm the environment in more serious ways. If so, we could apply:

> *The indirect harm principle*: We have a moral obligation not to perform an act that causes harm to others indirectly by causing someone to carry out acts that cause harm to others.

This principle would explain why it is morally wrong to drive a gas guzzler just for fun if this act led to other harmful acts.

One problem here is that my acts are not that influential. People like to see themselves as more influential than they really are. On a realistic view, however, it is unlikely that anyone would drive wastefully if I did and would not if I did not. Moreover, wasteful driving is not that habit-forming. My act of driving this Sunday does not make me drive next Sunday. I do not get addicted. Driving the next Sunday is a separate decision.[16] And my wasteful driving will not undermine my devotion to environmentalism. If my argument in this chapter is correct, then my belief that the government has a moral obligation to fight global warming is perfectly compatible with a belief that I as an individual have no moral obligation not to drive a gas guzzler for fun. If I keep this compatibility in mind, then my driving my gas guzzler for fun will not undermine my devotion to the cause of getting the government to do something about global warming.

Besides, the indirect harm principle is misleading. To see why, consider David. David is no environmentalist. He already has a habit of driving his gas guzzler for fun on Sundays. Nobody likes him, so nobody follows his example. But David still has a moral obligation not to drive his gas guzzler just for fun this Sunday, and his obligation has the same basis as mine, if I have one. So my moral obligation cannot depend on the factors cited by the indirect harm principle.

The most important problem for supposed indirect harms is the same as for direct harms: even if I create a bad habit and undermine my personal environmentalism and set a bad example that others follow, all of this would still not be enough to cause climate change if other people stopped expelling greenhouse gases. So, as long as I neither intend harm nor do anything unusual, my act cannot cause climate change even if I do create bad habits and followers. The scale of climate change is just too big for me to cause it, even "with a little help from my friends."

Of course, even if I do not cause climate change, I still might seem to contribute to climate change in the sense that I make it worse. If so, another principle applies:

> *The contribution principle*: We have a moral obligation not to make problems worse.

This principle applies if climate change will be worse if I drive than it will be if I do not drive.

The problem with this argument is that my act of driving does not even make climate change worse. Climate change would be just as bad if I did not drive. The reason is that climate change becomes worse only if more people (and animals) are hurt or if they are hurt worse. There is nothing bad about global warming or climate change in itself if no people (or animals) are harmed. But there is no individual person or animal who will be worse off if I drive than if I do not drive my gas guzzler just for fun. Global warming and climate change occur on such a massive scale that my individual driving makes no difference to the welfare of anyone.

Some might complain that this is not what they mean by "contribute." All it takes for me to contribute to global warming in their view is for me to expel greenhouse gases into the atmosphere. I do that when I drive, so we can apply:

> *The gas principle*: We have a moral obligation not to expel greenhouse gases into the atmosphere.

If this principle were true, it would explain why I have a moral obligation not to drive my gas guzzler just for fun.

Unfortunately, it is hard to see any reason to accept this principle. There is nothing immoral about greenhouse gases in themselves when they cause no harm. Greenhouse gases include carbon dioxide and water vapor, which occur naturally and help plants grow. The problem of global warming occurs because of the high quantities of greenhouse gases, not because of anything bad about smaller quantities of the same gases. So it is hard to see why I would have a moral obligation not to expel harmless quantities of greenhouse gases. And that is all I do by myself.

Furthermore, if the gas principle were true, it would be unbelievably restrictive. It implies that I have a moral obligation not to boil water (since water vapor is a greenhouse gas) or to exercise (since I expel carbon dioxide when I breathe heavily). When you think it through, an amazing array of seemingly morally acceptable activities would be ruled out by the gas principle. These implications suggest that we had better look elsewhere for a reason for my moral obligation not to drive a gas guzzler just for fun.

Maybe the reason is risk. It is sometimes morally wrong to create a risk of a harm even if that harm does not occur. I grant that drunk driving is immoral, because it risks harm to others, even if the drunk driver gets home safely without hurting anyone. Thus, we get another principle:

> *The risk principle*: We have a moral obligation not to increase the risk of harms to other people.[17]

The problem here is that global warming is not like drunk driving. When drunk driving causes harm, it is easy to identify the victim of the particular drunk driver. There is no way to identify any particular victim of my wasteful driving in normal circumstances.

In addition, my earlier point applies here again. If the risk principle were true, it would be unbelievably restrictive. Exercising and boiling water also expel greenhouse gases, so they also increase the risk of global warming if my driving does. This principle implies that almost everything we do violates a moral obligation.

Defenders of such principles sometimes respond by distinguishing significant from insignificant risks or increases in risks. That distinction is problematic, at least here. A risk is called significant when it is too much. But then we need to ask what makes this risk too much when other risks are not too much. The reasons for counting a risk as significant are then the real reasons for thinking that there is a moral obligation not to drive wastefully. So we need to specify those reasons directly instead of hiding them under a waffle term like "significant."

4. Internal Principles

None of the principles discussed so far is both defensible and strong enough to yield a moral obligation not to drive a gas guzzler just for fun. Maybe we can do better by looking inward.

Kantians claim that the moral status of acts depends on their agents' maxims or "subjective principles of volition"[18]—roughly what we would call motives or intentions or plans. This internal focus is evident in Kant's first formulation of the categorical imperative:

> *The universalizability principle*: We have a moral obligation not to act on any maxim that we cannot will to be a universal law.

The idea is not that universally acting on that maxim would have bad consequences. (We will consider that kind of principle below.) Instead, the claim is that some maxims "cannot even be thought as a universal law of nature without contradiction."[19] However, my maxim when I drive a gas guzzler just for fun on this sunny Sunday afternoon is simply to have harmless fun. There is no way to derive a contradiction from a universal law that people do or may have harmless fun. Kantians might respond that my maxim is, instead, to expel greenhouse gases. I still see no way to derive a literal contradiction from a universal law that people do or may expel greenhouse gases. There would be bad consequences, but that is not a contradiction, as Kant requires. In any case, my maxim (or intention or motive) is not to expel greenhouse gases. My goals would be reached completely if I went for my drive and had my fun without expelling any greenhouse gases. This leaves no ground for claiming that my driving violates Kant's first formula of the categorical imperative.

Kant does supply a second formulation, which is really a different principle:

> *The means principle*: We have a moral obligation not to treat any other person as a means only.[20]

It is not clear exactly how to understand this formulation, but the most natural interpretation is that for me to treat someone as a means implies my using harm to that person as part of my plan to achieve my goals. Driving for fun does not do that. I would have just as much fun if nobody were ever harmed by global warming. Harm to others is no part of my plans. So Kant's principle cannot explain why I have a moral obligation not to drive just for fun on this sunny Sunday afternoon.

A similar point applies to a traditional principle that focuses on intention:

> *The doctrine of double effect*: We have a moral obligation not to harm anyone intentionally (either as an end or as a means).

This principle fails to apply to my Sunday driving both because my driving does not cause harm to anyone and because I do not intend harm to anyone. I would succeed in doing everything I intended to do if I enjoyed my drive but magically my car gave off no greenhouse gases and no global warming occurred.

Another inner-directed theory is virtue ethics. This approach focuses on general character traits rather than particular acts or intentions. It is not clear how to derive a principle regarding obligations from virtue ethics, but here is a common attempt:

> *The virtue principle*: We have a moral obligation not to perform an act that expresses a vice or is contrary to virtue.

This principle solves our problem if driving a gas guzzler expresses a vice, or if no virtuous person would drive a gas guzzler just for fun.

How can we tell whether this principle applies? How can we tell whether driving a gas guzzler for fun "expresses a vice"? On the face of it, it expresses a desire for fun. There is nothing vicious about having fun. Having fun becomes vicious only if it is harmful or risky. But I have already responded to the principles of harm and risk. Moreover, driving a gas guzzler for fun does not always express a vice. If other people did not produce so much greenhouse gas, I could drive my gas guzzler just for fun without anyone being harmed by global warming. Then I could do it without being vicious. This situation is not realistic, but it does show that wasteful driving is not essentially vicious or contrary to virtue.

Some will disagree. Maybe your notions of virtue and vice make it essentially vicious to drive wastefully. But why? To apply this principle, we need some antecedent test of when an act expresses a vice. You cannot just say, "I know vice when I see it," because other people look at the same act and do not see vice, just fun. It begs the question to appeal to what you see when others do not see it, and you have no reason to believe that your vision is any clearer than theirs. But that means that this virtue principle cannot be applied without begging the question. We need to find some reason why such driving is vicious. Once we have this reason, we can appeal to it directly as a reason for why I have a moral obligation not to drive wastefully. The sidestep through virtue does not help and only obscures the issue.

Some virtue theorists might respond that life would be better if more people were to focus on general character traits, including green virtues, such as moderation and love of nature.[21] One reason is that it is so hard to determine obligations in particular cases. Another reason is that focusing on particular obligations leaves no way to escape problems like global warming. This might be correct. Maybe we should spend more time thinking about whether we have green virtues rather than about whether we have specific obligations. But that does not show that we do have a moral obligation not to drive gas guzzlers just for fun. Changing our focus will not bring any moral obligation into existence. There are other important moral issues besides moral obligation, but this does not show that moral obligations are not important as well.

5. Collective Principles

Maybe our mistake is to focus on individual persons. We could, instead, focus on institutions. One institution is the legal system, so we might adopt.

> *The ideal law principle*: We have a moral obligation not to perform an action if it ought to be illegal.

I already said that the government ought to fight global warming. One way to do so is to make it illegal to drive wastefully or to buy (or sell) inefficient gas guzzlers. If the government ought to pass such laws, then, even before such laws are passed, I have a moral obligation not to drive a gas guzzler just for fun, according to the ideal law principle.

The first weakness in this argument lies in its assumption that wasteful driving or gas guzzlers ought to be illegal. That is dubious. The enforcement costs of a law against joy rides would be enormous. A law against gas guzzlers would be easier to enforce, but inducements to efficiency (such as higher taxes on gas and gas guzzlers, or tax breaks for buying fuel-efficient cars) might accomplish the same goals with less loss of individual freedom. Governments ought to accomplish their goals with less loss of freedom, if they can. Note the "if." I do not claim that these other laws would work as well as an outright prohibition of gas guzzlers. I do not know. Still, the point is that such alternative laws would not make it illegal (only expensive) to drive a gas guzzler for fun. If those alternative laws are better than outright prohibitions (because they allow more freedom), then the ideal law principle cannot yield a moral obligation not to drive a gas guzzler now.

Moreover, the connection between law and morality cannot be so simple. Suppose that the government morally ought to raise taxes on fossil fuels in order to reduce usage and to help pay for adaptation to global warming. It still seems morally permissible for me and for you not to pay that tax now. We do not have any moral obligation to send a check to the government for the amount that we would have to pay if taxes were raised to the ideal level. One reason is that our checks would not help to solve the problem, since others would continue to conduct business as usual. What would help to solve the problem is for the taxes to be increased. Maybe we all have moral obligations to try to get the taxes increased. Still, until they are increased, we as individuals have no moral obligations to abide by the ideal tax law instead of the actual tax law.

Analogously, it is actually legal to buy and drive gas guzzlers. Maybe these vehicles should be illegal. I am not sure. If gas guzzlers morally ought to be illegal, then maybe we morally ought to work to get them outlawed.

But that still would not show that now, while they are legal, we have a moral obligation not to drive them just for fun on a sunny Sunday afternoon.

Which laws are best depends on side effects of formal institutions, such as enforcement costs and loss of freedom (resulting from the coercion of laws). Maybe we can do better by looking at informal groups.

Different groups involve different relations between members. Orchestras and political parties, for example, plan to do what they do and adjust their actions to other members of the group in order to achieve a common goal. Such groups can be held responsible for their joint acts, even when no individual alone performs those acts. However, gas-guzzler drivers do not form this kind of group. Gas-guzzler drivers do not share goals, do not make plans together, and do not adjust their acts to each other (at least usually).

There is an abstract set of gas-guzzler drivers, but membership in a set is too arbitrary to create moral responsibility. I am also in a set of all terrorists plus me, but my membership in that abstract set does not make me responsible for the harms that terrorists cause.

The only feature that holds together the group of people who drive gas guzzlers is simply that they all perform the same kind of act. The fact that so many people carry out acts of that kind does create or worsen global warming. That collective bad effect is supposed to make it morally wrong to perform any act of that kind, according to the following:

> *The group principle*: We have a moral obligation not to perform an action if this action makes us a member of a group whose actions together cause harm.

Why? It begs the question here merely to assume that if it is bad for everyone in a group to perform acts of a kind, then it is morally wrong for an individual to perform an act of that kind. Besides, this principle is implausible or at least questionable in many cases. Suppose that everyone in an airport is talking loudly. If only a few people were talking, there would be no problem. But the collective effect of so many people talking makes it hard to hear announcements, so some people miss their flights. Suppose, in these circumstances, I say loudly (but not too loudly), "I wish everyone would be quiet." My speech does not seem immoral, since it alone does not harm anyone. Maybe there should be a rule (or law) against such loud speech in this setting (as in a library), but if there is not (as I am assuming), then it does not seem immoral to do what others do, as long as they are going to do it anyway, so the harm is going to occur anyway.[22]

Again, suppose that the president sends everyone (or at least most taxpayers) a check for $600. If all recipients cash their checks, the government deficit will grow, government programs will have to be slashed, and severe economic and social problems will result. You know that enough other people will cash their checks to make these results to a great degree inevitable. You also know that it is perfectly legal to cash your check, although you think it should be illegal, because the checks should not have been issued in the first place. In these circumstances, is it morally wrong for you to cash your check? I doubt it. Your act of cashing your check causes no harm by itself, and you have no intention to cause harm. Your act of cashing your check does make you a member of a group that collectively causes harm, but that still does not seem to give you a moral obligation not to join the group by cashing your check, since you cannot change what the group does. It might be morally good or ideal to protest by tearing up your check, but it does not seem morally obligatory.

Thus, the group principle fails. Perhaps it might be saved by adding some kind of qualification, but I do not see how.[23]

6. Counterfactual Principles

Maybe our mistake is to focus on actual circumstances. So let us try some counterfactuals about what would happen in possible worlds that are not actual. Different counterfactuals are used by different versions of rule-consequentialism.[24]

One counterfactual is built into the common question, "What would happen if everybody did that?" This question suggests a principle:

> *The general action principle*: I have a moral obligation not to perform an act when it would be worse for everyone to perform an act of the same kind.[25]

It does seem likely that if everyone in the world drove a gas guzzler often enough, global warming would increase intolerably. We would also quickly run out of fossil fuels. The general action principle is, thus, supposed to explain why it is morally wrong to drive a gas guzzler.

Unfortunately, that popular principle is indefensible. It would be disastrous if every human had no children. But that does not make it morally wrong for a particular individual to choose to have no children. There is no moral obligation to have at least one child.

The reason is that so few people want to remain childless. Most people would not go without children even if they were allowed to. This suggests a different principle:

> *The general permission principle*: I have a moral obligation not to perform an act whenever it would be worse for everyone to be permitted to perform an act of that kind.

This principle seems better because it would not be disastrous for everyone to be permitted to remain childless. This principle is supposed to be able to explain why it is morally wrong to steal (or lie, cheat, rape, or murder), because it would be disastrous for everyone to be permitted to steal (or lie, cheat, rape, or murder) whenever (if ever) they wanted to.

Not quite. An agent is permitted or allowed in the relevant sense when she will not be liable to punishment, condemnation (by others), or feelings of guilt for carrying out the act. It is possible for someone to be permitted in this sense without knowing that she is permitted and, indeed, without anyone knowing that she is permitted. But it would not be disastrous for everyone to be permitted to steal if nobody knew that they were permitted to steal, since then they would still be deterred by fear of punishment, condemnation, or guilt. Similarly for lying, rape, and so on. So the general permission principle cannot quite explain why such acts are morally wrong.

Still, it would be disastrous if everyone knew that they were permitted to steal (or lie, rape, etc.). So we simply need to add one qualification:

> *The public permission principle*: I have a moral obligation not to perform an act whenever it would be worse for everyone to know that everyone is permitted to perform an act of that kind.[26]

This principle seems to explain the moral wrongness of many of the acts we take to be morally wrong, since it would be disastrous if everyone knew that everyone was permitted to steal, lie, cheat, and so on.

Unfortunately, this revised principle runs into trouble in other cases. Imagine that 1,000 people want to take Flight 38 to Amsterdam on October 13, 2003, but the plane is not large enough to carry that many people. If all 1,000 took that particular flight, then it would crash. But these people are all stupid and stubborn enough that if they knew that they were all allowed to take the flight, they all would pack themselves in, despite warnings, and the flight would crash. Luckily, this counterfactual does not reflect what actually happens. In the actual world, the airline is not stupid. Since the plane can safely carry only 300 people, the airline sells only 300 tickets and does not allow anyone on the flight without a ticket. If I have a ticket for that flight, then there is nothing morally wrong with me taking the flight along with the other 299 who have tickets. This shows that an act is not always morally wrong when it would (counterfactually) be disastrous for everyone to know that everyone is allowed to do it.[27]

The lesson of this example applies directly to my case of driving a gas guzzler. Disaster occurs in the airplane case when too many people do what is harmless by itself. Similarly, disaster occurs when too many people burn too much fossil fuel. But that does not make it wrong in either case for one individual to perform an individual act that is harmless by itself. It only creates an obligation on the part of the government (or airline) to pass regulations to keep too many people from acting that way.

Another example brings out another weakness in the public permission principle. Consider open marriage. Max and Minnie get married because each loves the other and values the other person's love. Still, they think of sexual intercourse as a fun activity that they separate from love. After careful discussion before they got married, each happily agreed that each may have sex after marriage with whomever he or she wants. They value honesty, so they did add one condition: every sexual encounter must be reported to the other spouse. As long as they keep no secrets from each other and still love each other, they see no problem with their having sex with other people. They do not broadcast this feature of their marriage, but they do know (after years of experience) that it works for them.

Nonetheless, the society in which Max and Minnie live might be filled with people who are very different from them. If everyone knew that everyone is permitted to have sex during marriage with other people as long as the other spouse is informed and agreed to the arrangement, then various problems would arise. Merely asking a spouse whether he or she would be willing to enter into such an agreement would be enough to create suspicions and doubts in the other spouse's mind that would undermine many marriages or keep many couples from getting married, when they would have gotten or remained happily married if they had not been offered such an agreement. As a result, the society will have less love, fewer stable marriages, and more unhappy children of unnecessary divorce. Things would be much better if everyone believed that such agreements were not permitted in the first place, so they condemned them and felt guilty for even considering them. I think that this result is not unrealistic, but here I am merely postulating these facts in my example.

The point is that even if other people are like this, so that it would be worse for everyone to know that everyone is permitted to have sex outside of marriage with spousal knowledge and consent, Max and Minnie are not like this, and they know that they are not like this, so it is hard to believe that they as individuals have a moral obligation to abide by a restriction that is justified by other people's dispositions. If Max and Minnie have a joint agreement that works for them, but they keep it secret from others, then there is nothing immoral about them having sex outside of their marriage (whether or not this counts as adultery). If this is correct, then the general permission principle fails again.

As before, the lesson of this example applies directly to my case of driving a gas guzzler. The reason Max and Minnie are not immoral is that they have a right to their own private relationship as long as they do not harm others (such as by spreading disease or discord). But I have already argued that my driving a gas guzzler on this Sunday afternoon does not cause harm. I seem to have a right to have fun in the way I want as long as I do not hurt anybody else, just like Max and Minnie. So the public permission principle cannot explain why it is morally wrong to drive a gas guzzler for fun on this sunny Sunday afternoon.[28]

One final counterfactual approach is contractualism, whose most forceful recent proponent is Tim Scanlon.[29] Scanlon proposes:

> *The contractualist principle*: I have a moral obligation not to perform an act whenever it violates a general rule that nobody could reasonably reject as a public rule for governing action in society.

Let us try to apply this principle to the case of Max and Minnie. Consider a general rule against adultery, that is, against voluntary sex between a married person and someone other than his or her spouse, even if the spouse knows and consents. It might seem that Max and Minnie could not reasonably reject this rule as a public social rule, because they want to avoid problems for their own society. If so, Scanlon's principle leads to the same questionable results as the public permission principle. If Scanlon replies that Max and Minnie *can* reasonably reject the antiadultery rule, then why? The most plausible answer is that it is their own business how they have fun as long as they do not hurt anybody. But this answer is available also to people who drive gas guzzlers just for fun. So this principle cannot explain why that act is morally wrong.

More generally, the test of what can be *reasonably* rejected depends on moral intuitions. Environmentalists might think it unreasonable to reject a principle that prohibits me from driving my gas guzzler just for fun, but others will think it reasonable to reject such a principle, because it restricts my freedom to perform an act that harms nobody. The appeal to reasonable rejection itself begs the question in the absence of an account of why such rejection is unreasonable. Environmentalists might be able to specify reasons for why it is unreasonable, but then it is those reasons that explain why this act is morally wrong. The framework of reasonable rejection becomes a distracting and unnecessary sidestep.[30]

7. What is Left?

We are left with no defensible principle to support the claim that I have a moral obligation not to drive a gas guzzler just for fun. Does this result show that this claim is false? Not necessarily.

Some audiences[31] have suggested that my journey through various principles teaches us that we should not look for general moral principles to back up our moral intuitions. They see my arguments as a "reductio ad absurdum" of principlism, which is the view that moral obligations (or our beliefs in them) depend on principles. Principles are unavailable, so we should focus instead on particular cases, according to the opposing view called particularism.[32]

However, the fact that we cannot find any principle does not show that we do not need one. I already gave my reasons for why we need a moral principle to back up our intuitions in this case. This case is controversial, emotional, peculiarly modern, and likely to be distorted by overgeneralization and partiality. These factors suggest that we need confirmation for our moral intuitions at least in this case, even if we do not need any confirmation in other cases.

For such reasons, we seem to need a moral principle, but we have none. This fact still does not show that such wasteful driving is not morally wrong. It only shows that we do not *know* whether it is morally wrong. Our ignorance might be temporary. If someone comes up with a defensible principle that does rule out wasteful driving, then I will be happy to listen and happy if it works. However, until some such principle is found, we cannot claim to know that it is morally wrong to drive a gas guzzler just for fun.

The demand for a principle in this case does not lead to general moral skepticism. We still might know that acts and omissions that cause harm are morally wrong because of the harm principle. Still, since that principle and others do not apply to my wasteful driving, and since moral intuitions are unreliable in cases like this, we cannot know that my wasteful driving is morally wrong.

This conclusion will still upset many environmentalists. They think that they know that wasteful driving is immoral. They want to be able to condemn those who drive gas guzzlers just for fun on sunny Sunday afternoons.

My conclusion should not be so disappointing. Even if individuals have no such moral obligations, it is still morally better or morally ideal for individuals not to waste gas. We can and should praise those who save fuel. We can express our personal dislike for wasting gas and for people who do it. We might even be justified in publicly condemning wasteful driving and drivers who waste a lot, in circumstances where such public rebuke is appropriate. Perhaps people who drive wastefully should feel guilty for their acts and ashamed of themselves, at least if they perform such acts regularly; and we should bring up our children so that they will feel these emotions. All of these reactions are available even if we cannot truthfully say that such driving violates a moral *obligation*. And these approaches might be more constructive in the long run than accusing someone of violating a moral obligation.

Moreover, even if individuals have no moral obligations not to waste gas by taking unnecessary Sunday drives just for fun, governments still have moral obligations to fight global warming, because they can make a difference. My fundamental point has been that global warming is such a large problem that it is not individuals who cause it or who need to fix it. Instead, governments need to fix it, and quickly. Finding and implementing a real solution is the task of governments. Environmentalists should focus their efforts on those who are not doing their job rather than on those who take Sunday afternoon drives just for fun.

This focus will also avoid a common mistake. Some environmentalists keep their hands clean by withdrawing into a simple life where they use very little fossil fuels. That is great. I encourage it. But some of these escapees then think that they have done their duty, so they rarely come down out of the hills to work for political candidates who could and would change government policies. This attitude helps nobody. We should not think that we can do enough simply by buying fuel-efficient cars, insulating our houses, and setting up a windmill to make our own electricity. That is all wonderful, but it does little or nothing to stop global warming and also does not fulfill our real moral obligations, which are to get governments to do their job to prevent the disaster of excessive global warming. It is better to enjoy your Sunday driving while working to change the law so as to make it illegal for you to enjoy your Sunday driving.

Acknowledgments

For helpful comments, I would like to thank Kier Olsen DeVries, Julia Driver, Bob Fogelin, Bernard Gert, Rich Howarth, Bill Pollard, Mike Ridge, David Rodin, Peter Singer, and audiences at the University of Edinburgh, the International Society for Business, Economics, and Ethics, and the Center for Applied Philosophy and Public Ethics in Melbourne.

Notes

1. For skeptics, see Lomborg 1998, chap. 24 and Singer 1997. A more reliable partial skeptic is Richard S. Lindzen, but his papers are quite technical. If you do not share my bleak view of global warming, treat the rest of this chapter as conditional. The issue of how individual moral obligations are related to collective moral obligations is interesting and important in its own right, even if my assumptions about global warming turn out to be inaccurate.

2. See Mahlman 2005, Schlesinger 2005, and Weatherly 2005.

3. See Shukla 2005.

4. See Bodansky 2005.

5. See Shue 2005.

6. See Jamieson (chap. 15 in this volume).

7. See Toman 2005.

8. See Driver 2005.

9. If I have an obligation to encourage the government to fulfill its obligation, then the government's obligation does impose some obligation on me. Still, I do not have an obligation to do what the government has an obligation to do. In short, I have no parallel moral obligation. That is what is at issue here.

10. I do not seem to have the same moral obligation to teach my neighbors' children when our government fails to teach them. Why not? The natural answer is that I have a special relation to my children that I do not have to their children. I also do not have such a special relation to future people who will be harmed by global warming.

11. See Claussen 2005.

12. Indeed, I am worried about how my environmentalist friends will react to this chapter, but I cannot let fear stop me from following where arguments lead.

13. For more on why moral intuitions need confirmation, see Sinnott-Armstrong 2005.

14. Another difference between these cases is that my failure to donate to famine relief is an inaction, whereas my driving is an action. As Bob Fogelin put it in conversation, one is a sin of omission, but the other is a sin of emission. But I assume that omissions can be causes. The real question is whether my measly emissions of greenhouse gases can be causes of global warming.

15. See Parfit 1984, pp. 75–82.

16. If my act this Sunday does not cause me to drive next Sunday, then effects of my driving next Sunday are not consequences of my driving this Sunday. Some still might say that I can affect global warming by driving wastefully many times over the course of years. I doubt this, but I do not need to deny it. The fact that it is morally wrong for me to do all of a hundred acts together does not imply that it is morally wrong for me to do one of those hundred acts. Even if it would be morally wrong

for me to pick all of the flowers in a park, it need not be morally wrong for me to pick one flower in that park.

17. The importance of risks in environmental ethics is a recurrent theme in the writings of Kristin Shrader-Frechette.

18. Kant (1785) 1959, p. 400, n. 1.

19. Ibid., p.424. According to Kant, a weaker kind of contradiction in the will signals an imperfect duty. However, imperfect duties permit "exception in the interest of inclination" (p. 421), so an imperfect obligation not to drive a gas guzzler would permit me to drive it this Sunday when I am so inclined. Thus, I assume that a moral obligation not to drive a gas guzzler for fun on a particular occasion would have to be a perfect obligation in Kant's view.

20. Ibid., p. 429. I omit Kant's clause regarding treating others as ends because that clause captures imperfect duties, which are not my concern here (for reasons given in note 19).

21. Jamieson 2005.

22. Compare also standing up to see the athletes in a sporting event, when others do so. Such examples obviously involve much less harm than global warming. I use trivial examples to diminish emotional interference. The point is only that such examples share a structure that defenders of the group principle would claim to be sufficient for a moral obligation.

23. Parfit (1984, pp. 67–86) is famous for arguing that an individual act is immoral if it falls in a group of acts that collectively cause harm. To support his claim Parfit uses examples like the Harmless Torturers (p. 80). But torturers intend to cause harm. That's what makes them torturers. Hence, Parfit's cases cannot show anything wrong with wasteful driving, where there is no intention to cause any harm. For criticisms of Parfit's claims, see Jackson 1997.

24. See Sinnott-Armstrong 2003 and Hooker 2003.

25. See Singer 1971.

26. See Gert 2005. Gert does add details that I will not discuss here. For a more complete response, see Sinnott-Armstrong 2002.

27. The point, of course, depends on how you describe the act. It would not be disastrous to allow everyone "with a ticket" to take the flight (as long as there are not too many tickets). What is disastrous is to allow everyone (without qualification) to take the flight. Still, that case shows that it is not always morally wrong to do X when it would be disastrous to allow everyone to do X. To solve these problems, we need to put some limits on the kinds of descriptions that can replace the variable X. But any limit needs to be justified, and it is not at all clear how to justify such limits without begging the question.

28. The examples in the text show why violating a justified public rule is not sufficient for private immorality. It is also not necessary, since it might not be disastrous if all parents were permitted to kill their children, if no parent ever wanted to kill his or her children. The failure of this approach to give a necessary condition is another reason to doubt that it captures the essence of morality.

29. Scanlon 1998.

30. Scanlon's framework still might be useful as a heuristic, for overcoming partiality, as a pedagogical tool, or as a vivid way to display coherence among moral intuitions at different levels. My point is that it cannot be used to justify moral judgments or to show what makes acts morally wrong. For more, see Sinnott-Armstrong 2006, chap. 8.

31. Such as Bill Pollard in Edinburgh.

32. Developed by Dancy 1993, 2004. For criticisms, see Sinnott-Armstrong 1999.

References

Bodansky, D. 2005. "The International Climate Change Regime." In *Perspectives on Climate Change: Science, Economics, Politics, Ethics*, ed. W. Sinnott-Armstrong and R. Howarth (Amsterdam: Elsevier), pp. 147–180).

Claussen E. 2005. "Tackling Climate Change: Five Keys to Success." In *Persectives on Climate Change: Science, Economics, Politics, Ethics*, ed. W. Sinnott-Armstrong and R. Howarth (Amsterdam: Elservier), pp. 181–187.

Dancy, J. 1993. *Moral Reasons*. Oxford: Blackwell.

———. 2004. *Ethics without Principles*. New York: Oxford University Press.

Driver, J. 2005. "Ideal Decision-Making and Green Virtues." In *Perspectives on Climate Change: Science, Economics, Politics, Ethics*, ed. W. Sinnott-Armstrong and R. Howarth (Amsterdam: Elsevier), pp. 249–263.

Gert, B. 2005. *Morality: Its Nature and Justification*, rev. ed. New York: Oxford University Press.

Hooker, B. 2003. "Rule Consequentialism." In *The Stanford Encyclopedia of Philosophy*. Available at http://plato.stanford.edu/entries/consequentialism-rule.

Jackson, F. 1997. "Which Effects?" In *Reading Parfit*, ed. J. Dancey. (Oxford: Blackwell), pp. 42–53.

Jamieson, D. 2005. When Utilitarians Should Be Virtue Theorists. *Utilitas*. In press. Also chapter 17 in this volume.

Kant, I. 1959. *Foundations of the Metaphysics of Morals*, trans. L. W. Beck. Indianapolis: Bobbs-Merrill. (Original work published in 1785.)

Lomborg, B. 1998. *The Skeptical Environmentalist*. New York: Cambridge University Press.

Mahlman, J. 2005. "The Long Timescales of Human-Caused Global Warming: Further Challenges for the Global Warming Process." In *Perspectives on Climate Change: Science, Economics, Politics, Ethics*, ed. W. Sinnott-Armstrong and R. Howarth (Amsterdam: Elsevier), pp. 3–29.

Parfit, D. 1984. *Reasons and Persons*. Oxford: Clarendon Press.

Scanlon, T. 1998. *What We Owe to Each Other*. Cambridge, Mass.: Harvard University Press.

Schlesinger, W. H. 2005. "The Global Carbon Cycle and Climate Change." In *Perspectives on Climate Change: Science, Economics, Politics, Ethics*, ed. W. Sinnott-Armstrong and R. Howarth (Amsterdam: Elsevier), pp. 31–53.

Shue, H. 2005. "Responsibility to Future Generations and the Technological Transition." In *Perspectives on Climate Change: Science, Economics, Politics, Ethics*, ed. W. Sinnott-Armstrong and R. Howarth (Amsterdam: Elsevier), pp. 265–283.

Shukla, P. R. 2005. "Aligning Justice and Efficiency in the Global Climate Change Regime: A Developing Country Perspective." In *Perspectives on Climate Change: Science, Economics, Politics, Ethics*, ed. W. Sinnott-Armstrong and R. Howarth (Amsterdam: Elsevier), pp. 121–144.

Singer, M. 1971. *Generalization in Ethics*. New York: Atheneum.

Singer, S. F. 1997. *Hot Talk, Cold Science*. Oakland, Calif. Independent Institute.

Sinnott-Armstrong, W. 1999. "Some Varieties of Particularism." *Metaphilosophy* 30: 1–12.

———. 2002. "Gert contra Consequentialism." In *Rationality, Rules, and Ideals: Critical Essays on Bernard Gert's Moral Theory*, ed. W. Sinnott-Armstrong and R. Audi (Lanham, M.D. Rowman and Littlefield), pp. 145–163.

———. 2003. "Consequentialism." In *The Stanford Encyclopedia of Philosophy*. Available at http://plato.stanford.edu/entries/consequentialism.

———. 2005. "Moral Intuitionism and Empirical Psychology." In *Metaethics after Moore*, ed. T. Horgan and M. Timmons (New York: Oxford University Press), pp. 339–365.

———. 2006. *Moral Skepticisms*. New York: Oxford University Press.

Toman, M. 2005. "Climate Change Mitigation: Passing Through the Eye of a Needle?" In *Perspectives on Climate Change: Science, Economics, Politics, Ethics*, ed. W. Sinnott-Armstrong and R. Howarth (Amsterdam: Elsevier), pp. 75–97.

Weatherly, J. 2005. "Watching the Canary: Climate Change in the Arctic." In *Perspectives on Climate Change: Science, Economics, Politics, Ethics*, ed. W. Sinnott-Armstrong and R. Howarth (Amsterdam: Elsevier), pp. 55–72.

Index